Wissenschaftliche Untersuchungen
zum Neuen Testament · 2. Reihe

Herausgeber / Editor
Jörg Frey

Mitherausgeber / Associate Editors
Friedrich Avemarie · Judith Gundry-Volf
Martin Hengel · Otfried Hofius · Hans-Josef Klauck

216

Young S. Chae

Jesus as the Eschatological Davidic Shepherd

Studies in the Old Testament, Second Temple Judaism,
and in the Gospel of Matthew

Mohr Siebeck

YOUNG SAM CHAE, born 1964; 1988 B.A. Yonsei University, Seoul; 1992 M.Div. Chongshin Seminary, Seoul; 1997 Th.M. Calvin Theological Seminary, Michigan, USA; 2004 Ph.D. Trinity Evangelical Divinity School, Illiois, USA.

BT
295
.C434
2006

ISBN 3-16-148876-8
ISBN-13 978-3-16-148876-4
ISSN 0340-9570 (Wissenschaftliche Untersuchungen zum Neuen Testament, 2. Reihe)

Die Deutsche Nationalbibliothek lists this publication in the Deutsche Nationalbibliographie; detailed bibliographic data is available in the Internet at *http://dnb.d-nb.de*.

© 2006 by Mohr Siebeck, Tübingen, Germany.

This book may not be reproduced, in whole or in part, in any form (beyond that permitted by copyright law) without the publisher's written permission. This applies particularly to reproductions, translations, microfilms and storage and processing in electronic systems.

The book was printed by Gulde-Druck in Tübingen on non-aging paper and bound by Buchbinderei Held in Rottenburg/N.

Printed in Germany.

Preface

The thesis of this book originated from a paper I submitted to a class held for the study of Matthew's Christology. As I analyzed Matt 9:36, the reference (Ezek 34:5) to the verse in my Nestle-Aland New Testament caught my attention. I looked up Ezekiel 34 and began reading the entire chapter. As my eyes reached the phrase, "I will search for the lost and bring back the strays. I will bind up the injured..." (v. 16), a thought came to my mind to view Jesus as the eschatological Shepherd, not of John 10, but in the entire narrative of the First Gospel. As I examined the plausibility of the case in view of the current debates on Matthew's Christology, the thesis left me with something to say concerning, particularly, the puzzling link between the Son of David and Jesus' healing activity in the First Gospel.

In this book, which represents the unabridged form of my doctoral dissertation, I present Jesus as the therapeutic Davidic Shepherd who fulfils the role of YHWH the eschatological Shepherd of Israel and also, that of the Davidic Shepherd-Appointee over the one eschatological flock. I call this fundamental framework of Matthew's Christology and mission "two Shepherds schema" according to the Davidic Shepherd tradition in the Old Testament and Second Temple Judaism.

It was a journey to finish writing a thesis. I surely believe it is God's grace that led all of the insights, researches, writings, discussions, even struggles and prayers, to this publication of my dissertation. Meeting a good teacher also reminds me of his careful guidance. I feel blessed whenever I recall Dr. Eckhard J. Schnabel, my Doktorvater. With such humility and competence, he served me as he diligently corrected the manuscript, guided the research, and stretched my ability to reach my best. I thank him for his encouragement to publish my dissertation.

I also thank Dr. Jörg Frey who, after having read the whole thesis for several weeks, gladly recommended it for the WUNT 2nd series. I appreciate his advice and cheerful encouragement for the publication of this thesis. Well, I have many more teachers to mention at this moment. Especially, I remember the late Dr. Donald Verseput (1952-2004) who loved the story of Matthew's Gospel. He had a special gift to make the students believe they can be their best. His legacy of excellence as scholar and teacher will stay with me for a long time. I should also thank Dr. Carson, my second reader, for lots of helpful comments and advices he offered to me. Dr. VanGemeren's encouragements always lightened up my heart. For a few lunches and good conversations, I thank Dr. Scott McKnight for his kindness. As I look back a few more years, I recall the classes held by Dr. David E. Holwerda and Dr. Jeff Weima at Calvin. I appreciate their work for me.

To write a thesis is one thing, yet managing a life to write it is another. Especially I thank Rev. Deuk-sil Jung (Toledo Korean Church, Ohio) for his financial help and friendship. He and his small church provided me with the tuition for the first two years of my doctoral program. Also, the members of Boondang Central Church (Rev. Jong-chun Choi, Korea) graciously supported my study for several years. My home church, Seo-dae-moon Presbyterian Church, prayed for me and supported my family.

Having sent her only son abroad, my mother, Jong Hee Kim, always encouraged me with her prayers. Her dedication and sacrifice always led me into a deeper appreciation of the love of God. For my affection for the Scripture, I owe much to my father, Seung Woo Chae. To me, his life is living proof of an authentic faith. I am truly grateful for their love. Also, I should thank my father-in-law, Nam Young Ahn and my mother-in-law, Chung Ja Chang for their prayers and support. Many other people contributed to the production of this book. I express my thanks to Mrs. Arlene Maass for her proofreading and Dr. Fred VonKamecke for his checking the manuscript once again for this book. I do not wish to forget Mrs. Jackie Pointer's interlibrary loan service and Mr. Kevin Compton's checking the format for the dissertation. For the book format, Jana Trispel at Mohr Siebeck offered me a careful and patient guidance. I also thank Dr. Henning Ziebritzki at Mohr Siebeck for his general editorial guidance.

Lastly, I love to mention my son, Daniel Ji-woong Chae. Not knowing how much strength he provided me on a daily basis, this little boy prayed for his father's thesis which often took away his due time with his dad. That is another grace. Until now, I feel I have not expressed enough appreciation to my wife Yang Hee for her love, patience, and support. With joy and gratitude, I gladly dedicate this book to my wife and my son.

Table of Contents

Preface V

Introduction 1
  1. The Issue 1
    1.1 Matthew's Shepherd Motif in Current Discussion 2
    1.2 The 'Therapeutic Son of David' in Matthew: Building a Case for Research 4
  2. The Thesis 5
  3. The Methodology 6
  4. Charting the Course 17

Chapter 1
The Davidic Shepherd Tradition in the Old Testament 19
  1.1 Shepherd Imagery in the Ancient Near East 19
    1.1.1 Shepherd Imagery and Kingship 20
    1.1.2 Shepherd Imagery and the Shepherd's Crook 23
    1.1.3 Summary 24
  1.2 Shepherd Imagery in the Old Testament 25
    1.2.1 Characteristics 25
    1.2.2 The Shepherd of Israel: Redemptive Leadership and David 27
    1.2.3 Summary 31
  1.3 The Davidic Shepherd Tradition in Micah 2-5 32
    1.3.1 Micah 2:12-13: The Shepherd as the Breaker 32
    1.3.2 Micah 4:11-13: The Eschatological Battle and the Gathered Flock 34
    1.3.3 The Davidic Shepherd in Mic 5:1-4 [4:14-5:3] 36
    1.3.4 Summary 37
  1.4 The Davidic Shepherd Tradition in Ezekiel's Vision of the Restoration 38
    1.4.1 The Structure of Ezekiel's Vision of Restoration in Chapters 34-48 40
    1.4.2 Restoration Themes Recapitulated in Chapters 34-37 46
    1.4.3 The Structure of Ezekiel 34-37 47
    1.4.4 The Traditions behind Ezekiel's Concept of Restoration 49
    1.4.5 God's Zeal for His Name 52
    1.4.6 The Shepherd in Ezekiel 34-37 57
    1.4.7 Summary and Conclusion 74
  1.5 The Davidic Shepherd Tradition in Zechariah 9-14 76
    1.5.1 The Good and the Worthless Shepherds 77
    1.5.2 Zechariah 11:4-17 and Ezekiel 34, 37:15-28 79
    1.5.3 The Shepherd Imagery in Zech 11:4-17 and the Temple 82
    1.5.4 The Smitten Shepherd and the House of David 83
    1.5.5 Like the Angel of the Lord (12:8): More Than David? 88
    1.5.6 Summary 90
  1.6 Conclusion: A Profile of the Eschatological and Davidic Shepherds 90

Chapter 2
The Davidic Shepherd Tradition in Second Temple Judaism 95
2.1 The *Animal Apocalypse* (1 Enoch 85-90) 97
   2.1.1 Animal Symbols 97
      2.1.1.1 The Sheep, the Rams, and Blindness 97
      2.1.1.2 The Wild Beasts, the Birds, and the Exile 101
   2.1.2 The Shepherds and the Judgment of the Lord of the Sheep 103
   2.1.3 The Lord of the Sheep and the Restoration 106
   2.1.4 The Eschatological White Bull and the Final Transformation 109
   2.1.5 Summary 113
2.2 The *Psalms of Solomon* 17:1-18:12 115
   2.2.1 The Davidic Shepherd as Leader and Teacher 117
   2.2.2 The [Iron] Rod and the New Davidic Regime 121
   2.2.3 The Compassion of the One Shepherd 123
   2.2.4 Summary 125
2.3 Qumran Writings 126
   2.3.1 4Q504 (4QDibHam^a) 128
   2.3.2 4Q252 (4QpGen) 131
   2.3.3 4Q174 (4QFlorilegium) 134
   2.3.4 4Q161 (4QIsaiah Pesher^a) 137
   2.3.5 4Q285 (4QSefer ha-Milhamah) 139
   2.3.6 1Q28b/1QSb (The Scroll of Blessings) 141
   2.3.7 4Q521 (4QMessianic Apocalypse) 143
   2.3.8 Damascus Document 146
      2.3.8.1 CD-A 13:7-12 146
      2.3.8.2 CD-B 19:7-13 (CD-A 7:10-21) 149
   2.3.9 Summary and Conclusion 151
2.4 The *Targum of Ezekiel* 153
2.5 The *Fourth Book of Ezra* 157
2.6 The Hellenistic Monarchy and the Shepherd Imagery 160
2.7 Conclusion: The Profile of the Shepherd Revisited 168

Chapter 3
Matthew's Textual Interaction with the Davidic Shepherd Tradition 173
3.1 Quotations 174
   3.1.1 Matthew 2:6 (Micah 5:1-4) 174
      3.1.1.1 Matthew's Reading of מוֹשֵׁל in Mic 5:2 176
      3.1.1.2 The Meaning of οὐδαμῶς in Matt 2:6 183
      3.1.1.3 Matthew 2:6 in Light of Micah 2-5 184
   3.1.2 Matthew 26:31 (Zechariah 13:7) 189
      3.1.2.1 Matthew's Passion Narrative and Zechariah 9-14 189
      3.1.2.2 Characteristics of Matthew's Version of Zech 13:7 191
      3.1.2.3 Προάγω in Matt 26:32 and the Shepherd Image 198
   3.1.3 Summary 204
3.2 Allusions 205
   3.2.1 Matthew 9:36: The Shepherd with Compassion 205
   3.2.2 Matthew 10:6, 16; 15:24: The Lost Sheep of Israel 212
   3.2.3 Matthew 25:31-46: The Shepherd as the Final Judge 219
      3.2.3.1 The Identity of the Shepherd-Judge 222
      3.2.3.2 The Criterion of Judgment and the Shepherd-Judge 226

  3.2.3.3 The Identity of 'The Least of My Brothers' 227
 3.2.4 Summary 232
3.3 Shepherd/Sheep Images in Matt 7:15; 12:11-12; 18:12-14 233
 3.3.1 Matthew 7:15: False Prophets in Sheep's Clothing 234
 3.3.2 Matthew 12:11-12: Saving a Sheep on Sabbath 236
 3.3.3 Matthew 18:10-14: A Shepherd Seeking One Lost Sheep 239
 3.3.4 Summary 244
3.4 Conclusion 244

Chapter 4
Seeking the Lost and Healing the Sick:
Jesus as the Eschatological Davidic Shepherd 247

4.1 Conflict as Divine Reversal: Jesus' Seeking the Lost (Matt 9:10-13) 248
 4.1.1 The Question of the Setting for Matt 9:10-13 249
 4.1.2 A Controversy Story: The Literary Context of Matt 9:10-13 252
 4.1.3 The Sinners and the Lost Sheep of Israel (Matt 9:10) 256
 4.1.4 The Pharisees before 70 and the Shepherds of Israel (Matt 9:11) 261
 4.1.5 Jesus' Seeking the Lost and Conflict as Divine Reversal (Matt 9:11-13a) 264
 4.1.6 More Than a Prophet (Matt 9:13b) 274
 4.1.7 Summary 278
4.2 The Davidic Shepherd as the 'Therapeutic Son of David" 279
 4.2.1 Jesus as the Son of David: Controversies 279
 4.2.2 Various Proposals 285
  4.2.2.1 The Servant: Early Christian Redaction 286
  4.2.2.2 Solomon as 'Son of David,' A Jewish Legend 288
  4.2.2.3 The Divine Man: A Greek Mediator 291
 4.2.3 The Therapeutic Davidic Shepherd: Ezekiel's Vision 292
  4.2.3.1 Echoes from Second Temple Judaism 293
  4.2.3.2 The Use of the Verb θεραπεύω and Restoration Motifs 296
 4.2.4 The Healing Son of David in Healing Contexts 302
  4.2.4.1 The Two Blind Men (Matt 9:27) 304
  4.2.4.2 The Blind and Dumb Demoniac (Matt 12:23) 308
  4.2.4.3 The Canaanite Woman (Matt 15:22) 313
  4.2.4.4 Two Other Blind Men (Matt 20:30-31) 316
  4.2.4.5 Healing in the Temple and the Children's Praise (Matt 21:14) 319
 4.2.5 Summary 323
4.3 Conclusion 324

Chapter 5
The Rise of the One Davidic Shepherd:
Echoes in Matthew 27:51b-53 and 28:16-20 327

5.1 The Rise of the Shepherd and the Resurrection of the Saints (Matt 27:51b-53) 328
 5.1.1 Historicity and Sources 328
 5.1.2 Traditions behind the Texts 330
 5.1.3 The Davidic Shepherd Tradition 334
 5.1.4 Summary 340
5.2 The Divine Presence and the One Davidic Shepherd-Appointee (Matt 28:16-20) 340
 5.2.1 Tradition and Setting 341
 5.2.2 The Significance of Matt 28:16-17 345

    5.2.3 An Analysis of Matt 28:18-20 347
    5.2.4 The Authority of the Davidic Shepherd 351
    5.2.5 One Shepherd over Israel and the Nations 353
    5.2.6 The Davidic Shepherd as the Prince-Teacher 359
    5.2.7 Jesus as the Eschatological Locus of Divine Presence 364
    5.2.8 Summary 369
   5.3 Conclusion 370

Chapter 6
Matthew's Narrative Strategy and the Davidic Shepherd Tradition 372
   6.1 Matthew's Grand Scheme 373
   6.2 The Preamble and Epilogue of the Gospel 377
   6.3 Preaching, Healing, and Teaching 378
   6.4 The Ezekielian Pattern: Two Shepherds Schema 380
   6.5 Conclusion 385

Conclusions 387
   1. Aspects of Jesus as the Davidic Shepherd 387
    1.1 The Christology of the Davidic Shepherd 387
    1.2 The Shepherd's Teaching and Mission 390
    1.3 The Shepherd's Mission and Salvation-History 392
   2. Concluding Statements 394

Bibliography 396

Reference Index 419

Author Index 438

Subject Index 443

Introduction

1. The Issue

"Is not Jesus the messianic Shepherd, whose responsibility is to gather eschatological Israel?" W. D. Davies and D. C. Allison leave this question unanswered as they comment on Matt 9:36 in their monumental commentary.[1] Supposedly, the designation of Jesus as the Shepherd rings familiar to many ears, though probably not in the context of Matthew's christology. Usually the tenth chapter of the Fourth Gospel is credited for that particular picture of Jesus. But even among the synoptics, it is Matthew who shows the greatest interest in that imagery. Francis Martin observes that "the first Gospel develops an increasingly inclusive image of Christ as shepherd."[2] Highlighting Jesus as the shepherd, he suggests, is one way Matthew mediates an understanding of who Jesus really is.[3]

Indeed, the shepherd image appears relevant to Jesus' various messianic activities. For instance, after Matthew summarizes Jesus' deeds in terms of teaching, preaching, and healing in 9:35, the Evangelist points readers to the identity of Jesus saying, "as he saw the crowd, he had compassion on them, because they were harassed and downtrodden, like sheep without a shepherd" (9:36). If the summaries in 9:35 along with 4:23-5 outline "the actual *programme* of Jesus' active ministry" as B. Gerhardsson suggests,[4] then Matt 9:36 yields crucial insight into the Evangelist's understanding of Jesus' identity.

Further, throughout the Gospel, the compassion/mercy motif, as illustrated by the term σπλαγχνίζομαι (9:36; 14:14; 15:32; 20:34), may point to the relevance

---

[1] W. D. Davies and D. C. Allison, *A Critical and Exegetical Commentary on the Gospel According to Saint Matthew* 2 (Edinburgh: T. and T. Clark, 1988), 148, make the link between the shepherd imagery and Moses typology: "the readers should perhaps think that Jesus the shepherd is taking up a Mosaic office when he seeks out the lost sheep of the house of Israel." Cf. John P. Heil suggests that Matthew presents Jesus as the true shepherd king, but fails to present any substantial evidence in his book, *The Death and Resurrection of Jesus: A Narrative Critical Reading of Matthew 26-28* (Minneapolis: Fortress, 1991), 9-10.

[2] For the passages involving the shepherd image in Matthew, see Martin, "The Image of Shepherd in the Gospel of Saint Matthew," *Science et esprit* 27 (1975): 298, lists 2:6, 9:36, 10:6, 12:9-14, 12:22-30, 14:14, 15:21-28, 18:12-14, 20:29-34, 21:1-12, 24:30. 25:32, 26:15, 26:31,56,58, 27:3-10. Cf. R. E. Bracewell, "Shepherd Imagery in the Synoptic Gospels" (Ph.D diss.; Southern Baptist Theological Seminary, 1983), 163.

[3] Martin, "The Image of the Shepherd," 270, 283, 299.

[4] B. Gerhardsson, *The Mighty Acts of Jesus According to Matthew* (Lund: CWK Gleerup, 1979), 23 (italics his).

of the shepherd image to Jesus' messianic activities.⁵ Besides Matt 2:6, 25:32 and 26:31, where the presence of the metaphor of the shepherd is explicit, this catchword of the shepherd's compassion in 9:36 frequently occurs in the Evangelist's description of Jesus' motivation for ministering to the people (14:14; 15:39; 18:27; 20:34). The Evangelist of the First Gospel appears to use various ways of describing Jesus as the eschatological Shepherd, though there is little, if any, substantial investigation in this direction.

1.1 Matthew's Shepherd Motif in Current Discussion

While the shepherd is a familiar image in Scripture, studies of this image are surprisingly few.⁶ In particular, its significance in Matthew's Gospel has yet to receive serious attention. In addition to Martin's article, "The Image of the Shepherd in the Gospel of the Saint Matthew" (1975), other major works include: J. Thompson, "The Shepherd-Ruler Concept in the OT and Its Application in the NT" (1955);⁷ Wilfred Tooley, "The Shepherd and Sheep Image in the Teachings of Jesus"(1964);⁸ W. J. Vancil, "The Symbolism of the Shepherd in Biblical, Intertestamental, and New Testament Material"(1975);⁹ R. E. Bracewell, "Shepherd Imagery in the Synoptic Gospels" (1983);¹⁰ Scot McKnight, "New Shepherds of Israel: An Historical and Critical Study of Matthew 9:35-11:1" (1986);¹¹ and Johannes Beutler, *The Shepherd Discourse of John 10 and Its Context* (1991).¹²

Yet it is only recently that John P. Heil made the connection that Ezekiel 34 contains the entire semantic field needed for the implied reader to fully appreciate Matthew's shepherd metaphor. According to Heil, Ezekiel 34 unifies Matthew's shepherd metaphor by supplying readers with key terms, concepts, and images.¹³ Nevertheless, Heil's thesis has not yet been adequately tested. Does Matthew truly draw various shepherd images from Ezekiel 34?

---

⁵ The verb σπλαγχνίζομαι in those texts occur in conjunction with Jesus' healing and feeding ministry, with the exception of 18:27. Further, five of the eight instances of the verb 'to have mercy' (ἐλεέω; 9:27; 15:22; 17:15; 20:30; 20:31; cf. 5:7 [x2; Jesus' teaching]; 18:33) occur in the similar context as well.

⁶ Cf. J. G. Thompson, "The Shepherd-Ruler Concept in the OT and Its Application in the NT," *SJT* 8 (1955): 406-418; J. Jeremias, *TDNT* 6:485-502.

⁷ Thompson, *SJT* 8 (1955): 406-418.

⁸ Tooley, *NovT* 7 (1964): 15-25.

⁹ Vancil, Ph.D. diss. (Dropsie University, 1975).

¹⁰ Bacewell, Ph.D. diss. (Southern Baptist Theological Seminary, 1983).

¹¹ McKnight, Ph.D. diss. (University of Nottingham, 1986).

¹² Beutler, Cambridge: Cambridge University Press, 1991.

¹³ John P. Heil, "Ezekiel 34 and the Narrative Strategy of the Shepherd and Sheep Metaphor in Matthew," *CBQ* 55 (1993): 698-708, see esp. 699, 708. Heil finds the Ezekiel shepherd metaphor in the passages like 2:6, 9:36, 10:16, 14:14, 15:24, 32, 18:12-14, 25:31-46, 26:31-32.

Indeed, Ezekiel 34 is a rich text for that particular image in the OT. Francis Martin, having noticed that the whole tone of Ezekiel 34 prepares the way for seeing the shepherd as a healer, remarks: "the imagery drawn from Ezekiel is expanded and orchestrated in many different ways; it is put in connection with the Son of David, inserted into many healing contexts."[14] Further, Martin underscores the significance of Matthew's constellation of Ezekiel's shepherd image with the shepherd-servant-king of Zechariah.

Yet it is striking that Matthew does not quote even a single verse from Ezekiel 34.[15] Neither Martin nor Heil explains why Matthew heavily uses, if in fact he does, the shepherd image. Nevertheless, their observation of Matthew's failure to cite Ezekiel leaves open the possibility that for some reason Matthew deliberately uses the shepherd image in a careful, selective, and systematic way. By citing Micah concerning the coming of the messianic Shepherd in 2:6 and Zechariah concerning the suffering of the shepherd in 26:31, Matthew intentionally sets up the framework of this particular christological picture at the beginning and at the end of the narrative structure. In the middle of the narrative, Jesus' messianic activities as the long-awaited Shepherd are meant to be deciphered by the shepherd language, which could readily be supplied by a text such as Ezekiel 34. In this way, the whole narrative can be viewed through this angle, that is, Jesus as the messianic Shepherd.

If the quotations from Micah 5 and Zechariah 13 are taken as the narrative framework and Ezekiel 34 is not quoted in the Gospel, then the source of the image in the Gospel should not be confined to Ezekiel. If it is true that Matthew drew, both directly and indirectly, the shepherd images from all these texts, i.e., Micah 5, Zechariah 13 and Ezekiel 34, then what constitutes the common background or a broad scheme behind that they all involve the shepherd image? For the case of Ezekiel, chapter 34 is an integral part of the Ezekielian vision for the restoration of Israel that stretches to chapter 37. Also, Ezekiel 40-48 as a unit can be taken as the revision of Ezekiel 34-37 in terms of restoring the eschatological Temple.[16] Further, in Micah 2-5 and Zechariah 9-14, the expectation of Davidic kingship converges with that of the eschatological visitation of the true Shepherd of Israel who is expected to finally restore the nation. A bigger picture may be detected behind Matthew's use of the shepherd image.

---

[14] F. Martin, "The Image of Shepherd," 275, 299.

[15] Robert H. Gundry, *The Use of the Old Testament in St. Matthew's Gospel* (Leiden: Brill, 1967), 27, 33, observes that Ezekiel's shepherd background is found only in Matt 9:36 (Ezek 34:5) and 26:31(Ezek 34:31).

[16] Moshe Greenberg, "The Design and Themes of Ezekiel's Program of Restoration," *Interpretation* 38 (1975), 182. Note also that Ezekiel 34-37, 40-48 and Zechariah 1-8 had a profound influence upon the New Testament's messianic and eschatological thoughts. See Walter Harrelson, "Messianic Expectation at the Time of Jesus," *Saint Luke's Journal of Theology* 32 (1988): 32-33.

The shepherd image in Matthew's Gospel is certainly neither accidental nor merely for the sake of rhetorical effects. If Jesus can be seen as the eschatological Shepherd as promised in the OT tradition, then its christological implications merit close attention. The following section presents an illuminating case for Matthew's christology that legitimates our research on the shepherd image as it is applied to Jesus in view of the Davidic expectation in the restoration context.

1.2 The "Therapeutic Son of David" in Matthew: Building a Case for Research

One christological issue in Matthean study tightly connects with our study. John J. Collins addresses this issue when he states, "how Jesus [who does miracles in the Gospel] came to be viewed as the Davidic Messiah remains something of a mystery."[17] It is known that the royal messiah in Jewish tradition does not quite fit the 'Son of David' picture in the Gospels, in which the title is associated primarily with "a figure who is so addressed by people in need of exorcism or healing." [18] In Matthew's Gospel, the association of the title 'Son of David' with healing has puzzled many.[19]

Why is this title used in its association with, particularly, healing in the Gospel? Besides the traditional answer that it is the product of early Christian redaction, two other competing suggestions have been put forward. Christopher Burger argues that the figure of the Son of David who heals results from the correlation of the royal Son of David with the Hellenistic 'divine man' in early Christianity.[20] The other line of conjecture involves the speculations about Solomon. For example, Dennis C. Duling argues that the Son of David was a popular religious concept among the first-century Jews of Palestine and that it was associated with Solomon as a renown exorcist and healer.[21] The title Son of David has been recognized as perhaps the most distinguished of the christological titles in Matthew.[22] Particularly, Duling affirms that Matthew's Son of David is

---

[17] J. J. Collins, *The Apocalyptic Imagination* (Grand Rapids: Eerdmans, 1998), 263.

[18] Dennis C. Duling, "Solomon, Exorcism, and the Son of David," *HTR* 68 (1975): 235.

[19] For example, G. Stanton, in his memorable research on Matthean scholarship from 1945 to 1980, utters, "I am still puzzled by the evangelist (Matthew)'s association of 'Son of David' with healing;" see G. Stanton, "Origin and Purpose of Matthew's Gospel," in *Aufstieg und Niedergang der römischen Welt* II. 25.3., 1889-1951 (ed. W. Hasse; Berlin: De Gruyter, 1985), 1924.

[20] Burger, *Jesus als Davidssohn: Eine traditionsgeschichtliche Untersuchung* (Göttingen: Vandenhoeck & Ruprecht, 1970), 169, says, "the Davidic Messiah takes the function of the hellenistic *theios aner* upon himself, which doesn't fit him according to Jewish expectation."

[21] "Solomon, Exorcism, and the Son of David," 235.

[22] Birger Gerhardsson, *Mighty Acts of Jesus According to Matthew* (Lund: Gleerup, 1979), 86, 88, 91, states that "none of the evangelists shows such interest in the Son of David theme as Matthew." As regard the significance of the title he says, "Son of David ... only says a superficial part of the truth about Jesus."

distinctively the 'therapeutic Son of David.'[23] As to the association of the title with healing activity, he contends that it is drawn from Mark 10:46-52, which reflects the popular Solomon-as-exorcist tradition.[24]

In Matthew's Gospel, however, the healing activities of the Son of David are better associated with the shepherd image. Jesus as the Son of David heals not because he is recognized as Solomon the exorcist, but because he is the messianic Shepherd. The shepherd language is clearly present in nearly all of the passages that Duling addresses for the association of the Son of David with healing activities.[25] Peter M. Head notes that there is definite evidence within the OT that the Davidic messiah as shepherd is associated with healing.[26] Matthew's use of the shepherd image can be key to understanding the association of the Son of David and healing in the Gospel.

In the end, Davies and Allison's unanswered question as to whether Jesus is the messianic shepherd, Heil's unproven thesis concerning Ezekiel 34 as the unifying source of the shepherd image, and Duling's dubious suggestion that the Solomon tradition is the background of 'therapeutic Son of David,' taken together call for a thorough and comprehensive examination of the shepherd image in the First Gospel.

2. The Thesis

Matthew presents Jesus as the eschatological Shepherd and as the Davidic Shepherd-Appointee according to the pattern of the OT Davidic Shepherd tradition, while echoing some significant developments of the tradition during the Second Temple period. Underlying this thesis is the basic idea that Matthew's presentation of Jesus in his narrative is a result of his intense dialogue with the Davidic Shepherd tradition (esp. Mic 2-5; Zech 9-14; and Ezek 34-37). Matthew (and Jesus) was conversant with this tradition, which is what this study aims to prove.

---

[23] Matt 9:27-31; 12:22-24; 15:21-28; 20:29-34; 21:1-16; 22:41-46. "The Therapeutic Son of David: An Element in Matthew's Christological Apologetic," *NTS* 24 (1977-1978): 392-410.

[24] Duling, "The Therapeutic Son of David," 409.

[25] For instance, 'compassion, sheep without the shepherd' in 9:27-38; 'gathering' in 12:22-30 (cf. 19:28); 'the lost sheep of Israel, compassion' in 15:21-31; 'compassion' in 20:29-34. Interestingly, there is the 'gentle king' motif in 21:1-16 and 22:41-45 introduces Jesus' confrontation with the false leaders of Israel in chapter 23.

[26] Head, *Christology and the Synoptic Problem* (Cambridge: Cambridge University, 1998), 185-186, mentions Ezek 34:23-24, which presents the Davidic Messiah who "would include, as part of his role as Shepherd, healing the sick, binding up the injured and strengthening the weak (Ezek 34:4, 6)"; in addition, he also recalls 4Q521, which makes a connection between messianic shepherding and healing in contexts that allude to Isa 61.

This aim will be achieved by presenting a fresh and comprehensive explanation of the shepherd image beyond the current state of research touched upon particularly by Martin and Heil. This study is also expected to contribute to unraveling of the puzzle of the association of the Son of David with healing in the Gospel. An awareness of the connection of Jesus as both the Son of David and the Shepherd, in turn, will open up new insights into Matthew's understanding of the OT promises of Israel's restoration. The OT vision of the restoration of Israel, along with the promise of blessings to the nations, finally begins to be realized through the mission of compassion of the eschatological Shepherd and his Davidic Shepherd-Appointee. An analysis of the Davidic Shepherd tradition in the context of the literary structure of the Gospel, unlike many previous studies of Matthew's structure, will take serious both the opening and the closing of the Gospel. The return of God's presence in terms of the kingdom of God and his Davidic Appointee is central to the First Gospel.

3. The Methodology

The method adopted in this study will not be limited to one specific approach to the exclusion of others. Recent developments in the methodological debate seem to suggest that being consistent with an employed methodology often means to neglect one or another integral dimension of the text. At minimum, the Gospel text is comprised of at least three integral dimensions: the historical, the literary and the theological.

Tradition historical criticism has often been noticed with its tendency to overlook the literary nature of the text.[27] Ever since James Muilenburg commented on the new direction toward literary and rhetorical analysis, the study of the literary/rhetorical dimension of the text flourished.[28] Yet it is also true that various literary/rhetorical analyses often operate without due attention to history or reality outside the text.[29] Further, in both cases mentioned above, the theological/ideological dimension of the text is often neglected or even eliminated.[30] While the traditional historical-grammatical and theological interpretation of Scripture has been challenged, its three categories are still defensible. The persistence of these basic three categories is self-evident even in the

---

[27] For instance, Graig Bartholomew et al., *"Behind the Text": History and Biblical Interpretation* (Grand Rapids: Zondervan, 2003), rethinks the historical-critical approach in the context of postmodern narrative-oriented hermeneutics.

[28] J. Muilenburg, "Form Criticism and Beyond," *JBL* 88 (1969): 1-13.

[29] Refer to G. Osborne, *The Hermeneutical Spiral* (Downers Grove: IVP, 1991), 153.

[30] For example, Roy A. Harrisville and Walter Sundberg, *The Bible in Modern Culture: Theology and Historical-Critical Method from Spinoza to Käsemann* (Grand Rapids: Eerdmans, 1995).

recent development of rhetorical criticism in terms of its socio-political, rhetorical, and ideological dimensions.[31]

In our study, with the emphasis upon the tradition-historical approach, attention will be paid to all three basic, yet integral dimensions of the text. Likewise, in our first two chapters we dedicate our efforts to the tradition-historical research of the Davidic Shepherd tradition in the OT and the Second Temple Judaism as well prior to the First Gospel. This will cover the research of the historical dimension of Matthew's texts relevant to our topic. In Chapters 3, 4 and 5, we will engage in an intertextual study. Here, the concept of 'intertextuality' is adopted while dismissing its deconstructive ideological baggage.

Intertextuality, according to Anthony C. Thiselton, was first coined as a technical term by Julia Kristeva, and since then, it has become a complex and technical term both in literary theory and in poststructuralist theories of signs. It is more than allusion.[32] Jonathan Culler's is instructive here. "Any prior body of discourse in terms of which a given text becomes intelligible; that which the text implicitly and explicitly takes up, prolongs, cites, refutes, transposes"; M. Riffaterre uses the term more specifically to indicate "self-referring or intralinguistic relations."[33] Yet if intertextuality is regarded as "an all-encompassing and infinitely expanding system or systems of signification," as it is for some theorists, then it becomes problematic.[34] While appreciating the usefulness of 'intertextual' dimension, we move forward cautiously as we adopt this term especially as intertextuality denounces the notion of 'author-ity' of the text; the Scriptures function as the constraining context for God's special revelation; and this should not be confused with other texts.[35]

Particularly in the case of the NT's use of the OT,[36] the concept of intertextuality seems to fit the notion of Scripture as God's unified yet diverse progressive revelation.[37] In fact, the definition of 'intertextual' is used by many in various

---

[31] Vernon K. Robbins' *Jesus the Teacher* (Minneapolis: Fortress, 1992) illustrates a case in point. Similarly, W. Randolph Tate, *Biblical Interpretation: An Integrated Approach* (rev. ed.; Peabody: Hendrickson, 1997), presents an integration the tripartite dimensions of the text: The world behind the text (history); the world within the text (literary); and the world in front of the text (rhetorical).

[32] Anthony C. Thiselton, *New Horizons in Hermeneutics* (Grand Rapids: Zondervan, 1992), 38.

[33] Ibid.

[34] Ibid., 41.

[35] Cf. Kevin J. Vanhoozer, *Biblical Narrative in the Philosophy of Paul Ricoeur: A Study in Hermeneutics and Theology* (Cambridge: Cambridge University Press, 1990).

[36] For example, Michael Fishbane, *Biblical Interpretation in Ancient Israel* (Oxford: Clarendon Press, 1985), esp. 1-19, proposes 'inner-biblical exegesis' as 'reinterpretation of the scriptural text within OT canon '; for useful distinctions between 'inner-biblical' study and 'inter-textual' study, see Sylvia C. Keesmaat, "Exodus and the Intertextual Transformation of Tradition in Romans 8:14-30," *JSNT* 34 (1994): 30-32.

[37] Similarly, James A. Sanders, "Paul and the Theology of History," in *Paul and the Scriptures of Israel* (JSNTSup 83; Sheffield: Sheffield Academic Press, 1993), 54, proposes the concept of 'theological history.'

ways.[38] Yet, neither the term nor the methodology involved in it (as we shall see in the following concerning Paul's use of Scripture) has reached any consensus, especially for the study of the Gospels. For the case of Paul's use of Scripture, Richard B. Hays' seminal volume, *Echoes of Scripture in the Letters of Paul*, appears to remain the most influential recent work in this area,[39] and which may have spawned a number of critiques, appraisals, and developments.[40]

Hays' proposal for using an approach to 'intertextuality' to investigate Paul's use of Scripture reflects another step taken toward the literary/rhetorical approach;[41] Hays himself acknowledges that he derives the concept 'intertextuality' from literary criticism.[42] That is, instead of focusing on "technical questions about the textual form of Paul's citations, or the historical background of Paul's interpretive techniques" (thus, contending for or against the legitimacy of Paul's hermeneutical practices), Hays proposes that one should rather be attuned to 'echoes' of the citations and allusions; thus seeking to describe "the system of codes or conventions" rather than to find "the genetic or causal explanations to specific texts."[43] Hays launches his proposal by addressing the current problem in the area of the study of Paul's use of Scripture:

The Pauline citations and allusions have been catalogued, their introductory formulas classified, their relation to various Old Testament text-traditions examined, their exegetical methods compared to the methods of other interpreters within ancient Christianity and Judaism...they have,

---

[38] For examples, refer to William S. Green, "Doing the Text's Work for It: Richard Hays on Paul's Use of Scripture," in *Paul and the Scripture*, 60-63; Keesmaat, "Exodus and the Intertextual," 33.

[39] New Haven, Conn. and London: Yale University, 1989; cf. J. Ross Wagner, *Heralds of the Good News: Isaiah and Paul "In Concert" in the Letter to the Romans* (Leiden: Brill, 2002), adopts Hays' proposal, along with Hays' seven criteria, without serious modifications.

[40] Craig A. Evans and James A. Sanders present a collection of the reviews of Hays' *Echoes of Scripture* and Hays' responses to these critiques as well, also with case studies of this method in *Paul and the Scriptures of Israel* (JSNTSup 83; Sheffield: Sheffield Academic Press, 1993); see also, Stanley E. Porter, "The Use of the Old Testament in the New Testament: A Brief Comment on Method and Terminology" in *Early Christian Interpretation of the Scriptures of Israel* (ed. C. A. Evans and James A. Sanders; JSNTSup 148; Sheffield: Sheffield Academic Press, 1997), 79-96; Kenneth D. Litwak, "Echoes of Scripture?: A Critical Survey of Recent Works on Paul's Use of the Old Testament" *CR:BS* 6 (1998): 260-288, who presents a series of succinct summaries of a number of developments worked out triggered by Hay's proposal.

[41] As noted earlier, the shift is known to have been addressed by J. Muilenburg's SBL presidential address as it was published in "Form Criticism and Beyond" in *JBL* 88 (1969): 1-13; C. D. Stanley, *Paul and the Language of Scripture: Citation Technique in the Pauline Epistles and Contemporary Literature* (Cambridge: Cambridge University Press, 1991), 28, n.86, calls Hays' approach "a laudable first step in this direction."

[42] Particularly, Hays, *Echoes of Scripture*, 5-10, mentions John Hollander's work, *The Figure of the Echo: A Mode of Allusion in Milton and After* (Berkeley: University of California Press, 1981).

[43] Hays, *Echoes of Scripture*, 15.

as it were, unpacked and laid out the pieces of the puzzle. But how are the pieces to be assembled? Most of the 'unpacking' of the Pauline citations was complete more than a generation ago, yet we still lack a satisfying account of Paul's letters as *hermeneutical events, discourse in which Paul is engaged in the act of reinterpreting Scripture* to address the concerns of his communities.[44]

The core of Hays' approach thus lies in the application of the concept of 'dialectic,' 'discourse,' or 'narrative' to the inter-textual relation, that is, the continuing engagement of the text in the dialogue with the tradition.[45] The discourse is not then confined within the text [of Paul or Matthew] but existing in the 'intertextual space,' or 'discursive space'[46] i.e., the space created by 'metalepsis' echoing between the tradition and the text. To explain the function of 'intertextual echo,' Hays introduces the literary device, 'metalepsis' or 'transumption,' borrowing it from Holland's usage.[47] An author evokes another text in such a way [by metalepsis] that significant points of contact between the new text and its precursor remain unexpressed, thereby inviting readers to interpret a citation or allusion by recalling aspects of the original context that are not explicitly quoted, i.e., the transumed connections.[48] Hays' analogy of the antithesis between metalepsis and metaphor explains well the heart of his methodological stance:

Allusive echo functions to suggest to the reader that text B should be understood in light of a broad interplay with text A, encompassing aspects of A beyond those explicitly echoed. This sort of *metaleptic* figuration is the antithesis of the *metaphysical* conceit, in which the poet's imagination seizes a metaphor and explicitly wrings out of it all manner of unforeseeable significations. Metalepsis, by contrast, places the reader within a field of widespread or unstated correspondences.[49]

The device of metalepsis is a literary means by which one is invited to engage in 'intertextual discourse,' and which characterizes the textual interaction that lies neither (primarily) between the text and its contemporary historical settings, nor between the text and the meta-physical conceit, but rather between the text and the tradition. In this sense, a study of citations and allusions may no longer be a

---

[44] Hays, *Echoes of Scripture*, 1-2 (italics mine).
[45] Hays, *Echoes of Scripture*, 14. Hays continues, "Paul repeatedly situates his discourse within the symbolic field created by a single, great textaul precursor: Israel's Scripture "; thus emphasizing the intertextual interplay within the discourse; that is, "the narrative framework for interpretation" (157-158); Hays, "Echoes of Scripture in the Letter of Paul: Abstract," 46, likewise sums up, "Paul's strategy of intertextual echo is neither 'sacramental' nor 'eclectic' nor 'heuristic,' but 'dialectical'."
[46] Cf. Litwak, "Echoes of Scripture?," 262-263, says that Hays' emphasis lies on "actual textual connections, citations and allusions" rather than on "the sense meant by writers like discursive space."
[47] J. Hollander, *The Figure of Echo*, 115.
[48] Wagner, *The Heralds of the Good News*, 9-10; cf. Hays, *Echoes of Scripture*, 19-21.
[49] Hays, *Echoes of Scripture*, 20 (italics mine).

science of examining one-to-one correspondence of rigid bits and pieces of the texts, as if an archeologist analyzes his finds of fossils. It is rather like an art of musician, which is to read the notes, eventually hearing the ongoing story or the (invisible) symphony of the echoing voices, from the past and the present, that reply to one another in that continuing discourse. The reader is thus able to hear the echo, and participate in it.

Prior to assessing Hays' proposal as to its benefits and weaknesses, an inevitable question arises: Can Hays' approach be applied to the Gospels' use of the OT?[50] In particular, regarding Matthew's use of the OT, how can the study of the Gospel's citations and allusions be assisted by this approach?[51] Or, how is Hays' approach to the NT's use of the OT different, specifically from typological study, or from the promise and fulfillment motif? In fact, it has been often argued that Matthew's (or the New Testament's in general) quotations from the OT are not to be taken as arbitrary *midrashic* proof-texting, although *midrashic* exegesis could still imply, more or less, a sort of inner-biblical exegesis.[52] Indeed, many have recognized the significance of the entire OT context from which the NT writers take the quotation.[53]

Likewise, Craig A. Evans argues that what operates "the underlying thinking which lies behind the metalepsis that Hays has observed in Paul's letters" is, in fact, 'typological thinking.'[54] Yet, James A. Sanders claims more fundamental differences:

"There is indeed but one God at work throughout Scripture. As Hays rightly notes, Paul's reading of Scripture is not typological *as that term is normally understood*; Paul does not fret about

---

[50] Litwak, "Echoes of Scripture?," 264, would expect this question: "Hays offer a departure from the usual way of approaching Paul's use of the Scriptures of Israel, or indeed any New Testament writer's use of the Scriptures of Israel."

[51] If Hays' approach is applied, then perhaps the product would take quite a different shape, for instance, something quite unlike what is presented in R. H. Gundry's *The Use of the Old Testament in St. Matthew's Gospel* (Leiden: Brill, 1967), which deals mainly with formal and allusive quotations as to Matthew's use of the OT.

[52] Cf. D. Boyarin, *Intertextuality and the Reading of Midrash* (Bloomington: Indiana University Press, 1990).

[53] C. H. Dodd, *According to the Scriptures: The Sub-Structure of the New Testament Theology* (Lodon: SCM, 1952), 126, 133; also, B. Lindars, *New Testament Apologetic: The Doctrinal Significance of the Old Testament Quotations* (Philadelphia: Westminster, 1961); recently, Martin C. Albl, *"And Scripture Cannot Be Broken": The Form and Function of the Early Christian Testimonia Collections* (Leiden: Brill, 1999). Max Wilcox, "On Investigating the Use of the Old Testament in the New Testament" in *Text and Interpretation* (ed. E. Best and R. McL. Wilson; Cambridge: Cambridge University Press, 1979), 231-243, argues that citation, as opposed to allusion, does not assume the reader's familiarity with the original context (esp. 37).

[54] Evans, "Listening for Echoes of Interpreted Scripture," 47.

correspondences between types and anti-types. Rather, Paul's argument, like Isaiah's and Luke's, and indeed much else in the Bible, is *from theological history.*"[55]

Sanders/Hays' critique of typology is based on their understanding of the term as mere "correspondences between types and anti-types." Admittedly, the legitimacy and usefulness of typology are not currently questioned in the study of the NT writers use of the OT (cf. Heb 8).[56] Further, Evans defines typology differently, "not as a method but as a presupposition" that "the biblical story (of the past) has some bearing on the present, or, to turn it around, that the present is foreshadowed in the biblical story." [57]

Nevertheless, as a case for this typological study in Matthean scholarship, Dale C. Allison's *The New Moses: A Matthean Typology* can illustrate the realistic difference that the respected methodological approach might bring about.[58] More pointedly, the concept of 'intertextual echo' as applied to the investigation of the NT's use of the OT seems to enable one to reach the OT context beyond the level of quotations or (historical; i.e., earthly and temporal) *types.*

In the case of our study, granted that "only '*historical* facts' (figures, events, and institutions) are materials for typological interpretation,"[59] this criteria (only '*historical* facts'; e.g., the historical figures such as Moses, David, and Elijah) would eliminate, by principle, the possible correspondence between YHWH as the Shepherd (Ezek 34:1-16) and Jesus as the shepherd in Matthew's Gospel. Our study investigates the interaction between *the texts* (with the emphasis on the level of discourse) not merely *the (historical) types.*

Obviously, a full treatment of these questions would stretch the limits of this introduction to the methodology for our study. In fact, Hays' proposal is still subjected to experiments that equally reveal its strengths and its weaknesses as well. In the following, we will discuss their relevance to our study in order to justify our methodology while pointing to both benefits and precautions as to Hays' approach.

(1) Beyond Citations: Hays' approach enables us to detect certain portions of the NT writers' interaction with the OT, which otherwise would be overlooked if approached from traditional methods. Hays' example of Phil 1:19 (cf. Job 13:16)

---

[55] Sanders, "Paul and Theology of History," 53-54 (italics mine); as to Hays on typology, see, *Echoes of Scripture,* 100-102.

[56] It is important to note that typology involves the heavenly and/or eschatological 'realities' in their correspondences with the *historical* facts (as 'shadows') in the context of *Heilsgeschichte.* Yet, typology as a study of the prototypes/types/antitypes, by definition, sets the limit of the scope and nature of the correspondences.

[57] Evans, "Listening for Echoes of Interpreted Scripture," 47-48.

[58] Allison, *The New Moses* (Minneapolis: Fortress Press, 1993).

[59] Cf. W. Dunnett, *The Interpretation of Holy Scripture* [New York: Thomas Nelson, 1984], 51.

illustrates a case in point.[60] If we limit the NT's use of the OT only to quotations introduced with an explicit citation formula and exclude the instances of allusion and echo, we may fail to hear the complete intertextual discourse, and perhaps even mistakenly claim that certain portions of the NT do not contain any OT references (e.g., Pastorals).[61]

Yet, as Wagner aptly notices in the case of Paul, intertexual echoes nearly always functions in tandem with more obvious references to scripture, including citations marked by introductory formula and more explicit modes of allusion.[62] This tendency can be true in the case of the Gospels. In our study of Matthew's interaction with the OT texts from the Davidic Shepherd tradition, the echoes of a few shepherd/sheep images and certain themes related to the Shepherd figure can hardly become intelligible if set apart from overt citations of Mic 5:2 (Matt 2:6), Zech 13:7 (Matt 26:31), and relatively explicit allusions found in Matt 9:36; 10:6, 16; 15:24 and 25:31-46.

Thus, our investigation of the restoration themes of the Davidic Shepherd tradition (ch. 4) and the Ezekiel pattern in Matthew (ch. 5) will be supported by our examination of the more explicit textual interactions in Chapter 3. Importantly, the study of the NT's use of the OT in terms of echo can be conducted effectively only if other more explicit references are available.[63]

(2) Definition: How are we then to define 'citation,' 'allusion,' and 'echo' for our study? Hays does not give specific definitions and scholars vary on this issue.[64] Yet, these terms can likely be distinguished according to the explicitness of their interactions with the tradition, whether referring to it (citation), alluding to it (allusion), or echoing it (echo). Quotation is the most *explicit*, allusion is more *implicit*, and echo is the most *implicit*.[65] However, it still remains uncertain where to draw the line among these three terms. Further, no definition for these terms is airtight. Having said that, some workable definitions can be made. In the

---

[60] Hays, *Echoes of Scripture*, 21-22; cf. Stanley E. Porter, "The Use of the Old Testament," 91-92, 96, despite his critiques of Hays' criteria, acknowledges this benefit: "As the example from Phil 1:19 illustrates, failure to be explicit in defining terms, or defining terms in overly restrictive ways for the task which is being undertaken, has pre-empted full and complete analysis of the use of the Old Testament in Philippians" (96).

[61] On this point, J. Christiaan Beker's critique deserves attention. Beker, "Echoes and Intertextuality: On the Role of Scripture in Paul's Theology," 64-69, critiquing Hays' criteria as too subjective and warning of its "ahistorical" tendency, also underlines the significance of contingency for the cases of the rarity/frequency of the NT's use of OT (e.g., Rom 5-8; Pastorals).

[62] J. R. Wagner, *The Heralds of the Good News*, 10.

[63] Some of Hays' criteria set out in *Echoes of Scripture*, 29-32, especially, "recurrence" and "thematic coherence" or perhaps even "satisfaction," would be helpful if kept in mind for this line of research.

[64] Litwak, "Echoes of Scripture?," 285.

[65] Keesmaat, "Exodus and the Intertextual," 32, n.8; cf. T. E. Morgan, "Is There an Intertext in this Text?: Literary and Interdisciplinary Approaches to Intertextuality," *American Journal of Semiotics* 3, no. 4 (1985): 1-40.

following, we begin with the most explict – the citation/ quotation. Then we proceed to the *implicit* – the allusion – followed last by the most implicit – the echo.

First, citation/quotation. Michael Thompson sees 'quotation' as referring to instances in which the writer uses direct quotation 'with an explicit citation formula.'[66] Under the category of quotation, Christopher D. Stanley presents a three-pronged proposal: those introduced by an explicit quotation formula, those accompanied by a clear interpretive gloss (e.g., 1 Cor 15:27 on Ps 110:1), and those that stand in demonstrable syntactical tension with their present Pauline surroundings (e.g., Rom 9:7; 10:18; Gal 3:12).[67] Stanley's major concern for these criteria is the plausibility for these varied sorts of quotations to be recognized as clear references to Scripture by the implied readers; that is, not to be Pauline formulations.

On the other hand, Stanley E. Porter, in seeking a more expansive definition than Stanley's, suggests, "the focus would be upon formal correspondence with actual words found in antecedent texts," though he admits there is the problem of determining how many words would qualify as a quotation.[68] While not dismissing the possibility of the less rigid definition (Porter), thus leaving further discussion to the experts, we safely offer that Mic 5:2 in Matt 2:6 and Zech 13:7 in Matt 26:31 are citations in keeping with the more rigid definitions (Thompson's and Stanley's). The former case, regardless of the disputable question as to the exact sources of the citation, is introduced by οὕτως γὰρ γέγραπται διὰ τοῦ προφήτου (Matt 2:5b); and the latter is justified as well by the introductory formula γέγραπται γάρ (Matt 26:31c).

Second, allusion and echo. According to T. E. Morgan, relations between texts can occur in a number of ways: they may be 'explicit or implicit' (a criterion which is useful for defining quotation) and they may also be 'intentional or unintentional.'[69] In the case of citation, the intentionality is indubitable if the introductory formula is present, or if a reasonable amount of matching words leave the readers without doubt. In the case of allusion and echo, however, the criterion of intentionality plays a significant role, for presumably there will be no clue offered by means of the introductory formula. Thus, intentionality rather than explicitness is considered the key to recognizing allusions and echoes.

Thompson makes a two-pronged proposal. First, allusion "refers to statements which are *intended* to remind an audience of a tradition they are presumed to

---

[66] M. Thompson, *Clothed with Christ: The Example and Teaching of Jesus in Romans 12:1-15:13* (JSNTSup 59; Sheffield: Sheffield Academic Press, 1991), 30.

[67] C. D. Stanley, *Paul and the Language of Scripture*, 37, applies these criteria to non-Pauline materials, while for support refers to Michael V. Fox, "The Identification of Quotations in Biblical Literature," *ZAW* 92 (1980): 416-431.

[68] Porter, "The Use of the Old Testament," 95.

[69] Morgan, "Is There an Intertext in This Text?," 5.

know as dominical; clear examples by this definition are 1 Cor 7:10 and 9:14." [70] Second, an echo or a reminiscence (in the case of Paul's use of the OT) refers to "the cases where the influence of a dominical tradition upon Paul seem evident, but where it remains *uncertain whether he was conscious* of the influence at the time of dictating."[71]

Others draw a similar distinction between allusion and echo. While including echoes in the category of allusions, Porter criticizes Hays' seven criteria as 'large and complex' and 'unworkable,' [72] and suggests 'rather a simpler' criterion. Porter does not distinguishes between allusions and echoes, and argues that allusion (thus, echoes as well), "could refer to the non-formal invocation by an author of a text (or person, event, etc.) that the author could reasonably have been expected to know (for example, the OT in the case of Paul)." [73] Again, while informality is presupposed, the intentionality is assumed as a criterion for allusions/echoes (for Porter). Similarly, Keesmaat takes a simpler approach: while allusion is intentional, echo is unintentional.[74]

As we consider in particular the broad range of methodological implications that can be projected by the term 'echo," it seems reasonable and useful to distinguish between an allusion and an echo. Yet the criterion of intentionality hardly suggests any scientific system of measurement; how can we be sure of an author's intentionality for allusion or echo? What would be the signs? Stanley's definitions of allusion/echo exhibit the same difficulty: how can we recognize and even distinguish them? Again, Hays' proposal of seven criteria for discerning intertextual echoes is useful if one seeks helpful guidelines rather than a 'rigorous set' of criteria.[75] In the case of an echo, especially as the explicitness and intentionality dims, these criteria would prove to be effective.

Conclusively, as to the definitions of an allusion and an echo, the criterion of intentionality may be helpful but is not critical. With allusion, intentionality might

---

[70] Thompson, *Clothed with Christ*, 30 (italics his).

[71] Ibid.,(italics mine).

[72] Hays, *Echoes of Scripture*, 29-32, suggests seven criteria: (i) Availability – "was the proposed source of the echo available to the author and/or original readers?"; (ii) Volume – "the degree of explicit repetition of words or syntactical patterns"; (iii) Recurrence – "how often does Paul elsewhere cite or allude to the same scriptural passage?"; (iv) Thematic Coherence – "how well does the alleged echo fit into the line of argument that Paul is developing?"; (v) Historical Plausibility – "Could Paul have intended the alleged meaning effect?"; (vi) History of Interpretation – "how other readers, both critical and pre-critical, heard the same echoes?"; (vii) Satisfaction – "with or without clear confirmation from the other criteria listed here, does the proposed reading make sense? Does it illuminate the surrounding discourse?"

[73] Porter, "The Use of the Old Testament," 95.

[74] Keesmaat, "Exodus and Intertextual," 32.

[75] Wagner, *Heralds of Good News*, 11, n. 44, harks back against Porter's critiques of Hays' criteria, arguing that Porter's demand for a right methodology for "such metaphorically-charged literature as Paul's letter," would lead to "methodological *rigor mortis*."

be clearer than it is with echo, but does echo completely rule out an author's intentionality? Hays states that echo as an 'intertextual space' often denotes 'the rhetorical effect' occurring "between the readers and the tradition, which is evoked by the text, and is not necessarily within the author." [76] Viewing echo as 'rhetorical effect,' however, does not necessarily rule out intentionality. Although echo is a more subtle literary device than allusion, nevertheless, that subtleness, as Hays would acknowledge, does not have to be seen as incompatible with intentionality.

Taking an example from our study, Jesus feeding the thousands on the mountains in Matthew 14 and 15, as we shall argue later in the appropriate section, echoes the pastoral activity of the promised eschatological Shepherd. Although we cannot find explicit citations or implicit, allusive texts, the event as a whole functions as a code or a convention echoing the Davidic Shepherd tradition. The implied readers with ears attuned to the tradition are supposed to realize that this is, so to speak, an acted-out text of the tradition as to the promised Shepherd.

In this case, if the intentionality is dismissed, then the core message in the intertextual discourse, invoked by the miracle, is deprived as well, i.e., the message concerning the identity and mission of Jesus. The echo must have its source(s), direction, and intention. It is a voice reverberating from the wall, that is, tradition. But it is not a resounding, clanging cymbal created by merely clashing tradition and readers together. What we maintain is that the speaker and audience, in the cave of the text and the tradition, can hear the same echo.

As illustrated in this case, the echo can refer to the intertextual discourse invoked by the codes and conventions as well as by minimal linguistic clues, intended by the author, which enable both author and readers to engage in dialogue as they participate in reinterpreting the tradition. Hearing the echo likewise involves catching the notes of the underlying codes, conventions, patterns, motifs, and themes, which play significant roles in inviting the readers into the intertextual discourse. While allusion represents a more limited and isolated yet still apparently 'textual' interaction with the tradition, echo, with minimal linguistic clues, reaches much farther into the intertextual space particularly using those various mediums as illustrated above.

For the case of allusions in our study, i.e., Matt 9:36, 10:6, 16; 15:24, and 25:31-36, we argue that the invocations of the Davidic Shepherd tradition – sometimes specific OT texts and contexts – are intentional by the Evangelist as well as by the historical Jesus (if the authenticity of the relevant text is defended). As to the echoes of the Davidic Shepherd tradition in the Gospel, we classify three categories: (i) the texts involving shepherd/sheep images, such as we find in Matt 7:15; 12:11-12; 18:12-14; (ii) the thematic studies involving Matt 9:10-13, and the texts that center on the 'therapeutic Son of David' (Matt 9:27; 12:23;

---

[76] See Hays, *Echoes of Scripture*, 33; Keesmaat, "Exodus and Intertextual," 29-33.

15:22; 20:30-31; 21:14); and (iii) the Ezekielian pattern in Matt 27:51b-53 and 28:16-20.

For the case of the shepherd/sheep images in the texts listed above, it is true that intentionality is not explicit in the invocations of specific OT texts from the Davidic Shepherd tradition. As Hays notes on the 'vanishing point' problem,[77] as we come to these case of mere images of shepherd/sheep, it becomes difficult to be assured of hearing the echo of the tradition. Yet, importantly, the narrative context of Matthew's Gospel and Jesus' usages and practices of those images in other cases where these images are clearly associated with more explicit and intentional textual interactions, may help amplify the echoes. In this respect, Hays' criteria such as 'recurrence' and 'thematic coherence' prove helpful.

For the allusions and echoes in the select texts, the questions as to the sources and historicity of the text, and if necessary, as to the historical Jesus will be addressed. The synoptic problems remain open and unsolved. To illustrate, David L. Dungan recently issued a reminder of the extent to which scholars are influenced by their 'cultural assumptions' as to the varied issues of the synoptic problem.[78] While appreciating the helpful contributions of the redactional approach, we compare the synoptic parallels as we remain open to various theories of literary dependency, including the two-source hypothesis or Markan priority.[79]

(3) Christological Hermeneutics: Those texts of Matthew that involve the picture of Jesus as the therapeutic eschatological Son of David (ch. 4) and the One Davidic Shepherd (ch. 5) will be examined, so that we can clarify the pattern of echoes implied in the Davidic Shepherd tradition, which the Gospel adopts, follows, and modifies. While this can be an intertextual study, our study also touches upon the significant aspect of Matthew's christology. Later in Chapter 6, with the examination of two complementary texts (Matt 27:51b-53; 28:16-20) in Chapter 5, we investigate this christological link between Matthew's presentation of Jesus and the Davidic Shepherd tradition in view of Matthew's narrative strategy.

It has been pointed out that Hays' proposal must be complemented with the incorporation of the exegesis of Scripture in late antiquity, in addition to the

---

[77] Hays, *Echoes of Scripture*, 23, notes, "As we move farther from overt citation, the source recedes into the discursive distance, the intertextual relations become less determinate, and the demand placed on the reader's listening powers grows greater; as we near the vanishing point of the echo, it inevitably becomes difficult to decide whether we are really hearing an echo at all, or whether we are only conjuring things out of the murmurings of our own imaginations."

[78] Dungan, *A History of the Synoptic Problem: The Canon, the Text, the Composition and the Interpretation of the Gospels* (New York: Doubleday, 1999); also see, N. T. Wright, *The Meaning of Jesus: Two Visions* (New York: HarperCollins, 1999), 20-22, summarizes this current state of the issue.

[79] As Matthean studies that adopted this approach, see Yang Yong-Eui's *Jesus and the Sabbath in Matthew's Gospel* (JSSNTSup139, Sheffield: Sheffield University Press, 1997); A.F. Segal, "Matthew's Jewish Voice," in *Social History of the Matthean Community: Cross-Disciplinary Approaches* (ed. D.D. Balch; Minneapolis: Fortress Press, 1991), 3-37.

Hebrew Bible, or as Evans calls it, a 'multilayered intertextual echo.'[80] Thus to decode this intertextual discourse in which Jesus and Matthew receive and respond to the tradition, it will be necessary to trace the later developments of the tradition during the Second Temple period. Then the tradition-historical research of Chapters 1 and 2 applied to the analysis in chs. 3, 4 and 5 will prove useful and might actually complement Hays' literary-oriented approach. Some of the aspects of Jesus as the Shepherd will be clarified immensely and amplified if those other echoes in the Second Temple period can be heard.[81]

Finally, ecclesiology and christology are inseparable in our study, as in the case of Hays' study of Paul's use of the OT. Yet, unlike the ecclesiology that plays a key role in Hays' analysis of the echoes in the Pauline corpus,[82] our intertextual study will lead eventually to Matthew's use of OT citations (selective), allusions and echoes. His approach serves primarily his christological concerns; it is a christological/theological heremeutic. Morever, as J. Christiaan Beker astutely indicates, Hays' approach exhibits theological imbalance with its emphasis on the continuity between the NT and the OT neglecting the aspect of discontinuity. Thus we will not endorse one particular 'methodology,' but will keep attuned with ears open, to hear, as it were, the whole symphony.

4. Charting the Course

Our research proceeds through two main parts. In the first part, we will research the Davidic Shepherd tradition, first in the OT (ch. 1), and second, in its development in Judaism (ch. 2). In the second part, we will investigate the relevant texts of the First Gospel to examine how Matthew's Gospel interacts with this tradition. To analyze the Gospel's interaction with this tradition, we will first exegete the citations, allusions and shepherd/sheep images in the Gospel (ch. 3). Second, we will trace the echoes following two critical themes of the tradition, i.e., the Shepherd seeking the lost, and healing the sick in the context of Israel's restoration (ch. 4). Third, as we focus on Ezekiel's pattern in Matt 27:51b-53 and 28:16-20 in the following chapter (ch. 5), we will investigate the picture of Jesus as the One Davidic Shepherd-Appointee. Finally, in the last chapter (ch. 6), we

---

[80] See Evans, "Listening for Echoes of the Interpreted Scripture," 50-51.

[81] A few pointed out that Hays' proposal must be complemented with the incorporation of the exegesis of Scripture in late antiquity; see Evans, "Listening for Echoes of the *Interpreted* Scripture," 50 (italics mine), suggests "multilayered intertextaul echo" (51); Hays, "On the Rebound: A Response to Critiques of Echoes of Scripture in the Letters of Paul" in *Paul and the Scriptures of Israel*, 70-96, acknowledges the significance to hear 'polyphony' (71); also, Litwak, "Echoes of Scripture?," 285.

[82] See J. Christiaan Beker, "Echoes and Intertextuality," 68-69, on his theological questions as to Hays' argument for Paul's ecclesiocentric hermeneutic; cf. Hays, "On the Rebound," 94.

will assess how Matthew's re-reading of the tradition shapes his story of Jesus in the Gospel, in view of the grand scheme of the narrative structure.

In the conclusion of the dissertation, we will briefly summarize critical aspects of Jesus as the Davidic Shepherd in view of Matthew's theology, in particular, the christology of the First Gospel.

Chapter 1

The Davidic Shepherd Tradition in the Old Testament

Davidic expectation (cf. 2 Sam 7:12-16) and shepherd imagery in the OT merge together in some portions of the prophetic literature, such as in Micah 2-5, Ezekiel 34-37, and Zechariah 9-14. In all of these passages, the theme of the restoration of Israel combines these separate motifs.[1] It is our thesis that Matthew's Gospel embraces the Davidic Shepherd tradition of the Hebrew Bible not only at the Gospel's critical junctures (Mic 5:2 in Matt 2:6; Zech 13:7 in Matt 26:31), but also throughout the narrative as a whole (Ezek 34-37, for instance, in Matt 4:23; 6:9; 9:10-13, 36; 10:6; 15:12-28, 32-39; 25:31-46; 27:51-53; 28:16-20).

In this chapter, our objective is to examine the profile of the eschatological and Davidic Shepherd(s) envisioned in Micah 2-5, Ezekiel 34-37, and Zechariah 9-14. While Matthew does not explicitly refer to the passage, it is beneficial to examine Ezekiel 34-37 more closely than the other two passages since we build from the thesis that in significant ways the eschatological Shepherd found in Ezekiel 34-37 authenticates Jesus as the promised Shepherd. Before turning to the Davidic Shepherd tradition in the OT, we will briefly survey the usage of shepherd imagery in the Ancient Near East (ANE). Then we will compare ANE shepherd imagery with that in the OT. Last, as we turn our attention to the OT tradition, we will be able to assess the modifications and development of Davidic Shepherd imagery during the exile and before Jesus/ Matthew interact with the tradition in the First Gospel.

1.1 Shepherd Imagery in the Ancient Near East

Shepherd imagery is often associated with kingship in the ANE where it was commonplace for shepherds to care for flocks. Kings and gods alike were described repeatedly as shepherds because of their ruling position; thus kingship is rooted deeply in the portrayal of rulership as typified in the figure of a

---

[1] Passages such as Isa 40:10-11, 61, Jer 23:1-6, 31:1-37 and Ps 78:68-72 demonstrate the partial merger of those three motifs.

shepherd. Likewise, to speak of YHWH particularly as a shepherd is to speak of YHWH's kingship and his kingdom.

1.1.1 Shepherd Imagery and Kingship

Regardless of the epoch or region, in the ANE examples abound that indicate how shepherd imagery was applied to both human kings and to gods. In Ugaritic texts, the supreme god El is described in terms of a shepherd-king motif, e.g., El calls himself a 'shepherd' (*r'y*). G. R. Driver translates it ruler, and suggests this may be intended as a title of a god.² Whether considered a god or a king as a god's son, especially in Egypt, a ruler is expected to protect his people and provide them with what is necessary for their well-being, such as food, land, and justice. In the same way, a shepherd is supposed to care for his flock.³ The god-man Gilagamesh, a hero of the Hittite Empire from the middle of the second millenium B.C., is recognized as a shepherd: "Two-thirds of him is god [one-third of him is human] ... Is this [our] shepherd, [bold, stately, wise]?"⁴ Bel-Marduk, the fertility god of Babylon, is called the shepherd of the black-headed ones, that is, his creatures. Marduk is also believed to be the shepherd of all the gods like sheep.

Many other god-man figures are depicted as shepherds, most notably, Shamash, Mullil, Hammurabi, and Enlil: "Faithful shepherd, faithful shepherd, god Enlil, faithful shepherd, Master of all countries, faithful shepherd. The Lord who drew the outline of his land." ⁵ In Ugaritic texts, it appears that the god *Ba'lu* or Haddu is the only god referred to as a 'shepherd.' According to Marjo C. A. Korpel, *Ba'lu* is the great savior believed to save the spirits of the dead so they can return to the earth; "He, the Shepherd (*r'h*), will revive the ghosts."⁶

E. O. James suggests that the first usage of the shepherd-king motif in reference to an earthly, historical entity occurs with Lugi-zaggissi, the king of Umma (ca. 2500 B.C.). This king attacked and subdued Lagash, and proclaimed himself 'king of the Land' with the blessing and sanction of Enlil. As he assumed his reign over the entire country, he prayed that he might fulfill his destiny and always be 'the shepherd at the head of the flock.'⁷ Furthermore, a search for the ideal kingship in terms of the (good) shepherd is exemplified in

---

² G. R. Driver, *Canaanite Myths and Legends* (Edinburgh: T. & T. Clark, 1956), 67, n. 9.
³ Jack W. Vancil, " The Symbolism of the Shepherd in Biblical, Intertestamental, and New Testament Material" (Ph.D. diss., Dropsie University, 1975), 78; cf. J. Jeremias "Ποιμην, αρχιποιμην, ποιμαινω, ποιμνη, ποιμνιον," *TDNT* 6:485-502.
⁴ James B. Pritchard, ed. *Ancient Near Eastern Texts Relating to the Old Testament*, vol. 1: *Anthology of Texts and Pictures* (Princeton: Princeton University Press, 1973), 41.
⁵ Pritchard, *ANET*, 69, 72, 337.
⁶ Marjo Christina Annette Korpel, *A Rift in the Clouds: Ugaritic and Hebrew Descriptions of the Divine* (Münster: UGARIT-Verlag, 1990), 448-449.
⁷ E. O. James, *The Ancient Gods* (London: Weidenfeld and Nicolson, 1960),118.

Ipu-wer, an Egyptian prophet (second millennium B.C.) who looked forward to a day when the Pharaoh would live up to his expectation: "The Lord of eternity .... abiding like the heavens ... Exalted above millions to lead on the people forever ... His eye is the sun ... His stride is swift, a star of electrum ... The good shepherd, vigilant for all people, whom the maker thereof has placed under his authority, lord of plenty."[8]

The Egyptian king Amenhotep III (1411-1374 B.C.) is also called a 'shepherd,' supposedly being in close relationship with a deity with whom he shared authority: "The good shepherd, vigilant for all people, whom the maker thereof has placed under his authority, lord of plenty."[9] A Sumerian poem written on the occasion of Shulgi assuming the throne addresses the king as a shepherd. When Shulgi beseeches Nanna, the tutelary deity of Ur, for the favor of the great god Enlil of Nippur, Nanna enters the assembly of the gods and pleads on behalf of Shulgi, saying: "Father Enlil, lord whose command cannot be turned back...Bless the just king whom I have called to my holy heart, the king, the shepherd Shulgi, the faithful shepherd of full of grace, let him subjugate the foreign land for me."[10]

The Assyrian king Shalmaneser I (ca. 1280 B.C.) is referred to as a shepherd, and Tukulti-Urta I (1244-1208 B.C.) describes himself as the rightful ruler, meaning the true shepherd.[11] Similarly, the Babylonian king Sannecherib (705-687 B.C.) appears in the image of a shepherd: "Sannecherib, the great king, the mighty king, king of Assyria, king without a rival; prayerful shepherd [ruler], worshipper of the great gods; guardian of the right, lover of justice, who lends support, who comes to the aid of the needy, who turns [his thoughts] to pious deeds." Just as the Sumerian Shulgi, the Assyrian king Assurbanipal (668-626 B.C.) pleads with Enlil, addressing himself as Enlil's [under]shepherd: "For Enlil, lord of the lands, Assurbanipal, the obedient ruler [shepherd], the powerful king, king of the four regions [of the world], has rebuilt the brick [work] of Ekur, his beloved temple."[12]

From these examples, it is apparent that gods and kings in the ANE are pictured as shepherds of their people. Furthermore, kings perceived themselves as under-shepherds to the gods whose permission, guidance, and blessings were sought equally for themselves and for their flock. While shepherd imagery seems similar throughout the ANE, it has been observed that there are some distinct differences in its usage in Egypt, Mesopotamia, and Greece.

---

[8] James H. Breasted, *Egypt* (New York: Russell & Russell, 1962), 3:195, 3:243.
[9] Breasted, *Egypt*, 2:365-66.
[10] Samuel N. Kramer, *History Begins at Sumer: Thirty-Nine Firsts in Man's Recorded History* (Philadelphia: University of Pennsylvania Press, 1981), 278.
[11] Daniel D. Luckenbill, *Ancient Records of Assyria and Babylonia 1: Historical Records of Assyria from the Earliest Times to Sargon* (New York: Greenwood Press, 1968), 49-68.
[12] Daniel D. Luckenbill, *Ancient Records of Assyria and Babylonia 2: Historical Records of Assyria from Sargon to the End* (New York: Greenwood Press, 1968), 405.

According to J. W. Vancil, "Mesopotamian kingship is to be seen in great contrast to that of Egypt. Whereas the god-king of Egypt ruled by divine right and in virtue of his own divinity, the Mesopotamian ruler was made king through the tenuous thread of selection by his peers."[13] In Egypt, the idea of a king as shepherd conveys a 'divine' quality; in the minds of the people this means a god resides within the king. In terms of being the god's shepherd, the king in Egypt is thereby invested with the authority of a god. This is not the case in Mesopotamia where the king emerges without any divine quality ascribed to him.

In Greece it was believed that before men were put in charge of creatures inferior to them, the god in charge of mankind was called 'shepherd.' Thus when the god shepherds men, it is thought to be the ideal age.[14] Later, the imagery of shepherd appears particularly in military contexts. For instance, Nestor, 'king of sandy Pylos' and 'master of persuasive speech' succeeds in convincing the Greeks to battle against the city of Troy: "He spake, and led the way forth from the council, and the other sceptered kings rose up thereat and obeyed the shepherd of the host."[15] This raises the question of how 'shepherd' [of the host] is used here. The Homeric phrase, 'shepherd of the host' (ποιμένα λαῶν) refers mostly to 'a chief or commander of military forces.'[16] In Egypt and Mesopotamia, the concept of kindness turns up repeatedly as an attribute of the shepherd, but this attribute, according to Vancil, is absent from the Greek understanding of the shepherd.[17]

In summary, it may be a mistake to exaggerate the differences of shepherd imagery used in ancient Egypt, Mesopotamia, and Greece. Yet distinct characteristics exist. In Egypt, shepherd imagery is characterized by the power and authority of the god-kings. In Mesopotamia, shepherd imagery is characterized by the caring leadership of the king, while in Greece, shepherd imagery is often used in a military context, and is noticeably devoid of kindness. We have examined the relationship of shepherd imagery with the idea of kingship. Now we turn our attention to a subsidiary motif, the shepherd's crook.

---

[13] Vancil, *Symbolism,* 39-40, 42-43, 97.

[14] Plato, *Statesman* (H. N. Fowler, LCL), 271d -272a, says through the mouth of Eleatic: "For then, in the beginning, God ruled and supervised the whole revolution ... moreover, the animals were distributed by species and flocks among inferior deities under his own care, so that no creature was wild, nor did they eat one another, and there was no war among them, nor any strife whatsoever ... God himself was their [mankind's] shepherd, watching over them, just as man, being an animal of different and more divine nature than the rest, now tends the lower species of animals."

[15] Homer, *Iliad* (A. T. Murray, LCL), 2.78-85 (esp. 2.85).

[16] Homer, *Il.* 5.13-14 (esp. 5.14), alludes to the military chiefs (Astynous and Hyperim) as shepherds who fight against a lion to protect their sheep, i.e., soldiers.

[17] Vancil, *Symbolism*, 127, 249, argues that Jeremiah employed the idea with militant connotations (ch. 23) because of a consciousness that the kings and military leaders of the nations were called by the title.

1.1.2 Shepherd Imagery and the Shepherd's Crook

We have established how shepherd imagery conveys the idea of kingship. This is further discerned by the use of multiple, effective metaphors such as the shepherd's crook. A shepherd carries a crook and a king carries a scepter. Numerous ancient art objects signify connections between crooks and scepters.[18] For instance, a late Egyptian pre-dynastic (3400-3200 B.C.) pictorial record shows men approaching the palace of their overlord with a shepherd's crook over the shoulder of each local chief.[19]

From earliest times, the shepherd's crook is a badge of princely, and later of royal, leadership; it is preeminently the symbol of the power of leadership. The Egyptian deity Osiris is usually depicted as a human male clothed in a white garment. In his hands he holds a shepherd's crook and a flail, while on his head he wears a crown. In the 'Book of the Dead' he is named 'Osiris, the Shepherd.'[20] The Egyptian kings were exalted to the level of deity. John Wilson notes how pharaohs were regarded as shepherds sent by gods to tend mankind: "This is perhaps the most fitting picture of the good Egyptian ruler, that he was the shepherd for his people."[21] King as shepherd involves the role of 'feeder' who secures food, protection, and justice for the people under his care. Vigilance for the flock was one of the qualities of the ideal shepherd, the king.[22]

The link between kingship and shepherd imagery is vividly evidenced by the shepherd crook as the earliest insignia of the Pharaoh.[23] Likewise in Mesopotamia (second millennium B.C.) the crook is an instrument reserved for a male deity as an apparent representation of authority.[24] Just as in Egypt, the Mesopotamian god or god-king possesses a shepherd's crook.[25] In Greece, the

---

[18] For the items such as palettes and vases that have revealed something appearing to be a shepherd's crook from this prehistoric period of Egypt, see Elise J. Baumgartel's *The Culture of Prehistoric Egypt* (London: Oxford University Press, 1960), 2:146; or Mesopotamia, see John Wilson, *Before Philosophy* (Baltimore: Penguin Books, 1971), 151-153; Henri Frankfort, *Cylinder Seals* (London: Gregg Press, 1939), 164, presents the picture of Amurru, one of the first dynasties of Babylon holding assumably a shepherd's crook; E. D. Van Buren, *Symbols of the Gods in Mesopotamian Art* (Rome: Pontificum Institutum Biblicum, 1945), 142-144.

[19] William C. Hayes, *The Scepter of Egypt, Part 1* (New York: The Metropolitan Museum of Art, 1953), 28, 286; see also, A. B. Mercer, *The Pyramid Texts in Translation and Commentary* IV (New York: Longmans, Green, 1952), 58-60.

[20] E. A. Wallis Budge, *Osiris and the Egyptian Resurrection* (London: Medici Society, 1911), 2:16; also, Korpel, *Rift in the Clouds*, 449; Fikes, *Shepherd-King*, 52.

[21] J. Wilson, *Before Philosophy*, 88.

[22] Ibid., 89.

[23] Hayes, *The Scepter 1*, 285 (fig. 187), illustrates various kinds of crooks: (i) *mekes* is called a scepter, frequently represented in the hand of a king; (ii) *was*-scepter is believed to endow its owner with prosperity; (iii) *dja'm* is referred to in the Pyramid texts as the 'scepter of heaven' representing authority.

[24] Vancil, *Symbolism*, 31.

[25] Van Buren, *Gods*, 142-144.

connection between the shepherd's crook and kingship is less clear. In Homeric material King Agamemnon's staff is referred to in certain militaristic contexts evidenced in the title 'shepherd of the host.'[26]

The king's scepter might have been derived from a military weapon, i.e., a lance or spear. Since shepherds and their flocks probably preceded military figures, however, and early on the shepherd's crook was recognized as a symbol of authority, it is likely that the crook is the precursor to the king's scepter. J. Wilson points out that one of the words meaning 'to rule' finds its origins in the shepherd's crook.[27] As the shepherd's crook is used to protect the flock from wild beasts and guide them, it becomes obvious that concepts like justice and fairness are inherent in shepherd imagery. Hammurabi (1792-1750 B.C.) as the shepherd-king is introduced as the one who "makes law prevail; who guides people aright," as he says, "I (Hammurabi, the shepherd) have governed them in peace."[28]

Other attributes of kingship, such as showing mercy toward the people and securing the peace of the land, while not used explicitly and frequently, were also associated with the shepherd.[29] An ancillary question is raised if any healing activity is evidenced in the shepherd-king imagery used in Ezekiel 34-37. In Greece, Machaon was a kind of surgeon general, yet it is not certain whether he is called 'shepherd of the host' because of his position as an army commander, or because of his position as army surgeon.[30]

1.1.3 Summary

To sum up, shepherd imagery in the ANE is a means of expressing kingship. Protection and provision are key characteristics of shepherd imagery applicable to both kings and gods, although the usage of the imagery differs in emphasis according to regions. As we have just seen, the shepherd's crook or staff is a symbol of authority, leadership, and rule. When examining the shepherd imagery used in the ANE, a question arises concerning similarities and differences evidenced in the shepherd imagery used in the OT.

Indeed, the image of the caring shepherd of Mesopotamia, the authoritative shepherd-king of Egypt, and the shepherd figure associated with the military are found in the OT (Jer 23:1-4; Ps 2:9; cf. Isa 11:4; Mic 2:12-13) as well. Our research thus far, however, has failed to uncover any indication of healing in the

---

[26] Homer, *Il. ii*.100-108.
[27] Wilson, *Before Philosophy*, 88.
[28] Pritchard, *ANET*, 164-165, 177-178.
[29] Vancil, *Symbolism*, 343; Pritchard, *ANET*, 440, comments on a cuneiform acrostic poem that was written before 700 B.C. which reads: "May the goddess Ishtar, who ..., have mercy upon me! May the shepherd, the sun of the people, [have mercy]."
[30] Homer, *Il. ii*.505-598.

ANE use of shepherd imagery. Thus to emphasize this aspect of shepherd imagery seems arbitrary.

In contrast, as we shall examine later more fully in the next section, the OT shows YHWH as the true Shepherd of Israel (Ezek 34-37) seeking the lost among the flock, healing the sick, and strengthening the weak. There is a proliferation of shepherd imagery in the OT, especially pronounced in the prophets' promise of the restoration of Israel. It is noteworthy how the imagery is interrelated to restoration and resembles YHWH's redemption in the exodus. As the course of Israel's history unfolds, Davidic expectation plays an increasingly significant role as it employs the shepherd/sheep image to draw blueprints for the future.

1.2 Shepherd Imagery in the Old Testament

1.2.1 Characteristics

Before we trace shepherd imagery in the OT, we must first consider its characteristics in light of the surrounding cultures. In the ANE, shepherd imagery is applied to both gods and kings; in the OT shepherd imagery is applied to YHWH, human rulers in Israel, and also to Gentile nations.[31] One particular aspect of the OT's use of the term shepherd is that unlike the gods in the ANE, YHWH is believed to be the one, true, and transcendent God supreme over all creation, including the pantheons of gods regarded as idols and thought to be nothing but part of the creation itself.[32] Significantly, in the OT, the opposite holds true: these gods are referred to as lifeless (e.g., Isa 46).[33]

In the OT, the term shepherd is rarely used to refer to pagan kings, and the term is never used to refer to idols.[34] Though shepherd imagery is plentiful in the OT, there seems to be a specific development of its usage as Israel's history progresses. For instance, no patriarch is called a shepherd of Israel. Clearly the

---

[31] For YHWH himself: Gen 48:15; 49:24 (cf. Ps 80:2); Mic 2:12-13; 4:6-8; 7:14-15; Jer 23:2; 31:10; 50:19; Isa 40:10-11; 49:9b-13; Pss 23; 28:9; 80:1; 77:20; 78:52-55; 74:1; 79:13; 95:7; Ezek 34:31. For Israelite leadership: Num 27:17; 1 Kgs 2:17; 1 Sam 21:8; 2 Sam 5:2; 7:7-8; Ps 78:70-72; Jer 2:8; 3:15; 10:21; 22:22; 23:1-4; 25:34-36; 50:6; Ezek 34:2-10; Isa 56:11; Zech 10:3; 11:5-8. For Gentile rulers: Jer 6:3; 12:10; 25:34-36; 49:19; 50:44; Nah 3:18; Isa 44:28.

[32] For the comparison of monotheistic religion with the myth of the nations around her, see J. Wilson, *Before Philosophy*, 236-237 (esp. 'The Emancipation from Myth').

[33] Fikes, *Shepherd-King*, 68.

[34] One exception is the case of the Persian King Cyrus in Isa 44:28, who is called, 'my shepherd'(רֹעִי, cf. Zech 13:7). This is an anomaly, an unexpected use of 'shepherd' as a title as we consider that none of the kings of Israel is called by this title except the new David in Ezek 34:23 (cf. J. J. Collins, *The Scepter and the Star* [New York: Doubleday, 1995], 28). The case of Cyrus may characterize YHWH's free and sovereign exercise of his shepherd rulership.

pre-eminent shepherd of Israel at the time of the patriarchs was YHWH himself (Gen 48:15, 49:24). Even Moses is remembered as being under the shadow of YHWH, Israel's prime shepherd: "You led your people like a flock by the hand of Moses and Aaron" (Ps 77:20; cf. Isa 63:11).

The inception of the monarchy in Israel is pivotal in terms of the development of the OT shepherd imagery. J. Vancil asks why shepherd imagery is reserved exclusively for YHWH until the onset of the kingship in Israel,[35] for at a later date kings themselves were designated shepherds (Jer 22:22, 23:2, 4; Ezek 34:2, 7; cf. 1 King 22:17). Yet no specific king in Israel is described in shepherd imagery as YHWH's royal representative, with the exception of David before he assumed the throne (cf. 2 Sam 5:2; 1 Chr 11:2; Ps 78:71). ANE shepherd imagery reveals an intimate connection between a human king and deity in terms of title. In Israel, the title 'shepherd' squarely belongs to YHWH, and thus indicates his particular rulership. The OT tends to reserve shepherd imagery for YHWH and, significantly, extends its use only for YHWH's Davidic Appointee (Mic 5:2-4 [5:1-3]; Jer 3:15; 23:4-6; Ezek 34:23-24; 37:24-25; cf. Zech 13:7).

Shepherd imagery is extensively articulated around the period immediately preceding, during, and following the exile, as J. Jeremias comments: "It is to be noted that the references are not spread evenly over the whole of the OT. It is true that in Exodus-Deuteronomy shepherd terms are used in the exodus stories ('to lead,' 'to guide,' 'to go before'), but in general it is hard to determine whether there is any conscious feeling for the shepherd metaphor. More commonly, and with details which show how vital the concept is, the figure of speech is found in the Psalter and in the consoling prophecy of the Exile."[36] J. L. Mays agrees that "the title 'shepherd' belongs to YHWH's identity as ruler of his people. The role of shepherd became prominent especially in exilic salvation prophecy when the predicament of the dispersion appeared to offer no future beyond the dissolution of Israel among the nations (e.g., Jer 23:3; 31:8-10; Isa 40:11; Ezek 34)."[37]

Both exodus and exile settings are replete with shepherd imagery. While in Exodus-Deuteronomy YHWH is the prime Shepherd, in the context of the restoration of Israel from exile, YHWH along with his Davidic Appointee are the Shepherds. During the monarchy, many under-shepherds appear and as noted previously, the titular use of shepherd imagery is reserved until YHWH's Davidic Appointee is announced with a view to the restoration of Israel. J. Jeremiah is correct that shepherd imagery undergoes 'a unique and, from the NT

---

[35] Vancil, *Symbolism*, 340-343. One should note, however, that the leaders of Israel in the wilderness were also designated shepherds by YHWH (2 Sam 7:7//1 Chr 17:6; Isa 63:11).

[36] Jeremias, *TDNT* 6:487.

[37] J. L. Mays, *Micah: A Commentary* (Philadelphia: Westminster, 1976), 75.

standpoint, final development' in the later part of Zechariah.³⁸ It is to the task of investigating this development of shepherd imagery in the OT that we now turn.

1.2.2 The Shepherd of Israel: Redemptive Leadership and David

Shepherd image is used extensively, both in in the OT and in the ANE, though it is difficult to pinpoint when and how the imagery found its way into the biblical accounts.³⁹ The first occurrence of the imagery and probably the earliest text in which YHWH is called 'shepherd' is Gen 49:24-25 (cf. 48:15-16, 49:10).⁴⁰ Jacob blesses Joseph calling YHWH 'the shepherd (רֹעֵה), the Rock of Israel' who leads and blesses Joseph. Similarly we find this thought expressed in Ps 80:1, "Hear us, O Shepherd of Israel, you who lead (נֹהֵג) Joseph like a flock." The prime role of the shepherd of Israel is to lead his flock.

Interestingly, we discover in Gen 49:10 that the scepter (שֵׁבֶט) or the ruler's staff, which may refer to the shepherd's crook, belongs to Judah, and its dominion will extend over the nations. While YHWH is the Shepherd of Israel, the scepter belongs to Judah, and the nations will eventually obey the authority coming from Judah. YHWH is called the Shepherd of Israel, yet the need of and the search for the human representative of YHWH's leadership emerges as a crucial issue in the history of the nation.

In Num 27:16-17, Moses' request for a new leader for Israel contains the metaphor 'sheep without a shepherd' (כַּצֹּאן אֲשֶׁר אֵין־לָהֶם רֹעֶה; LXX, ὡσεὶ πρόβατα οἷς οὐκ ἔστιν ποιμήν; cf. 1 Kgs 22:17; 2 Chr 18:16, Ezek 34:5, 8). It is intriguing how the role of a shepherd is described: "a man over the community to go out and coming in before them (LXX, ὅστις ἐξελεύσεται πρὸ προσώπου αὐτῶν καὶ ὅστις εἰσελεύσεται πρὸ προσώπου αὐτῶν; cf. Matt 26:31-32), one who will lead them out (יָצָא; LXX, ἐξάγω) and bring them in (בּוֹא, LXX, εἰσάγω)."⁴¹ It is possible that the expressions, ἐξάγω and εἰσάγω, may echo YHWH's role in leading them out of Egypt and bringing them into the promised land. Numerous uses of this expression illustrate YHWH's redemptive activity for his people. For example, in Exod 3:8 (LXX) YHWH rescues (נָצַל) them from Egypt to bring them out (LXX, ἐξαγαγεῖν) from that land and to bring them into (LXX, εἰσαγαγεῖν) the good land of milk and honey (cf. Exod 3:10, 11, 17; 6:6, 13, 26; 7:5; 12:42; 17:3; 23:23; Num 20:5; Deut 4:38). Deut 6:23 (LXX) also repeats this expression and applies it to YHWH leading his people from Egypt

---

³⁸ Jeremiah, *TDNT* 6:488.

³⁹ Vancil, *Symbolism*, 248, suggests an Egyptian connection as a strong possibility, that is, the early Hebrew group's experience in Egyptian society.

⁴⁰ Korpel, *A Rift*, 449-450.

⁴¹ The language here is often considered to depict military movements led by the shepherd as military leader (Deut 20:1, 21:10, 28:25, Judg 4:14, 2 Kgs 19:9, 2 Sam 5:24). Refer to Baruch A. Levine, *Numbers 21-36* (*AB*; New York: Doubleday, 2000), 42-43; Philip J. Budd, *Numbers* (WBC 5; Waco: Word Books, 1984), 306.

into the promised land: καὶ ἡμᾶς ἐξήγαγεν ἐκεῖθεν ἵνα εἰσαγάγῃ ἡμᾶς δοῦναι ἡμῖν τὴν γῆν ταύτην ἣν ὤμοσεν δοῦναι τοῖς πατράσιν ἡμῶν. 'To bring out' and 'to bring in,' therefore, can be viewed as technical language for the shepherd's role that echoes YHWH's own redemptive leadership.

As we reflect on Israel's history, we see that the OT usage of shepherd imagery for human leaders finds its ideal type in David the son of Jesse (2 Sam 5:2; 7:5-7, 12-16). YHWH's promise of an eternal throne for David's house (2 Sam 7:12-17) coincides with YHWH entrusting David as the shepherd of Israel: "For some time, while Saul was king over us, it was you who led out (LXX, ἐξάγω) Israel and brought it in (LXX, εἰσάγω); the LORD said to you, 'It is you who shall be shepherd (רָעָה; LXX, ποιμαίνω) of my people Israel, you who shall be ruler over Israel '" (2 Sam 5:2; cf. 1 Chr 17:6). Once again we discover the language 'leading out' and 'leading in' clearly associated with shepherd image. YHWH entrusted others to lead his people as shepherds of Israel (2 Sam 7:5-7), but David's leadership is promised to be perpetual.

Psalm 2 depicts the heavenly king enthroning his son, the anointed, on Zion, declaring, "You are my son, today I have become your Father"(v.7, NIV), which recalls the promise given to David in 2 Sam 7:12-17. The Lord's anointed will "rule [the nations] with an iron scepter" (LXX, ποιμανεῖς αὐτοὺς ἐν ῥάβδῳ σιδηρᾷ) and will "dash them to pieces like pottery" (v.9).[42] Surprisingly LXX reads MT's תרעם (רָעָה, 'to break') as ποιμανεῖς ('to shepherd')! While the imagery of shepherding evokes military action, the concept here speaks of an irreversible, final judgement in the manner pottery is broken into pieces.[43] The shepherd imagery used here is implied by the metaphor of the shepherd's staff that symbolizes the authority of YHWH, the shepherd-king of Israel.

In the Psalms, shepherd imagery reveals one of the deepest aspects of the relationship between YHWH and his people, and that is demonstrated in the covenant formula: "For he is our God, and we are his people of his pasture, and the flock of his hand" (95:7; cf. 100:3). Several Psalms describe YHWH as the Shepherd of Israel who cares for, saves, and guides them (Pss 23; 28:8-9; 77:20; 19:13; 80:1-3; 95:6-7; 119:176), and who gathers and even heals them (Pss 147:2-3; cf. 44:11, 22). Particularly in Psalm 78:68-72 David's leadership is idealized as YHWH's own (v.72). The last verse preceding Psalm 78 (Ps 77:20)

---

[42] The LXX reads MT's שֵׁבֶט as ῥάβδος, the shepherd's staff, not 'tribe.' A. A. Anderson, *Psalms 1-72* (Grand Rapids: Eerdmans, 1972), notes that the 'rod of iron' here refers to the royal scepter in the form of a long staff or a short handled battle mace.

[43] Hans-Joachim Kraus, *Psalms 1-59* (trans. Hilton C. Oswald; Minneapolis: Augsburg, 1989), 132 suggests Egyptian coronation and jubilee rituals when "the king demonstrated his worldwide power by symbolically smashing earthen vessels that bore the names of foreign nations," and in a similar sense "the Mesopotamian texts frequently mention the fact that a ruler smashes nations 'like pottery'." Therefore, shepherd imagery may not be irrelevant to the picture of the ruler/king in those texts.

1.2 Shepherd Imagery in the OT

makes apparent that it is YHWH himself and not simply Moses who led his flock Israel.

In contrast, Psalm 78 ends with the confirmation of YHWH choosing 'the tribe (שֵׁבֶט; LXX, φυλή) of Judah' (v.68).[44] A later parallel structure with the same verb means YHWH chooses 'his servant, David' (v.70) to be 'the shepherd of his people' (v.71). YHWH's choice of David is to be understood in terms of Israel's unfaithfulness to the laws and decrees of their Lord (vv. 51-56). In this respect, David is depicted as the ideal shepherd leading his people 'with integrity of heart' (v.72). The impression is, therefore, that the coming restoration of YHWH will be like the exodus, yet the prime agent of this new exodus will be a new David – that is, *one not like Moses but one like David.*

In the book of Jeremiah, we observe that shepherd imagery is closely tied to the hope of restoration and to YHWH in his role as the Shepherd of Israel gathering his scattered flock. Though lacking the details found in Ezekiel 34-37, Davidic expectation is undeniably linked to the hope for the future time of the promised restoration: "I will place shepherds over them who will tend them, and they will no longer be afraid or terrified; 'The days are coming,' declares the Lord, 'when I will raise up to David a righteous King who will reign wisely'" (23:4-5; cf. 33:12-18). The theme of YHWH's confrontation with the failed shepherds of Israel probably refers to political local leaders as distinguished from priests and prophets (cf. Jer 2:8): "Weep and wail, you shepherds; roll in the dust, you leaders of the flock. For your time to be slaughtered has come; you will fall and be shattered like fine pottery" (Jer 25:34; cf. Ps 2:9).

In the context of the restoration and the new covenant set out in Jeremiah 30-31, YHWH healing his flock (30:13-17; cf. 31:10) is associated with forgiveness of their sins and guilt (30:15; cf. 31:31-34). The emphasis is on the eschatological Shepherd who gathers and heals the scattered flock, thus forgives their sins. However, the establishment of the Davidic kingship is deemed the result of the restorative act of YHWH, the Shepherd of Israel (30:9).

We notice an intriguing pattern in the use of shepherd imageries found in Jeremiah 30 and 31. In Jer 30:8-9, we find the promise of the Davidic king at the time of YHWH's restoration of Israel. David himself is mentioned: "In that day, I will break the yoke off their neck; instead, they will serve the Lord their God and David their king, whom I will raise up for them" (cf. Matt 1:1-17). Then an explicit shepherd-saying appears in Jer 31:10: "He who scattered Israel will gather them and will watch over his flock like a shepherd" (cf. Matt 2:6). After delivering the promise of the future David, and assuring that the shepherd would gather scattered Israel, Jer 31:15-17 announces that comfort will come

---

[44] The term שֵׁבֶט here means 'tribe,' while it can also mean the shepherd's staff (cf. Gen 49:10; Num 24:17; further, Isa 14:5, 29; Ezek 19:14, 15; Amos 1:5, 8). The term means 'tribe' mostly in the Pentateuch (e.g., Exod 28:21; 29:14; Num 4:18; 18:12; Deut 10:8; 29:7; Jos 22; 1 Sam 10:20; 1 Chro 5) rarely in the Prophets. The philological connection is intriguing.

to the weeping Rachel (cf. Matt 2:8). This is set in the context of hope for the future restoration. In the restoration is an implied process. With the coming of the new David, YHWH himself will be the eschatological Shepherd for afflicted Israel, who will then be restored to their land.⁴⁵

The book of Isaiah is rich with shepherd imagery associated with the Davidic expectation in the context of the restoration of Israel. The vision described in Isa 11:1-16 expands to universal dimensions in its portrayal of future renewal as YHWH brings up and gathers his exiled remnant from Egypt (cf. Isa 63:10-14). The Davidic ruler to whom the eternal throne is promised (Isa 9:7, cf. 2 Sam 7:12-17) will be called "wonderful counselor, mighty God, everlasting father, prince of peace"(Isa 9:6).

Furthermore, this Davidic figure is 'a branch' coming from the stump of Jesse (v. 1).⁴⁶ The coming new ruler, markedly described in shepherd imagery, is the Davidic branch who will "strike the earth with the rod (שֵׁבֶט; LXX, λόγος) of his mouth; with the breath (רוּחַ; LXX, πνεῦμα) of his lips he will slay the wicked"(v. 4; cf. Ps 2:9). The shepherd imagery is invested with authority and imbued with power to judge in the eschatological context when the branch judges the nations. The term שֵׁבֶט here refers to a shepherd's staff. Rendering שֵׁבֶט as 'rod' fits the expression, 'he will strike.' Interestingly, LXX translates it as λόγος since it is to come out of 'his mouth.' In Ps 23:4, 6, the Lord's rod and staff are paralleled to his 'goodness'(טוֹב) and 'faithfulness'(חֶסֶד).⁴⁷ The Lord's staff is already seen as his ruling principles in the OT.

Isaiah 40:1-11 highlights a distinctive aspect of shepherd imagery in the context of Israel's restoration (esp. vv. 9-11): the compassion of the Shepherd. After the preparation of the way of the Lord in the wilderness (v. 3; cf. Matt 3:3// Mk 1:3//Lk 3:4), the Lord who comes i s depicted as the compassionate Shepherd of Israel: "He tends the flock like a shepherd; He gathers the lambs in his arms and carries them close to his heart; he gently leads those that have young"(v. 11; cf. Isa 49:1-12, 22). YHWH's compassion as the Shepherd for Israel at the time of the exodus is emphasized again in Isa 61:1-14 (cf. Isa 14:1).⁴⁸

---

⁴⁵ The restoration process in Jer 30-31 reveals a remarkable resemblance to the flow of the motifs found in Matt 1-2: from 'new David' (1:1-17), 'new ruler who will shepherd Israel' (2:6), to 'the end of Rachel's weeping' (2:18).

⁴⁶ Brevard S. Childs, *Isaiah* (Louisville: Westminster John Knox Press, 2001), 97-106, succinctly notes that "chapter 11 begins with the end of the old ... the proud and corrupt house of David"(102) and "the prophetic picture [vv.6-9] is not a return to an ideal past, but the restoration of creation by a new act of God through the vehicle of a righteous ruler"(104).

⁴⁷ Korpel, *Rift*, 450, says the abstract qualities of the ideal shepherd are the "names of the shepherd's staffs," and also mentions Zech 11:7.

⁴⁸ In the OT, besides the texts mentioned above, there are few passages in which shepherd imagery is adapted in different contexts. In Ps 49:14, it is in death that those shepherds (יִרְעֵם) are appointed to Sheol. Jer 43:12 uses the shepherd metaphor to relay that Babylonian King

1.2.3 Summary

To sum up, shepherd imagery applies to deity and human leaders both in the ANE and OT, though there is a noticeable difference. The monotheistic reservation of the titular use is characteristic of the OT usage. YHWH's leadership as the Shepherd of Israel can be described as 'redemptive' and restorative especially in Exodus-Deuteronomic and in the exilic context. In this regard, YHWH's healing his flock (Jer 30:13-17; cf. 31:10) associated with forgiveness of Israel's sins and guilt (30:15; cf. 31:31-34), and the eschatological Shepherd's compassion (Isa 40:11; 49:1-12, 22; 61:1-14), are the features of kingship almost unique to the OT against the ANE background. Leadership defined by OT shepherd imagery is not merely a projection of an idealized kingship, for YHWH himself reveals his shepherd leadership in the course of Israel's history. The promised Davidic kingship is thus characterized by YHWH's shepherd-leadership demonstrated in the exodus, which is yet to be revealed in the coming restoration.

Davidic leadership is couched in shepherd imagery. In the OT, the only Shepherd of Israel is YHWH who will eventually initiate his reign over Israel and the nations. As the ideal under-shepherd in the OT was David, the ideal under-shepherd to come will be one like David. As YHWH's Appointee in view of Israel's restoration, the new David will be distinctive in his shepherd-leadership as he leads the flock to follow YHWH. Israel's restoration is pictured in the coming new exodus indicated by the language of YHWH's shepherding activities of 'bringing out' and 'bringing in'; however, the major figure is not Moses but the Davidic Shepherd (cf. Ps 78).

In the following sections, we will examine three OT passages (Mic 2-5, Ezek 34-37, and Zech 9-14) to see how shepherd imagery is imbued with Davidic expectation related to the restoration of Israel. To neglect an examination of these texts is to undermine our comprehension of the background of Jesus as the Davidic Shepherd as employed in the First Gospel. Micah 2-5 and Zechariah 9-14 are no less important than Ezekiel 34-37 as they pertain to Matthew's Gospel.

Nevertheless, Ezekiel's shepherd imagery requires closer attention. Even though Ezekiel is never quoted directly in Matthew, it remains one of the major resources for the Gospel, and provides it with a profile of the eschatological and Davidic Shepherd(s) and a critical pattern for the restoration of Israel.

Nebuchanezzar's conquest of Egypt will be as easy as a shepherd wrapping around himself his garment. Ecc 12:11 introduces the expression, רֹעֶה אֶחָד ('one shepherd') by whom the words of the wise are given.

1.3 The Davidic Shepherd Tradition in Micah 2-5

Micah's vision of Israel's restoration unfurls with YHWH confronting false prophets (2:6-11), followed by YHWH's eschatological gathering his flock, and opening the way as their leader who goes before them (2:12-16). YHWH rebukes the false leaders and prophets (3:1-9) once again; the lack of the divine presence among the flock, and the consequential barrenness in Zion are highlighted (3:10-12). All is not lost, however, for 'in the last days,' the Lord's temple will be established (4:1), and the nations will rush to the mountain of the Lord (4:2). YHWH's kingship will be restored (4:7-8). YHWH will antithetically gather the nations that will be defeated by the 'Daughter of Zion,' described dramatically as being outfitted with horns of iron and hoofs of bronze (4:11-13). Micah 5:1-4 is thus placed in the context of Israel's future restoration, Davidic expectation, and the return of YHWH's kingship to Zion. As we shall see in this series of prophetic visions, shepherd imagery in various ways plays a significant role.

1.3.1 Micah 2:12-13: The Shepherd as the Breaker

Micah 2:12-13 is a salvation oracle sandwiched between YHWH's indictments issued against certain leading groups in Israel, including false prophets and leaders (2:1-11 and 3:1-12). The sudden transition from verse 11 to 12 is not careless editing, or Micah's allegedly one-sided oracle of judgment.[49] Nor may it be said that YHWH is capricious. The lack of reasons given for this abrupt shift leaves us with the singular explanation that YHWH will accomplish the restoration only because of his compassion, his חֶסֶד (cf. Mic 7:18).

---

[49] One of the arguments for an exilic or post-exilic dating of Mic 2:12-13 is that the motif of YHWH as a shepherd leading his sheep is often found in exilic and post-exilic texts, allegedly in Isa 40:11; Jer 23:3; 31:7, 10; Ezek 34:10ff; where YHWH as king goes before his people in Isa 43:15; 52:7, 12. See, Hans W. Wolff, *Micah: A Commentary* (trans. Gary Stansell; Minneapolis: Augsburg, 1990), 74, 76; also, R. Manson, *Micah, Nahum, Obadiah* (Sheffield: JSOT press, 1991), 43. Contra Charles S. Shaw, *The Speech of Micah: A Rhetorical-Historical Analysis* (JSOTSup; Sheffield: JSOT press, 1993), 77, argues against dating Mic 2:12-13 later than the eighth century B.C. for lack of conclusive evidence, and insists there are sufficient connections with themes, images, and ideas to justify considering vv.12-13 part of the rhetorical unit of vv. 1-13; Francis I. Andersen and David N. Freedman, *Micah: A New Translation with Introduction and Commentary* (AB; New York: Doubleday, 2000), 343, present a more convincing argument in this case: "To say that vv. 12-13 cannot be an integral part of the discourse of chapters 2-3 is to miss the complexity of the pradoxical, ambivalent, dialectical relationship between YHWH and Israel in the ongoing covenant (at once conditional and unconditional) as it was appreciated by the classical prophets." Why should one expect Micah or YHWH to follow one-sided linear logic as would modern scholars as they write their dissertations, particularly when YHWH/Micah are filled simultaneously with anger and compassion as they speak against/for the people?

1.3 The Davidic Shepherd Tradition in Micah 2-5

In Mic 2:12, YHWH as the eschatological Shepherd proclaims that he will reverse the fate of his scattered people (vv. 4-5) by gathering (LXX, συνάγω) them like sheep in affliction (LXX, ὡς πρόβατα ἐν θλίψει) gathered in a pen. In verse 13, the speaker appears to be the prophet. The shepherd figure becomes the king (מֶלֶךְ) who will pass through the gate with his gathered people going 'before' (לִפְנֵיהֶם, cf. Exod 13:21; Num 27:16-17; Zech 12:8) them at their head (LXX, καὶ ἐξῆλθεν ὁ βασιλεὺς αὐτῶν πρὸ προσώπου αὐτῶν ὁ δὲ κύριος ἡγήσεται αὐτῶν). How verse 13 relates to verse 12 remains problematic.

Rather than the imperfect, in verse 13 the perfect and imperfect consecutive verbs are used, though this does not confine what is said to Israel's past history. The function of the perfect in verse 13 may be to assure the results of YHWH's eschatological gathering of Israel.[50] Moreover, the image of the shepherd going 'in front of them (the flock)' [x2] in verse 13 closely intertwines with such concepts such as 'gathering,' 'breaking through,' and the 'king' applied in the context of the eschatological restoration.

The context indicates that verse 13 is both the chronological and logical continuation of verse 12, though the image changes from pastoral (v. 12) to militaristic (v. 13). The main figure in verse 12 is YHWH as the Shepherd, whereas in verse 13 the main figure is the king, the breaker (הַפֹּרֵץ). The roles are not incompatible since shepherd and kingship are inseparable in the OT as well as in the ANE. It is likely that in verse 13 YHWH steps back, and the verse reflects on what has transpired as a result of his intervention (v. 12).[51]

In verse 13, the role of 'the breaker,' that is, 'the one who makes a breach' (הַפֹּרֵץ) is closely tied to the movement of the flock. The breaker leads and goes before them in order to break through the gate. Then they are able to go out and follow him. The identity of the breaker is unclear, and whether the breaking is totally dependent on the breaker or not in verse 13 is obscure. The subject of הפרץ can easily be seen as YHWH, yet there is a careful distinction between the first person singular in verse 12 and the third party in verse 13. The מֶלֶךְ also remains unidentified. Moreover, it is unclear whether the breaker alone or whether the flock participates with the breaker in causing the breach. Of the five verbs in verse 13, only the first and last are singular, while the other three verbs are attributed to the movement of the flock with their shepherd at their head, namely, their leader.[52] Even though the shepherd-king is the breaker, the picture that emerges is of a gathered flock with the shepherd at its head breaking

---

[50] See Wolff, *Micah*, 84-85.

[51] Carol J. Dempsey, "Micah 2-3: Literary Artistry, Ethical Message, and Some Considerations about the Image of Yahweh and Micah," *JSOT* 85 (1999): 123.

[52] Cf. Andersen and Freedman, *Micah*, 340, state that "the passage through the gate could then be that of an army, not necessarily of a flock of sheep." If the shepherd can be compatible with the kingship, then the compatibility between the flock and the army is not impossible as well.

through the gate together, as a whole, to go out (v.13a, עָלָה הַפֹּרֵץ לִפְנֵיהֶם פָּרְצוּ).[53] In verse 12, the task of gathering is clearly YHWH's as the Shepherd of Israel. Yet in verse 13, YHWH leading his flock will be revealed in the king breaking through the gate, thereby opening the way out for his flock.

When considering the role of the eschatological Shepherd, it is intriguing to discover what it means to break open the way through the gate. J. L. Mays notes that YHWH breaks out to slay anyone who violates the limits set by the sanctity of his presence, especially in connection with the ark (2 Sam 6:8; 1 Chr 15:13; Exod 19:22, 24; cf. similarly, 2 Sam 5:20).[54] But the verb פָּרַץ has YHWH as its subject in other texts where it is used for breaking through the wall of a fortified city (Pss 80:13 [12]; 89:41 [40]) or for breaking through enemy lines, employing the simile of a bursting dam in 2 Sam 5:20 (cf. 1 Chr 15:13; 2 Sam 6:8).[55] Also, the imagery of the 'fold' and the 'pen' does not suggest Jerusalem but a foreign city, perhaps Babel (Isa 48:20; 52:11f) from whence all of the exiles are to return to their own country. This line of interpretation makes sense because one goes down, not up (עָלָה) from Jerusalem. Moreover, the salvation hoped for would not be a departure from, but a return to, Jerusalem (1 Kgs 12:27f.; Isa 40:9f; 52:7-12; Jer 31:6; Ps 122:4).[56]

The shepherd imagery, the exodus motif, and the hope of the restoration all intermingle in Mic 2:12-13. It is the shepherd who will break through what imprisons the remnant flock of YHWH; it is the king who will shepherd YHWH's flock by breaking the gate and going ahead of them. He will be their Lord. Clearly the shepherd imagery is closely tied to YHWH's future action in the time of the restoration.

1.3.2 Micah 4:11-13: The Eschatological Battle and the Gathered Flock

The second unit in which shepherd imagery resumes is Mic 4:6-5:4. YHWH as the Shepherd of Israel will gather (קָבַץ) the exiles (v. 6; cf. Jer 31:10; Ezek 34:13; Zech 10:10). The imagery is implied also in the phrase, מִגְדַּל־עֵדֶר ('citadel of the flock,' v. 8), that is, YHWH is the Shepherd of Israel who watches over Jerusalem (cf. עֵדֶר in Isa 40:11; Jer 13:17, 20; Zech 10:30). Yet, it is the gathered flock and not the shepherd that defeats the nations in verses 11-13, in which the eschatological battle between Zion and the nations is depicted. The main concern of Mic 4:11-13 is that YHWH's restored people will face the

---

[53] Among the five verbs in v. 13, with the exception of the first and last being singular, the rest are third-person masculine plural. The impression is that the Shepherd, although he is leading the flock, is counted one among the flock. Thus they break through together (against Wolff, *Micah*, 85-86).

[54] Mays, *Micah*, 75.

[55] Shaw, *Speeches*, 95, notes that 'to go' (עָלָה) is often used of a military campaign (2 Kgs 17:3, 5).

[56] Wolff, *Micah*, 85.

nations gathered by the Lord to expose their rebellion against YHWH's sovereign rule. A prototype of this scene is found in the dramatic exodus story (Exod 14). YHWH leads them through the Red Sea, and the Israelites cross through, carrying what they had plundered from Egypt (Exod 12:33-36). The Egyptian army pursues the Iraelites, only to find itself buried underneath the sea (Exod 14:1-31). Thus, the final battle following the eschatological restoration of the remnant of Israel can be visualized in terms of the exodus.

The context of Mic 4:11-13 presents that theme of YHWH's return to Zion. The kingship will be restored to the people of Zion. But interestingly, here it is they not YHWH *per se* who will defeat the nations gathered by YHWH to fight against Zion. The destruction of YHWH's enemies usually appears in oracles of doom delivered against other nations (Isa 17:12-14; 29:5-8; Zech 14:1-3; 12-15;12:2-9; Joel 3:1-3; 9-12; Pss 46:6, 48:8-14; 76:3-6; especially Ezek 38-39). In all of the prophetic texts, Mic 4:11-13 may very well be the only passage where the summons to battle is addressed to Israel.[57] In all other instances when the assault of Zion motif occurs, it is YHWH who intervenes directly and mysteriously to vanquish those gathered in opposition to his sovereign rule.

If Mic 4:11-13 is compared to Ps 2:7-9 which refers to the anointed one who will rule the hostile nations with an iron scepter, ultimately dashing them into pieces like pottery, we notice how both texts employ shepherd imagery. Each passage may relate a different aspect of the same eschatological confrontation between YHWH's Appointee/ his restored people and the nations. However, it is unclear whether it is YHWH's Appointee (Ps 2:9) alone, the restored people, or both (Mic 2:11-12?) as to the subject who will defeat the hostile nations (Mic 4:11-13).

It is also unclear what is meant exactly by the description of breaking them into pieces 'with the iron scepter' (בְּשֵׁבֶט בַּרְזֶל; LXX, ἐν ῥάβδῳ σιδηρᾷ, Ps 2:9; cf. Hos 6:5). It seems that LXX's rendering 'the rod of his mouth' used in a similar context in Isa 11:4 as 'the word of his mouth' illuminates the characteristic of the eschatological battle set out in Mic 4:13 and Ps 2:9.[58] Nevertheless, at this stage it is uncertain how the shepherd and his gathered flock will confront the hostile nations.

---

[57] May, *Micah*, 108, states, "Micah 4:11-12 presents the assault of 'many nations' as a strategy initiated by YHWH to break their power. In this respect it seems to have some relation to Ezek 38-39."

[58] Cf. J.-G. Heintz, "Royal Traits and Messianic Figures: A Thematic and Iconographical Approach," in *The Messiah: Developments in Earliest Judaism and Christianity* (ed. J. H. Charlesworth with J. Brownson, M. T. Davis, S. J. Kraftchick, and A. F. Segal; Minneapolis: Fortress Press, 1992), 62-64, while studying ANE (esp. Mesopotamian and Syro-Palestinian) royal iconography of 'a storm god,' underlines the close relation between 'the sword' and 'the word' from the mouth of god-figure (often, a roaring lion or thunder imageries; cf. 2 Sam 22:14 = Ps 18;14; Amos 3:8).

1.3.3 The Davidic Shepherd in Mic 5:1-4 [4:14-5:3]

All three motifs – Davidic expectation, shepherd imagery, and Israel's restoration – interconnect in Mic 5:1-4. Mic 2:12-13 and 4:11-13 certainly focus on the tasks that the shepherd (2:12; 4:6-8), the king (2:13; 4:8), the breaker (2:13) and the gathered flock (4:13) will carry out at the time of the restoration. Yet, the integrating theme in Mic 5:1-4 is the coming Davidic Shepherd.

While in the text the name of David is not referred to explicitly, many expressions in verse 2 intentionally recall a figure of Davidic origin. 'Bethlehem-Ephratha' refers to the hometown of David's father Jesse (1 Sam 17:12; 1 Sam 16:18; 17:58; cf. 1 Sam 16:1, 2); therefore, both names clearly are associated with David's origins before he assumed the kingship in Hebron and Jerusalem. Also significant is YHWH's choice of ruler from a small, insignificant, and even despised rank (cf. צָעִיר in Ps 119:41) among the clans of Judah. This points to the divine mystery that marvels and provokes human expectation in their finding the ruler, as is well attested in the examples of Gideon (Judg 6:15), Saul (1 Sam 9:21) and David (1 Sam 16:11-13). Moreover, the momentous expressions 'from of old' and 'since ancient days' in verse 2 occur in connection with the time of David in Neh 12:46 and Amos 9:11 (cf. Mic 7:14; Isa 45:21; 46:10; 63:9, 11; Mal 3:4). In our text, מוֹצָאָה ('origin') in verse 2 picks up the root יָצָא in verse 1, and the twice-occurring מִן in מִקֶּדֶם ('from of old') and מִימֵי עוֹלָם ('since ancient days') in verse 2 echoes מִן in מִמְּךָ לִי ('from you') in verse 1.[59] Thus, the time of David is viewed as an era of the distant past belonging to 'ancient days,' while the new ruler comes from people sharing David's origins, if not the exact Davidic royal linage.[60]

YHWH intervenes to raise up the new ruler. As this becomes the emphasis, the shepherd imagery found in verse 4 expresses both subordination to YHWH's sovereignty and faithful representation of YHWH's leadership. As verse 2 indicates, the coming Davidic Shepherd, described in terms of מוֹשֵׁל, is the one who will bring back the former מֶמְשָׁלָה ('dominion') in Mic 4:8. The term מֶלֶךְ ('king') is avoided in order to underline YHWH's theocratic rule, likely because Micah accepts the end of Israel's monarchic period and looks beyond her failure. The shepherd imagery is sharper in verse 4: וְעָמַד וְרָעָה ('He will stand and shepherd'; LXX, καὶ στήσεται καὶ ὄψεται καὶ ποιμανεῖ τὸ ποίμνιον αὐτοῦ). Further, the imagery is quite graphic. As the shepherd stands up, the flock sits down (וְיָשָׁבוּ, v. 4b). Yet, the rulership of the shepherd will not be restricted within Israel; it will extend even to עַד־אַפְסֵי־אָרֶץ ('to the end of the earth,' v. 4b). As the rulership reaches to the ends of the earth, there will come שָׁלוֹם (5:5; v. 4b LXX). The vision of the Davidic Shepherd is followed by YHWH raising up

---

[59] Wolff, *Micah*, 145.
[60] May, *Micah*, 115-116.

'the (under) shepherds' (רֹעִים, 5:6; v. 5 LXX) in vv. 6-15 as YHWH expands his rulership among the nations.

Thus the restoration under the leadership of the coming shepherd looks beyond the restoration of the nation, signaling the time of the shepherd's universal reign (v. 3). In Mic 2:12-13, the image of the eschatological Shepherd is associated with the breaking open of the way and the going before them at the head of the flock, recalling the exodus and the redemptive role of YHWH as their Shepherd. On the other hand, in Mic 4:11-13 (cf. vv. 6-8), shepherd imagery is used to depict the final battle against the nations, with the flock depicted in terms of warriors breaking the nations into pieces. The culmination is found in Mic 5:1-4, where shepherd imagery is employed to center on the figure of the Davidic ruler who will stand and feed YHWH's flock, and שָׁלוֹם begins within Israel and issues forth to the nations.

1.3.4 Summary

Pulling together a coherent restoration program from select passages in Micah 2-5 might prove to be challenging. Admittedly the shepherd imagery is scattered, lacking clear evidence of its interconnectedness. Still we reaffirm the three main elements embedded in Micah's vision of restoration: Davidic expectation, shepherd imagery, and the hope of the restoration. Overall, Micah's vision of the eschatological Shepherd moves from the breaker who goes before the gathered flock (Mic 2:12-13), to the scene of the final battle between YHWH's flock and the nations (4:11-13), and finally to the coming Davidic Shepherd figure whose reign will extend to the ends of the earth (Mic 5:1-4).

It is noteworthy how Mic 5:1-4 presents the order of these three motifs: Davidic expectation (v. 2), shepherd imagery (v. 4a), and the restoration (v. 4b-5a). The oracle in Mic 5:1-2 is often interpreted as a 'revision' of Nathan's oracle (2 Sam 7:6-7), namely, Davidic expectation.[61] Yet, the echo of Mic 5:1-2 to Nathan's oracle is not limited to a Davidic motif. In 2 Sam 7:7-16, especially vv. 8-11, is a parallel display of the motifs and their order as found in Mic 5:1-4, for Nathan's prophecy begins with the figure of David whom YHWH took 'from the pasture and from following the flock' (v. 8). In verse 7 we find the shepherd imagery that depicts Israel's leaders as 'the shepherds of his people,' and what is implied in verse 8 is YHWH chose David to shepherd the flock. Then YHWH promises he will make his name great (שֵׁם גָּדוֹל, v. 9, cf. Mic 5:4), and rest will be given (v. 11, cf. Mic 5:5, 'peace').

It appears that we are able to establish a similar pattern both in Micah's vision and Nathan's prophecy: Davidic expectation is followed by shepherd imagery introducing the theme of the restoration that brings rest or peace. Yet, in Micah's vision, the consequence of the coming of the Davidic Shepherd is the

---

[61] Shaw, *Speeches*, 146; cf. Wolf, *Micah*, 144; Mays, *Micah*, 115.

restoration of Israel and the eschatological confrontation with the nations. Now, we turn to Ezekiel's vision of the restoration to examine the development of this sort of pattern with the shepherd imagery related to the Davidic expectation.

1.4 The Davidic Shepherd Tradition in Ezekiel's Vision of the Restoration

To assess the various aspects of the shepherd motif in its historical, literary, and theological contexts, it is necessary to first define Ezekiel's concept of the restoration of Israel in the broad framework of Ezekiel 34-48.[62] We can detect a progression in his vision of the new Israel. We will pay attention to the various themes of the restoration motif that are repetitious and recapitulated in Ezekiel 34-37. Ezekiel's critical and creative reflection on the Mosaic covenant, and his much emphasized theological understanding of God's own zeal for the sake of his name, lay the foundation for his concept of the restoration of Israel in chapters 34-37. Finally, we will focus on the shepherd imagery used in these chapters, exegeting those critical passages germane to our study.

The authorship and date of the Book of Ezekiel are not our immediate concerns. Yet I assume, along with the majority of contemporary scholars, that the major part of the book originates from Ezekiel, a priest in Jerusalem carried off to Babylon, together with King Jehoiachin in 597 B.C. (Ezek 1:1).[63] What

---

[62] Ezekiel's vision of restoration is often called a 'program' as illustrated by Jon D. Levenson, *Theology of the Program of Restoration of Ezekiel 40-48* (Harvard Semitic Monographs 10; Scholars Press, 1986). The word 'program,' however, involves detailed instructions and rules to be carried out at a definite time following definite procedure under definite conditions; thus it seems an overstatement which serves no justice to what Ezekiel communicates in chs. 34-37 as well as 40-48. For instance, Yehezkel Kaufmann,*The Religion of Israel: From Its Beginnings to the Babylonian Exile* (trans. abridged by Moshe Greenberg; New York: Schocken, 1960), 443, points out that Ezek 40-48 is "not a program to action, and anticipates nothing of what was actually carried out at the Restoration." Terms such as vision, idea, or concept would better represent what is actually described by Ezekiel concerning the future which God will actualize 'for the sake of His name,' which Ezekiel consistently emphasizes throughout the book of Ezekiel.

[63] While Ezekiel's emphasis on chronology is nearly exceptional in the OT, its historicity has been doubted. For instance, G. Hölscher (*Hesekiel, der Dichter und das Buch*, Giessen: A. Töpelmann, 1924) and V. Herntrich (*Ezekielprobleme,* Giessen: A. Töpelmann 1933) raised questions about the authorship and date of the book based mainly on stylistic ('the poetic, authentic message of the doom' vs. 'the prose, phseudographic interpolated message of hope') and geographical objections ('a bearer of the prophetic message working in Palestine' vs. 'a prophet without people in Babylon'). Nevertheless it seems the majority have less difficulty accepting the traditional view. Plausible later alterations may be acknowledged either due to the activities of 'the school of Ezekiel' (W. Zimmerli) or what is typical with other OT writings (E. Hammershaimb). Many hold to the traditional unity of the book for its remarkable consistency of style and content; cf. G. A. Cooke, *A Critical and Exegetical Commentary on*

can be said with more certainty is that the fall of Jerusalem, the captivity, the exile and the resultant cessation of the Davidic dynasty (Ezek 33:21; 40:1) form the background of the indictments and the hopes set out in chapters 34-37. The report of Jerusalem's fall (33:21) provides the historical and literary contexts for Ezekiel's condemnation of the shepherds, particularly the kings of the Davidic dynasty considered largely responsible for Jerusalem's calamity (vv. 1-10).[64] In this respect, chapters 34-37 serve as a part of divine communication in this historical setting where "the prophet is described as addressing a post-disaster audience ... living after the end of the old Israel, but prior to the creation of the new Israel." [65]

Ezekiel's vision of restoration is to be understood as the result of a divine communication in this particular time in history.[66] The prophet might have faced painful yet unavoidable questions concerning his people, their future, and their God: What went wrong? What brought the chosen nation into exile? Where is the hope for Israel? What happened to the covenants YHWH made with his people in the past? From here on in, what is God planning to do with his people, the temple, and the nations? What would motivate God to do this?[67]

The prophet's search for the root causes of the exile is illustrated in such themes as the departure of God's glory from the temple (10:3-18), the incurably adulterous hearts of the people (16:1-34), their impotence to obey the laws,

---

the Book of Ezekiel (Edinburgh: T. & T. Clark, 1985), xx-xxxi; Walter Zimmerli, "The Message of the Prophet Ezekiel," *Interpretation* 23 (1969): 131-132; E. Hammershaimb, *Some Aspects of Old Testament Prophecy from Isaiah to Malachi* (Rosenkilde Og Bagger: 1966), 52-52; Walter Eichrodt, *Ezekiel: A Commentary* (Philadelphia: Westerminster, 1970), 1-7; Moshe Greenberg, *Ezekiel 1-20* (Anchor Bible 22; New York: Doubleday, 1983), 20-27.

[64] Gerald van Groningen, *Messianic Revelation in the Old Testament* (Grand Rapids: Baker, 1990), 771. Nevertheless, it is impossible to be precise as to which kings and leaders the indictments were directed. Perhaps it was directed at the whole of Israel as Walter Zimmerli suggests from "house of Israel" or "the mountains of Israel" (*Ezekiel 2* [trans. James D. Martin; Philadelphia: Fortress, 1983],185). Ronalds M. Hals, *Ezekiel: The Forms of The Old Testament Literature* (Grand Rapids: Eerdmans, 1989), 250, states that the standard genre pattern of a prophecy of punishment cautions readers not to draw precise correspondence between metaphoric false shepherds and historical figures.

[65] Thomas Renz, *The Rhetorical Function of the Book of Ezekiel* (Leiden: Brill, 1999), 103.

[66] Hals, *Ezekiel,* 264, expresses a similar idea with his "divine reflection on history" in his comments on Ezek 36:16-32; yet it is not simply a reflection or monologue in the mind of the author; rather it is communication through which the meaning or intention can be detected through the spoken text as a whole or as, better put, a dialogue. Communication has become a focal point in the recent development of hermeneutics. Cf. Kevin J. Vanhoozer, *Is There a Meaning in This Text?* (Grand Rapids: Zondervan, 1998), 281-366; also, T. Renz, *Rethorical Function,* 1-18.

[67] Renz, *Rhetorical Function,* 104, summarizes the fundamental issue in the prophet's rhetoric addressed to his post-disaster audience: "If Yahweh denies the Jerusalemites the rights to live in the land, he will hardly grant the exiles restoration if they continue to be the same 'Israel of the past.'" The challenge is to create a new Israel.

perhaps even the problem of the law itself (20:15) and the absence of true leadership (34:1-16). These questions come to face divine solutions illustrated in the major themes of the book: God's zeal for his holy name (36:20, 21; 39:7, 25; 43:7, 8), the new covenant (36:16-32), the resurrection of Israel (37:1-14); YHWH's theocratic shepherding (34:11-22), the Davidic Shepherd (34:23-24; 37:24), and God's sheltering presence among his people (37:26-27; 48:35).

On the other hand, Ezekiel's vision of the restoration of Israel reveals also the prophet's interaction with traditions that most likely influenced his vision in chapters 34-37. For example, Ezekiel's vision reveals the impact of such passages as Hos 3:5 and Jer 30:9 regarding the Davidic ruler in Ezek 34:23-24; 37:23-24, Jer 23:1-8 regarding the Shepherd's tasks in Ezek 34:1-16, Lev 26:4-13 regarding the return of the blessings in Ezek 34:25-30, 36:30-32; 37:26-28, and Jer 31:31-34 regarding the new covenant in Ezek 36:16-38.

Ezekiel's interaction with these traditions are often distinguished by his own unique formations. For instance, he introduces new dimensions such as new exodus language, a new Shepherd, a new David; Ezekiel heralds a new covenant, speaks to a new residence in the land, and points to a new sanctuary. Ezekiel's vision of restoration is historically situated in the time of the temple's destruction and the subsequent exile, set forth with the intention to address his fellow exiles. What he presents is yet a blueprint couched in the language and metaphors of his time (esp. Ezek 40-48) that looks to the near future of restoration, the rebuilding of the temple, and the return of God's glory, even anticipating the full consummation of the covenant of peace, as history nears its conclusion.[68]

1.4.1 The Structure of Ezekiel's Vision of Restoration in Chapters 34-48

Our discussion of chapters 34-37 begin with preliminary remarks about the macro-structure of Ezekiel's restoration vision in 34-48. What draws our attention here is the inner connectivity within these chapters, especially the relationship between chapters 34-37 and 40-48.[69] The book as a whole displays

---

[68] Groningen, *Messianic Revelation*, 782, adds, "From the exile to the very end of time, during this entire span of time there is to be no interruption in their dwelling under the reign of the king and his shepherding care."

[69] There is a consensus on the demarcation of these units, perhaps with the exception of ch. 33. But there have been many questions on the authenticity of chs. 40-48 and their connection to 34-37, particularly raised by form critical approaches. Levenson, *Program of Restoration*, 7, rightly points out: "Even if we denied Ezekiel's authorship of any of the pericopae of chs. 40-48, we should still be obliged to see in that later version the work of a school which believed itself to be expounding the fulfillment of Ezekiel's promise." Increasingly, however, scholars see a more substantial coherence or pattern of thoughts in the whole block from 34-48 in the context of canonical and rhetorical approaches. For instance, see L. Boadt's efforts to prove the various connective links between chs. 33-37 and 40-48; cf. L. Boadt, "The Function of The Salvation Oracles in Ezekiel 33 to 37," *HAR* 12 (1990): 19-21.

1.4 The Davidic Shepherd Tradition in Ezekiel's Vision

symmetrical balance. It opens with the exile and the departure of Yahweh's glory from Jerusalem (chs. 1-11), and closes with a vision of restoration from exile, and the return of God's glorious presence among his people (chs. 33-48). In this symmetry, the motif of God's presence forms an inclusio within the entire book, as Willem VanGemeren illustrates. What is relevant to our study from the symmetry structure appears below:[70]

B. The departure of Yahweh's glory and Israel's guilt, chs. 1-11
    C. Reasons for God's judgment, chs. 12-34
    C'. The vision of the restoration, chs. 34-39
B'. The return of Yahweh's glory, the new temple, transformation, chs. 40-48.

The literary structure indicates that the presence motif indeed gives an overarching sign of Israel's restoration. This presence motif appears at the two-fold climax of the vision of restoration in 37:28, "my sanctuary is among them forever"; and in 48:35, "the name of the city shall be from now on, 'The Lord is There.'" The vision of restoration is placed within this framework, nestled between the opening and closing motif of God's presence. Ezekiel proclaims God's resolve to reestablish his presence again in the renewed land (cf. 20:40), and then sets out the blueprint for its implementation in chapters 34-48.[71]

Daniel I. Block calls these chapters "the gospel according to Ezekiel," and this message of salvation, restoration, and hope takes its turn in chapter 33.[72] Few would disagree that chapter 33 plays a pivotal role in the book. The prophetic word formula is used in 33:1 to mark the inception of YHWH's instruction to the prophet to address his own community (אֶל־בְּנֵי־עַמְּךָ in v. 2; cf. vv. 12, 17, 30; also Ezek 3:11; 13:17; 37:18). Thomas Renz aptly states: "In summarizing the first part of the book and in presenting the situation just after the collapse of the nation in 587 BC, chapter 33 gives the readers the perspective from which the following chapters are to be read." [73] All that has preceded chapter 33 neatly divides into two major subsections: YHWH's judgment of Jerusalem (3-24) and YHWH's judgment of other nations (25-32).[74] The latter

---

[70] W. A. VanGemeren, *Interpreting the Prophetic Word* (Grand Rapids: Academie, 1990), 326, 337.

[71] The appearance and departure of YHWH's glory in chs. 1, 8-11, and the announcement in 20:40 that Israel will serve YHWH 'on my holy mountains, the mountain height of Israel' are both linked up in the final section, chs. 40-48.

[72] D. I. Block, *The Book of Ezekiel: Chapters 25-48* (Grand Rapids: Eerdmans,1997), 272.

[73] Renz, *Rhetorical Function*, 105. Likewise, Hals, *Ezekiel*, 230, asserts that "Chapter 33 deliberately aims at serving as a transition from the message of doom to that of hope." It seems a minor issues whether one includes ch. 33 as an introduction to the message of hope in the following chapters, or excludes it as a transitional section. Suffice it to say that the indefinite nature of the chapter itself may well explain its transitional function.

[74] The message of judgment in these section, however, is often concluded with the oracles of hope (11:14-21; 16:53-63; 17:22-24 and 20:39-44).

half of the book (33/34-48) contains narratives, oracles of hope, and salvation aimed at future rebuilding.[75] According to Renz, the main issue of chapters 34-48 is how to create the 'Israel of the future' as a 'community of character' enabled to obey the law for the sake of God's name.[76] That answer is found in God's restoration of his kingship/presence among them (34-37; 40-48).

In Ezekiel, Israel's restoration is utterly dependent upon God who will restore his own kingship and honor, resulting in the return of his glory to and presence among his redeemed and renewed people.[77] God's presence, in effect, means to *reverse* the wrongdoings of the past rulers of Israel in chapter 34, where the heart of the matter is the reestablishment of God's kingship. The theme of reversal is one of the main emphases in Ezekiel's vision of restoration. L. Boadt, for example, finds this series of reversal in chapters 34-37: the reversal of the disordered structures in kingship and possession of the land (ch. 34; 35:1-36:15); and, the reversal of the history of infidelity (36:16-38).[78]

In chapters 40-48, God's presence requires the building up of the redeemed society centered in the Lord's sanctuary among them. Chapters 34-48 no longer focuses on the past (chs. 1-33) but on how the transformation will take place (chs. 34-37), with an emphasis on how the already redeemed society will take shape. Many scholars view chapters 40-48 as a consequential vision that follows the restoration in chapters 34-37.[79] As a whole, chapters 40-48 present a vision of new Israel, a transformed society.

This transformation, according to W. Eichrodt, reaches back to *Urzeit*, the paradise regained as yet a part that stands for the whole.[80] The whole redeemed society is depicted primarily as a temple surrounded by the land. R. Stevenson notes that the two-dimensional blueprint for this temple draws a distinction between the holy and the common, while using the term 'YHWH's territorial claim' as the king of his people (cf. 43:6). The architectural center is the altar, though not the Holy of Holies. Likewise, this community is meant to be a cultic-centered society. Having received cleansing from this altar, societal and cosmic

---

[75] For the nearly unanimous agreement on the division presented above, see Greenberg, *Ezekiel 1-20*, 3-6; Hals, *Ezekiel*, 3; Block, *Ezekiel 25-48*, 235; Boadt, "Salvation Oracles," 5, who takes chs. 25-32 as 'indirect words of hope,' thus belonging to the message of salvation.

[76] Renz, *Rhetorical Function*, 105, 130.

[77] Block, *Ezekiel 25-48*, 272, presents YHWH's restoring process, beginning with the restoration of Yahweh's kingship of Israel (34:1-31), Yahweh's land (35:1-36:15), Yahweh's honor (36:16-38), Yahweh's people (37:1-14), Yahweh's covenant (37:15-28), Yahweh's supremacy (38:1-39:29), Yahweh's presence among his people (40:1-46:24), and Yahweh's presence in the land (47:1-48:35).

[78] Boadt, "Salvation Oracles," 13.

[79] In recent articles, the unity and possible authenticity of chapters 40-48 are more widely accepted. See e.g., Levenson, *Program*, 1976; M. Greenberg, "The Design and Themes of Ezekiel's Program of Restoration," *Interpretation* 38 (1975): 181-216; Boadt, "Salvation Oracles," 1990; Renz, *Rhetorical Function*, 1999; contra Cooke, *Ezekiel*, 425-427.

[80] Eichrodt, *Ezekiel*, 585.

1.4 The Davidic Shepherd Tradition in Ezekiel's Vision

well-being (cf. 'covenant of peace' in 37:26) become a reality.[81] Further, this transformation is initiated by YHWH himself along with the Davidic Appointee in chapters 34-37. In the relationship between chapters 34-37 and chapters 40-48, chapters 40-48 can be seen as the *'proleptic corroboration* of these promises' where the theme of the restoration of God's presence in Israel is more fully developed.[82]

How does Ezekiel's concept of restoration progress as chapters 34-37 and 40-48 are interpreted in a sequence? L. Boadt traces a careful progression from the condemnation of the mountains of Israel in chapter 6, their restoration in chapter 36, to their full blessing in the later sections of chapters 38-39 and 40-48.[83] T. Renz sees a thematic inclusio of YHWH's sovereign kingship in chapter 34 with YHWH's territorial claim as king in chapters 40-48. In between are prophecies of YHWH securing the land of Israel and opposing foreign nations (35:1-36:15; 38-39), with visions of spiritual (36:16-38) paired with political (37:15-28) transformation. At the center of these visions lies the vision of Israel's transformation that combines the political with the spiritual (37:1-14).[84] The following diagram gives us a glimpse of the thematic progression of Ezekiel's concept of restoration, with the focus on the Shepherd figure set forth in chapters 34-48:

Progression in Ezekiel's Concept of Restoration

Redemption/Restoration	Establishment of New Leader	Settlement of New Community
YHWH as Shepherd →	Davidic Shepherd-king/Prince →	princes (shepherds)
theocratic	spiritual/political/cultic	cultic/spiritual/political
chs. 34, 36 (37:1-23)	chs. 37:24-28 (34:23-24)	chs. 40-48

Twice in Ezek 45:8 and 9, the plural constr. form of נְשִׂיאֵי ('princes') appears, both expressing the negative aspect of the office. The context is that YHWH, who has ultimate authority over his restored people, warns his princes against their potential oppression and the exploitation of the members of this newly gathered eschatological community. This is also an expression of one of Ezekiel's consistent concerns to undo the past failures of the unfaithful

---

[81] R. Stevenson, *The Vision of Transformation: The Territorial Rhetoric of Ezekiel 40-48* (SBLDS 154; Atlanta: Scholars Press, 1996), 160-164.

[82] Moshe Greenberg, "The Design and Themes of Ezekiel's Program of Restoration," *Interpretation* 38 (1975): 182 (italics mine). It has been observed that 37:26-28 forms a more natural transition to chs. 40-48 than to chs. 38-39, and at least one ancient MS (Papyrus Codex 967) testifies to this arrangement. Cf. Floyd V. Filson, "The Omission of Ezekiel 12:26-28 and 36:23b-38 in Codex 967," *JBL* 62 (1943): 27-32. Note also that Ezek 34-37, 40-48 and Zechariah 1-8 had a profound influence upon New Testament messianic and eschatological thought. See Walter Harrelson, "Messianic Expectation at the Time of Jesus," *SLJT* 32 (1988): 32-33.

[83] Boadt, "Salvation Oracles," 12.

[84] The demarcation of these units can be generally accepted, yet Renz's too neat chiastic structure does not quite seem convincing. T. Renz, *Rhetorical Function*, 128.

shepherds who fed themselves with the sheep (ch. 34:1-14). The implication of the plural form most likely denotes the offspring or successors of that office.

This, however, does not diminish Ezekiel's emphasis on 'one' (אֶחָד) shepherd in 34, 37.[85] Jon D. Levenson rules out the possibility of the prince's role as a political leader by defining him as an 'a-political leader.'[86] Yet certain passages demand consideration that the prince possibly possesses some political authority. For instance, Ezek 46:12 shows the prince having access through the east gate, which could mean his royal authority; also Ezek 46:16-18 may imply that the prince has the right to handle people's possession of their land. The progression from transformation to settlement is unmistakable.

There is a noticeable change of emphasis on the role of the agent, shifting from YHWH, to the Davidic Appointee, to prince(s). Ezekiel's vision progresses from God redeeming and restoring his people Israel, to appointing a new Davidic Shepherd over this regathered people, and finally to settling them under the leadership of the entrusted princes to secure them and ensure they will not be scattered again. Thus might the nations acknowledge God's holy name. The scattering of Israel among the nations is a sign of disgrace, possibly fueling an accusation of God's impotence (Ezek 12:15; 20:23; 22:15; 29:12; 30:23; 30:26). In Ezekiel's vision of restoration, God intends to reverse this past disgrace. To gather and secure them in their land (so they no longer will be dispersed), is to restore God's Name in the eyes of the nations (34:28, 29, recapitulated in 36:20, 30; 37:21, 22, 23, 28).

Unlike chapters 40-48 where maintaining the community is the major concern, chapters 34-37 presents YHWH sanctifying his people before the eyes of the nations for the sake of his holy name, creating from dry bones a new people, restored from exile (cf. 34:23-24; 37:24-28). The focus of chapters 34-37 is YHWH's redemptive act, and the pinnacle is YHWH's appointment of the Davidic Shepherd. YHWH will save his people just as he did in the Exodus. Ezek 37:24-25 recapitulates the announcement in 34:23-34, that is, at the time of the restoration YHWH will appoint a new Shepherd over his newly redeemed people. This new Davidic Shepherd is called king (37:24) and prince (34:24; 37:25). It is noteworthy that in chapter 34 YHWH's own saving activities are emphasized and twice in vv. 23, 24 the Davidic Shepherd is referred to as עַבְדִּי ('my servant'). Redeemed and restored by YHWH the true Shepherd, and ruled by YHWH's Servant the Davidic Shepherd, the new community is described in chapters 40-48 as cult-centered, and is to be settled around the sanctuary.[87] The

---

[85] Cf. Iain M. Duguid, *Ezekiel and the Leaders of Israel* (Leiden: Brill, 1994), 38, 50.

[86] Levenson, *Program of Restoration*, 75-101.

[87] Thus Greenberg, *Ezekiel,* 760, states: "The conclusion to the prophecies of restoration prepares the way for the 'constitution' of the new Israel in the sequel. The sudden mention of the temple-sanctuary (מִקְדָּשׁ, 37:26b) presages the extensive description of the future building, its personnel and its rites, the duties of the chief, and the settlement of the restored tribes that will be set out in chs. 40-48"; cf. Cooke, *Ezekiel*, 377.

1.4 The Davidic Shepherd Tradition in Ezekiel's Vision

center of this new Israel is the temple rather than a city, as politics are subordinated to religion. The main responsibility of the king, characterized by the designation נָשִׂיא as leader of a single unit, is to represent Israel by providing offerings (ch. 46).[88] There is no comparison between the previous false shepherds of the Davidic dynasty and the princes in this eschatological community. This is a far cry from any typical nation; even the name David does not occur at all in chapters 40-48.

After passing the stages of YHWH's redemptive and restorative acts, the vision comes to the point where YHWH appoints the Davidic Shepherd who is to play a crucial role as God unifies the divided nation (37:15-28). While this hints at a political dimension, the Davidic Shepherd is also called 'prince,' which as a title is less political than king. In chapters 40-48 where the constitution and settlement of the restored community becomes the issue, the designation 'king' and 'prince' appears. In these eight chapters, the cult itself becomes the main focus of the vision.[89]

Interestingly, when Ezekiel describes the eschatological regathering of the house of Israel, the main focus is inner, religious renewal and outer, environmental renewal, i.e., the land (34:1-22; 36:16-38; 37:15-23). Thus the two main concerns are the spiritual regeneration of Israel, and the renewal of the land (= return to the state of Eden), which appears to occur at the far end of the whole vision. The spiritual and political reformation is followed by a detailed account of the cultic reconstitution. Of significance to our study is that both YHWH and his Servant in the future are depicted in shepherd imagery. In chapters 34-37, the activities and mission of YHWH the Shepherd and the Davidic Shepherd, with their respective roles, characterize Ezekiel's concept of restoration. We will now turn our attention from the macro-structure of Ezekiel's vision to multiple themes of the restoration.

---

[88] The city is not even referred to as 'Jerusalem,' let alone as the 'city of David.' On the contrary, the enumeration of the gates (one gate for each tribe) emphasizes that all tribes will have equal access to the city, whose name from that time on shall be: יְהוָה שָׁמָּה (48:35). T. Renz points out that in the ANE temple building was a privilege reserved for kings and was an expression of their power. Thus the pre-exilic temple built by Solomon was a royal sanctuary. In Ezekiel's vision, however, no human king is responsible for building the temple; Ezekiel does not receive a building plan, but perceives a temple already built. This suggests that YHWH's new sanctuary will solely legitimize his own kingship and not the rule of any human monarch. Therefore, Renz, *Rhetorical Function*, 124-126, concludes that the vision carefully reconstructs aspects of the Zion tradition, but resists any attempt to include in it a Davidic component.

[89] For instance, Karl Begrich, "Das Messiasbild des Ezechiel," *ZWT* 47 (1904): 453-454, argues here for a sharp contrast between the political-national and the religious-ethical, and states: "Der Fürst (נָשִׂיא) spielt eben in Cap. 40-48 eine so untergeordnete Rolle, dass er nicht seiner Persönlichkeit wegen, sondern nur gelegentlich als oberster administrativer Beamter, als 'Kirchenpatron' in Betracht kommt."

1.4.2 Restoration Themes Recapitulated in Chapters 34-37

There are numerous themes of restoration found in chapters 34-37. Cooke nicely represents the implied stages of restoration in these chapters: (i) YHWH himself will feed his flock, gathered and safe in their native land, 34:1-16; (ii) the land will be transformed, 36:8-15; (iii) the re-assembled nation will be purified in heart and spirit, 36:16-38; (iv) Israel, as good as dead, will rise to new life, 37:1-14; (v) the old division of the kingdoms will vanish, and (vi) a David will rule over a united nation, in the midst of which YHWH's sanctuary will be set for evermore, 37:15-28.[90] It is difficult to determine, however, one coherent program since many themes are recapitulated, i.e., repeated along with the addition of other aspects or dimensions to the previous vision.

In Ezek 34:23-24, the appointment of a single shepherd over Israel is recapitulated in 37:25-28 where more details are provided as to this undershepherd. What is striking is the connection between the Davidic Shepherd and the people following the laws. This connection is unclear in 34:23-24, which simply relates that the Davidic Shepherd is to shepherd the restored people. While 37:24b is uncertain as to the future David becoming the Shepherd over new Israel, it is stated with certainty that they will follow my laws.

The context of 37:23-34 presupposes what is to happen in 36:27: the Spirit is given to the people so they will follow the laws. Yet the role played by this future Davidic ruler regarding the people's obedience to the laws is implicit. Ezekiel 34:25-31, a seemingly creative re-adaptation of the covenant blessings of Lev 26:4-13, is recapitulated in 36:9-11 and again in 37:26-28.[91] According to the context of these passages, the return of blessing to the land seems allied to the covenant of peace. The fulfillment of the covenant of peace presupposes all that precedes it: God's redeeming, restoring process, and the giving of his Spirit enables the people to follow the laws. The renewed covenant in 36:24-28 is followed by the returning of the wasteland to the garden of Eden (36:35). What is the relationship between the renewed covenant in 36 and the resurrection of Israel in 37:1-14?

Leslie C. Allen argues that Ezek 36:27a ("I will put my Spirit in you") is recapitulated and interpreted in 37:1-14, and that 36:27b ("and move you to follow my decrees and be careful to keep my laws") is recapitulated and interpreted in 37:25-28.[92] If Allen is correct, then the scene of God revivifying the hopeless exiles explains on a different level how the new heart and new spirit will be given to the house of Israel: When God brings them back to the

---

[90] Cooke, *Ezekiel*, 372.

[91] Renz, *Rhetorical Function*, 107.

[92] L. C. Allen, "Structure, Tradition and Redaction in Ezekiel's Death Valley Vision" in *Among the Prophets: Language, Image and Structure in the Prophetic Writings* (Sheffield: JSOT Press, 1993), 127-142.

land, his Spirit will revive them to follow his laws. Moreover, this implies that God's redeeming and restoring activities (34:1-16) involve the enablement of his people to obey the laws, which entails pouring out his Spirit (36:24-32). The vision reaches its climax at God's revivifying or re-creating Israel (37:1-14). All these belong to the Shepherd's mission.

Ezekiel 37:23-28 is a recapitulation of the entire range of restorative acts extended throughout chapters 34 and 36.[93] Furthermore, this pericope foreshadows chapters 40-48 where the situation following the restoration is described: How the new future David will guarantee obedience and security (37:24-25).[94]

Boadt rightly states, "Ezek 37:23-28 recapitulates the major themes of the preceding four chapters. It clearly is a closing. It combines the promise of the purification of the people with the restoration to the land, under a new David, and in a covenant of peace when God's dwelling is reestablished in their midst. It sums up the various stages outlined in chapters 34-37, and at the same time looks ahead to the new order described in chapters 40-48." [95]

1.4.3 The Structure of Ezekiel 34-37

The absence/presence of YHWH as the true Shepherd of Israel forms an inclusio at the beginning (34:1-10; 11-16) and the end (37:27-28) of the unit. Ezekiel 34:1-10 is a series of indictments directed at the unfaithful shepherds of Israel, and also sets up the inevitable necessity of YHWH's intervention as the Shepherd: מֵאֵין רֹעֶה (v. "because there is no shepherd").

The oracle intentionally postpones YHWH's proclamation of his intervention (v. 10) by confirming the recurring failures of Israel's past leaders (vv. 2-6, 8,10). The composite connective particle לָכֵן in verse 7 ("this being so"; LXX: διὰ τοῦτο) refers to the absence of one shepherd repeated twice and thus emphasized at the end of verse 6: וְאֵין דּוֹרֵשׁ וְאֵין מְבַקֵּשׁ ("there is no one searching; there is no one seeking"). Verse 6 already makes clear how the shepherds are neither searching nor seeking. Verse 8 repeats the desperate situation of צֹאנִי ("my flock") becoming the prey of beasts because of the absence of 'one shepherd,' and because the current shepherds (הָרֹעִים) are shepherding (וַיִּרְעוּ)

---

[93] In Ezek 37:23, Israel shall be purged and delivered from sin/pollution = 36:25, 29a; YHWH their God, Israel his people = 36:28 (cf. 34:24a); from 37:25 to the end each restored condition is said to be eternal, a new emphasis. In 37:25, they shall dwell in their ancestral land = 36:28a; my servant David, their chief = 34:24a; a covenant of peace (salvation; well-being) = 34:25; they shall be numerous = 36:10f, 37f. In 37:26b-28, God's sanctuary and his sheltering presence in the midst of people; cf. Moshe, *Ezekiel 21-37*, 785.

[94] Ezekiel 37:26-28 for the first time mentions a sanctuary in new Israel and thus prepares the ground for chs. 40-48, with chs. 38-39 serving as the interim. The purpose of chs. 38-39 is to affirm that Israel's living in the land is indeed 'forever' (37:25). See Renz, *Rherical Function*, 117.

[95] Boadt, "Salvation Oracles," 15.

themselves and not the flock. The second appearance of לָכֵן in verse 9 maximizes the significance what transpires in v. 10 (אָמַר אֲדֹנָי יְהוִה הִנְנִי אֶל־הָרֹעִים כֹּה) and v. 11 (הִנְנִי־אָנִי, "behold, I"): YHWH, he himself will come to shepherd.[96]

This presence motif is most visible and is used most confidently at the end of the unit (37: 26-28). YHWH will give sanctuary (מִקְדָּשׁ) to the restored Israel forever (vv. 26, 28). This sanctuary to be given by God at the time of restoration is described in v. 27 as his מִשְׁכָּן over them. This is his tabernacling presence over his renewed people.[97] The entire semantic matrix of vv. 24-28 helps clarify what will be the nature of this presence. In this critical section, לְעוֹלָם occurs five times (vv. 25 [2x], 26 [2x], 28). This adverbial noun connects vv. 25-28 and accentuates the climax of the unit. Four items are promised to last 'forever': (i) peaceful and abundant inhabitancy in the renewed land (v. 25b); (ii) the new David's reign (v. 25c); (iii) the covenant of peace (v. 26b); and (iv) the sanctuary where God sanctifies his people and the land while his sheltering presence dwells among them (vv. 26e, 28).[98]

This appears to be a vision of the fulfillment of the Davidic covenant (2 Sam 7:12-14). The first occurrence of לְעוֹלָם in v. 25, where living in perpetual peace in the land is promised, is sandwiched between the appearance of the future Davidic ruler in v. 24 ("My servant David will be king over them and one shepherd will be for them") and v. 25 ("David my servant will be their prince forever"). The appearance of the Davidic Shepherd is not immediate in chapters 34-37, however. Verse 2 introduces the motif, and the larger part of chapter 34 (vv. 1-22) describes YHWH's mission of reversing the failures of the shepherds. In fact, the unfaithful shepherds of Israel reversed their roles and responsibilities: "The Lord YHWH says, alas! Israel's shepherds, they have been 'shepherding' only themselves!"[99] Verse 3 explains v. 2c, i.e., in shepherding themselves

---

[96] In Ezek 34:10 the threat introduced in v. 8 by אִם־לֹא is, after a long parenthesis, resumed with הִנְנִי. E. Kautzsch, *Genesius Hebrew Grammar* (Oxford: Clarendon, 1910), 149c.

[97] The word מִשְׁכָּן ('residence, dwelling place') occurs only here in Ezekiel with reference to the house of God (cf. 25:4, used of human dwellings). YHWH's residence in these passages is identified also by מִקְדָּשׁ (from קָדַשׁ, 'to be holy') as indicated above. D. I. Block, *Ezekiel: 25-48*, 421, suggests that these two expressions reflect the opposite dimensions of the divine character, one for his transcendent nature and the other for his condescending immanent presence. It is interesting to note that מִשְׁכָּן goes with the idea of 'over (them)' as in עֲלֵיהֶם while מִקְדָּשׁ goes with the idea of 'in the midst of (them)' as in בְּתוֹכָם. Does this not have anything to do with the expressions that the future king will reign 'over' them and the prince will be 'in their midst'?

[98] The perpetual presence of the sanctuary is emphasized (twice in vv. 26, 28). Obviously, this is meant to be the highlight. Later, God's dwelling presence in the new temple is reassured in 43:7: "This is where I will live among the Israelites 'forever' (לְעוֹלָם); the house of Israel will never again defile my holy me" (also twice; in v. 9).

[99] There is a word play with רָעָה (translated as 'to tend, feed'). This idea of role reversal is explicitly repeated in vv. 2, 8, 10: וַיִּרְעוּ הָרֹעִים אוֹתָם (v. 8d), רֹעֵי־יִשְׂרָאֵל אֲשֶׁר הָיוּ רֹעִים אוֹתָם (v. 2d);

and not the flock, the sheep fed and clothed the shepherds. This reversal motif is most visible in v. 16 with the reversal of the shepherd's task illustrated in v. 4. Until the Davidic Shepherd is introduced in vv. 23-24, the entire section displays a symmetrical structure supported by the theme of reversal:

a. vv. 1- 3: unjust shepherding: the false shepherds feed themselves
  b. v. 4: the task of the shepherd that Israel's shepherds failed to do
    c. v. 5: results - flock scattered and became food/prey
      v. 6: scattered, absence of the one who seeks
      vv. 7-8 : sum (v.8a - scattered/prey; v. 8b - false shepherds)
        d. vv. 9-10 : therefore, "I will seek and deliver"
        d´. vv. 11-12: "I will seek and deliver"
    c´. v. 13: reversal - new exodus (bring out, gather, lead into)
      v. 14: "I will feed them in good pasture"
      v. 15: "I will make them lie down"
  b´. v. 16 - the task of the shepherd that YHWH reverses
a´. vv. 17-22 - restoring of the just shepherding within the community

This reversal motif appears to continue in the vision as YHWH renews the heart of his people by pouring out his spirit (36:24-32). Thus the fate that led them from their land into exile is reversed, and they are empowered to keep the laws they failed to keep in the past (36:27b, cf. 37:24b). The apex is reached in the valley of the dry bones and graves (37:1-14), where the utter hopelessness of the exiles in Babylon is graphically illustrated. The vision of the dry bones and the raising of the dead from their graves may be the most profound aspect of the vision, in which YHWH, as the eschatological Shepherd (34:1-22 and also 36:24-32), reverses death to life.

The theme of the Davidic Shepherd occurs at the closing of YHWH's shepherding activities. The flow of the text suggests that God, as he reverses Israel's plight, will appoint the new David. At the end of chapters 34, 36-37 the Davidic Shepherd appears, though the nature of his mission is not explicit. It is clear that he will reign over the renewed people (34:23-24; 37:24-25), while any other implications about his task are addressed in the semantic interconnections with other restoration themes and motifs.

1.4.4 The Traditions behind Ezekiel's Concept of Restoration

When considering the traditions behind chapters 34-37, there is a growing recognition of the significance of Ezekiel 20 and Leviticus 26 in this respect. Moshe Greenberg argues, "The main oracle (34:16-21; vv. 22-32) carries on the grand theme of chapter 20: the indissoluble link between the fate and destiny of

---

and conclusively לא־יִרְעוּ עוֹד הרעים אוֹתָם (v. 10d). In all three of these occurrences, the marked emphatic אוֹתָם intensifies the tone of indictment (cf. E. Kautzsch, *Genenius Hebrew Grammar*,135 k).

Israel and the general recognition of YHWH God of Israel as true God."[100] Leviticus 26 clearly presents the blessings and curses that result from Israel's response to God's decrees and laws. All kinds of idolatry are prohibited (vv. 1-2). If Israel obeys, the land is blessed (vv. 3-13); if Israel does not obey, the result is exile, with a view toward humbling their pride (vv. 14-39). If Israel confesses their sins, God will remember the covenant 'for their sake' (v. 45) since he is the Lord their God (vv. 40-46).

Similarly, the progression of the vision in Ezekiel 20, according to R. Rendtorff, unfolds in three stages: a history of disobedience (20:8-31), followed by exile from the land as judgment on evildoers and the purging of their guilt (20:33-38), and restoration for the sake of God's name (20:44).[101] L. Boadt argues that this three-stage outline of the divine plan closely resembles the pattern found in Lev 26:14-32, 33-39, and 40-46, and claims that the vision reflected in Leviticus 26 and Ezekiel 20 controls the development of chapters 33-37, and most likely all of chapters 33-48 in Ezekiel.[102]

The question remains why Ezekiel's concept reflects the tradition of the Mosaic covenant. Dieter Baltzer suggests that Ezekiel 20 represents counter-theology to the prevailing David-Zion theology of the Jerusalem establishment.[103] Likewise, Boadt suggests that Ezekiel rejects the name Zion or God's holy mountain; this is reflected preliminarily in 17:22-24, and carried out completely in 40:2, 43:12. Further, the prophet seems reluctant to speak of any major role for a king, eschewing the title מֶלֶךְ in favor of a mere נָשִׂיא (leader or prince), in 34:23-24, 37:24-25, 43:6-12. By doing this, Boadt conjectures that Ezekiel modifies the existing royal theology of the Davidic covenant as known in pre-exilic Judah in favor of a fuller integration of the northern tradition of the Mosaic covenant.[104] Yet it is uncertain what type of David-Zion theology of the Jerusalem establishment Ezekiel might have faced in his time.

As argued above, Ezekiel 20 and Leviticus 26 are likely reflected in Ezekiel's blueprint in chapters 33-37. Then Ezekiel's vision of the future of Israel can be seen as an attempt to synthesize and overcome the past Mosaic paradigm and

---

[100] Greenberg, *Ezekiel 21-45*, 735.

[101] Rolf Rendtorff connects ch. 20 to ch. 36 as a 'threat' to the 'fulfillment of the threat,' concluding that 36:16-28 (plus vv. 29-32) is part of the completion of the program laid out in ch. 20, and not an independent addition to the book. "Ez 36, 16ff im Rahmen der Komposition des Buches Ezechiel," *BETL* 84 (1986), 260-265, cited from L. Boadt, "Salvation Oracles," 19-20.

[102] Boadt, "Salvation Oracles," 8, 20, states: "Especially chapters. 33 to 37 show many connective links to each other and to the larger schematic program put forth in ch. 20; this in turn connects to a wider priestly reformation program tied into the Holiness Code and especially Leviticus 26."

[103] D. Baltzer, "Literarkritische und literarhistorische Anmerkungen zur Heilsprophetie im Ezechiel-Buch," *BETL* 76 (1986): 166-181.

[104] Boadt, "Salvation Oracles," 20.

the contemporary David-Zion theology. There are particular connotations to the use of the shepherd metaphor in chapters 34-37. The shepherd metaphor, used to describe the kingship over Israel, seems to modify the existing royal theology of the Davidic expectation by Ezekiel.

Further, the shepherd metaphor in Ezekiel 33-37 is closely related to the Mosaic tradition, and it is apparent that the law of God is fundamental to Ezekiel's understanding of restoration. Israel's failure to keep God's law deeply distressed Ezekiel, a failure that subsequently led to the departure of God's glory from the temple. Obedience to the law is key to understanding the failures of the nation in the past, and the shaping of the new vision in the future. The visions of the restoration in chapters 34-37 can be seen as the reversal of the curses of the covenant documents – particularly Leviticus 26. The blessings promised for obedience to the covenant laws are turned into unconditional prophecies of future bliss.[105]

In Ezekiel 34, 37, the Davidic Shepherd leads the renewed people to obey the laws and decrees of YHWH, but this presupposes the new solution in Ezekiel 36 to the plight of Israel's impotence to obey the Law: "I will put my Spirit in you and move you to follow my decrees and be careful to keep my laws" (36:26). The tradition of Leviticus 26 as background to Ezekiel 20 and 33-37 shows Ezekiel reflecting the plight *under* the law, that is, the people's impotence to obey the laws and decrees of YHWH, not *of* the law itself (cf. Ezek 20:25).

In this respect, Ezek 20:25 frames the issue: "I also gave them (Israel) over to statues that were not good and laws they could not live by." Zimmerli and Cooke understand 'the statues' and 'the laws' here referring to the Law of Moses. Zimmerli suggests that Ezek 20:25 expresses "the mystery of a divine punishment contained in the law itself (cf. Rom 5:20; 7:13; Gal 3:19)."[106] Corrine Patton argues that Ezekiel 40-48 is the new law and Ezekiel 20 sets up the possibility of the giving of this new law by the decree that the laws of the wilderness were no good. He claims that "for Ezek 20:25 the problem is not one of disobedience to the law. Israel is literally 'damned if they do and damned if they don't.' Such is Yahweh's justice for the nation of Israel. Israel has been set up for failure."[107]

The key issue is whether the laws and decrees in Ezek 20:25 refer to the Law of Moses or the pagan laws. Patton claims that the people "correctly understood the law of child sacrifice [as one of the laws that were not good] as something prescribed by Yahweh." On the other hand, M. Greenberg points out that the laws and decrees here refer to the the deadly laws of pagans as countering God's law. The context of Ezekiel 20 supports this view.[108]

---

[105] Greenberg, *Ezekiel 21-37*, 705.

[106] Zimmerli, *Ezekiel 1*, 410, mentions Rom 5:20; 7:13; Gal 3:19, suggesting that St. Paul reflects the similar notion regarding the nature of the law in this respect; Likewise, Cooke *Ezekiel*, 218.

[107] C. Patton, "'I Myself Gave Them Laws That Were Not Good': Ezekiel 20 and the Exodus Traditions," *JSOT* 69 (1996): 73-90 (78-79).

[108] Greenberg, *Ezekiel 1-20*, 369.

A negative outlook on the law appears incompatible with Ezekiel's view of the Mosaic Law in light of his vision of the new Israel. Ezekiel assumes that YHWH's law still ought to be fulfilled, yet in a new way under the leadership of the Davidic Shepherd (Ezek 36:27; 37:24). In Ezekiel's vision, the restoration (chs. 34-37) precedes the regulations for the new community (chs. 40-48), not the other way around. YHWH's tasks of rescue (ch. 34), revivification (ch. 36), and restoration (chs. 34-37) call for the inevitable accommodation of the laws and decrees in a new situation as YHWH's glory and presence return among them (chs. 40-48). Yet this does not necessarily create a new law. For Ezekiel, the plight under the law is the adulterous human heart, not the law itself.

The imagery of YHWH as the eschatological Shepherd and the Davidic Shepherd in Ezekiel 34-37 makes obsolete Ezekiel's contemporary David-Zion theology as a repetition of the failed monarchy, but at the same time provides a means by which Ezekiel's vision moves beyond the Mosaic system. More importantly, Ezekiel highlights YHWH's ultimate motivation to push forward Israel's history to transcend their failures: YHWH's own zeal for his holy name before the nations.

1.4.5 God's Zeal for His Name

God's sovereignty is more explicit in Ezekiel 20 than in Leviticus 26. The Lord YHWH desires to be known by Israel and the nations, and this desire is the driving objective of Israel's exile and her restoration (Ezek 20:32-37). The recognition formula, 'you (they) will know that I am YHWH,' is one of the most striking characteristics of Ezek 20:12, 20, 26, 38, 42, 44. There appears to be a hint of the recognition formula already in Lev 26:45: "in the sight of the nations to be their [Israel's forefathers] God" (cf. Ezek 20:22). But God's desire to be known by the nations as well as by Israel is more explicitly stated in Ezekiel 20. Overall in the book of Ezekiel, we find expressions of God's concern for his holy name defiled by Israel among the nations. It is noteworthy that the texts containing God's concern for his name are mostly found in blocs such as chapters 20, 36, 39, and 44.[109]

---

[109] YHWH acts 'for the sake of his name' (20:9, 14, 22, 44; 36:22). He is also said to 'have concern for' his name (36:21), to 'vindicate the holiness of' his name (36:23), to 'make known' his name (39:7), and to 'be jealous for' his name (39:25). It must be noted that the verb with which the divine 'name' most frequently appears is 'to profane' (חלל; 20:9, 14, 22, 29; 36:20, 21, 23; 39:7). The divine name takes a variety of forms, 'my name' (20:9, 14, 22, 44), 'my holy name' (20:39; 36:20, 21, 22; 39:7 [x2]; 39:25; 43:7, 9), 'my great name' (36:23). See Paul Joyce, *Divine Initiative and Human Response in Ezekiel* (JSOTSup 51; Sheffield: JSOT Press, 1989), 101; cf. Elsewhere in the Hebrew Bible, לְמַעַן שְׁמִי ('for the sake of my name') appears 18 times: Exod 9:16; 1 Kgs 8:41; 2 Chr 6:32; Isa 48:9, 66:5; Jer 14:7, 21; Pss 23:3, 25:11, 31:4, 79:9, 106:8, 109:21, 143:11. Its equivalent, לשמו, occurs in Jos 7:9.

Judgment as well as salvation oracles often testify to God's implicit concern about the nations and Israel's relationship with them. In rebuking Israel, the sovereign God is explicit: "This is Jerusalem; I have set her in the center of the nations, with countries all around her" (Ezek 5:5). What God does to Israel is always done before the eyes of the nations, and the sovereign God even compares Israel's deeds to the nations' deeds. He rebukes Israel because she "has rebelled against my ordinances and my statutes, becoming more wicked than the nations and the countries all around her"(5:6).[110] It comes as no surprise that Israel's punishment will be meted out in the sight of the nations (5:8). Clearly the nations are in view whenever God deals with his own people. While the nations primarily function as on-lookers, their presence nevertheless takes on more significance.

The openness of the knowledge of God to the nations in the book of Ezekiel, however, has not been suggested without facing serious objections. Paul Joyce presents a chart that shows that the nations will know God most when God punishes the nations and often when God delivers Israel, but says, "there seems to be no case in which it is said that the nations will come to know that 'I am YHWH' when YHWH punishes Israel."[111] By this, he suggests that God is the God of Israel, not of the nations, in the book of Ezekiel. Joyce's chart shows also that Israel will know YHWH most when YHWH punishes her (6:7, 10, 13, 14; 7:4, 27; 11:10, 12; 12:15, 16, 20; 20:26; 22:16; 23: 49; 24:24, 27; 39:28a) or delivers her (16:62; 20:42, 44; 28:24, 26a; 29:21; 34:27; 36:11, 38; 37:6, 13; 39:28b), but rarely when Yahweh punishes the nations (28:26b; 39:22). There is no occurrence, however, of YHWH delivering the nations. On the other hand, according to Joyce, the nations will know YHWH most when YHWH punishes the nations (25:5, 7, 11, 17; 26:6; 28:22, 23; 29:6, 9, 16; 30:8, 19, 25, 26; 32:15; 35:4, 9, 15; 38:23; 39:6) or delivers Israel (36:23; 39:7); but never is there any case of Yahweh punishing Israel or delivering the nations.

Suffice it to present several examples to challenge Joyce's view. In 29:6-21, God treats Egyptians as he deals with Israel. The language employed of God's scattering (because of the sin of pride) and gathering of the Egyptians (vv. 12, 12-24) is strikingly similar to the language used for Israel. Here, the nations will know YHWH when YHWH delivers the Egyptians. Also, it is noted that 'on that day' Israel will have a horn grown and God will open their mouth and they (Israel) will know YHWH (v. 21). For the instance that testifies the nations will know God when he punishes Israel, see 39:21-23: "the nations will know that the people of Israel went into exile for their sin, because they were unfaithful to me." To conclude, it is not an overstatement to say that the nations are in view as God deals with Israel's exile and restoration in the book of Ezekiel.[112]

---

[110] Some Hebrew manuscripts do not have לא conforming to the same phrase in 11:12 where God accuses Israel of having done in accordance with the manners of the nations. The accusation that Israel did 'worse' than the nations is a recurring theme throughout the book (cf. 'more depraved than they,' or 'worse than Sodom' in 15:45-48).

[111] Joyce, *Divine Initiative*, 90-91.

[112] Joyce, *Divine Initiative*, 154, n.7, takes 29:13-16 only as a case of humiliation.

The relationship God desires to have with his people is laid down in the covenant formula, 'They will be my people and I will be their God.'[113] This exclusive relationship, in terms of Israel being God's own possession, turns out to be a terror to herself. One clear reason why God had to purge Israel is that she failed to follow the laws and decrees of her God whose name is to be known among the nations (6:8-14). Her detestable practices before God were even worse than those of Sodom (7:4-9; 16:45-48).

Moreover, Israel is accused of having oppressed and mistreated the aliens and the fatherless within her own community (22:7; cf. 47:22-23). Yet all of these practices are summed up in Israel's pursuit: "We want to be like the nations, like the peoples of the world, who serve wood and stone" (20:32; 11:12). This is the exact reversal of what God intended for Israel from the beginning (cf. Exod 19:5-6). Nevertheless, it is God's zeal for his name – 'for the sake of his name,' says Ezekiel – that would prevent this from happening (20:32-38). Herein lies the hope for Israel's future. M. Greenberg succinctly states, "the indissoluble link between God's reputation and Israel's fortune guarantees that Israel shall be restored; but so that God's name never again suffer disgrace, Israel's restoration must be irreversible."[114]

Thus the motivation for and guarantee of God's restoration of Israel is God's zeal for his name before the nations. This motif of God's concern for his name as the only ground for his restoration of Israel is reaffirmed repeatedly throughout the vision for the restoration (Ezek 36:20, 21; 39:7, 25; 43:7, 8). Few would disagree that the recognition formula motif weaves and holds together the entire matrix of the book: וְיָדְעוּ כִּי־אֲנִי יְהוָה ("then they [Israel or the nations or both] shall know that I am the Lord")[115] Zimmerli argues that "the ידע - formulation belongs to the sphere of legal examination in which a sign of truth was demanded" (cf. Gen 42:34).[116] Ezekiel's use of the formula intended an experiential, confessional knowledge of YHWH on the part of the subject.[117]

---

[113] In this hopeful future context, they will surely obey his laws: 11:20; 14:11; 34: 30; 36: 28; 37: 23, 27. 'My people' against the nations appears in 13: 9, 10, 18, 19, 21, 23.

[114] Greenberg, *Ezekiel 21-45*, 735.

[115] The following is a comprehensive list of the occurrences of the recognition formula in its various forms: 5:13; 6:7, 13, 14; 7:4, 9, 27; 11:10, 12; 12:15, 16; 13:9, 14, 21, 23; 14:8; 15:7; 16:62; 17:21, 24; 20:12, 20, 26, 38, 42, 44; 21:10; 22:16, 22; 23:49; 24:24, 27; 25:7, 9, 11, 17; 26:6; 28:22, 23, 24, 26; 29:6, 9, 16, 21; 30:8, 19, 25, 26; 32:15; 33:29; 34:27, 30; 35:4, 9, 12, 15; 36:11, 23, 36, 38; 37:6, 13, 14, 28; 38:23; 39:6, 22, 28. Associated with the above formula is another that deserves close attention: "then they shall know that a prophet has been among them" (2:5; 33:33); cf. John Strong, "Ezekiel's Use of Recognition Formula in His Oracles Against the Nations," *PRS* 22 (1995): 118.

[116] Zimmerli, *Ezekiel 1*,37.

[117] W. Zimmerli, "Knowledge of God According to the Book of Ezekiel," in *I Am Yahweh* (ed. Walter Brueggemann; Atlanta: John Knox, 1982), 97.

1.4 The Davidic Shepherd Tradition in Ezekiel's Vision

If the formula denotes an experiential, confessional knowledge of YHWH, then the subject becomes very significant. The subject of the main clause may be varied depending on the context. It could be 'all the trees of the field' (17:24), 'the house of Israel' (39:22), 'the inhabitants of Egypt' (29:6), or 'the nations' (36:23, 36; 37:28; 39:23), yet the most comprehensive description is found in 21:10, 'all flesh' (כָּל־בָּשָׂר).[118] Then it is concluded that in Ezekiel God desires all people to have some kind of confessional knowledge of him.

According to H. M. Orlinsky, 'universalism' denotes the belief that YHWH was the creator and ruler over all creation; 'internationalistic' can be defined as the belief that the foreign nations would be included within Israel as the covenantal people.[119] John S. Strong doubts whether Ezekiel's concept of restoration is 'international,' i.e., whether that knowledge was open to the peoples in the nations. He raised a series of questions that are one in essence: "What kind of knowledge of YHWH did Ezekiel intend when he placed the foreign nations as the subject of the recognition formula: Judgment?"; "Will Egypt also be a covenantal partner of YHWH?"; "Was Ezekiel an internationalist?" Strong posits that Ezekiel's use of the recognition formula in his foreign nation oracles is wholly nationalistic.

Strong's argument seems ineffectual, however. First, according to Ezekiel, God's concern for his name is not confined only to Israel. There are fourteen references to the divine 'name' in Ezekiel: chapter 20 [5x], 36 [4x], 39 [3x], 43 [2x]. The divine 'name' always appears in association with the recognition formula, regardless of the subject of that formula is either the nations or Israel. Wherever God's name appears, the nations are always in the picture: 'in the eyes of the nations' (20:9, 14, 22), 'before their eyes' (36:23), 'in the sight of many nations' (40:25), or 'among the nations' (36:20, 21, 22, 23). Second, the particular phrase, 'opening the mouth among the nations' (29:21), affirms YHWH's determination to make his name known to the nations despite the failure of his chosen nation. Third, among chs. 20, 36, 39, and 43 in which God's name appears, judgment on Gog as the representative of the nations' rebellious power (39) sounds the opening of the new history, beginning with the new temple and reaching to the new land. The nations will be witnessing the Holy One among Israel (v. 7), and also the nations and Israel will witness God's redemption of Israel out of the nations: "And the nations will know that the people of Israel went into exile for their sin ... I will show myself holy through them [Israel] in the sight of many nations" (39:21-27).

Peter Ackroyd rightly suggests that here the place occupied by the nations as witness is to be understood not in the narrow sense of spectators as non-participants in the action of God, but as witnesses who are themselves involved in what happens because "they must assess their own position relative to it." [120]

---

[118] Though the term, בָּשָׂר denotes simply a human being, it is frequently taken in a negative context (11:19; 36:26; 37:6; 44:7,9).

[119] Harry M. Orlinsky, "Nationalism-Universalism in Ancient Israel," in *Translating and Understanding the Old Testament* (Nashville: Abingdom, 1970), 206-236.

[120] H. G. Reventlow (in "Die Völker als Jahwes Zeugen bei Ezechiel," *ZAW* 71 [1959]: 33-43, cited in Ackroyd, *Exile and Restoration* [Philadelphia: Westminster, 1968], 15-17), points out that witnesses in the OT sense are not indifferent but involved: "they are to assess their own position relative to what takes place."

For the sake of this opportunity, God's name should not be profaned among the nations. God must act for his name's sake. The formula discloses that God reveals himself in his name among the nations as well as within the people of Israel. Thus W. Zimmerli states:[121]

> YHWH is always the subject, even when it is mentioned that his action is mediated by men, and fulfils the function of a sign of proof that he is who he claims to be in his name. All that which is preached by the prophet as an event which is apparently neutral in its meaning has its purpose in that Israel and the nations should come to recognition, which in the Old Testament also always means an acknowledgment of this person who reveals himself in his name. All YHWH's action which the prophet proclaims serves as a proof of YHWH among the nations.

As we argued earlier, it is 'before the eyes of the nations' that God does what he does with Israel. Conversely, it is 'before the eyes of Israel' that God deals with the nations. God's purpose is to be known by all. God's desire to be known by the nations as well as by his people Israel has unmistakable universal dimensions. Even Ezekiel's vision of resettlement in the land amazingly expands to include peoples from the nations.

This is hinted at in 47:13-23 where we find the allotment of the new land as the inheritance, which is the result of God returning to the new temple. From the beginning, the principle of treating aliens nearly on par with the status afforded the Israelites has been included in the law (Lev 19:34; Deut 24:17). In the OT, the land was allotted only to the tribes of Israel (Num 26:55-56), while aliens were not to hold any property in the land, similar to what was accorded to the Levites (Deut 14:21). Listed among Jerusalem's sins, Ezekiel points to the oppression of aliens (22:7). Now the context of Ezek 47:13-23 is the allotment of the new land as the inheritance, which is the result of God returning to the new temple. Here Ezekiel is taking 'a great and daring step' toward placing the alien on equal footing with the native born.[122]

The principle exists in the laws, yet it is Ezekiel's vision that actualizes the principle: "So you shall divide this land among you according to the tribes of Israel. You shall allot it as an inheritance for yourselves and *for the aliens* who reside among you and have begotten children among you. They shall be to you as citizens of Israel; with you they shall be allotted an inheritance among the tribes of Israel. *In whatever tribe aliens reside, there you shall assign them their inheritance*, says the Lord God" (47:21-23). In this respect, Zimmerli is correct when he makes the general observation that "the majority of the statements are concerned with the recognition that is to take place within the people of Israel. Beyond that, however, we see that this same recognition is expected from the

---

[121] Zimmerli, *Ezekiel 1*, 38.
[122] The quote is W. Eichrodt's own expression; see *Ezekiel*, 592.

rest of the world's nations. On this point Ezekiel is similar to Deutro-Isaiah." [123] Taking into consideration the recapitulations, syntactic and semantic connections, and the traditions behind the concept, the procedure in Ezekiel's vision of restoration in chapters 34-37 may be depicted as follows:[124]

a. Indictment of False Shepherds of Israel - absence of the true Shepherd
  b. God's concern for his name defiled by Israel before the nations
    (the cause of the exile; also the problem of disobedience of the law)
    c. The Mission of the Divine Shepherd
       - reversing the failed shepherding: seeking, saving, gathering
       - reversing the impotence to keep the law: renewed covenant
         (thus, bringing the Mosaic covenant into fulfillment [Lev 26])
       - reversing death to life: the revivification of Israel
    c'. The Mission of the Davidic Shepherd
       - shepherding the redeemed people to follow God's laws
       - bringing the fulfilment of the covenant of peace in the land
       - securing God's sanctuary that sanctifies the people/ the land
  b'. Acknowledging God's holy Name by the nations - the ultimate goal
    (God's own zeal for his name is the ultimate cause of the restoration)
a'. The Presence of God the Shepherd and the Davidic Shepherd in their midst

Many have deemed the book of Ezekiel the most nationalistic among the major prophetic writings, though its scope and intention are fundamentally open to universalism. Ezekiel 37:24-28, which closes Ezekiel's vision of restoration in chapters 34-37 and foreshadows what follows in chapters 40-48, underscores this theme: "Then the nations shall know that I the Lord make Israel holy, when my sanctuary is among them forever" (v. 28; cf. 36:1-23). Ezekiel's vision of restoration expands to embrace people from the nations.

1.4.6 The Shepherd in Ezekiel 34-37

Of the various motifs, shepherd imagery distinguishes these chapters from chapter 33 and chapters 38-39, providing literary and thematic links that bind together the bloc. Chapter 34 begins with YHWH indicting the unfaithful shepherds of Israel (vv. 1-10), and proclaiming himself the Shepherd who will perform their unfulfilled job 'on a day of cloud and darkness' (v. 12; vv. 11-19). This chapter closes with the promise of the Davidic Shepherd (vv. 23-24) and

---

[123] Zimmerli, "Knowledge of God," 88; cf. Philip R. Davies, "The Judaism(s) of the Damascus Document" in *The Damascus Document A Centennial of Discovery* (ed. J. M. Baumgrarten, E. G. Chazon and Avital Pinnack; Leiden: Brill, 2000), 32, comments on Ezek 47:22, "a non-Israelite resident of the land is to be reckoned among the tribes of Israel, similarly in CD 14:3-4."

[124] This does not necessarily represent a chronological sequence, nor is it ever clear in this regard. Yet the diagram can be seen as a possible conceptual sequence of Ezekiel's vision, though often recapitulated, yet is definitely progressive.

the claim of YHWH that his people are his 'sheep' (v. 31). The shepherd motif resurfaces at the end of chapter 36: YHWH as the Shepherd of the house of Israel promises to increase the sheep, i.e., his people (v. 37). Chapter 37 is no exception. Besides the possible shepherd imagery in verses 12-14, the motif clearly reappears in verses 24-25, taking various forms, and further developing as the oracle unfolds. As mentioned previously, the reversal motif is used throughout the whole of chapter 34. To start the process of the reversal, the first thing to do is to establish the inevitable intervention of YHWH as the righteous Shepherd (vv. 1-10).

As was demonstrated, chapter 34 displays a symmetric structure with the emphasis on YHWH's claim placed at its center: אֲבַקֵּר אֶת־צֹאנִי וְהִצַּלְתִּי אֶתְהֶם ("I will seek my flock and deliver them," vv. 10, 12). The idea that 'YHWH himself' is the one who 'will seek' (בָּקַר, 34:10, 12, בקשׁ, 34:16) the lost and wounded sheep is characteristic feature of Ezekiel's description of the restoration, especially when compared to Jer 23:1-8.[125] W. Zimmerli notes that in Lev 13:36 in the diagnosis of leprosy, בָּקַר ('to care for, seek, examine') describes the precise examination of the rash of the disease; in Lev 27:33 it describes the examination of an animal with a view to its suitability for sacrifice. In the context of Ezek 34:10, 12, however, the idea of examination is unclear until later in verses 16 (בקשׁ). Here YHWH seeks the lost as one of his rescue activities. The main task implied in YHWH's seek-and-deliver mission (vv. 10, 12) is illustrated in a series of redemptive actions set out in verse 13:

I will cause them to come out (וְהוֹצֵאתִים; LXX, ἐξάγω);
I will gather them (וְקִבַּצְתִּים; LXX, συνάγω) from the peoples and the countries;
I will lead them into (וַהֲבִיאֹתִים; LXX, εἰσάγω) their own land;
I will shepherd them (וּרְעִיתִים; LXX, βόσκω) on the mountains of Israel.

The four main verbs in verse 13 give a coherent perspective of the Shepherd's tasks. The meaning of רָעָה ('to shepherd') is limited 'to feed,' and the preceding three action stages, 'to bring out, gather, and bring into' are expressed in terms of ἄγω cognates. The first step is to reverse the final stage of the consequences of the evil shepherding, i.e., 'they have become prey' (vv. 5, 8, 10). The reversal begins when YHWH rescues the flock from the claws and mouths of the beasts (v. 10). At the center of the symmetrical structure of 34:1-24, YHWH solemnly proclaims: "I will seek (אֲבַקֵּר) out my flock; and I will deliver (וְהִצַּלְתִּי) them" (v.

---

[125] According to Hals, *Ezekiel*, 251, some of the differences between the two oracles (Ezek 34 and Jer 23:1-8) are: (1) the idea of God himself as the good shepherd in Ezek 34 is missing in Jer 23:1-6; (2) Jeremiah 23:1-4, 5-6, and 7-8 are separate units connected to the end of a series of prophecies about specific Judean kings, achieved by means of a wordplay on the root 'righteousness' in the name Zedekiah. Ezekiel 34, however, is more of a programmatic exploration of various facets of shepherding. The word 'righteousness' never appears, while 'to seek' appears prominently. In Jer 23:1-6 it does not appear at all.

12). If Ezek 34:13 is seen as a detailed expansion of what is proclaimed in verse 12, all consecutive stages of 'bringing out, gathering, bringing in, and feeding' can be seen as implied stages of YHWH's rescuing activity. If taken in a narrower sense, however, הִצַּלְתִּי (graphically "I will snatch away") could indicate the first half of a series of YHWH's redemptive action. The verb form of הִצַּלְתִּי (*hiphil*) is found in Exod 6:6, 7:4, 5, and with the third person plural suffix in Ezek 34:13, 27. This points to God manifesting his mighty power on the battlefield to rescue his people from oppressive, external powers.[126]

It is quite possible that verses 12-13 echo the exodus motif, seemingly more evident in verse 13: 'in the day of cloud and darkness' (v. 12c), and the imagery of YHWH 'bringing out, gathering and bringing into (v. 13).' [127] Ezekiel 20:32-38 is a prophecy of the day when YHWH will deliver his people, before he leads them into the land of Israel (v. 38). YHWH will purge them with judgment in the land of the nation – 'as in the desert of the land of Egypt' (v. 36; cf. Exod 32:15-29; Num 11; 14:10-19; 16:31-50).[128] Here YHWH is also depicted as the Shepherd who will make them pass under the 'rod': וְהַעֲבַרְתִּי אֶתְכֶם תַּחַת הַשָּׁבֶט (v. 37). Again, the LXX's rendering of וְהַעֲבַרְתִּי as διάξω suggests the presence of the shepherd metaphor. In fact, the shepherd motif is implicit already in Ezek 20:32-38. All the verbs that describe YHWH's redemptive and restorative actions are identical with those used in 34:12-13 (MT). Even here, the LXX mainly uses the cognates of ἄγω:

20:34a, "I will bring you out (וְהוֹצֵאתִי; LXX, ἐξάγω) from the peoples" - 34:13a
20:34b, "I will gather you (וְקִבַּצְתִּי); LXX, εἰσδέχομαι) from..."- 34:13b
20:35a, "I will bring you into (וְהֵבֵאתִי; LXX, ἄγω ὑμᾶς εἰς) ..." - 34:13c
20:35b, "I will execute judgement (וְנִשְׁפַּטְתִּי) there face to face" - 34:17

The Shepherd's premier task is to rescue his sheep. What surprises us in Ezek 34:10 is the fact that evil shepherds also come under the category of 'wild

---

[126] Cf. Ezek 34:27, "I will deliver them (וְהִצַּלְתִּים; 'save' NRSV) from the hands that enslave them."

[127] The LXX renders בְּיוֹם־הֱיוֹתוֹ בְתוֹךְ־צֹאנוֹ נִפְרָשׁוֹת (v. 12a) as ἐν ἡμέρᾳ ὅταν ᾖ γνόφος καὶ νεφέλη ἐν μέσῳ προβάτων διακεχωρισμένων, thus making an inclusio with v. 12c. Historically, on the basis of 30:3, one might see here a motif of the description of the day of Yahweh: "the day of the destruction of Jerusalem and the final collapse of the political entity 'Israel'(Judah)" (Zimmerli, *Ezekiel 2*, 216). While it is certain that this temporal clause refers to 'Yahweh's day of judgment,' no comment is made about the judgment day. W. Eichrodt, *Ezekiel*, 471, doubts whether Ezekiel here looks beyond the judgment day of Jerusalem. Cooke, *Ezekiel*, 375, suggests Zeph 1:15 as the tradition that is also the source of Joel 2:2.

[128] The same sequence is found in Ezek 34: rescue (vv. 12-13), restoration (vv. 14-16), and judgment (vv. 17-22). That this process of purging/cleansing is required before the rescued people enter into the land is more vividly described in 36:24-32; cf. Zimmerli, *Ezekiel 2*, 217; see also Levenson, *Program of Restoration*, 89-90, for the exodus motif used here (cf. Pss78:52-54).

beasts' from whose hands and mouths YHWH will snatch away his flock.[129] The shepherds of Israel are the enemies of YHWH's flock; it is they who plundered and devoured it. Who then is denoted by the term 'shepherds'?

If 'the fat and strong' in verse 16, 17-22 are directed at non-royal persons who possess power in Judah, that is, the ruling class, then the term 'shepherds' in verses 1-10 seems to refer to the former kings of Judah.[130] Later in verses 23-24, however, the new David whom YHWH appoints as 'one' shepherd over the people, is also called 'prince' (v. 24) as well as 'king' (v. 23). This cautions us to refrain from narrowing the range of the 'shepherds' exclusively to the kings of Judah. It is unclear if we can exclude any other ruling groups within the society. Throughout the book of Ezekiel, there is a hint that the focus of YHWH's indictment and judgment shifts from the kings of Judah, to the leaders in general, and eventually to the people themselves!

This shift is evidenced in the following passages: (i) The kings are under judgment: Ezek 17:2-10 (Zedekiah); 19:1-14, "No strong branch is left on it fit for a ruler's scepter" (v. 14); "O profane and wicked prince of Israel" (Zedekiah as prince, 21:25-27); (ii) the princes come under judgment: "See how each of the princes of Israel who are in you uses his power to shed blood" (22:6, cf. v. 25); 34:17-22; and (iii) the people come under judgment: "The house of Israel shall no more defile my holy name, neither they, nor their kings, by their prostitution, and the lifeless idols of their kings at their high places" (43:7,8). Certainly, YHWH holds accountable 'the leadership' of the nation.[131]

Opposing these evil shepherds, YHWH shows remarkable compassion for his flock. The phrase, צאני ('my flock') occurs at least ten times within chapter 34.[132] YHWH fiercely confronts the shepherds of Israel out of compassion for his flock, who have strayed on the mountains (v. 6) since they did not bother to search for his flock (v. 8). Therefore, YHWH will deliver his flock from the mouths of the evil shepherds (v. 11); He will search for his flock (v. 11); He will seek out his flock (v.12); and He will shepherd his flock by himself (v. 15).

---

[129] In Ezek 22:25, the land is barren and YHWH's word reveals the reason for it: "there is a conspiracy of her (land) princes within her like a roaring lion tearing its prey; they devour people, take treasures and precious things and make many widows within her."

[130] Duguid, *The Leaders*, 39-40, prefers this option based on the new David's appearance as a kingly figure in vv. 23-24. Jeremiah looks forward to a new group of shepherds (25:34-38) in contrast to Ezekiel's 'one' shepherd. Cooke, *Ezekiel*, 373, includes leading members of the community as well as kings. Zimmerli, *Ezekiel* 2, 214, notes that the oracle here is directed to the history of Israel as a whole; therefore the term shepherds of Israel is meant to represent God's people as a whole. Zimmerli refuses to pose the question to identify any specific group either before or after the exile.

[131] Cf. Duguid, *The Leaders*, 33-43, misses this transition of the focus of YHWH's indictment.

[132] Outside Ezekiel, Jer 23:1-3 retains a stock of similar images: "sheep of my pasture" (v. 1; cf. Ezek 34:31); "you have scattered my flock" (v. 2); "I will gather them the remnant of my flock" (שארית צאני, v. 3).

1.4 The Davidic Shepherd Tradition in Ezekiel's Vision 61

Then YHWH will judge his flock to purge them and establish justice within the flock (v. 17). In verse 19, YHWH's flock is identified with the downtrodden and the outcast. YHWH declares to save (יָשַׁע; LXX, σώσω) his flock in verse 22, and in verse 31, YHWH affirms his unfailing covenantal relationship saying, "You are my flock, the flock of my pasture (cf. Jer 23:1), people (הָאָדָם) you!"[133] Consistent in the use of צֹאנִי in Ezekiel 34 is that צֹאנִי refers mostly to YHWH's remnant of the covenant people victimized by the oppression of the evil leaders and the robust of their society.[134] YHWH particularly identifies this group as his flock since no one seeks and saves them.[135] The tripartite initial action of the shepherd, i.e., 'bringing out, gathering from, and bringing into' in Ezek 34:13 (cf. Ezek 20:34-35) seems to be recapitulated in 37:12, yet from a different aspect and in different imagery.[136] One cannot dismiss possible indications of the shepherd image, however. This passage might be describing a decisive moment of YHWH's shepherd's redemptive act (37:12):

(Thus, prophesy, and you speak to them, here I say the Lord YHWH, behold),
a. I will open (פתח; LXX, ἀνοίγω) your graves;
b. and I will raise (וְהַעֲלֵיתִי; LXX, ἀνάγω) you, from your graves, my people;
c. and I will bring (וְהֵבֵאתִי; LXX, εἰσάγω) you in the land of Israel.

It is not difficult imagining in verse 12a-b a shepherd forcefully opening (פתח) the mouth (פֶּה) of graves, causing those buried within to rise up. This is also what the verb נָצַל ('to snatch/take away [i.e., from the 'mouths' of the wild

---

[133] The phrase, 'people, you' seems enigmatic. To what does it refer? Is there any universal dimension applied to this verse? Does this refer only to Israel, or to humanity in general, thus expanding on 'my flock'? One rabbinic commentator applies it only to the Israelites. R. Simeon b. Yohai thinks the predicate הָאָדָם is limited to Israel: "You (Israelites) are called men but the idolaters are not called man" (*B. Yebad* m. 61a; *b. B. Mes.* 114b; *b. Ker.* 6b; cited from W. Zimmerli, *Ezekiel* 2, 221). Since few manuscripts omit the word הָאָדָם, G. A. Cooke suggests it may have crept in from below, 35:2, 'son of man' (*Book of Ezekiel*, 379). Perhaps the intention may be to notify the cessation of the usage of the metaphor shepherd and sheep, or YHWH revealing himself as God, not man, who alone accomplishes what man can only dare to imagine, thereby emphasizing the gap between the holy God and mere man.

[134] Likewise, Renz, *Rhetorical Function,* 107, comments, "It is striking that the phrase 'my sheep' is applied only to the victims." Also, Cooke, *Ezekiel,* 376-377, notes that "the divine Shepherd will care for precisely those who have no one else to care for them; Jahveh's sheep are the helpless and weak, as distinguished from the fat and the strong."

[135] The flock (in 'my flock') that YHWH supports is referred to in what follows as feminine (vv. 19, 21-22, see also 'you' [fem.] in v 17) while the wicked thoughtless animals are referred to in vv. 18, 19 ('your [masc.] feet'), 21 as masculine. Thus two groups are separated, in v. 20, the two feminine forms שֶׂה בְּרִיָה ('fat beasts') and שֶׂה רָזָה; cf. Zimmerli, *Ezekiel* 2, 217.

[136] Not necessarily as Cooke, *Ezekiel,* 377, states, "In Ezekiel, the emphasis lies upon restoration rather than salvation." Moreover, Ezekiel repeatedly projects the image of YHWH's great salvific act through recapitulating the aspect of 'delivering' that is to be *presupposed* for the subsequent restoration process.

beasts],' 'rescue') graphically illustrates in 34:10, 12, 27.[137] The LXX certainly leans in this direction by adopting 'ἀν/εἰσάγω' to project the shepherd's activity in this particular passage. The following scene re-introduces the shepherd image: "and I will cause you to rest (וְהִנַּחְתִּי; LXX, fut. τίθημι) in your land" (37:14). What is meant by the opening of the graves and the bringing up of the exiles remains to be dealt with separately.[138] The point is the shepherd metaphor appears continually. In this passage, YHWH raising the dead from their graves should be understood as one of the eschatological Shepherd's tasks.

From Ezek 34:13 to 34:14-15, there is a shift of focus in terms of the shepherd's task. Verses 14-15 proleptically project the picture of YHWH guiding, lying down, and feeding his flock that he delivered and brought in the mountains of Israel in verse 13. Similarly in the bloc of Ezek 36:24-32, there appears t be a shift in terms of the progression of YHWH's restoration.

Compared with נָצַל (34:10, 12, 27), the verb יָשַׁע while used in a different context, still conveys deliverance: YHWH will 'save' (LXX, σώσω) them from being further plundered (34:22); from uncleanness (36:29), and he will save (LXX, ῥύσομαι) them from all apostasies (37:23). The root יָשַׁע is rare in Ezekiel, and occurs only in these three texts. In the cases of 36:29 and 37:23, its meaning indicates rescue from Israel's own moral or religious troubles. Indeed, the use of יָשַׁע seems to be differentiated from נָצַל in the OT.[139]

---

[137] Of course, the word 'mouth' is not present in 37:13. A phonetic or verbal link between the verb פתח and the noun פֶּה as can be illustrated in Ezek 3:27 (אֶפְתַּח אֶת־פִּיךָ) and 16:63 (פֶּה פִּתְחוֹן). In addition, 'to lead in' (הֵבֵאתִי, 37: 13c) is the same verb that precedes YHWH's acts of delivering and gathering in Ezek 34:13 and 20:34.

[138] To link the *hiphil* form of קום ('raise up,' also in 37: 12; cf. Jer 30:9-10, Hos 3:5) with the idea of resurrection of David, i.e., *David redivivus* is possible but not necessary. It is used for the raising up of judges (Judg 2:16), deliverers (Judg 3:9), prophets (Jer 29:15), and kings (1 Kgs 14:14). Thus, 'to raise up' is used for the appointment of a plurality of future shepherds, but in contrast to Jer 23:4, the singular interpretation is preferred in Ezek 34:23 (Zimmerli, *Ezekiel 2*, 219). It may be one of David's legitimate offspring or *David redivivus*, or less likely rulers according to the Davidic line. M. Greenberg, *Ezekiel 21-40*, 760, assesses these two options as too clear-cut between the Davidic offspring or David *redivivus*. He notes that "the text requires no more than a new David who is not the old one resurrected or merely one of David's line." It might be that it emphasizes the sure continuation of what Yahweh is to do. Cf. Groningen, *Messianic Revelation*, 773.

[139] Likewise, Robert L. Hubbard Jr., *NIDOTTE* (ed. Willem VanGemeren; Grand Rapids: Zondervan, 1997) 2:556, says: "In general, the root *yš‛* implies bringing help to people in the midst of their trouble *rather than in rescuing them from it* " (italics mine). The *BDB* also gives a sense of the *hiphil* form of יָשַׁע as "God's saving his people from external evils," placing Ezek 34:22 in this category, but failing to list Ezek 36: 29 (446). Another sense conveyed is 'to save from moral troubles.' In this category, *BDB* includes Ezek 37:23 (447). It is interesting to see that יָשַׁע can carry different nuances in different contexts.

Block notes that in 36:29 Israel is held captive 'not by human enemies but by their own uncleanness.'[140]

These three occurrences share a common setting, that is, YHWH judges (34:22), purges, (36:29) and cleanses (37:23) within the regatherd community. Conceptually, this is to take place after they are taken out of the nations and gathered from all countries, but before they enter the new land (cf. 36:24). Ezek 34:16 presents a list of YHWH's tasks as the Shepherd, which reads as the exact reversal of the failed tasks of the shepherds of Israel's past, failures of which they are accused in verse 4, yet with much more. It is with pronounced harshness that the Shepherd confronts the shepherds of Israel; YHWH determines to perform every single task the evil shepherds had neglected to fulfill for his flock. In verse 16 these tasks are set in stark contrast:[141]

v. 4: You (Israel's shepherds) have not:
  a. strengthened the weak,
  b. healed the sick
  c. bound up the injured,
  d. brought back the outcast
  e. sought to find the lost
("You treaded them violently and harshly")

v. 16: I (YHWH) will:
  e´. seek to find the lost
  d´. bring back the outcast
  c´. bind up the injured
  b´. the sick
  a´. strengthen the weak
("I'll destroy the fat; shepherd with justice")

Six types of sheep are identified who will be cared for by the Shepherd: (i) the weak (הַנַּחְלוֹת; LXX, τὸ ἠσθενηκός, v. 4); (ii) the sick (הַחוֹלָה; τὸ κακῶς ἔχον, v. 4; τὸ ἐκλεῖπον, v. 16); (iii) the injured (נִשְׁבֶּרֶת; τὸ συντετριμμένον, vv. 4, 16); (iv) the outcast (הַנִּדַּחַת; τὸ πλανώμενον, vv. 4, 16); and (v) the lost (הָאֹבֶדֶת; τὸ ἀπολωλός, vv. 4, 16); in contrast, (vi) the fat and robust sheep form another group. YHWH as the shepherd responds to each group respectively. In verse 16, YHWH's restorative actions invert the sequence of the failed tasks of the evil shepherds.[142] Importantly, YHWH's remedy for his flock begins with the most victimized group: the lost. Likewise, the eschatological Shepherd's reversal begins at the least thing the evil shepherds wanted to do: to seek. Thus to seek the lost (אֶת־הָאֹבֶדֶת אֲבַקֵּשׁ; LXX, τὸ ἀπολωλὸς ζητήσω, vv. 4, 16), is the priority of YHWH, the eschatological Shepherd. I believe this inverted order is

---

[140] Block, *Ezekiel 25-48*, 357.

[141] In v. 16, וְאֶת־הַחוֹלָה אֲחַזֵּק sums up and replaces 'strengthened (חָזַק) the weak and healed the sick (הַחוֹלָה)' of v. 4. Instead the first half of the last part of v. 16, 'I will destroy the robust,' stands in contrast to 'you tread down them with harshness and violence' (v. 4), and the second half is added: 'I will shepherd with justice.' The customary expression, 'harshly and brutally,' is exactly the way Egyptians treated the Israelites in the land of slavery, behavior that is prohibited for redeemed Israel (Exod 1:13-14; Lev 25: 43, 46); see W. Lemke, "Life in the Present and Hope for the Future," in *Interpreting the Prophets* (ed. J. L. Mays and P. J. Achtemeier; Philadelphia: Fortress Press, 1987), 173.

[142] The implication surely involves what Daniel I. Block states in the *Book of Ezekiel: Chapters 25-48*, 291: "the tragedies will be reversed."

64 Chapter 1. The Davidic Shepherd Tradition in the OT

significant, for it renders the task of seeking the lost a catch-phrase topping the list of the shepherd's tasks. Seeking the lost is how the restoration will be inaugurated.

The next group consists of outcasts who strayed or were driven away by a storm or an attack of wild beasts. The metaphoric expression itself makes it difficult for us to determine exactly what type of group is being referred to in an exiled society at the time of Ezekiel. While the outcasts are not totally lost or exterminated, their very survival remains at stake. It is likely they are an extremely marginalized people driven beyond the fringe of society. YHWH will bring them back (אָשִׁיב; LXX, ἐπιστρέφω, vv. 4, 16) into the community to restore their rightful place. Later Ezek 34:21 describes graphically a situation in which the robust and strong ones (v. 16) push away all the diseased/sick (הַחוֹלָה) with their horns till they have scattered them to the outside. The outcast, then, denotes the sick driven out of the community. Bringing and gathering them back is the second principal task of the shepherd.

In verse 16 we note a gradual movement from the most urgent task to actions of a more restorative nature on the list of the shepherd's mission. Binding up (חָבַשׁ; LXX, καταδέω, in vv. 4, 16) the crushed or injured, healing (רָפָא; LXX, σωματοποιέω, v. 4) the sick, and strengthening (חָזַק; LXX, ἐνισχύω, vv. 4, 16) the weak, can be seen as a series of restorative work for the suffering people remaining within the community. Thus YHWH the eschatological Shepherd will build a new community. He will reach out to the lost and the outcast by healing the broken and sick, and strengthening the weak.

Therefore, YHWH as the Shepherd, at the time of the restoration will be recognized by this antithetical attitude toward the tasks originally entrusted to the leaders of Israel. Consequentially, the leaders of the nation are held accountable, and YHWH proceeds to confront them point by point. The inverting of the order of the tasks in verse 4 not only announces the time of blessings, but also means – or historically has already signified for Ezekiel – the time of embarrassment and judgment meted out to the leaders of the nations. Eventually, they appear to be replaced by YHWH himself and his servant David, רֹעֶה אֶחָד ('one shepherd') over his people (vv. 23-24).

In this respect, the rendering of the MT in the last part of verse 16 seems critical: וְאֶת־הַחֲזָקָה אַשְׁמִיד. Either it means 'I will destroy (the robust/strong ones)' or 'I will police (them).' W. Zimmerli and D. I. Block emend אַשְׁמִיד ('I will destroy; wipe out') to אֶשְׁמֹר ('I will watch over; police').[143] R. H. Alexander dismisses this emendation in favor of YHWH's implied negative evaluation of

---

[143] Zimmerli, *Ezekiel 2*, 208. Hiphil form of שָׁמַד ('to destroy, exterminate') is read שָׁמַר with LXX (φυλάσσω), Vulg., and Syr., though many delete the word as a gloss. Block, *Ezekiel 25-48*, 287-288, argues that the parallel between 'I will police' and 'I will shepherd with justice' supports this emendation.

the robust and strong in the same verse 16.¹⁴⁴ Interestingly, G. A. Cooke preserves MT's reading, despite the strong impression of the gentile, healing shepherd thus far in verse 16. He bases his argument on the grounds that it better agrees with what follows – 'I will feed them (the strong ones?) with judgment' (אֶרְעֶנָּה בְמִשְׁפָּט).¹⁴⁵ Verse 6 reads 'I will shepherd with judgment (or justice)' without 'them.'

Nevertheless, Cooke observes precisely: "The Divine shepherd, if merciful, is also just and can be stern."¹⁴⁶ This point should not be missed. Besides possessing compassion and mercy, the shepherd possesses authority in verses 16, 17-22. This is a stern and powerful figure imbued with authority who takes control of the community with justice. YHWH will do what the past leaders of Israel were supposed to do, but this time the tasks will be performed with compassion and authority (cf. vv. 4 and 16). In this regard, Ian M. Duguid rightly states:¹⁴⁷

> According to Ezekiel 34, the expected change in Israel's governance will be accomplished not so much through a change of nature of the *office* but through a change in the nature of the *occupant* ... Only the one who is stronger can protect against the strong. Strength and service are the two features emphasized in the description of the new ruler.

It is hardly surprising that the shepherd's task of judgment is the motif by which verse 16 introduces what follows in verses 17-22. Conceptually in this section, the restoration process reaches the point where YHWH is concerned about justice within the community. Not only are the royal figures of Judah held accountable, but also the leading members of the community.¹⁴⁸ Morever, verses 17-22 along with verse 16 provides the immediate context the oracle uses to introduce the Davidic Shepherd and what transpires in 22-23.

As G. A. Cooke argues, it is an acceptable interpretation that Ezekiel has no respect for the contemporary kings (17:1-21; 19:10-14; 21:30-32), though he betrays some sympathy for the fate of Jehoahaz and Jehoachin (19:2-9). Ezekiel tends to share Jeremiah's opinion of Zedekiah's immediate predecessors (Jer 22:10-30).¹⁴⁹ Regarding the historical background of Ezekiel's disappointment with the kings of Israel, Duguid focuses on Zedekiah's impotent rule and inability to control the political situation.

---

¹⁴⁴ R. H. Alexander, *Ezekiel* (Expositor's Bible Commentary 6; Grand Rapids: Zondervan, 1986), 915.

¹⁴⁵ Cooke, *Ezekiel*, 376.

¹⁴⁶ For this use of רָעָה, Cook, *Ezekiel*, 376, lists Mic 5:5; Zech 11:4-7, Ps 2:9.

¹⁴⁷ Duguid, *Leaders*, 47 (italics his)

¹⁴⁸ The passage, which denotes its addressee with the opening וְאַתֵּנָה צֹאנִי, suggests that YHWH's judgement is not limited to Israel's past rulers, but also includes the exile community, and thus links with the separation judgment in 20:33-34. See, Renz, *Rhetorical Function*, 107.

¹⁴⁹ Cooke, *Ezekiel*, 373.

For instance, Zedekiah failed to uphold the covenant of freedom for slaves in the face of the opposition of the royal officials (הַשָּׂרִים) and the people (Jer 34:8-11). When the officials demand the death of Jeremiah, Zedekiah replies: "Behold he is in your hands; for the king can do nothing against you" (38:5). Here is a man unable to control his own officers. Duguid concludes: "In light of past history, what Judah requires is not a weak, depoliticized king but a strong shepherd."[150] Since Zedekiah is also called נָשִׂיא in 21:25 ('And you, O profane and wicked one, prince of Israel,' cf. Ezek 17, 19), it is possible to trace the hope that the Davidic Shepherd, also called נָשִׂיא (34: 24), is expected to replace Zedekiah, and thus reverses what Zedekiah had failed to do as YHWH's vassal king.

As noted in the discussion of Ezekiel's progressive concept of restoration in 34-48, however, the title נָשִׂיא should also be read in light of the last part of the grand vision set out in 40-48. The Davidic Shepherd, called נָשִׂיא in chapters 34, 37, connects with the נְשִׂיאִם both of the past (cf. 20:6, 25) and the future (cf. 43:7-8). The shepherd imagery employed in Ezek 34:1-22 ceases to be simply imagery of YHWH at the eschatological moment of Israel's restoration, but seems to transform to a title that recalls the Davidic covenant (2 Sam 7:12-14) in 34:23-24 and 37:24-25: especially, רֹעֶה אֶחָד (34:23; 37:24). The focus shifts from YHWH himself as the Shepherd who will reverse the evil done by the shepherds of Israel, to David, who is YHWH's servant and the Shepherd-Appointee over YHWH's eschatological flock gathered on the terrible day of clouds and darkness.

The paramount question remains: Who is this David the shepherd? What is his relationship with YHWH? What is he supposed to do in this restoration process? To answer these questions, we must first look at the literary structure of 34:23-24 and 37:24-25. Here we find a particular development of the strands of conceptions found in these two crucial portions regarding the Davidic Shepherd:

	34:23-24	37:24-25
A	I will raise up over them one shepherd my servant David	A′ My servant David will be king over them and they will all have one shepherd
B	and he will shepherd them he will shepherd them and be their shepherd	B′ (a) they will follow my ordinances and be careful to keep my statutes (b) and they will live in the land...

---

[150] Duguid, *The Leaders,* 47-48, further comments, "This clear affirmation of the power of the coming ruler should serve as a balance against the common tendency to focus exclusively on the servant nature of this coming king."

I YHWH will be their God C my servant David will be prince in their midst (my sanctuary in their midst forever, v. 28)	C´ and David my servant will be their prince forever (I will be their God, v. 27)

Similarities abound in these two passages as well as differences too. Notably, the latter passage repeats nearly all the components of the former, yet displays other aspects of the same concepts. If chapters 34-37 are seen as a continuum of the author's communication, then the concepts in 37:24-25 can be seen as repetitions of 34:23-24 with more nuanced implications. First of all, both passages say neither 'prince *over* them (עֲלֵיהֶם)' nor 'king *in their midst* (בְּתוֹכָם; LXX, ἐν μέσῳ αὐτῶν).' A subtle distinction exists. It is always '*shepherd* (34:23; A; cf. 37:24) or *king* (37:24, A´) over them,' and '*prince* (34:24, C) or sanctuary (37:28, C´) in the their midst.' Only מֶלֶךְ or רֹעֶה is said to be placed 'over' the people.[151]

From this we can infer that both king and shepherd are the figures possessing authority. Their authority as rulers is emphasized. On the other hand, YHWH places נָשִׂיא (34:24, C; 37:25, C´) and also 'my sanctuary' (מִקְדָּשִׁי; LXX, τὰ ἅγιά μου, 37:28, C´) 'in the midst of' the new Israel. Markedly, the prince is never described to be 'over' people. This possibly indicates that the authority and role of the נָשִׂיא are considered on a different level than the king and shepherd as kingly figures. In the case of the shepherd, however, his function is more flexible. While the shepherd parallels the king in both A and A´, he is also depicted as one closely associated with the people in 34: 23 ('their shepherd,' B) as the prince is expected to be (נָשִׂיא לָהֶם, 37:25, C´).[152] YHWH's servant, the future David, is therefore designated by three functional titles: shepherd, king, and prince.

We observe that no matter with what designation it is combined, the phrase 'my servant (עַבְדִּי) David' remains the same, and is repeated four times (34:23, 24; 37:24, 25). As the Shepherd Appointee, this future king of the new Israel, and this prince in the midst of them forever, is preeminently the Son of David. Here we find a clear association of ideas among the Son of David, the shepherd, and the prince. The main function of this Davidic ruler is only indirectly implied in terms of the literary structure. The phrase 'my servant David' in 37:24-25 [2x] forms an inclusio (A´ and C´) with B´ section in the middle. This suggests, as Renz and many others rightly point out, that "Israel's obedience and security are

---

[151] Note that the LXX does not show such sensitivity to the subtle differences between נָשִׂיא and מֶלֶךְ. In these passages, the LXX renders both without exception as ἄρχων. Thus, YHWH's servant David, according to LXX, will be ποιμήν and ἄρχων.

[152] The term רֹעֶה can functionally represent the positions of both מֶלֶךְ and נָשִׂיא, that is, רֹעֶה will be placed 'over' people and also will be 'their' shepherd (34:24); cf. also, 'their God' (לָהֶם אֶהְיֶה) in 34:24 and 37:27.

closely connected with the reign of the Davidic king."[153] The expression 'my servant David' has powerful overtones in OT historiography. Repeatedly YHWH's servant David is presented as the standard by which every king of Israel should be measured (1 Kgs 14:8; 15:3; 2 Kgs 14:3).[154] As king, David was remembered as the archetypal man of integrity, the man 'after YHWH's own heart' (1 Sam 13:14). We are struck by the memorable picture of David in 1 Kgs 9:4 where YHWH addresses Solomon and reminds him of his father David: As for you, if you will walk before me, as David your father walked, with integrity of heart and uprightness, doing according to all that I have commanded you, and keeping my statues (חֹק) and my ordinances (מִשְׁפָּטַי).

David is YHWH's servant, primarily because of the integrity of his heart. He was faithful to walk before God, following the ordinances and keeping the decrees of YHWH. In Ezek 34:23-24, it is quite remarkable that after YHWH denounced all of the shepherds of Israel as unfaithful, and after YHWH established himself as the Shepherd and explains what is to be done, he finally appoints David as the one Shepherd over his flock, his people. Since the future David is appointed as 'One Shepherd,' his task is supposed to be similar to or perhaps identical with that of YHWH as the shepherd illustrated in Ezek 34:1-22.[155]

However, the impression is that YHWH is the prime agent in the scenes described in Ezek 34:1-22, and the role of the Davidic Shepherd is set down later in vv. 23-24, conveyed implicitly in vv. 25-31, in 37:24-25 and implicitly

---

[153] Cf. Renz, *Rhetorical Function*, 116.

[154] Cf. G. von Rad, *Old Testament Theology* (Edinburgh: Oliver and Boyd, 1962) 1:345. The negative counterpart to King David is Jeroboam, son of Nebat. His sin of setting up altars in Bethel and Dan eventually caused the kingdom to be divided in two, thus becoming a scale by which the sins of the kings are often measured (cf. 2 Kgs 10:29). Similarly, 1 Kgs 6: 12 and 2 Ch 7: 17. In 1 Kgs 11:33, against the sins of Jeroboam, YHWH speaks, "I will do this because they have forsaken me and worshiped Ashtoreth ... (they) have not walked in my ways, nor done what is right in my eyes, nor kept my decrees and ordinances as David, Solomon's father, did."

[155] Van Groningen, *Messianic Revelation*, 771, raises a question, "Will he, Yahweh the divine One, enter directly, visibly, and personally into the lives of the flock?" He points out that Ezekiel does not explain how this rescue and redemption are to occur and how Yahweh will actually carry out this shepherding work. Groningen explains this so-called problem, i.e., "Yahweh declares to be the Shepherd and he appoints another to be a shepherd," by alluding to Yahweh speaking the word through Ezekiel the prophet as if he communicates his words with whom he had to deal. Thus Groningen, *Messianic Revelation*, 773, suggests, "this is how Ezekiel functions as well as the one whom YHWH as Shepherd will appoint to serve as shepherd." But the distinction between YHWH's tasks and the Davidic Shepherd's tasks are clear enough not to assume this interchangeability. It seems better to leave this tension as it is in the text at this stage.

in vv. 26-28.[156] The usage of 'shepherd' as a title in the ANE reveals a flexibility applicable to a particular deity and to a human king as the representative of that deity.[157] Yet in the OT, the titular use of this term never applies to any kings, but applies exclusively to YHWH (Jer 2:8; 23:2, 31:10; Ezek 34:1-22; cf. Mic 4:6-8) and his Davidic Appointee (34:23-24 and 37:24-25).[158] From this we assume the existence of a close relationship between YHWH and David in terms of their sharing the same title 'shepherd,'[159] though once again, the passages above appear to divide the restoration stages respectively for YHWH (34:1-22; cf. 36:24-32; 37:1-14) and for his servant the Davidic Shepherd (34:23-24; cf. 37:24-28).

'My servant' appears fragmentary and enigmatic in that it serves as a humiliation motif, or designates exaltation through humiliation because in these two passages the title "servant seems to be a primarily high and lifted up designation."[160] Surely it points to 2 Sam 7:12-14[161] in which 'my servant David' is both king and shepherd 'over' the people. It is also true that this servant will be in their midst as their prince, fulfilling his main role to shepherd YHWH's flock. It is possible that in 34:23 the title 'shepherd' refers back to the compassionate (vv. 11-16) yet stern YHWH as the shepherd (vv. 1-10; 17-22). In 37:24, however, it is noticeable that when David will be shepherd-king over them (A') and prince (C') among them, he will reverse past failures (B') in two major

---

[156] Zimmerli, *Ezekiel 2*, 216, comments on the emphatic 'I myself' in 34:11: "No human figure comes as mediator any longer between YHWH and his flock. Only YHWH's activity is mentioned"; Greenberg, *Ezekiel 21-37*, 707, likewise observes the sequence of the "divine act of restoration followed by the appointment of human rulers" as in Jer 23:3-4; also, Cooke, *Ezekiel*, 337.

[157] Refer to ch. 1.1. "The Shepherd Imagery in the Ancient Near East."

[158] J. Jeremias, "ποιμήν," 485-502; the formulaic double statement, 'YHWH their God and David their king' has been developed as traditions as seen in Hos 3:5 and Jer 30:9f. Jeremias also remarks, "With the title 'shepherd' Ezekiel seeks to guard against a one-sided political understanding of the figure of the future ruler, and also to leave the manner of the fulfillment of the promise of God" (ibid., 488).

[159] I believe D. Block's points made about the new ruler's status, *Ezekiel 25-48*, 298, are the minimum: (1) this ruler will be chosen, not self-appointed nor elected, by YHWH himself; (2) the shepherd will be alone, singular (cf. Jer 23:4, 'shepherds'), thus probably not the restoration of the dynasty, i.e., a series of kings; and (3) the shepherd will be *David* without any indication of *David redivivus*.

[160] The idea of the 'suffering servant of the Lord' can be detected only indirectly through YHWH's compassion toward his flock victimized by the evil and strong ones in 34:1-22. Hals, *Ezekiel*, 254, comments: "this David, though, is not king, but 'prince' and even 'servant' ... functioning here to describe a task which is indeed in the form of a servant, exercised not in ruling but in feeding."

[161] VanGroningen, *Messianic Revelation*, 776, brings up the humiliated yet exalted One of whom Isaiah had prophesied (Isa 52:13-53:12) in connection with the 'servant' in Ezek 37:24. Likewise, J. B. Taylor, *Ezekiel* (Downers Grove, Ill.: Inter-Varsity, 1969), 46, 240, refers to the Messiah king in the phrase 'my servant' (37:24) and states it is 'a clear messianic title.'

realms: (a) "they will follow my ordinances and be careful to keep my statutes"; and (b) "they will live in the land I gave to my servant Jacob ... forever." With the reign of the shepherd Davidic Appointee over them and his presence among them, the people will finally follow my ordinances (מִשְׁפָּטַי) and keep my statues (חֻקֹּתַי), exactly as King David did. The Davidic Appointee will also secure blessings on the land where the sanctuary will be placed in the midst of the people (37:28).

The picture of the Davidic Shepherd in 34:23 (B) reveals little about his task; we are simply told twice that 'he will shepherd them.' In 37:24 (B´), however, the task of the Davidic Shepherd seems to be clarified. In the discourse of chapters 34-37, Ezekiel presents the Davidic Appointee again in 37:24-25, though this time we garner more details in chapter 36 that enlarge the Davidic picture presented in 34:23-24. We discover in 36:24-32 that YHWH gives a new heart and new spirit as he rescues and gathers his people from the nations (v. 24), as YHWH proclaims, "I will put my Spirit in you, and move you to follow my statues and to keep my ordinances" (v. 27; 37:24, B´). It is not surprising to see YHWH give the land and secure the blessings of the land as fulfillment of the covenant formula found in the next verses (36:28-38) recapitulated in exact order in 37:25 (B´[b]) and 27 (C´). Further, רֹעֶה אֶחָד carries the connotations definitely colored by 37:15-23.[162]

The future David is not merely 'one' shepherd opposing 'many' evil shepherds as illustrated in 34:1-10, but as we come to 37:24 ('single shepherd'), we see distinctly that he is the one who will be the ruler of the unified nation (37:15-23), reigning over all twelve tribes of Israel.[163] This notion is continued in terms of the role of נָשִׂיא in Ezekiel 40-48; there he is "*a single individual who rules all twelve tribes* without being identified with any."[164] This poses the question why the kingdom, divided or unified, is expressed in the imagery of a 'tree, stick or staff (עֵץ; LXX, ῥάβδος).'

K. Nielson notes that the connotation of the word עֵץ is fundamentally associated with life, which makes sense in the present context of YHWH

---

[162] Zimmerli, *Ezekiel 2*, 218, rightly comments that the emphatic אֶחָד is to be understood in terms of 37:15ff with the consideration of the historical plight of the division of Israel into two separate political entities. However, the information relayed in 37:24-28 may be progressive in nature as is the case with 34:23-24; chs. 40-48 will probably shed more light on him and his task.

[163] In Ezek 37:15-23, the shepherd motif is unmistakable. The unification of the divided nations at the time of restoration is figuratively to form one long stick (עֵץ, 37:16-17; LXX, ῥάβδος). Some comment that if a different word had been used such as מַטֶּה (7:10, 11; 19:11, 12, 14 [2x]) or שֵׁבֶט (19:11, 14; 20:37; 21:15, 18), it would make a nice play on words with 'tribes of Isarel' (שִׁבְטֵי יִשְׂרָאֵל) in 37:19 or מַקֵּל (39:9; cf. Zech 11:7). But עֵץ in 37:15ff might be related to עֶצֶם ('bones') in 37:1-14. Cf. See T. Renz, *Rhetorical Function*, 114, fn.138; M. Greenberg, *Ezekiel 21-37*, 758-759; R. Hals, *Ezekiel*, 275.

[164] Duguid, *Leaders*, 50 (italics mine).

1.4 The Davidic Shepherd Tradition in Ezekiel's Vision

reviving the divided Israel.[165] On the other hand, there seems to be a play on words with שֵׁבֶט ('tribe' in 'tribes of Israel'), which also means 'stick or staff,' in 37:19.[166] Davis M. Fouts explains that both שֵׁבֶט and מַטֶּה refer to parts of a tree (עֵץ) from which a staff or a weapon could be made. He observes that while שֵׁבֶט when meaning rod or scepter normally involves a symbol of the figure in authority such as a father, a king, or God, מַטֶּה refers to a shepherd's staff or the staff of a leader of lower status than God or king, such as priest, prince, or tribal leader.[167] Furthermore, Fouts notes that שֵׁבֶט symbolizes the authority and discipline normally reserved for the leader of the tribe, *especially one of the tribes of Israel*. Later it depicts the rod of discipline or the scepter of the Messiah (Pss 2:9; 45:7).[168]

The picture implied here may be that YHWH brings the divided kingdoms together in his hand as his scepter or staff, and it is likely that the Davidic Shepherd is to hold this scepter or staff, i.e., the unified new Israel, so that no longer will they defile themselves with idols. Thus will they be saved from their sinful apostasies (37:22-24).[169] Is the nation Israel meant to be a staff or scepter in the hand of YHWH the Shepherd and in the hand of the Davidic Shepherd? If so, for what purpose? These two questions are not answered in our passages.[170] What is clear is that as the shepherd YHWH unifies the divided kingdoms, and the Davidic Shepherd rules over this new Israel as a single shepherd over them – the twelve tribes/staffs of Israel. Morever, as YHWH

---

[165] K. Nielson, *TDOT* 11:265-277.

[166] Zimmerli, *Ezekiel 2*: 273. Zimmerli also suggests that עֵץ alludes to 'scepter' in line with the LXX rendering of the word (ῥάβδος). The LXX renders both עֵץ in Ezek 37:17 and מַקֵּל in Zech 11:7 as ῥάβδος. The context of Zech 11:7 indicates that Zechariah understands Ezekiel's passage in light of the shepherd imagery. For the influence of Ezek 37:16-17 on Zech 11:7, see Cooke, *Ezekiel*, 401.

[167] Fouts, *NITDOTT* 2:924 (#4751).

[168] Fouts, *NITDOTT* 4:27 (#8657; italics mine).

[169] The text is not explicit in that the unified Israel as one stick will be 'in the hand of the Davidic ruler' in the future; it is said to be 'in My (YHWH's) hand' (v. 19). Symbolically, the joined stick was held in the prophet, the son of man's, hand, i.e., 'in your hand' (בְּיָדֶךָ, 37: 17). Yet Renz, *Rhetorical Function*, 115 [n.144], comments: "the LXX seemingly interprets בְּיָדֶךָ as an abbreviation for 'in the hand of Judah' thus highlighting the rule of the future David."

[170] Here, 'one' in the phrase וְהָיוּ לַאֲחָדִים בְּיָדֶךָ is plural (37: 17). Cooke says it denotes either 'a few' (Gen 27:44, 29: 20; Dan 11:20) or as used here 'one and the same' (Cooke, *Ezekiel*, 401). Zimmerli and Hals point out that the Davidic ruler will serve to protect the newly regathered people of God from 'the danger of a new schism' (Zimmerli, *Ezekiel 2*, 276; Hals, *Ezekiel*, 275); Renz emphasizes the comprehensiveness of the Davidic ruler's reign including 'all former tribes' (*Rhetorical Function*, 115). However, the context of 37:15-25 indicates that the unification under the Davidic Shepherd will produce righteousness as a result of their following the laws and decrees of God. Quite interestingly, Ps 2:9 refers to the Lord's Anointed One (v. 2) and this figure rules the nations with an iron scepter (LXX, ῥάβδος) in his hand, dashing them into pieces like pottery. Cf. Rev 2:27 quotes Ps 2:9; here the scepter serves as a symbol of authority over the nations.

holds the unified Israel as a staff or scepter in his hand, it is implied that the Davidic Shepherd's rule over the new Israel will be akin to holding them as a staff or scepter in his hand.

In 34:23-24 and 37:24-25, 'my servant David,' who is introduced as 'one shepherd' and 'king' over the people, is now introduced as נָשִׂיא in their midst, a term recurring at the end of each section. Clearly there is a movement from YHWH as the shepherd, to David the shepherd and king over the people, and finally to David as prince among the people in Ezekiel's concept of restoration. Notably נָשִׂיא is the predominant figure in the chapters following 34 and 37, namely, in chapters 40-48.[171] The Davidic Shepherd is placed in the middle, between the salvific figure of YHWH as the shepherd and the restorative figure of נָשִׂיא among the people, who is also identified as the representative of YHWH's kingship over the people in chapters 34-37. The role of the Davidic Shepherd follows YHWH's salvific act as the eschatological Shepherd, and as the prince his role is obviously crucial for the settlement of the new community (chs. 40-48). In this regard, the figure of the Davidic Shepherd can be seen as transitional.

On the other hand, נָשִׂיא is always said to be 'in their midst' as YHWH's sanctuary will be (C'). This does not strike us as an arbitrary composition, as we also notice that both נָשִׂיא and מִקְדָּשִׁי (YHWH's sanctuary) are said to be forever in their midst (for prince, C and C'; for sanctuary, C'). In chapters 40-48, the נָשִׂיא is mainly the representative of the people, who is basically one among them.[172] At the same time, the role of נָשִׂיא closely relates to the task of maintaining the cultic life of the people settled and centered around the sanctuary. A pivotal question is raised: What does it mean that 'my servant David' will be 'their prince' in their midst forever? In what sense is the Davidic Shepherd-king described as prince?

Neither King David nor the Davidic Appointee is called נָשִׂיא outside of Ezek 34:24 (C) and 37:25 (C').[173] The identity of נָשִׂיא in these passages has been

---

[171] The absence of terms like 'David,' 'shepherd,' or 'king' in chs. 40-48 convinced many scholars of the discontinuity of the נָשִׂיא figure in chs. 34, 37 used synonymously with king and shepherd from what is used in chs. 40-48. Thus chs. 40-48 has been regarded as an interim arrangement awaiting the fulfillment described in chapters 33-39 (J. Boehmer, O. Procksch), as the work of different authors (J. B. Harford, G. R. Berry), or even as a result of a change of mind in the prophet (K. Begrich), cf. Duguid, *The Leaders*, 10.

[172] R. W. Klein emphasizes בְּתוֹכָם in defining the role of נָשִׂיא over against the past shepherds of Israel: "This prince ... will rule *among*, rather than over, the people. The oppressive shepherds, of course, had ruled *over* the people" (*Ezekiel: the Prophet and the Message*, Columbia, S.C.: University of South Carolina Press, 1988, 123, italics his).

[173] Outside of Ezekiel, King David was called נָגִיד ('prince') in 2 Sam 6:21 and 7:8, with the phrase 'over the people,' used before he became the king of Israel. Neither 'my servant David' nor 'David' occurs in Ezekiel 40-48. This is puzzling since nowhere in Ezekiel 40-48 is prince directly identified as the Davidic Appointee. As we note the considerable difference between

1.4 The Davidic Shepherd Tradition in Ezekiel's Vision

interpreted in many different ways. E. Hammersheimb assumes that Ezekiel's thought closely resembles what is presented in chapters 40-48, which allegedly rejects the monarchy. The term נָשִׂיא in chapters 40-48 might have influenced chapters 1-39 and superseded מֶלֶךְ in several places as demonstrated here.[174] Yet considering that Zedekiah, who broke the covenant of YHWH, is also called נָשִׂיא (17:12ff; 21:30, cf. 12:10ff), it is legitimate to view נָשִׂיא in chapters 34 and 37 as primarily belonging to Ezekiel's concept of the reversal of Israel's past failures.

Further, the term נָשִׂיא does not necessarily indicate later influence since it is a heavily used term in the pre-monarchic period; it occurs 72 times in the Hexateuch (in Numbers alone it appears 60 times). The princes during this period were tribal leaders heading up a patriarchal group that represented the chief political authority. They had the right to declare war (Jos 22:14), make treaties (Jos 9:15), or decide on property issues (Num 36:1). As figureheads of the community, they acted on behalf of their people (Num 7). During the monarchy, the office of נָשִׂיא was eclipsed. Authority gradually transferred from the representatives of the tribes to the representatives of the king.[175]

It may be an exaggeration contrasting מֶלֶךְ as a political leader and נָשִׂיא exclusively as a cultic patron (allegedly in Ezek 40-48). According to Duguid, in Ezekiel's scheme the נָשִׂיא is still held responsible for doing justice and righteousness (45:9), is assumed to exercise certain political authority (cf. 44:24), is occasionally treated as a royal figure (44:1-3), or is warned against their political powers (45:8; 46:18). Yet, unlike the pre-monarchic tribal princes or the later kings, Ezekiel's נָשִׂיא is no longer responsible for the administration of the cult. Still the נָשִׂיא takes upon himself representative responsibility for the sake of the people. This representative responsibility involves gathering the gifts and offerings required of the people in order to provide the cereal, sin, burnt,

---

מֶלֶךְ and נָשִׂיא in the OT in general (1 Kgs, 11: 34; Hos 8:3-4, 13:10), R. Hals' comment, *Ezekiel*, 252, helps understand the uniqueness of this link: "When Ezekiel refers to David as נָשִׂיא rather than king, we can see how bold Ezekiel could be in reinterpreting this old tradition in view of his own particular heritage."

[174] He argues for the authenticity of Ezekiel's Davidic hope in chs. 34 and 37 noting that it must have been easier for Ezekiel to omit than to insert these terms. This does not conflict with his own thought, i.e., a rejection of the monarchy, in chs. 40-48, especially in light of certain circles who reject the monarchy and stress YHWH's assumption of power in his own person, whereas others saw the high-priest as the one who inherits the monarch's functions; cf. E, Hammersheimb, "Ezekiel's View of the Monarchy" in *From Isaiah to Malachi*, 51-62. Hammerscheimb's conjecture is based on the LXX's tendency to render מֶלֶךְ as ἄρχων and to make no distinction with נָשִׂיא which frequently is translated also as ἄρχων, as in Ezek 37:24 (55-56). He does not explain, however, why in chs. 40-48 the term נָשִׂיא is rendered by the LXX in most of the cases as ἀφηγούμενος, and in a few exceptional cases as ἡγούμενος (44:3, 45:7).

[175] Duguid, *The Leaders*, 14-15.

and fellowship offerings 'to make atonement [לְכַפֵּר] for the house of Israel' (Ezek 45:16-17).[176]

On the other hand, since Ezekiel's נָשִׂיא differs significantly from the old tribal princes, the prophet's choice of the term נָשִׂיא is not necessarily evidence of one-sided antagonism against the monarchy or against the Davidic hope.[177] As described in Ezekiel 40-48, Ezekiel's נָשִׂיא differs from מֶלֶךְ, and neither is it identical with the pre-monarchic non-royal tribal leader. Even in Ezekiel 34 and 37, the nature of this נָשִׂיא is not further clarified or described. Zimmerli rightly acknowledges:[178]

> Suffice it to know in this 'prince' there exists the shepherd who no longer lives off his sheep and 'looks after himself,' but the shepherd who lives for his sheep; suffice it to know that in him there exists the shepherd who in his flock brings to victory the true 'righteousness of God,' whereby the powerful member of the flock gives to the weak his right to life and no longer thrusts the other away with fist and elbow.

1.4.7 Summary and Conclusion

In the framework of his concept of restoration, Ezekiel's Davidic Shepherd is indeed a way of modifying the Davidic expectation in the light of the historical experiences during the monarchy. The theocracy is justified (34:1-22). At the same time, it is not simply a return to the pre-monarchical past. In Ezekiel's concept of restoration, it is imperative to fulfill the Mosaic covenant. Thus the matter of disobedience is reflected upon (20:25), and the remedy of a new heart and the spirit is promised (36:24-32). But all of this will be accomplished through YHWH the Shepherd and his appointee the Davidic Shepherd.

Ezekiel's eschatological Shepherd and the Davidic Shepherd-Appointee can be characterized summarily as follows: (i) YHWH the eschatological Shepherd will confront the false shepherds of Israel; (ii) he will seek first the lost and outcast; (iii) he will heal, bind up the wounded, and strengthen the weak; (vi) he will judge the fat and robust sheep; (v) he will revive and reunify Israel from her exile. Then, after YHWH the eschatological Shepherd fulfills his rescue mission, (vi) the Davidic Shepherd-Appointee will teach restored Israel

---

[176] The princes of Leviticus were to make atonement that only extended to their own sins (Lev 4:26); only the Aaronic priesthood makes atonement (לְכַפֵּר) for others (Lev 10:17; 14:29; 16:33). There is an exception: One pre-exilic king, Hezekiah, 'had ordered' the provision of an offering to atone for all of Israel (2 Chr 29:24); see *The Leaders*, 50-55 (52).

[177] Levenson, *Program of Restoration*, 69, argues that Ezekiel's choice of the term נָשִׂיא points to a return to Sinai, i.e., bringing "the monarchy under the governance of the Sinaitic covenant." According to him, Ezekiel's נָשִׂיא as an 'a-political Messiah' democratizes the Davidic promise (99); thus, in Ezek 34 the Davidic Shepherd, described as נָשִׂיא, is to be seen, in fact, as a 'substitute king' without power of his own. This new David is the sole member of YHWH's 'shadow cabinet,' ready to assume office when the present administration falls (88).

[178] Zimmerli, *Ezekiel 2*, 223.

[through the outpouring of the Spirit into their hearts] to follow the laws of God in order to produce the righteousness; thus (vii) he will eventually bring peace to the land back and secure the presence of YHWH in the midst of restored Israel; likewise, (viii) he will accomplish all this for the sake of YHWH's name in view of all the nations.

Ezekiel's Shepherd will reverse and redefine the leadership of the monarchy by confronting and consequently dismissing them from their positions as the shepherds of Israel. This signals a type of return of the theocracy in which God's reign is unique, unlike that of a monarchy in Israel as a nation. A dominant feature of the shepherd's rule is the compassion of Israel's covenant God, and the righteousness of his servant David. Thus YHWH the Shepherd will first seek 'the lost' – the most victimized group of his little flock, the people most neglected by the failed shepherds of Israel.

As the Shepherd establishes the community of the new covenant, not only will he seek the lost and outcast, he will also heal the sick and strengthen the weak. Thus will his eschatological mission for the time of the restoration will be distinguished. Likewise the Davidic Shepherd will lead this rescued flock of YHWH the shepherd to finally produce the righteousness of the laws and decrees of YHWH. Thus will the people fulfill the demands of the Mosaic covenant. The task of the restoration of Israel involves the shepherd's judgment between the robust and those who are the down-trodden of his sheep. The recapitulated themes of YHWH's acts of rescue and salvation reach their climax as he revivifies exiled Israel and unifies the divided tribe ('tree or stick') of Israel. The revivified and unified twelve tribes of Israel at the time of YHWH's eschatological reversal will be placed in the hands of YHWH the Shepherd and of his appointee, the Davidic Shepherd.

The נָשִׂיא is not necessarily an a-political figure. According to Ezekiel's concept of restoration in general, the political and spiritual dimensions are inseparable (36:24-32; 33-38 and 37:1-14; 15-23). The title נָשִׂיא in chapters 34, 37 also enhances the modified concept of the future Davidic Shepherd: his righteous and caring intimacy with people (34:1-16, 24; 37:25), his role to provide the sin-offerings on behalf of the people (45:16-17), which secures the presence of God in the sanctuary in their midst (37:26-28; 40-48).

It is characteristic of Ezekiel's concept of Israel's restoration that the past and present failures of Israel will be overcome fundamentally because of YHWH's own zeal for his name before the nations. Ezekiel's vision of the restoration of Israel appears exclusively nationalistic, but in its nature and scope, it indeed remains open to the nations. According to Ezekiel's vision in 34-37, both Israel and the nations will recognize the arrival of the Davidic appointee, the future מֶלֶךְ and נָשִׂיא, when they witness God as the shepherd beginning to seek the lost, gather the outcast, heal the sick, feed the gathered, and revivify and unite the house of Israel. It will be the time when their evil leaders must face the harsh judgment of God himself, and their rule over his flock will cease. They will see the Davidic Shepherd leading God's gathered flock finally to follow God's laws and decrees, thus ushering in the time of peace in the land. Then God's indwelling sanctifying presence will be with them, and remain in their midst forever. Both Israel and the nations will eventually come to understand that the Lord is God whose zeal for his holy name finally hallows his own.

1.5 The Davidic Shepherd Tradition in Zechariah 9-14

In light of the Davidic Shepherd tradition in Micah 2-5 and Ezekiel 34-37, the shepherd images that appears in the oracles in Zechariah 9-14 takes a unique turn. Here we find images oblique and yet rich. The challenge here is to understand how the OT shepherd imagery develops in Zechariah 9-14. Assuming Zechariah is receptive to the Davidic Shepherd tradition, how then does he develop it in his writings? How can we explain the characteristics of Zechariah's use of the shepherd imagery? While shepherd imagery is widespread throughout the oracles in Zech 9-14, the imagery is especially pronounced in sections such as 9:9-15, 10:1-6, 11:4-17, 12:6-14, and 13:1-9.[179]

Zechariah 9:9-15 announces the meek king coming to Jerusalem riding on a donkey (v. 9). World dominion will belong to him; thus he will declare שָׁלוֹם to the nations, again 'to the ends of the earth' (v. 10), as seen in Mic 5:4 [5:5]. At that time, YHWH will appear and will be the eschatological Shepherd over them. The shepherd language is clearly reminiscent of the original Exodus, though used in the context of the coming restoration at the end: "YHWH their God will save them on that day as the flock of his people" (v. 15). There is a subtle distinction between YHWH and the meek king in Zech 9. It is distinctively YHWH who is the eschatological Shepherd figure while the meek king riding on a donkey is himself the object of YHWH's salvation.[180]

Compared to the warlike picture of the Davidic king in Psalm 72, the Davidic king in Zech 9:9-10 presents a different look, one possibly befitting the suffering shepherd imageries throughout the remainder of Zechariah 9-14. Nevertheless, the basic structure of the Davidic Shepherd tradition is confirmed once again: YHWH will be the eschatological Shepherd who will act decisively

---

[179] The shepherd imagery continues also in Zech 14:1-21, but only indirectly. For example, the expression, 'I will gather' (וְאָסַפְתִּי, v. 2), echoes YHWH as the shepherd; the motif of 'being struck,' a particular imagery associated with the shepherd in 13:7, continues to occur in 14:12, 13; last, the term כְּנַעֲנִי ('Canaanite' or 'traders,' 'merchants') found in the last verse of chapter 14 recalls the opening of 11:4-17, where sheep 'merchants' are mentioned.

[180] Note that וְנוֹשַׁע has a passive force. For the argument for this rendering, see Ian Duguid, "Messianic Themes in Zechariah 9-14," in *The Lord's Anointed* (ed. P. E. Satterthwaite, R. S. Hess and G. J. Wenham; Grand Rapids: Baker, 1995), 209.

1.5 The Davidic Shepherd Tradition in Zechariah 9-14

to restore his flock and bring a state of blessedness to the nations, and he will set up a new David over the restored flock.

Zechariah 10:2 contains an intriguing expression in this respect, 'like a flock oppressed because of not having a shepherd' (כְמוֹ־צֹאן יַעֲנוּ כִּי־אֵין רֹעֶה). The LXX translates it, 'like a flock mistreated – thus sickened – because there is no *healing*' (ὡς πρόβατα καὶ ἐκακώθησαν διότι οὐκ ἦν ἴασις). The translator of the LXX equates the presence of a shepherd with the possibility of healing the oppressed and sick ones. As the true Shepherd YHWH claims Judah as his flock; he will confront the leaders of Israel. This motif is closely tied to the absence of a shepherd in v. 2: "My anger is hot against the shepherds [הָרֹעִים] and I will punish the leaders/he-goats [הָעַתּוּדִים]" (v. 3; cf. Ezek 34:2, רֹעִים and v. 17, עַתּוּד).

As seen in Ezekiel 34-37, the absence of a (good) shepherd is characteristic. Likewise, YHWH's confrontation with the wicked shepherds in Zechariah 10 leads to renewed obedience to YHWH: "I will strengthen them in the Lord and in his name they will walk" (v. 12). In Ezekiel's vision, ultimately it is YHWH's concern about his reputation 'for the sake of my holy name' (20:39; 36:20, 21, 22; 39:7 [x2]; 39:25; 43:7, 9) that overcomes the failures of the shepherds in Israel's monarchic history. Yet instead, Zechariah underscores YHWH's compassion for the distressed flock as the major motif for the future restoration that would otherwise be impossible: כִּי רִחַמְתִּים (v. 6).

1.5.1 The Good and he Worthless Shepherds

Zechariah 11:4-16 is another rich text in which various shepherd images are amassed, with 11:1-3 serving as an introduction: "Listen to the wail of the shepherds" (v. 3).[181] Verses 4-5 seem to recall the problem of the shepherd's absence in 10:2. The contemporary shepherds do not spare their flock, and mark them for slaughter. Eventually these irresponsible shepherds evoke YHWH's anger upon the flock as well as upon themselves. In response, YHWH makes the prophet represent himself as the true yet mistreated shepherd (v. 4). Here, as a representative of YHWH, the prophet's identity as a shepherd converges with God's metaphorical role as Israel's shepherd.[182] The prophet's symbolic action signifies that even though YHWH as the true shepherd pastured them with the rod of 'Favor' and 'Union,' the flock detested the shepherd, that is, the prophet

---

[181] C. L. Meyers and E. M. Myers, *Zechariah 9-14* (*AB*; New York: Doubleday, 1993), 296-302, labels the section "shepherd narrative," suggesting the word 'shepherd' supplies the identifying focus, the motif whereby its surface as well as deep meaning are conveyed.

[182] The 'prophetic' connotation of the shepherd image, as some argue, is neither necessary nor required in reading the prophet's symbolic identification with YHWH as the shepherd of Israel. The point of the identification is the kingly aspect of the shepherd image. Zechariah is chosen to represent YHWH as the shepherd not to lead the people but to deliver the message of what it is like for YHWH to be the shepherd.

who represents YHWH (vv. 4-14). Thus, YHWH decides to stop caring for the dying and perishing ones among the flock (v. 9). YHWH revokes 'the covenant he made with all peoples' (v. 10), producing fatal consequences.

What is the covenant that was annulled? Various readings are offered. 'My covenant that I had made' can be read as 'the covenant of YHWH which he made,' or 'the peoples' can be read as 'the people,' though the latter is unlikely. For the former, neither option will affect the implication that it refers to the covenant which YHWH made, since in the verse, the 'I' of YHWH and the 'I' of the prophet are almost inextricable. The messenger's 'I' is also YHWH's 'I'.[183] To discover which covenant it refers to, however, remains problematic simply because there is not much information provided in the text. The context does seem to yield some suggestions, however. S. R. Driver and A. Plummer noted, "the covenant, if represented by the staff, can only be a covenant with peoples represented by sheep, and surely the Jews were among them."[184] The covenant in the verse is to be broken since the two staffs of 'Favor' and 'Union' are broken, and those two staffs representing such covenantal principles certainly involves the covenant YHWH made with Israel.

Then why do we find here the covenant with all 'the peoples' (הָעַמִּים) and not the people of Israel? First, the phrase can refer to the people of Israel even though it is plural. It is not unusual to refer to the people of Israel in the plural (Gen 27:29; 48:4, 49:10; Isa 3:13; Mic 1:2).[185] Similarly, 'all the peoples' could mean the mixed population of the territory of the old Northern kingdom, or might even lean toward a 'Samaritan' interpretation.[186] Further, 'all the peoples' can refer to non-Israelite nations (1 Chr 16:24; 1 Kgs 5:14; Hab 2:5), usually the larger group of ethnic or national groups out of which YHWH had chosen Israel to be his people (e.g., Deut 7:6, 14; 10:14; 10:15; 14:2). Especially in Zech 12:1-9, the nations from which YHWH will rescue Judah and Jerusalem are frequently designated 'all the peoples' (cf. הָעַמִּים in vv. 2, 3a, 4, 6, and הַגּוֹיִם in vv. 3b, 9).

If the phrase 'the peoples' is understood restrictively to mean non-Israelite peoples, then the covenant in question remains unidentified. At best, some suggest it be equated with 'the covenant of peace' (בְּרִית שָׁלוֹם) in Ezek 37:26, which is the everlasting covenant.[187] Zechariah 9 already announces the day

---

[183] Edgar W. Conrad, *Zechariah* (Readings: A New Biblical Commentary; Sheffield: Sheffield, 1999), 175; likewise, C. L. Meyers and E. M. Meyers, *Zechariah 9-14*, 270.

[184] S. R. Driver and A. Plummer, *Haggai, Zechariah, Malachi and Jonah* (ICC; New York: Charles Scribner's, 1912), 380.

[185] Meyers and Meyers, *Zechariah 9-14*, 270-271.

[186] Katrina J. A. Larkin, *The Eschatology of Second Zechariah: A Study of the Formation of Mantological Wisdom Anthology* (Kampen: Kok Pharos, 1994), 128, introduces this option, though she thinks this view too narrow.

[187] B. F. Batto, "The Covenant of Peace: A Neglected Ancient Near Eastern Motif," *CBQ* 49 (1987):187-211.

when YHWH the shepherd of Israel will save his flock (v. 16); however, and as we noted, the king who comes riding on a donkey will bring peace (שָׁלוֹם) to the nations (הַגּוֹיִם), to the ends of the earth (v. 10, cf. Mic 5:4). It is likely that 'the peoples' primarily refers to non-Israelites. Nevertheless, it does not exclude Israelites who can also be called 'peoples,' either as a part of peoples or the representative of peoples. Actually the instances of the plural 'peoples' being applied to Israel show from the beginning that it retains multi-ethnicity (cf. Exod 12:38).

Moreover, as previously indicated by S. R. Driver and A. Plummer, the broken staffs involve sheep tended by shepherds with staffs, and surely the flock includes the Jews. As we infer from these observations and suggestions, we can safely agree with R. A. Mason that the 'covenant I had made with all peoples' is the covenant YHWH made with Israel in order to enlighten the Gentiles: "I have given you as a covenant to the people" (Isa 42:6; cf. 49:8).[188] From the context of Zech 11:4-17, we conclude that YHWH's covenant with Israel, in view of bringing peace to all the people on the earth, inevitably is to be broken because the sheep refuse their shepherd YHWH. Thus it is quite likely that Zech 11:10-12 demonstrates the chosen flock's unfitness for the task of mediating YHWH's covenantal favor to all the people.[189] In verses 11-12, only the afflicted among them perceived the word of YHWH while the obstinate majority remained virtually ignorant of or even confused about YHWH's judgment.[190]

1.5.2 Zechariah 11:4-17 and Ezekiel 34, 37:15-28

Another issues arises when we come to recognize the similarities and differences between Zechariah 11 and Ezekiel 34-37. Frequently three main passages are singled out as possible influences on the thought, if not the language, of Zech 11:4-17: Ezek 34, 37:15-28, and Jer 23:1-8.[191] These texts center on the image of YHWH as the shepherd of Israel in the context of the hope of restoration mingled with Davidic expectations. How does Zechariah's shepherd imagery reflect and develop Ezekiel's vision of the eschatological Shepherd and the Davidic Appointee?

Zechariah 11:16 is apparently reminiscent of Ezekiel 34 (esp. vv. 1-6, 11-16). The language echoes those of Ezek 34:1-16 in which YHWH rebukes the wicked shepherds by listing the tasks they failed to perform, such as 'to care for

---

[188] R. A. Mason, "Some Examples of Inner-Biblical Exegesis in Zech IX-XIV," in *Studia Evangelica* VII (ed. E. Livingstone; Berlin: Akademie-Verlag, 1982), 350.

[189] Meyers and Meyers, *Zechariah 9-14*, 270; also, in agreement with Larkin, *Eschatology*, 129.

[190] John Calvin, *Commentaries on the Twelve Minor Prophets: Zechariah and Malachi* (Edinburgh: The Calvin Translation Society, 1849), 322-323.

[191] W. Rudolf, *Haggai, Sacharja 1-8, Sacharja 9-14, Maleachi* (KAT XIII/4; Gütersloh), 205; Larkin, *Eschatology*, 118-132.

the lost,' 'seek the young,' 'heal the injured,' 'sustain the healthy'; likewise they are rebuked for such deeds as 'to eat the meat of the choice of sheep' and 'tearing off their hoofs.' Taking into consideration the broader context of Zechariah 9-14, the similarity between Zechariah 11 and Ezekiel 34 and 37 becomes increasingly evident. Zechariah 10:2c can be taken as a prologue to Zechariah 11: "Therefore the sheep wander like sheep oppressed כִּי־אֵין רֹעֶה ('for the absence of a shepherd')." In Ezekiel 34, what necessitates YHWH's inevitable intervention as the eschatological Shepherd is the absence of a (good) shepherd who seeks, heals, judges, and so on: "So they were scattered because there was no shepherd (מִבְּלִי רֹעֶה, v. 5c); there is no one searching; there is no one seeking" (וְאֵין דּוֹרֵשׁ וְאֵין מְבַקֵּשׁ, v. 6b).

Although the lexical links between the two texts may not be crystal clear, the underlying issue for both Ezekiel 34, 37 and Zechariah 11 is indeed the quest of searching for the prime under-shepherd(s) who would actualize YHWH's own shepherd leadership over his flock. This hope for one shepherd in Zech 10:2c who would truly represent YHWH, in fact, underlies the expectation of the coming meek king riding on a donkey in Zech 9:9. The following chapters 11-13 then can be seen as a description of how this meek king would come, intimately involving shepherd imagery in the telling of the mysterious way in which YHWH will restore shepherd-leadership over his people.

Returning to Zechariah 11, shepherd imagery appears in two phases, namely, the good shepherd who represents YHWH in 11:4-14, and the worthless and smitten shepherd in 11:15-17. The prophet acts symbolically in both roles: YHWH the shepherd to be sold, and the worthless shepherd to be smitten. Similarly, the shepherd image in Ezekiel 34 and 37 can be distinguished in two figures, namely, YHWH the eschatological Shepherd in 34:1-22, 25-31 and 37:1-23, 26-28, and the Davidic Shepherd in 34:23-24 and 37:24-25. While YHWH the eschatological Shepherd in Ezekiel 34 and 37 overcomes the failures of the wicked shepherds and restores his flock, the good shepherd in Zechariah 11 is detested by the flock, and the covenant is broken. Likewise, the worthless shepherd in Zechariah 11 is representative of YHWH dealing with his flock in the respective discourse, while it is the Davidic Shepherd who replaces the wicked shepherds in Ezekiel 34, 37.

Besides these parallels and differences, another common feature shared between Ezek 34, 37 and Zech 11 is the metaphor of the two sticks or staffs. In Ezek 37:15-28, the two sticks represent the divided kingdoms of Israel, while the two staffs in Zech 11:4-14 represent YHWH's covenantal favor and the union promised to the nation. Yet, Zechariah's use of the metaphor of the two sticks seems constructed as the negative image of Ezekiel's joining the two together, which Larkin sees as the interpretation given in Zech 11:14 to the breaking of the second staff (the staff of Union) as the breaking of the brotherhood between 'Judah and Israel', exactly corresponds with the naming of Ezekiel's first stick, on which was written 'For Judah, and the Israelites associated with it.' [192]

Knowing that Zech 11:4-14 is engaged in the reception and development of the tradition of Ezekiel's earlier prophecies, the question is how to explain the use of the tradition in Zech 11:4-14. Is Zechariah harshly modifying the future vision of Ezekiel, thus 'criticizing the optimistic views of the hierocratic followers of Ezekiel'?[193] Besides the uncertain reconstruction of the social-political backgrounds of those responsible for the book of Zechariah, a comparison of the literary contexts of both texts may yield firmer ground for a more convincing argument.

In Ezekiel 37, the promise of the union of the divided kingdoms of Judah and Israel is presented as a consequence of YHWH revivifying the nation from the exile (vv. 1-13) and giving them his Spirit (v. 14). Even in the macro-structure of Ezekiel 34-37, the fact that the nation is united and settled in the land concurs with YHWH placing his Spirit in the hearts of the exiles, and their subsequent ability to obey the decrees and the laws of YHWH (esp. Ezek 36:22-38). This rationale for the united nation is completely absent in the shepherd discourse prior to Zechariah 11 (cf. Zech 12:10; 13:1-2).[194] Even the later opening of the cleansing fountain for the house of David and the inhabitants of Jerusalem 'on that day' (Zech 13:1-2) appears an abrupt and enigmatic consequence despite of Israel's detesting her shepherd (Zech 11:4-14).

While Ezekiel envisioned the union of the nation as a consequence of YHWH restoring his flock by placing his Spirit in them, Zechariah 11 describes the impossibility of the union of the nation, so to speak, apart from the work of the Spirit. For those who read the oracles of Zechariah as a unified discourse, Zech 4:6 appears to be relevant and helpfully sheds light on this: "So he said to me, 'This is the word of the Lord to Zerubbabel: Not by might nor by power, but by my Spirit, says the Lord Almighty.'" [195] The two staffs of 'Favor' and 'Union' under the covenant YHWH made with Israel in view of all the peoples in Zech 4:4-14 were broken, since the flock under the covenant came to detest the rulership of the shepherd. In light of Zech 4:6 (cf. Zech 13:1-2), the ignorant will be enlightened only after they will be given 'the spirit of grace and

---

[192] Larkin, *Eschatology*, 119.

[193] For this view, see P. D. Hanson, *The Dawn of Apocalyptic: The Historical and Sociological Roots of Jewish Apocalyptic Eschatology* (Philadelphia: Fortress Press, 1979), 343.

[194] Likewise, Larkin, *Eschatology*, 121, says, "only the Lord's gift of the 'new heart' (Ezek 36:25-26) can make unity become a reality."

[195] Conrad, *Zechariah*, 160, concerning the meek king riding on a donkey//on a colt, the foal of a donkey (Zech 9:9) in the light of Zech 4:6, states that the coming king will be 'humble' because Zerubbabel's victories will have been won 'not by might, nor by power, but by my spirit.'

supplication' (12:10). They detested their shepherd to the point of paying for a shepherd who did not bother to hire himself for the flock.

In short, Zechariah judges Ezekiel's vision of the reunion of Israel as an ideal unattainable in the face of the 'givens' of Israel's disobedience.[196] Zechariah reverses Ezekiel's vision of the two staffs only in his own context following the rebuilt temple, and without changing the original intention of Ezekiel's vision. Zechariah's interpretation, therefore, reinforces rather than negates the original vision. Ezekiel's vision of union and the consequential coming of the Davidic Shepherd's rulership over the whole of Israel was not fulfilled at the time of Zechariah. They still expect the outpouring of the renewing spirit of YHWH guaranteed both in Ezekiel 36-37 and as is foretold in Zech 12:10; 13:1.

Ezekiel's eschatological and Davidic Shepherds are triumphant (Ezek 34-37). They succeed when they confront the wicked shepherds of Israel. At the time of the restoration, they fulfill all the expected tasks on behalf of the scattered flock. YHWH places his Spirit on restored Israel so they may finally obey the laws and decrees of YHWH.

On the other hand, the shepherd imagery in Zechariah, employing the same metaphor of the two staffs found in Ezek 37:15-28, describes the suffering and rejected shepherd. This picture is a renovation of the typical image of the shepherd in related materials found both in the OT and the ANE. The shepherd imagery found in Zechariah 11 is deeply involved with the sins and blindness of the flock. By adapting shepherd imagery and the metaphor of the two staffs in Ezek 37:15-28, Zechariah thus points out that the greatly expected future of the united nation will not come about unless the flock is enabled to follow their shepherd.

Meanwhile, in contrast to Ezekiel's eschatological Shepherd who rescues his oppressed flock, in Zechariah, YHWH delivers his disobedient flock to the hands of the wicked shepherd who is yet to be judged. The struggle between YHWH and his flock becomes irrevocable.

1.5.3 The Shepherd Imagery in Zech 11:4-17 and the Temple

Why did the prophet-shepherd throw the thirty pieces of silver to the potter, or 'metalworker' in the temple? It is commonly remarked that the amount is the equivalent price of purchase offered for a slave (Exod 21:32).[197] The wage for YHWH as their shepherd – the selling price of a slave – and the thirty pieces of silver thrown into the temple generate sarcasm.[198] E. W. Conrad argues that the

---

[196] Larkin, *Eschatology*, 132.

[197] Larkin, *Eschatology*, 130, mentions several scholars who find Sumerian parallels meaning that the one who sold for thirty shekels is held in low esteem.

[198] Cf. D. R. Jones, "A Fresh Interpretation of Zechariah IX-XI," *VT* 12 (1962): 254; Meyers and Meyers, *Zechariah 9-14*, reads the tossing thirty pieces of silver to the temple treasury in light of 2 Chr 24:4-6 as a contribution, not necessarily as sarcasm.

oracles (Zech 9:1-11:17; 12:1-14:21; Mal 1:1-4:6) are an attempt to address the question of the difference between current circumstances and the envisaged future. Moreover, he suggests that the context of the sarcasm may be traced through Haggai and Zechariah where the temple construction is associated with increased wealth for which YHWH is responsible. Perhaps, he argues, the gold and silver collected from the exiles who returned from Babylon (Zech 6:9-15) were to be used for fashioning items for the temple.[199]

The tossing of the thirty shekels to the potter in the temple then may symbolize that building a temple is futile when YHWH himself is rejected. Thus YHWH is absent, and the covenant is broken. In this respect, Larkin is right when he states, "It is possible that just as the wages are supposed to pay the prophet for his mission to the whole community (Zech 11:4b-6), so his rejection of them symbolizes his rejection of the whole community, which however is centered upon the temple."[200] The sarcasm testifies that things are not what they are supposed to be. The expected outcomes from the construction of the temple are yet to eventuate.

In this context, Zechariah 11 modifies the shepherd imagery in a surprising way when compared to Ezekiel 34-37. Zechariah's unique use of shepherd imagery develops in 11:15-17 and 13:7-9, while these images are more clearly linked with some of the 'David' sayings 'on that day' in 12:6-9 and 13:1-6.

1.5.4 The Smitten Shepherd and the House of David

The saying concerning the striking of the worthless shepherd in Zech 11:14-17 is followed by the visions of the future restoration 'on that day' (12:3, 4, 6, 8, 9, 11; 13:1, 2, 4). Another saying follows that concerns the striking of the shepherd in13:7-9. This composes an inclusion of the shepherd imagery surrounding 11:15-13:9. Morever, the motif of the affliction reappears between those two sayings of smitten shepherds (11:14-17; 13:7-9) in 12:10. Indeed, the theme of YHWH's eschatological battle with the nations and the blessings bestowed upon 'the house of David and the habitants of Jerusalem' (12:7, 10; 13:1) is sandwiched between the two separate stricken shepherd sayings in 11:14-17 and 13:7-9, yet with another affliction motif in the middle. How do these sayings where YHWH strikes the shepherd relate to YHWH blessing the house of David?

---

[199] Conrad, *Zechariah*, 153, 175; cf. D. L. Peterson, *Zechariah 9-14 and Malachi: A Commentary* (Louisville: Westminster John Knox, 1995), 97.

[200] Larkin, *Eschatology*, 130; also, refer to R. F. Person, *Second Zechariah and the Deutoronomic School* (JSOTSS 67; Sheffield: Sheffield, 1993), 202-205, who associates the eschatological outlook of Zechariah 10-14 with disenchantment with the Jerusalem temple administration under Persian control.

A few scholars suggest relocating Zech 13:7-9 to the end of Zech 11:4-17 since these verses are believed to have no obvious connection with the rest of Zechariah 12-14.[201] Setting aside the significance of reading the text in its present literary context, however, is there not any rationale for the current location of YHWH striking the shepherd both in 11:14-17 and 13:7-9? In other places in the Davidic Shepherd tradition in the OT, such as in Ezekiel 34-37 and Micah 2-5, shepherd imagery applies both to YHWH and his Davidic Appointee. Yet, apart from Zech 9:9, the Davidic royal figure is not clearly addressed in Zech 11:15-13:9. Rather, YHWH's eschatological blessings and his restorative Spirit (12:10; 13:1) to be poured upon the house of David, coincide with YHWH striking the shepherds. As YHWH strikes the worthless shepherd and the shepherd who is YHWH's associate, and fights off the gathered nations surrounding Jerusalem (12:1-5) – enemies both within and without – blessings will fall upon the house of David (12:6-9).

From a literary context, we observe that the announcements of YHWH's salvation of Israel (12:1-9; 13:1-6; 14:1-21) follow each appearance of the motif of YHWH's judgment upon the shepherd and his flock (11:14-17; 12:10-14; 13:7-9).[202] At first glance, the shepherd saying in Zech 13:7-9 does not seem to fit what precedes in 13:1-8, but in the broader context of Zechariah 11-14, it does. First, YHWH striking the wicked shepherd (11:14-17) is followed by YHWH battling the nations (12:1-5) and bestowing blessing upon the house of the David and the inhabitants of Jerusalem (12:6-9).

Similarly, the saying of the pierced one (12:10-14) is followed by the Lord bestowing blessing upon the house of David and the inhabitants of Jerusalem (13:1-6). More precisely in the case of 11:14-17 and 12:1-9, YHWH striking his shepherd (13:7-9) is followed by YHWH battling the nations (14:1-15) and bestowing blessing upon the house of the Lord (14:16-21). Since the sections containing YHWH's eschatological blessings are said to be bestowed 'on that day' (12:6; 13:1; 14:8), the later may likely be the recapitulation of the former depiction of the same event.[203] Then all three scenes of the affliction of the shepherd or YHWH can be seen as closely related to one another, identical in essence.

Thus there is a progressive deepening of the vision, though all of the implications are not fully clarified as the prophecies move from the striking of

---

[201] Eibert J. C. Tigchelaar, *Prophets of Old and the Day of the End: Zechariah, the Book of Watchers and Apocalyptic* (Leiden: Brill, 1996), 112, names B. Stade, "Deuterozacharja: Eine kritische Studie," *ZAW* 1 (1881):1-96, as the scholar responsible for the suggestion, followed by many.

[202] M. Butterworth, *Structure*, 290-291, notes varying amounts of judgment and/or salvation for the enemies of and for Judah/Jerusalem in Zech 9-14.

[203] Tigchelaar, *Prophets of Old*, 111, 133, speaks to the movement from present, to near past, and to the near future in Zech 11-14; perhaps addressing Zech 9-11 as 'historical survey' and 12-14 as an 'announcement of the future.'

1.5 The Davidic Shepherd Tradition in Zechariah 9-14

the wicked shepherd, to the pierced one, and finally to the striking of YHWH's associate.[204] The idea expressed is the suffering of the shepherd. The one who suffers may vary, however, depending on the scenes as well as on how one interprets them: the good shepherd, the wicked one, the pierced one, YHWH, or YHWH's associate, and even taking into consideration of 'the house of David and the inhabitants of Jerusalem.'

Can we identify who is the stricken shepherd? The text of 13:7-9 does not help us be decisive about whether or not the smitten shepherd in 13:7-9 is identical with the wicked shepherd of 11:14-17. Clearly the motif of YHWH striking in 13:7 is reminiscent of the striking of the wicked shepherd in 11:14-17. Reading up to 13:7, we note the pierced one in וְהִבִּיטוּ אֵלַי אֵת אֲשֶׁר־דָּקָרוּ (12:10; LXX, ἐπιβλέψονται πρός με ἀνθ' ὧν κατωρχήσαντο). This figure refers probably to YHWH himself or perhaps his representative, though the implication may be the same.[205]

Furthermore, the shepherd in 13:7-9 is 'my shepherd,' 'my associate' may refer to the good shepherd in 11:4-14.[206] The text remains ambivalent, however. There is no indication that the wicked shepherd in 11:14-17 should be excluded as a possible referent, 'my shepherd' in 13:7 is rendered in the plural by the LXX, ἐπὶ τοὺς ποιμένας μου.[207] It would be wise to leave the expression 'my shepherd,' 'my associate' unidentified. That is to say, YHWH's representative either refers to the wicked or to the good shepherd, or some other alternative not indubitably revealed.

What can be known is the context of the shepherd's affliction. When we ask why YHWH strikes the shepherd, and consider the ramifications, the context implies that it is YHWH's judgment meted out on the leaders of Israel, and thus upon the whole community. In chapter 10, only 'shepherds' and 'leaders' are rebuked harshly by YHWH, and the Almighty himself will care for his flock (v.

---

[204] Rudolf, *Sacharja 9-14*, 201-223, relocates 13:7-9 before 12:1-14, and identifies 'the pierced one' in 12:10 with 'my shepherd' in 13:7-9 (213, 219).

[205] Does MT's case represent the omission of *waw*? M. Dahood, "A Note on the Third Person Suffix -y in Hebrew," *UF* 4 (1972): 163, emends 'upon me' (אֵלַי) to 'upon *him*' (אֵלָיו) claiming that the poetic form אֵלַי is a variant of אֵלָיו. Likewise, P. D. Hanson, *The Dawn of Apocalyptic*, 357, emends it; M. Butterworth, *Structure*, 215, understands it as a bold expression of YHWH's suffering through or with his representative, and he also suggests its association with the 'smitten shepherd' and sees the pierced one in some sense as YHWH's representative (291, n.1). The mourning of the whole community in vv.10-14 may indicate a royal figure. Yet, the situation recalls the people's rejection of the good shepherd in 11:4-14.

[206] Notes that even the Persian conqueror Cyrus was called רֹעִי (Isa 44:28).

[207] Antti Laato, *Josiah and David Redivivus: The Historical Josiah and the Messianic Expectations of Exilic and Postexilic Times* (Stockholm, Sweden: Almqvist & Wiksell International, 1992), 287, sees 13:7-9 as a reference to the good shepherd in 11:4-14; likewise, I. Duguid, "Messianic," 274; Yet Latto, 287, n.102, lists the scholars who see it as the wicked shepherd in ch.11 in such as B. Stade, J. Wellhausen, R. Mason, and also *The New English Bible* where 13:7-9 is added after 11:17.

3). In the next chapter the focus changes concerning the Lord's frustration. Now YHWH expresses his frustration with the entire flock. The thirty pieces of the silver tossed into the temple treasury (11:13) likely signifies YHWH's rejection of the whole community. Then, YHWH delivers his flock into the hands of the wicked shepherd (11:16), and later YHWH striking the shepherd signifies his judgment against as well as salvation for the whole community (11:17).

The wicked shepherd appointed and stricken in 11:14-17 is the end result of the broken covenant between YHWH and his flock that detested him (11:4-14). In 12:10, it is the whole community of Israel, 'the house of David and the inhabitants of Jerusalem,' who pierced (LXX, 'despised') YHWH (or, his shepherd).[208] Likewise, 'my shepherd' is stricken, the whole flock is scattered, destined to go through the tribulation set out in 13:7-9. Thus, the theme of the striking of YHWH's shepherd (or YHWH) is the eschatological judgment upon the community.

Then again, the texts do not say precisely who will be smitten or pierced. The imagery used is repetitive but complex, and we may leave untouched the obscurity presented in those texts. While the object of the piercing is uncertain, the outcome of such affliction upon YHWH's shepherd is certain. Those alternating salvation oracles emphasize the restoration of YHWH's own reign over Israel and the nations (12:1-9; 13:1-6; 14:1-21). YHWH striking the shepherd eventually leads to YHWH raising up leaders/shepherds like David, who will participate in YHWH's extension of his rulership over the nations (12:6-9).

Within Israel, YHWH striking the shepherd will result in the banishment of the names of the idols from the land. This means true worship of YHWH will be restored, genuine worship that will extend to the ends of the earth: "on that day, there will be one Lord, and his name will be the only name" (14:9). Here, YHWH presents himself as the only true shepherd who claims his shepherd-leadership by judging Israel's shepherd, and fighting off the nations. Interestingly, this suggests the vision of the inauguration of theodicy at the time of restoration, a theme characteristic of the Davidic Shepherd tradition in Ezekiel 34-37, and 40-48.

In Ezekiel's vision, YHWH's theodicy is justified by the dearth of true shepherd(s) for the sake of his oppressed flock (34:6-7, 10-11). Opposing the false shepherds of Israel, YHWH will prove that he, as the true eschatological Shepherd, will truly fulfill the tasks of the restoration of the flock back in the land, and will bring peace. In contrast, Zechariah's picture of the eschatological shepherd(s) is gloomy and dark. Why? In Zechariah 11-14 YHWH's theodicy arrives primarily as the result of YHWH dealing with the sins of the flock.

---

[208] Duguid, "Messianic," 275, rightly notes that the difference between Josiah who was pierced by an arrow at Megiddo (2 Chr 25:24), and the one pierced in Zech 12:10 is the latter was apparently pierced by the Jerusalemites themselves.

1.5 The Davidic Shepherd Tradition in Zechariah 9-14

YHWH's confrontation is aimed at the flock in Zechariah; the shepherds play the roles of the mediators through whom YHWH deals with his flock at the time of restoration.

All the shepherd figures in Zechariah 11-13, namely, the wicked in chapter 11, the good or wicked in chapter 13, and YHWH himself or his representative in 12, share a common feature: suffering. Zechariah certainly makes a peculiar contribution to the development of OT shepherd imagery. His shepherd imagery unfolds in the context of YHWH's conflict with the flock, not merely with the shepherds of Israel.[209] Noticeably, it is 'the flock' that detested the good shepherd (11:8); the whole community of Israel is represented as their temple is rejected (11:13).

Consequently, after the good shepherd is rejected, YHWH delivers his flock to the hands of the wicked shepherd to be smitten (11:16). The house of David and the inhabitants of Jerusalem pierce YHWH or his shepherd (12:10), and the flock is scattered as their shepherd is stricken (13:7). Yet the judgment of YHWH is laid solely upon the shepherd-figure who represents the wicked shepherd, the good shepherd, or YHWH himself.

There is no explicit indication that Zechariah's smitten shepherd is Davidic (cf. Zech 9:9), however. Ezekiel's Davidic Shepherd in 34 and 37, as in Mic 5:1-4, is apparently triumphant. If pressed to find a parallel picture of the Davidic Shepherd in Ezekiel's vision, we can point to the figure of נָשִׂיא who is supposed to provide the cereal, sin, burnt, and fellowship offerings 'to make atonement (לְכַפֵּר) for the house of Israel' (Ezek 45:16-17; cf. 34:1-16, 24; 37:25).

In fact, the pierced one in Zech 12:10-14 fits in the context of the rejection of the Shepherd/YHWH in 11:4-14 whose covenantal Favor and Union is utterly despised by his flock. Furthermore, if the pierced one in 12:10 can be associated somehow with the smitten shepherd(s) in 11 and 13, then it seems likely that 'my associate' in 13:7-9 matches some of the profiles of Ezekiel's נָשִׂיא (34:24; 37:25), who is identical with the 'shepherd,' 'David,' and the 'king' (34:23, 24; 37:24, 25). Only this נָשִׂיא will have to provide the people with sin-offerings to make atonement for them. As the covenant is broken, and the temple is rejected by the Lord (Zech 11:10, 13), no one remains who can provide sin-offerings except YHWH himself and 'my shepherd' or נָשִׂיא as is the case in Ezekiel's vision.

In addition, it is unlike Ezekiel's vision of restoration, 'for the sake of My name' (cf. Ezek 36:21, 23), where the fundamental motivation for YHWH's restoration in Zechariah 9-14 is expressed in terms of YHWH's compassion for the flock. In the programmatic oracle of Zech 10 for the remaining chapters,

---

[209] Rudolf, *Sacharja 9-14*, 219, likewise, highlights this aspect in 12:10: "Es ist wie in 13:7-9: Wie dort trotz des eschütternden Anfangs am Ende die Harmonie zwischen Gott und Volk in Aussicht gestaltet wird, so bricht hier nach dem aufziehnden Feindgewitter zuletzt doch die göttliche Gnadensonne durch."

YHWH declares, וְהוֹשִׁבוֹתִים כִּי רִחַמְתִּים (v. 6, "I will restore them because I have compassion on them," cf. also, 'the spirit of חֵן וְתַחֲנוּנִים' in 12:10). While this may seem marvelous to the remnant of the people at that time, it will not seem marvelous to YHWH their true shepherd (Zech 8:6, cf. 8:1-5).

1.5.5 Like the Angel of the Lord (12:8): More Than David?

Besides the possible link between 'my shepherd' of 13:7-9 and Ezekiel's נָשִׂיא, we do not see any clear indication of the Davidic Shepherd figure in Zech 11:15-13:9. Instead, we read the prophecy regarding the house of David in 12:7-8, "the feeblest among Jerusalem's inhabitants, on that day, will be like David, and house of David will be like God, *like the angel of the Lord [going] before them*" (כְּמַלְאַךְ יְהוָה לִפְנֵיהֶם). The context of Zech 12:6-9 is clearly the eschatological restoration of Israel 'on that day' (vv. 6, 8, 9). Here we have all three major motifs of the Davidic Shepherd tradition: David, the shepherd, and the restoration. Moreover, this is perhaps the only place in Zech 11:15-13:9, besides Zech 9:9, in which we find a parallel to the triumphant Davidic Shepherd of Ezek 34:23, 37:24-25 and Mic 5:2.

The prophet's vision of the house of David at the time of the restoration, however, seems to surpass the status and glory of the monarchic shepherds of Israel. Shepherd imagery can be detected in the phrase, לִפְנֵיהֶם in Zech 12:8, which is a semi-technical term that refers to a shepherd figure (cf. Ex 13:21). In Mic 2:12-13, YHWH is the eschatological Shepherd; v. 13 says, "their king will go in advance before them (וַיַּעֲבֹר מַלְכָּם לִפְנֵיהֶם), the Lord at their head," while v. 12 clearly affirms shepherd imagery: "I will surely gather; I will bring them together like sheep in a pen." Similarly, the picture of the messenger/angel of the Lord, מַלְאַךְ יְהוָה '[going] in front of them' (לִפְנֵיהֶם) in Zech 12:8 not only affirms the restored presence of YHWH among them (cf. Zech 8:23), but also suggests that this messenger can be seen as YHWH's shepherding activity.

In Zech 9-14, the Davidic Shepherd figure of Ezekiel 34-37, that is, the new David who will lead the restored flock to obey the laws and decrees of YHWH, is nearly out of sight.[210] Yet the messenger of the Lord is presented as a figure equivalent to the eschatological shepherd-king who would go ahead of the flock, fighting the eschatological battle with the nations (12:9; cf. Mic 4:11-13).

We then can ask about the relationship between the house of David and יְהוָה מַלְאַךְ who assumes the role of the eschatological Shepherd. For the angel, we can suggest various prototypes including the angelic presence leading the

---

[210] It is true that in Zech 9-14, the motif of the Davidic expectation is not as expressive or intense as in Mic 2-5 and Ezek 34-37 (cf. Zech 12:7, 10, 12; 13:1). Probably the reasons vary. Besides the widespread disillusionment with the monarchy of Israel, the king Cyrus' benevolent policy for the returnees may have suppressed any politically incorrect utterances. Still we read in Haggai and Zechairiah implicit expectations of and encouragement for Zerubbabel. Refer to Josiah Derby, "Prophetic Views of the Davidic Monarcv," *JBO* 2 (2000): 111-116.

Exodus (Exod 14:19), and the angel who defended Jerusalem against Sennacherib (2 Kgs 19:35). The angel in Zech 12:8 can be associated with the messenger/angel of God in 2 Sam 14:20 (cf. 2 Sam 14:17; 16:23) in which the phrase is used as a circumlocution of the deity.[211] Nathan's prophecy in 2 Sam 7:7-16 attests that the king could be addressed using language befitting a deity such as מַלְאַךְ יְהוָה (cf. Pss 2, 110). By using a simile, it is relayed that the house of David is to be elevated to a divine status of sorts, while in contrast Davidic royal privilege is democratized (cf. Isa 55:3).[212]

Regardless of the identity of the angel of the Lord,[213] it can hardly be denied that Zech 12:8, along with its use of a simile, affirms the idea that the house of David is to be elevated to a divine or semi-divine status.[214] Ironically, the idea of the house of David being elevated in Zech 12:8 points back to the hopelessness of the current Davidic shepherds of Israel (cf. Zech 10:2-3).

The whole series of the visions related to the affliction of YHWH's shepherd in Zechariah 11-13 opens the way for YHWH to intervene in the future. Clearly the future will be unlike the past and unlike the present. Despite elements of obscurity, it is certainly implied in Zech 12:8 that 'on that day' the eschatological shepherd's leadership of the house of David will be much more than merely Davidic monarchic shepherds.

There is a continuity between Zechariah's Davidic expectation and Micah's and Ezekiel's (cf. Mic 5:2; Ezek 34:23-24; 37:24-25): the new David will come.[215] Nevertheless, we also detect a discontinuity in Zechariah's vision. The figure of the eschatological shepherd, like the angel of the Lord, surpasses the glory of the house of David, for this hope transcends beyond the monarchy.

---

[211] Driver and Plummer, *Zechariah*, 326, on the comparison of the house of David to God, lists Ps 8:5-6, Isa 9:5-6, 1Sam 14:17; Contra. Larkin, *Eschatology*, 119, nevertheless, eventually suggests that the messenger of the Lord here refers to the prophet himself whose oracles can be seen as the wisdom of the angel of the Lord.

[212] Peterson, *Zechariah 9-14*, 119.

[213] A variety of answers present themselves to the question of the identity of the angel of the Lord. G. van Groningen, *Messianic Revelation*, 215, asserts that the angel of the Lord is an angel who is not YHWH but a special and specific representative of God; Stephen L. White, "Angel of the Lord: Messenger or Euphemism," *TynBul* 50 (1999): 299-305, argues that it is basically a euphemism for God; some believe the angel of the Lord refers to a plurality of the divine being added later to express the anthropomorphic operations of YHWH. White, "Angel of the Lord," 304. We focus, however, not on the identity of the figure, but on the shepherd imagery.

[214] Tigchelaar, *Prophets of Old*, 123, suggests the polemic theory to be convincing that the author responsible for the text belonged to apocalyptic movement opposing the Jerusalemite hierocratic party. However, whether this view is likely or not, the notions of democratizing the Davidic royal privilege and elevating the house of David can be affirmed without serious difficulties.

[215] Zech 12:8 is indeed an expression of the hopes for the reestablishment of the Davidic dynasty at the eschaton. See, Antti Laato, *David Redivivus*, 294, 299.

1.5.6 Summary

Zechariah 9-14 is an important text that revises Ezekiel's vision (34-37). The shepherd image, associated with the Davidic expectation, abounds in Zechariah 9-14. Zechariah 9:9-15, announcing the meek king riding a donkey, projects the vision of YHWH as the eschatological Shepherd who will eventually set up a new David over the restored flock. Significantly, we have noted that Zech 10:2 (LXX) equates the absence of a shepherd with the absence of healing. Zech 11:4-16 shows that YHWH's covenant with Israel, in view of bringing peace to all the people on the earth, inevitably is to be broken because the sheep refuse their shepherd YHWH. Zechariah's revision of Ezekiel 34 and 37 is characteristic in that the former underscores the givens of Israel's disobedience.

While Ezekiel 34, 37 and Mic 5:1-5 focus on the triumphant Davidic Shepherd figure who would lead the restored flock into the way of righteousness, Zechariah tackles the problem of the flock's sin in terms of the mysterious suffering of the various shepherd figures in chapters 9-12. Zechariah's emphasis upon the sin of the flock and the suffering of the shepherd figure are indeed a renovation of the typical image of the shepherd in related materials found both in the OT and the ANE. Zechariah underscores that the whole community of the flock is placed under the judgement; as YHWH is rejected by his flock, building a temple is futile (11:13). Nevertheless, those alternating salvation oracles emphasize the restoration of YHWH's own reign over Israel and the nations (12:1-9; 13:1-6; 14:1-21). YHWH's striking the shepherd eventually leads to his raising up leaders/shepherds like David, who will participate in YHWH's extension of his rulership over the nations (12:6-9). Zechariah presents 'the angel/messenger of the Lord' (12:6) whose appearance coincides with the renewal of the house of David; the figure is the closest parallel to the triumphant Davidic Shepherd of Ezek 34:23, 37:24-25 and Mic 5:1-4.

1.6 Conclusion:
A Profile of the Eschatological and Davidic Shepherds

Now we offer a profile of the eschatological and Davidic Shepherds as we raise a question: "How would Israel recognize the arrival of the promised eschatological Shepherd and God's Davidic Shepherd-Appointee?" In the OT, we find numerous indications spread out among various portions of text that follow the Davidic Shepherd tradition. Unlike what occurs in ANE materials, the titular use of the shepherd imagery, especially in the context of the promised restoration, is reserved solely for YHWH and his Davidic Shepherd. This may imply that the OT develops a specific notion of shepherd-leadership in the manner of YHWH's redemptive and restorative activities in Israel's history. The search intensifies for the shepherd after the heart of YHWH as Israel's history passes from the time of the wicked shepherds of the monarchy to consequential exile.

The first and foremost sign of the coming of the eschatological Shepherd is the introduction of YHWH's theocratic intervention in Israel's history: "I myself will search for my sheep" (Ezek 34:11). This involves two characteristic phenomena at the inauguration of the restoration: First, the return of YHWH's presence among his flock, and second, YHWH's fatal confrontation with the wicked shepherds of Israel (Ezek 34:2-16; Zech 10:1-6; cf. Jer 23:1-8). To secure YHWH's presence as the shepherd over his flock, YHWH must confront the wicked shepherds of Israel in order to judge them. This confrontations is irrevocable since it signals the arrival of the eschatological theocracy of YHWH governing his people in view of the nations. For Israel, it is the eschatological divine reversal, achieved by the eschatological Shepherd compassionately and powerfully shepherding the scattered flock.

We discover linguistic clues to the profile of the eschatological Shepherd. Ezekiel 34-37 in particular depicts the shepherd restoring his flock at the time of the restoration. YHWH as the eschatological shepherd will first and foremost be characterized by his 'seeking the lost' as he gathers the flock who had been scattered as the result of the wicked shepherd's deplorable shepherding.

To restore the community, YHWH will begin the process of restoration from the most remote fringe of the community 'by seeking the lost and outcast.' Afterward, he will 'bind up the wounded' and 'strengthen the weak,' 'feeding the hungry,' and judge between the robust and the downtrodden among the flock. The eschatological Shepherd healing the sick, an unusually compassionate shepherd unique when compared to ANE shepherd imageries, is characteristic in Ezekiel 34-37. Ironically, in Zechariah 11-13 the shepherd figures are replete with images of being detested, sold, outcast, smitten, and mourned.

YHWH's redemptive shepherding activities have been established early as evidenced from the Exodus. 'To bring out' (cf. LXX, ἐξάγω) and 'to brings in' (cf. LXX, εἰσάγω) and 'going before' his flock at their head (Exod 3:8; Deut 6:23) can be regarded as semi-technical shepherd language often found in redemptive and restorative contexts. These terms reveal the ultimate image of the shepherd: He is the leader who leads (cf. -άγω) his flock. As a result of that leading, the eschatological Shepherd '[going] before' (לִפְנֵיהֶם) his gathered flock signifies a significant breakthrough as the Breaker in Mic 2:13, the angel of the Lord (Zech 12:8), and YHWH in the Exodus (Exod 13:21; cf. 2 Sam 5:2; 7:5-7, 12-16).[216]

---

[216] Peter R. Ackroyd, *Exile and Restoration: A Study of Hebrew Thought of the Sixth Century B.C.* (Philadelphia: Westminster, 1968), 239, notes the difference between the Exodus and the exile stating, "the exile is not comparable with the period of the Exodus. For at no point in the Exodus narratives is it suggested that the people in Egypt were brought into subjection by reason of their own sinfulness."

One of the key outcomes of YHWH's eschatological shepherding will be the renewed obedience of the flock to YHWH's laws and decrees. To secure this, YHWH will set up one shepherd like David over the restored people. This is the consequence of the restorative act of YHWH, the true shepherd of Israel. This is a common pattern in the Davidic Shepherd tradition of Micah 2-5, Ezekiel 34-37, and Zechariah 9-14 (cf. Ezek 34:23-24; Zech 12:8-10). Under the leadership of the Davidic Shepherd, the ultimate problem existing between YHWH and his flock – their inherent inability to follow their shepherd's law under the covenant of Favor and Union – at last will be resolved by YHWH giving his Spirit and a new heart (Ezek 36:16-37:14; cf. Zech 12:10-13:6). Thus YHWH's presence among them will be perpetual, along with the representative kingship of the Davidic Shepherd governing over them.

As mentioned earlier, Zechariah 9-14 adds other significant signs to the profile of Ezekiel's eschatological shepherd. The idea of the detested or stricken shepherd is added by Zechariah to Ezekiel's healing shepherd. The key concept in Zechariah's new image of the shepherd finds its roots in the inner conflict between YHWH and his flock. In Ezkiel, the eschatological Shepherd rescues his flock from mouths of the wicked shepherds, but in Zechariah, YHWH delivers them to the wicked shepherd. This shows the deeper side of YHWH's confrontation with the shepherds in the Davidic Shepherd tradition in the OT.

Ezekiel's vision solves this problem in terms of YHWH giving his Spirit and a new heart, though Ezekiel does not detail the struggle and painful consequences endured as is recounted in Zechariah 11-13. Ezekiel comes closest to Zechariah's description of the suffering shepherd when he uses נָשִׂיא which is another name for the Davidic Shepherd and king. Ezekiel's נָשִׂיא is characterized by his intimacy with the flock (Ezek 34:24; 37:25) and his responsibility to provide the renewed people with sin-offerings (Ezek 45:16-17). When the rebuilt temple turned out to be a futile attempt to usher in the promised restoration, there would be no one left to be held responsible to offer sacrifices for the sins of the flock.

In the case of Zechariah's shepherd imagery with its emphasis on the suffering of the shepherd caused by the sins of the flock, the ultimate motivation for YHWH to restore his flock is his unfailing compassion toward them (Zech 10:6; 12:10; cf. Mic 7:18; Isa 40:9-11; 61:1-14). When the disastrous and irrevocable course of Israel's history leaves no alternative but the broken covenant (Zech 11:10), the compassion of Israel's true Shepherd intervenes. In Ezekiel's vision, however, the motivation differs. YHWH is concerned for his reputation among the nations: 'for the sake of my name' (Ezek 20:39; 36:20, 21, 22; 39:7 [x2]; 39:25; 43:7, 9). Despite the hopeless situation of the exiled, akin to an army of dead bones piled in the valley (Ezek 37:1-4), YHWH's zeal for his holy name among the nations overcomes Israel's uncontrollably adulterous hearts.

1.6 Conclusion: A Profile of the Eschatological and Davidic Shepherds 93

The one whom YHWH will establish over the restored flock is the new David (Mic 5:1-2). He is the new David, a shepherd according to the heart of YHWH (Ezek 34:23), empowered by the might of YHWH (Mic 5:4; Zech 12:8) and by YHWH's compassion. The new David will lead the restored flock to walk in the way of the Lord. Through this Davidic Shepherd, YHWH's sovereign shepherd-rule over the nations will extend 'to the ends of the earth' (Mic 5:4; Zech 9:10).

While not dominant, there is an image of the shepherd as the eschatological warrior who will defeat the nations gathered against the Lord and his flock. It is not clear whether it will be YHWH alone as the eschatological Shepherd (Zech 14:1-19; Ezek 38-39), the Davidic Shepherd (Mic 2:12-13; Zech 12:8-9; cf. Ps 2:9), or the restored and empowered flock with and commanded by the shepherd figure (Mic 4:11-13; Zech12:8-9) who will fight off the nations. Regardless, all of these texts involve shepherd imageries. It remains obscure what is meant by the imagery of 'breaking into pieces like pottery' with 'staff/scepter' (Ps 2:9) or 'hoofs' (Mic 4:13). Since we observed that the 'rod' of the Davidic Branch (שֵׁבֶט; cf. 'staff' in Ezek 37:19; Ps 2:9) in Isa 11:6 is translated by the LXX as λόγος that 'comes out of his mouth,' we can only imagine what the battle will be like.[217] The vision of YHWH's final battle with the nations often incorporates shepherd imagery in terms of the shepherd's staff, and extends the vision of the Davidic Shepherd tradition beyond the territory of Israel, and outside the realm of her history (Zech 14:1-19; Mic 4:11-13; cf. Ezek 38-39).

Last but not least, the Davidic Shepherd tradition in the OT presents a consistent pattern or flow of similar visions of the restoration. In the instance of Jeremiah 30, 31, the vision of the future begins with Davidic expectation (30:8-9) followed by shepherd imagery (31:10), and climaxes with the hope that the exile will come to an end (31:15-17). Micah follows suit. The example of Mic 5:1-4 exhibits the identical pattern: from Davidic expectation (vv. 1-2), shepherd imagery (v. 4a), to the climax – the hope of restoration (vv. 4bc, 5). Likewise, the pattern in Zechariah 9-14 is identical. The vision opens with the hope for the coming of the Davidic king (9:9), followed by various shepherd images in Zechariah 11-13. Here too the vision climaxes with the hope of the restoration and YHWH's reign over the nations in chapter 14.

Ezekiel's pattern is intriguing (Ezek 34, 37). It begins with YHWH confronting the wicked shepherds of Israel, serving as a precursor of the expectation of the 'one shepherd' pointing to the advent of the Davidic Shepherd (34:6, 23). The central section of the vision is comprised of rich imageries of

---

[217] Similarly, yet in different context of Ps 23:4, 6, the Lord's rod and staff are paralleled to his 'goodness'(טוֹב) and 'faithfulness'(חֶסֶד). Even the term שֵׁבֶט ('tribe, staff') evokes the imagery of the shepherd holding the tribes of Israel in his hand (cf. Ezek 37:15-28; Zech 11:4-14).

what the eschatological Shepherd would do at the time of the restoration. The pattern concludes with renewed obedience to YHWH and consequential restoration, including the united nation, peace in the land, and YHWH dwelling among them for everlasting (37:26-28; cf. Zech 14:20-21). Thus we can say that the Davidic Shepherd tradition begins with the motif of the absence or presence of the true shepherd and ends, eventually, with the returning presence of the shepherd among the flock: "And the name of the city from that time on will be: יְהוָה שָׁמָּה"('The Lord is there,' Ezek 48:35; cf. Ezek 37:26-28).

Chapter 2

The Davidic Shepherd Tradition in Second Temple Judaism

Thus far our research shows that Davidic expectation (esp. 2 Sam 7:12-15) went through various interactions with the prophets who combined it with the shepherd motif now in the context of the hope of Israel's restoration. Of these prophets, it may be Zechariah who reaches the climax of the development of the OT tradition as he receives and reinterprets Ezekiel's vision of the Davidic Shepherd. Ezekiel's vision of Israel's restoration depends largely on the rescue of the scattered flock by YHWH as the eschatological Shepherd, and the salvation wrought under the righteous leadership of the Davidic Shepherd.

Although Ezekiel underscores Israel's impotence to keep the law, it is Zechariah who concludes that the central question of Israel's restoration should fall upon the hopeless flock itself rather than on their need for the good shepherd. He recognizes the more tragic aspect of Israel's failure: Israel's hatred of her good shepherd. As a consequence of this conflict, some shepherd(s) – wicked, good, or both – will be pierced before the fountain bursts forth from Jerusalem to cleanse the people and restore the land.

Concerning the lingering hope of the revival of the Davidic dynasty, the long process of its development had already reached a conclusion when Zechariah modified Ezekiel's vision. This appeared to be the case until the appearance of the Hasmoneans triggered a resurgence of the Davidic hope. As John J. Collins observes, messianism (defined as royal Davidic expectation with military connotations) was virtually dormant from the early fifth to the late second century B.C; then, the renewed interest in monarchy in the first century B.C. was largely in reaction to "the flawed restoration of the Jewish kingship by the non-Davidic Hasmoneans."[1]

Whether supportive or critical of the Hasmoneans, they all would have read into their new situation the age-old Davidic expectation along with the old traditions. Thus we will set out to examine how the revived messianism, spawned in the Maccabean revolt (163 B.C.), interacts with the OT Davidic Shepherd tradition evidenced in a variety of texts from the Second Temple period.

One way to start is to trace the reworkings of any texts from the OT Davidic Shepherd tradition during this period. For instance, 4QPseudo-Ezekiel (4Q385c,

---

[1] John J. Collins, *The Scepter and the Star: The Messiah of the Dead Sea Scroll* (New York: Doubleday, 1995), 40.

4Q386), particularly 4Q386 1, 2:1-3, echoes YHWH revivifying the dead bones in Ezek 37:1-14: "And they will know that I am YHWH; and when will you assemble them?" (4Q386 1, 2:1a, 3a).[2] Another explicit reference to Ezekiel 34-37 is found in fragment 5 of the *Apocryphon of Ezekiel*: "Therefore he says by Ezekiel ..., 'And the lame I will bind up, and that which is troubled I will heal, and that which is led astray I will return, and I will feed them on my holy mountain ... and I will be,' he says, 'their shepherd and I will be near to them as the garment to their skin.'"[3]

This fragment is dated no later than the first century A.D.[4] Here, the image of the shepherd is bound up with various motifs such as the healing, gathering, and feeding of the sheep. The intimacy of the shepherd with the sheep recalls the role of the Davidic Shepherd as the 'prince' in the midst of the restored flock in Ezekiel 34 and 37. The fact that this fragment, though Jewish in origin, has been preserved only in Christian sources, suggests the familiarity of Ezekiel 34 and 37 among the early Christians.[5] These explicit references to the OT Davidic Shepherd tradition, however, are mostly fragmentary and often fail to involve all three criteria of the OT Davidic Shepherd tradition, i.e., the shepherd imagery, and Davidic expectation in the context of the restoration.

For a comprehensive research which meets these criteria, on the one hand, we will trace the occurrences of the shepherd imagery in the hopes of discovering any allusions to and echoes of Davidic expectation; and, on the other hand, we will examine those occurrences of David expectations to determine if there is any link at all with the shepherd imagery. Accordingly, a text reflecting only one of the three motifs will be dismissed for failing to meet our criteria.[6]

A case in point, which fails to meet the criteria, may be found in Sirach (before 163 B.C.). Some shepherd imageries can be traced in this passage: "the

---

[2] The Qumran texts will be based on *Discoveries in the Judaean Desert* (Monographic series, Oxford: Clarendon Press, 1955-2004); *The Dead Sea Scrolls: Hebrew, Aramaic, and Greek Texts with English Translations* (ed. James H. Charlesworth; Tübingen: Mohr Siebeck, 1994); and, *The Dead Sea Scrolls: Study Edition* (ed. F.García Martínez and Eibert J. C. Tigchelaar; Leiden: Brill, 1997, 1998); for the overall influence of Ezekiel in Qumran sectarianism, see Ben Zion Wacholder, "Ezekiel and Ezekielianism as Progenitors of Essenianism," in *The Dead Sea Scrolls: Forty Yeas of Research* (ed. Devorah Dimant and Uriel Rappaport; Brill: Leiden, 1992), 186-196.

[3] Translated by J. R. Mueller and S. E. Robinson in *The Old Testament Pseudepigrapha* (ed. James H. Charlesworth; New York: Doubleday:, 1983) 1:488; Hereafter, *OTP* 1.

[4] Mueller and Robinson, ibid., place the composition of the document between 50 B.C. and A.D. 50; cf. K. G. Eckart, "Das Apokryphon Ezekiel," *JSHRZ* 5, no.1 (1974), 45-54.

[5] Cf. W. D. Stroker, "The Source of an Agraphon in the Manichaen Psalm-Book," *JTS* 28 (1977): 114-118. The text appears in Clement of Alexandria (*Strom* 1.9), in Origen (Homilies on Jeremiah 18:9), and in the Manichaean Psalmbook (Psalm 239:5-6.) with slight variations.

[6] This approach differs from J. Vancil's (*The Symbolism of the Shepherd*) since Vancil traces only the shepherd imagery in the Second Temple period, and differs also from Collins' *The Scepter and the Star* because Collins focuses mainly on Davidic expectation, and seems unaware of the significance of its link with shepherd imagery.

compassion of man is for his neighbor, but the compassion of the Lord is for all living beings. He rebukes and trains, teaches them, and then turns them back, as a shepherd his flock"(18:13).

Certainly we find the compassion motif associated with YHWH as the shepherd who trains and teaches, but we do not find any association with Davidic expectation. Likewise, Sirach's positive comments about David are found in 45:25 in the context of retelling Israel's history, but which are noticeably lacking in shepherd imagery there. With this criteria in mind, we will now begin looking at selected texts that include both Davidic expectation and shepherd imagery in the context of Israel's restoration. If possible, we will put the investigation of these texts into chronological order.

2.1 The *Animal Apocalypse* (1 Enoch 85-90)

Without a doubt this text is imbued with 'hope for the restoration of Davidic line.'[7] Furthermore, the *Animal Apocalypse* [hereafter, *An. Apoc.*] presents itself as incredible evidence for the continuing validity of the shepherd imagery in the Second Temple period.[8] Using a complex allegorical system that depicts human beings as animals, disobedient watchers as descending stars, and seven archangels as men, the narrative attempts to summarize the whole events of history, beginning with the creation of Adam to some point of the Hellenistic period, and including in its grand scheme the final judgment and the onset of a new era of transformation.

The *An. Apoc.* can be understood as a history retold by employing the traditional sheep-shepherd images. An examination of this book should shed invaluable sights on the reception and use of the OT Davidic Shepherd tradition in this period.

2.1.1 Animal Symbols

2.1.1.1 The Sheep, the Rams, and Blindness

As the *Animal Apocalypse* describes the vision of the fallen stars and the Great Flood in 86:1-89:9, the fallen stars become black cows, and their mating with bovids produces elephants, camels, and donkeys (86:4). Seth, Abram, and Isaac

---

[7] Collins, *Scepter*, 34.

[8] J. T. Milik, *The Books of Enoch: Aramaic Fragment of Qumran Cave 4* (Oxford: Clarendon, 1976), 5, 225, 246, asserts that the Qumran Scrolls covers 26 percent of the Dream Visions dated at least before the first century B.C. Isaac, *OTP* 1:7, dates them to 165-161 B.C. close to the time of the Maccabean revolt. Likewise, George W. E. Nickelsburg, *1 Enoch 1: A Commentary on the Book of 1 Enoch, Chapters 1-36; 81-108* (Minneapolis: Fortress Press, 2001), 361, places them between 165 and 163 B.C.

are depicted as white bulls (86:6; 89:10-11). Most likely the color of the animal represents its religious disposition; for example, white tends to mean 'pure' while black tends to mean 'sinful.'[9] The sheep symbol is applied first to Jacob, the 'snow-white sheep,' who bore 'twelve sheep' (89:12). Particularly, the exodus setting for the first appearance of the sheep metaphor in *An. Apoc.* also harmonizes with the OT use of the image for YHWH's redemptive leadership (cf. Exod 3:10, 11, 17; 6:6, 13, 26; 7:5; 12:42; 17:3; 23:23; Num 20:5; Deut 4:38). The expression of Israel as 'twelve sheep' (89:12) is noteworthy; they "grew up in the midst of the wolves" [Egyptians] until "they multiplied and became many flocks of sheep" (89:12-14).[10]

Interestingly, leader-figures in Israel's history such as Caleb and Joshua (89:39), Zerubbabel, Ezra, and Nehemiah (89:72), are called 'sheep' rather than 'shepherds.' Moses is no exception: he is also seen as a 'sheep' saved from the wolves (89:16). Though Moses is credited for 'leading' Israel in the wilderness (89:31), he is later described as becoming a man (89:36, 38)[11] similar to Noah, who born as a bovid, later becomes a person (89:1).

Yet Moses is not a shepherd in the Vision. The 'ram' in *An. Apoc.* refers to an under-shepherd figure, but is not designated a 'shepherd.' This is true of Saul, and surprisingly, David as well. Saul is called a ram, strong enough to defend the sheep against their enemies, yet Saul is not called 'shepherd' (89:42).[12] David, when called by the Lord, was a sheep who later "became a ram" and "led the [little] sheep" (89:45, 48).

Yet we must clarify that for David, additional descriptions are included in the text. Samuel appoints David 'to rule' and to be the 'ruler and leader' of the sheep (89:46). The Aramaic root *marha* behind 'to rule' is used to denote the leadership function of many other figures such as God (89:22), Moses (vv. 28, 32, 38), Joshua and Caleb (v. 39), the judges (v. 14), Saul (v. 42), and David and Solomon (v. 48).[13] But the Greek root ἄρχειν is used only for David and Solomon, which is likely a translation of שלט.[14] Moreover, David is appointed

---

[9] Ida Fröhlich, "Symbolical Language of 1 Enoch 85-90," *RevQ* 14 (1990): 630, suggests that the color of the animal indicates the character of the given human figure: white for the elect, like Adam; black for sinners, like Cain; and red (neutral), like Abel and Ham.

[10] Note the use of a similar sheep metaphor found in Matt 10: '*twelve* disciples' (v. 1); "Go rather to the lost *sheep of Israel*" (v. 6); "I am sending you out like *sheep among the wolves*" (v. 6).

[11] Patrick A. Tiller, *A Commentary of the Animal Apocalypse of 1 Enoch* (SBL Early Judaism and Its Literature 4; Atlanta: Scholars Press, 1993), 295-296, suggests that the transformation of the sheep [Moses] into a man alludes to Moses' angel-like glorification (Exod 34:29-35).

[12] Cf. Vancil, *Symbolism*, 256.

[13] Nickelsburg, *Commentary*, 383.

[14] Ibid.

as the 'ruler and leader' (*mekwennan wamārāhe*; ἄρχων καὶ ἡγούμενος) in 89:46.[15]

Nickelsburg remarks that the double usage here corresponds to 11QPs[a] 28:11 in which the nouns נגיד ("prince") and מושל ("ruler") are paired to refer to David's functions as leader of God's flock.[16] Yet, it may be more accurate to say that the doubling of the titles is to serve the two-fold relationship of this leader-figure, that is, with the people and also with YHWH. The full text of 11QPs[a] 28:11-12a clarifies this: "and [YHWH] made me 'prince of his people' (לעמו נגיד) /and 'ruler/ over the sons of his covenant' (ומושל בבני בריתו)."[17]

If the double usage applied to David (89:46) is consciously used to convey this subtle distinction, one may then trace Davidic expectation in the phrase. This might explain why only Solomon is called 'a ruler and a leader' (89:48b) after the fashion initiated in the case of David (v. 46a). Indeed, no one else before David and no one after Solomon in *An. Apoc.* is designated by these double titles. After Solomon, all of the rulers of Israel and Judah until the destruction of Jerusalem, are not even mentioned as rams because the texts simply bypass them (89:51-54), and skip over even the reforms of Josiah and Ezra.[18] The double titles, 'prince [among the people]' and 'leader [appointed by God],' thus can be regarded as the functional equivalent to the Davidic Shepherd/Prince in the OT Davidic Shepherd tradition (esp. Ezek 34:23; 37:25).

Like Moses and David, Elijah is also regarded as one of the 'sheep.' The Vision does not call him a 'ram,' most likely because he is remembered not as an administrator among the people but as a prophet. While Moses 'becomes' a man and David is called the 'prince and leader,' *An. Apoc.* 89:52 reads that Elijah is taken up by the Lord of the sheep to settle with Enoch in heaven.[19] All three of these prominent figures are distinguished from others, yet not one of them is referred to as 'shepherd.' The *An. Apoc.* describes Moses as entering into angelic status; Elijah is taken up to the heavenly realm, and the Vision depicts David in terms of being the prince of the people and the leader appointed by God. Nevertheless, we must caution that David is counted among the sheep. Although he is esteemed, he is not the shepherd, and his special relationship with the Lord of the sheep is only implicit.

---

[15] Isaac, *OTP* 1:67, renders it 'a judge and a leader'; Tiller, *1 Enoch*, 304, translates it 'administrator and leader.'

[16] Cf. Nicklsburg, *Commentary*, 383, n. 77.

[17] J. A. Sanders, *The Dead Sea Scrolls v. 4a: Pesudepigraphic and Non-Masoretic Psalms and Prayers* (ed. J. H. Charlesworth; Tübingen: Mohr Siebeck, 1994), 164-165 renders נגיד as 'leader' as so do Martines and Tigchelaar, *Scrolls*, 1179; the rendition, 'prince' by Nickelsburg, ibid.

[18] Tiller, *1 Enoch*, 340.

[19] Fröhlich, "Symbolical," 630-631, notes for this reason that Elijah is the most important figure marking the end of the epoch of the Divided Kingdom.

The fundamental problem of the sheep is their blindness, the symbol of their apostasy. The *An. Apoc.* frequently underscores the blindness of the sheep (89:28, 32, 40, 41, 74). They are dim-sighted or blinded, they fail to see (the glory of) the Lord about which the Vision displays particular concerns. At the exodus, the sheep were following the Lord gazing at the face of the Lord: "His face was glorious, adorable, and marvelous to behold " (89:22); "they began to open their eyes; then I saw the Lord of the sheep bringing them to a pasture and giving them grass and water" (v. 28b). To have their eyes opened, and thus have the ability to gaze at the Lord and follow him, is critical. Characteristic of Israel's eschatological state involves this metaphor as well: "the eyes of all of them are opened" (90:35).

Nickelsburg is probably right when he points out that blindness in the Vision is particularly 'related to the cult' rather than to the lack of their religious and moral knowledge (89:32-35, 41, 44-45, 51, 73-74).[20] In this respect, the 'house' or the 'tower' (temple) closely relates to their blindness (89:40) as can also be seen in 89:54: "when they have abandoned the house of the Lord and his tower, they went astray completely, and their eyes became blinded."[21] This may be a significant point as we consider that in *An. Apoc.*, the blindness of the sheep is seen as the result of the exile not only as its cause just as 89:54 demonstrates.

If this is correct, then Nickelsburg's conclusion that the "acceptance of a new revelation will bring about Israel's final and permanent deliverance from their enemies" rings hollow in its generality.[22] It simply misses the point of the symbolism of blindness in the Vision. The solutions to the plight of Israel's blindness presented in *An. Apoc.* do not seem to urge the acceptance of new revelation with regard to God's law *per se*. It is rather likely that the blindness of the sheep is depicted as part of their inevitable nature, and thus, from the beginning, the exile does not take them totally by surprise (89:32, 41, 54).

To prove it, the restoration announced in 90:37 is completed by the sudden appearance of the eschatological snow-white cow. Up to this point, *An. Apoc.* elaborates on a series of judgment scenes, mostly focusing upon fallen stars and the seventy shepherds (90:20-27). In the extensive justification of these judgments, the seventy shepherds (89:59-90:17) are blamed for their wrongdoings. While the blindness of the sheep is the probable cause, it is certainly not *the* cause for the entire misery of the exile, and the excessive disaster that befell the nation. The Vision points to evil powers working behind

---

[20] Nickelsburg, *Commentary*, 380.

[21] Tiller, *1 Enoch*, 292, misses this point as he comments on 89:28: "The implication of seeing, then, seems to be possession of God's law and obedience to it." In v. 28, the sheep's eyes are open and they see the Lord pasturing them after the Exodus. In v. 32, however, their blindness refers to their idolatry.

[22] Nickelsburg, 381.

the blinded sheep, and seems to accentuate the hope of God's abrupt 'transcendental' intervention.

Certainly, the sheep metaphor is foundational for Israel, but in *An. Apoc.* 'sheep' is not the ultimate category for Israelites, especially as their final destiny is in view. Along with the Gentiles (the wild beasts), the sheep will be transformed into 'white bulls' at the time when the messianic age arrives (90:37-39). With an astonishing twist in imagery, 'the Lord of the sheep,' who rejoiced over the return of the sheep (v. 33), will likewise 'rejoice over all the bulls' (v. 38). Thus the sheep metaphor for Israel serves as the author's view of *historical Israel* as particularly led by 'the Lord of the sheep' (89:16, 26, 29, 30) only until the time of the eschatological transformation.

Thus to assess this material, we see that the sheep image, beginning with Jacob, sets up the basic characters and relationships with the nations, the shepherds, and the Lord of the sheep. Contrasts are set up between the sheep and the beasts, as well as between the sheep and the shepherds, while the Lord of the sheep tends his flock in the midst of the beasts.[23] Yet the shepherds are no longer the leader figures renowned in Israel's history. The fact that the Vision avoids calling any leader figure 'shepherd', accentuates God's status as the one and only 'Lord of the sheep' over Israel.

Furthermore, the Lord's shepherding is not limited to his flock, for inevitably the Lord of the sheep will shepherd all – Israel and the nations. The sheep metaphor is valid only as long as the sheep are set against the beasts. In the end, the eschatological white bull transforms both the sheep and the beasts into its own kind. Thus, the sheep is the category for Israel only for their historical journey up to the eschaton.

2.1.1.2 The Wild Beasts, the Birds, and the Exile

While the symbol of sheep is chosen for and applied to Israel, the role of the "beasts of the fields and the birds" (89:9) is assigned to the Gentiles. Emerging from Noah's three sons, these nations are personified as animals: "lions, leopards, wolves, snakes, hyenas, wild boars, foxes, squirrels, swine, hawks, eagles, kites, striped cows, and ravens" (89:10). Noticeably, these animals and birds are unclean scavengers. As we indicated earlier, a ram [Jacob] born to the while bull [Isaac] becomes the progenitor of the Israelite sheep that will be the victims of the beasts and the birds (89:11-12, 55-56).

Nickelsburg notes how this dichotomous imagery of predators and sheep focuses on "the account of this era on the history of Israel as a time when the Gentile beasts devour the sheep."[24] Further, according to David Bryan, this

---

[23] Fröhlich, "Symbolic," 632, states, "each period is a dichotonomic system" between two major symbols such as white and black, oxen and beasts, sheep and beasts, and sheep and shepherds.

[24] Nickelsburg, *Commentary*, 354.

symbolism means more than a dichotomous structure, and argues that the seer is one whose mentality was shaped and fully governed by the world-view and inherent symbolism of the kosher rules.[25] Truly, the animals are unclean that represent the Watchers'freak offsprings and the Gentile nations; in *An. Apoc.*, Israelites are never unclean animals or birds. Bryan suggests that the author was, after all, "part of an ultra-conservative Jewish party which was vigorously opposed to the Hellenizing policies of the priests who led the Jerusalem community at the beginning of the second century B.C."[26] If that is correct, that may explain why the seer adopted a positive outlook on the Maccabean revolt and its leaders (90:6-17). The main struggle ensuing in the Vision then takes place between the holy Israelite and the unclean nations, with the latter attacking the former.

However, Bryan's reconstruction of the central storyline, namely, the return to the *Urzeit* at the *Endzeit* in which "the original purity of the creation is re-established: 'and they all [both Israelite and the nations] became white bulls' (1 *En.* 90:38),"[27] simply do not match with the kosher mentality which does not articulate *the vision of the inclusion of the Gentile* at the *eschaton*. The more serious question that *An. Apoc.* raises may be that of theodicy: Where is the Lord of the Israelite when the beasts devour freely the sheep in the field? Bryan's scheme, based on the kosher mentality, never wrestles with this problem central to the seer's outlook in the Vision as he painfully witnesses the sheep suffering without their shepherd.

Particularly after the exile, the wild beasts "tear those sheep into pieces"(89:55), and "eat them" (v. 56). The horror of Israel's life in the exile is well articulated in 89:58:

"Then I [Enoch] began to cry aloud with all my strength and to call upon the Lord of the lions and to reveal to him concerning the sheep, for he had fed them to all the wild beasts; But he remained quiet and happy because they were being devoured, swallowed, and snatched; so he abandoned them into the hands of all the wild beasts for food."

Thus the flock remains without a shepherd who could protect it from wild beasts, leaving them exposed and vulnerable to danger and destruction.[28] The motif of the abandoned sheep without a shepherd invokes YHWH's serious indictments upon the wicked shepherds of Israel in Ezekiel 34: "They were scattered because there was no shepherd (v.5). Moreover, the expression, "devoured, swallowed, and snatched" does echo Ezek 34:2-3, 5: "You shepherds of Israel who have been feeding [shepherding] yourselves! Should

---

[25] David Bryan, *Cosmos, Chaos, and the Kosher Mentality* (JSPSupS12; Sheffield: Sheffield Academic Press, 1995), 168-169.
[26] Bryan, *Kosher Mentality*, 169.
[27] Bryan, *Kosher Mentality*, 170.
[28] See Judith 11:19 as a case in point; cf. Vancil, *Symbolism*, 262.

not shepherds feed [shepherd] the sheep? You eat the fat, you clothe yourselves with the wool, you slaughter the fatlings; they became food for all the wild animals." [29] The only difference between these two texts is the reference to the subject of the reversal of good shepherding. The implied subject by whom the sheep "were trampled and eaten" in *An. Apoc.* 89:74, in this regard, is unclear: "they were delivered to their shepherds for an excessive destruction, so that the sheep were trampled and eaten."[30]

Nevertheless, the suffering of the sheep in these texts is depicted in language reminiscent of Ezek 34: "You [Israel's shepherds] eat the fat, you clothe yourselves with the wool, you slaughter the fatlings" (vv. 2-4). Especially, the phrase, "trampled and eaten" in *An. Apoc.* 89:74 alludes to Ezek 34:4, 18: "You [the shepherds] treaded them violently and harshly; Must you [the fat and robust among the flock] also trample the rest of your pasture with your feet?"

The language is familiar, but the primary agent who causes the suffering of the sheep changed. It appears that the injustice within the community of Israel, as described in Ezekiel 34, is now fully extended to the sheep among the nations in *An. Apoc.* The question of how YHWH would rescue the sheep, then, may well be modified too. In Ezekiel 34-37, the eschatological shepherd focuses on the leadership within the community by rescuing the sheep primarily from the mouth of the negligent shepherds. In *An. Apoc.*, 'the Lord of the sheep' is to rescue the sheep primarily from the mouth of the wild beasts, the nations, and from the shepherds behind the nations. This might be one reason why the shepherds in *An. Apoc.* cannot simply refer to Israel's shepherds.

2.1.2 The Shepherds and the Judgment of the Lord of the Sheep

The shepherds in *An. Apoc.* are not part of the cause of the exile (cf. Ezek 34), nor are they responsible for the destruction of the house in 89:54. Rather, on behalf of the will of the Lord of the sheep, they execute punishment during the exile (v. 59). Later, however, they invoke the final judgment to befall upon themselves because their destruction of the sheep was excessive, going beyond the limits set by the Lord of the sheep (90:20-29). The main exegetical issue regarding the shepherds concerns identity: Who are they? The options vary, ranging from the kings of the Gentile nations, seventy years, or angelic beings.[31] J. Vancil suggests that the shepherds might represent "the reign of the nations over Israel during the years of her captivity."[32] It is not difficult, however, to

---

[29] Nickelsburg, *Commentary*, 395, notes only 89:75 reminiscent of Ezek 34:12.

[30] Tiller, *1 Enoch,* 314, remains undecided; Nieckelsburg, *Commentary*, 395, thinks they are the nations that trample and eat the sheep in this text.

[31] R. H. Charles, "The Book of Enoch," in *Apocrypha and Pseudepigrapha of the Old Testament* (Oxford: The Clarendon Press, 1913), names it 'the most vexed question' in *1 Enoch.*

[32] Vancil, *Symbolism*, 258.

find the traditional roots of the notion of the heavenly rulers or leaders such as those found in Deut 32:8 (LXX); Jub.15:32; Dan 10-11; Ps 89:72b-90:1.[33]

Most likely, the [seventy] shepherds in *An. Apoc.* represent the 'angelic' shepherds who 'pasture' behind the scene of the nations afflicting Israel during the exile. This position can be supported in a variety of ways. First, the shepherds are not the wild beasts, and the shepherds are placed over the nations in a hierarchical structure (89:54-59, 65-66) that draws them closer to the Lord's authority: "He summoned seventy shepherds and surrendered those sheep to them (v. 59); they [seventy shepherds] abandoned those sheep into the hands of the lions; So the lions and the leopards devoured the majority of the sheep (v. 66)."

Second, the literary parallelism between the beasts' devouring the sheep and the shepherds' grazing the sheep in 90:3 reflects the shepherds as the angelic power behind the historical events:

a. Then the sheep cried aloud
    b. because their flesh was being devoured by the birds
a'. I too cried aloud and lamented in my sleep
    b'. because of that shepherd who was pasturing the sheep

The first half of verse 3 depicts what is happening on the historical level – though of course symbolically – and the second half reveals only what Enoch can see through the vision. Similarly, at the final judgement scene (90:20-25), while the fallen angels are held responsible for the age before the Flood, most likely it is the 'angelic' shepherds who are blamed for the period following the Flood, i.e., Israel's history. S. B. Reid's analysis of the structure of 1 Enoch 83-90 helps us understand that in the entire Book of Dream Visions is one literary unit with one introduction serving the whole (83:1-2) and one conclusive judgment applied to both the 'Watchers and the shepherds' (90:20-27).[34]

Third, the Lord of the sheep calls another 'group of the shepherds' (89:61) to report the activity of the seventy shepherds, and the Lord receives the report in the heavenly realm as they "elevate all of it [the report] to me [the Lord]" (89:64; cf.90:22). The expression suggests these are the angelic beings and likewise they are the seventy shepherds.

The more puzzling question, however, is not the identity of the shepherds but the implications of *An. Apoc.* as it receives and modifies the Davidic Shepherd tradition. What is the reason for *An. Apoc.* replacing Israel's shepherds with the angelic shepherds in its vision of Israel's restoration? Certainly, 1 Enoch interacts with the Davidic Shepherd tradition, particularly Ezekiel 34, as Reid

---

[33] Nickelsburg, *Commentary*, 391.
[34] S. B. Reid, "The Structure of the Ten Week Apocalypse and the Book of Dream Visions," *JSJ* 16 (1985): 195-199.

names the section of 89:15-67 "Midrash on Ezekiel 34."³⁵ Nickelsburg also recognizes that Ezekiel 34 is the most obvious source for the negligent shepherds in *An. Apoc.* Yet he notices that the seeds of the idea of God delivering the flock into the hands of the destructive shepherd in *An. Apoc.* are found in Zechariah 11. In particular, Zech 11:15-17 offers a parallel; God speaks of appointing a destructive shepherd, but this is immediately followed by a proclamation of woe against the worthless shepherd who deserts the flock.³⁶ Both aspects of the Enochic picture are present.

Apparently, *An. Apoc.* is aware of all of the elements as well as the development of the Davidic Shepherd tradition in the OT. Even the figure of 'the Shepherd as the Breaker' of Mic 2:12-13 is echoed in 1 *En.* 90:18 and 'the eschatological battle' motif by the re-gathered flock in Mic 4:11-13 is alluded to in 90:19 as well. *Animal Apococalypse*'s reception and reinterpretation of the OT Davidic tradition contains its own theological reflections as the seer reviews the past, present, and foresees the future of the nation's history in view of the Lord of the sheep reigning over the nations.

Returning to Ezekiel's vision, the accusation is directed mainly at the wicked shepherds of Israel. Zechariah's modification of Ezekiel's Davidic Shepherd tradition is distinct in its emphasis on the hopelessness of the flock themselves, while his modification introduces the idea that the Lord hands over the flock to the wicked shepherd who will be afflicted later. In *An. Apoc.*, we observe that the blame for Israel's exile is placed on the shepherds, the disobedient and cruel shepherds to whom the Lord handed over the flock. In this regard, Reid is accurate when he states that "the structural key is the midrash on Ezekiel 34 (1 *En.* 89:15-67); the reason for the suffering of the community now becomes not its sins [of the flock] but rather the excesses of the rulers or shepherds."³⁷

However, Reid overlooks not only the relevance of Zechariah 11 but also the 'angelic' nature of the shepherds of *An. Apoc.* Indeed, the seventy shepherds are called on the carpet for the final judgment in a manner similar to Zechariah's afflicted shepherds. However, *An. Apoc.*'s seventy shepherds are not Israel's leaders but the heavenly angels in charge of Israel in exile just as the fallen Watchers are held accountable for the age that predates Israel.

It appears that *An. Apoc.* 89:59-90:39 is a midrash on Zechariah 11-13 rather than on Ezekiel 34. Note that Ezekiel 34 lacks the point that the Lord hands over the flock to the bad shepherds (89:59) and the consequential afflictions endured by the bad shepherds. To detect allusions to Ezekiel 34 in 1 *En.* 89:59-90:39, thus, would seem fitting only if Zechariah 11-13 itself can be seen as a modification of Ezekiel's Davidic Shepherd tradition. Basically, Zechariah finds

---

³⁵ Ibid., 199.
³⁶ Nickelsburg, *Commentary*, 390-391 (Excursus on "The Biblical Sources of the Idea of the Negligent Shepherds").
³⁷ Reid, "Structure," 200.

fault with the flock who detest the shepherd, resulting in the Lord handing them over to bad shepherds who end up being afflicted. To take up what is left by Zechariah's reading of Ezekiel 34, *An. Apoc.* accepts that the Lord handed the flock over to the bad shepherds. As in Zechariah's vision, it is accepted that the restoration would come as these shepherds are afflicted, though now it is said that the shepherds are mostly responsible for the destruction of the nation. More importantly, they are not historical figures but angelic powers.

In this regard, *An. Apoc.* departs radically from the OT Davidic Shepherd tradition as it finds outside Israel's historical arena the solution for the exile. Morever, in the context of the eschatological drama, the Lord's judgment of casting the angelic shepherds into the fiery abyss (90:25) in *An. Apoc.* differs from the afflictions of the shepherds found in Zech 11:7; 12:10; 13:7. The latter case coincides with the cleansing of the sins of the flock in Jerusalem (Zech 13:1) while the former is merely a step away from the transformation of the sheep initiated by the appearance of the eschatological white bull (1 *En.* 90:37).

Unlike what we find in Zechariah's and Ezekiel's visions, it is characteristic to find in *An. Apoc.* a separation in terms of the roles played by the shepherd imagery [angelic] and the messianic figure [bull = human]. The image of Zechariah's afflicted shepherds is carefully set aside in the instance of the seventy angelic shepherds destined for the abyss in order to avoid confusion with any leader figure in the history of Israel, both in the past and in the *eschaton*. Reserving the shepherd imagery for this select purpose in the Vision is supported also by the Vision's picture of the Lord of the sheep. When *An. Apoc.* uses the term 'shepherd,' it reveals its awareness of the heavenly powers behind the horrible mystery of Israel's exile. There is no hope to be found among the flock nor among the leaders of the nation. Blame is assigned neither to the flock nor to the leaders of the nation, but to the angelic powers behind their tragic history. Therefore, *An. Apoc.* envisions what lies behind and beyond the history of Israel to fulfill YHWH's promise to end their exile.

2.1.3 The Lord of the Sheep and the Restoration

Strikingly, the Lord of the sheep, frequently referred to in the OT as the 'shepherd' of Israel, is never addressed by that title in *An. Apoc.* As noted earlier, this is typical, for no historical leadership figure in Israel is ever called [under-] 'shepherd.' The term shepherd is singularly reserved for accusation and destruction similar to what is found in Zechariah 9-11, for the Lord of the sheep and his flock are reconciled only after the [angelic] shepherds are cast into the abyss.

The Lord of the sheep does not have any direct role in the destruction of the sheep, thus preserving his justice. By condemning the angelic shepherds for their excessive destruction of the flock, *An. Apoc.* succeeds in justifying the Lord of the sheep without blame for the misery of the exile. Moreover, by removing the shepherd's positive role in the historical realm, *An. Apoc.* keeps the role of the Lord of the shepherds in the heavenly realm and later at the *eschaton*. Only the Lord of the sheep, however, is expected to settle the score. Even the role of the eschatological white bull is so minimized that at the end of the drama, the Lord of the sheep takes the major part.[38]

The title, 'the Lord of the sheep,' appears along with the sheep in the Vision for the first time in the scene of the rescue of the flock in the Exodus: "the Lord of the sheep descended at their entreatment from a lofty place, arriving to visit them" (89:16). He not only rescues but also leads; as the sea is parted, 'their Lord [God as the shepherd of the sheep]' as 'their leader' stands between the flock and the wolves [Egyptians] (89:25). God, the 'owner' of the sheep, as Tiller renders it, owns and leads the flock by himself: "but the Lord of the sheep went with them *as their leader*, while all his sheep were following him" (89:22).[39]

This picture fits the characteristic of YHWH as Israel's shepherd/leader particularly in the Exodus (cf. Exod 3:10, 11, 17; 6:6, 13, 26; 7:5; 12:42; 17:3; 23:23; Num 20:5; Deut 4:38). In *An. Apoc.*, however, the Lord of the sheep withdrew into the heavenly realm when the sheep blindly wandered in the wilderness only to be devoured by beasts and birds (89:54-90:17). It is only at the end of the excessive destruction of the flock by their entrusted shepherds that the Lord of the sheep descends to visit again and leads the sheep:

I kept seeing until the Lord of the sheep came unto them [the shepherds?][40] and took in his hand the rod of his wrath and smote the earth; and all the beasts and all the birds of the heaven fell down from the midst of those sheep and were swallowed up in the earth, and it was covered upon them. Then I saw that a great sword was given to the sheep; and the sheep proceeded against all the beasts of the field in order to kill them and all the beasts and birds of heaven fled from before their face (1 *En.* 90:18-19).

This eschatological visit of the Lord of the sheep in verses 18-19 is followed by scenes of the last judgment (vv. 20-27), the building of a new house [temple] (vv. 28-29), the return of the sheep, with their eyes opened, into the house (vv. 30-36), and the final transformation of Israelites and the nations into 'white bulls' (vv. 37-39).[41] It is noteworthy that the appearance of the eschatological

---

[38] Only one verse addresses the role of this significant messianic figure (90:37).

[39] Tiller, *1 Enoch*, 280 (italics mine); he renders מר (4QEn^c 4:4) as 'owner,' which also means 'Lord.'

[40] It is not clear to whom 'them' refers. Tiller, *1 Enoch*, 364, thinks 'the sheep,' while Nickelsburg thinks, *Commentary*, 400, 'the Gentile nations.' Yet in the preceding verse (17), the subject is 'the twelve shepherds' and the object is 'their [the twelve shepherds'] predecessors.'

[41] J. A. Goldstein, "How the Authors of 1 and 2 Maccabees Treated the 'Messianic' Promises," in *Judaisms and Their Messiahs at the Turn of the Era* (ed. Jacob Neusner, William S. Green and Earnest S. Frerichs; Cambridge: Cambridge University Press, 1987), 72-73, thinks this figure is generally interpreted as a symbol for the Messiah.

Shepherd – here, the Lord of the sheep – to rescue the sheep precedes the coming of "the white bull," that is, the messianic figure at the eschaton in the Vision. This pattern of the eschatological drama is characteristic to Ezekiel's vision.

Among the concluding activities of the Lord of the sheep during his eschatological visit, the most obvious parallels to the Davidic Shepherd tradition are found in verses 18-19. The Lord of the sheep appears with 'the rod' of his wrath. Nickelsburg comments that here 'rod' or 'staff' (*batr*) denotes both a shepherd's implement (Ezek 20:37; Mic 7:14; Ps 23:4) and a royal scepter (Isa 11:4; Esth 4:11; Ps 2:9; Ep Jer 14 [13 LXX]), and the phrase 'the rod of my/his anger' is used twice in Scripture of God's judicial activity (Isa 10:5; Lam 3:1).[42] This clearly indicates that the Lord of the sheep is depicted by the languages of the shepherd imagery particularly developed in the Davidic Shepherd tradition. Shepherd imagery invested with authority and power is common to the passages depicting the eschatological battle, the outbreak of YHWH's force, and likewise with his people evident in such passages as Ps 2:7-9, Mic 2:12-13, and Mic 4:11-13 (cf. Zech 12:7-9).

The 'rod' or 'staff' in the hand of the Lord of the sheep in *An. Apoc.* 90:18 helps to identify him as the one and only shepherd of Israel. Yet in Ps 2:7-9, it is the Lord's Anointed who dashes the nations like pottery with the iron scepter. Likewise, it is the Lord's Appointee, the Branch from the stump of Jesse (Isa 11:1), who will "strike the earth with the rod (שֵׁבֶט; LXX, λόγος) of his mouth; with the breath (רוּחַ; LXX, πνεῦμα) of his lips he will slay the wicked" (v. 4). In 1 *En.* 90:18, however, it is the Lord of the sheep himself who takes the staff in his hand and strikes the earth. The messianic figure (90:37-38) is not engaged in this battle, and only appears after the temple is rebuilt (90:28-36).

The scene depicted in 1 *En.* 90:19 probably finds its closest parallel in Mic 4:11-13, though Mic 2:12-13 describes another aspect of the same eschatological outbreak of YHWH's force against the nations. The question of who participates in this final battle nets a variety of answers in the OT passages. In Ps 2:9, it is only YHWH's Anointed. In Mic 2:11-12, both the returning King of Zion and the re-gathered people advance to the battle together. In Mic 4:11-13 – the passage known to be the only place where the 'summons to battle' for the saints in all the occurrences in the OT prophetic texts[43] – the restored flock defeat the hostile nations. The flock's combat on the battle field in Mic 4:11-13 is particularly vivid and unambiguous in 1 *En.* 90:19; a "large sword" is given to the sheep. Nowhere in the OT Davidic Shepherd tradition is it explicitly mentioned that the sword will be given to the people to battle against the nations at the *eschaton*.

---

[42] Nickelsburg, *Commentary*, 401.
[43] May, *Micah*, 108.

There is a tendency in the OT to view the rod or the staff of YHWH or of his appointee as eventually transforming into or becoming something other than a military weapon, i.e., principles of the reign such as 'goodness'(טוֹב) and 'faithfulness'(חֶסֶד) in Ps 23: 4, 6 and 'favor' and 'union' in Zech 11:7. The LXX's rendering of 'the rod (שֵׁבֶט) of his mouth' in Isa 11:4 as 'the word (λόγος) of his mouth' is also intriguing. This may well illuminate the LXX's less militant view of the eschatological battle such as found in Mic 4:13 and Ps 2:9. But this tendency is clearly denied in 1 *En.* 90:18-19 in the demonstration that the rod for the Lord of the sheep and the sword for the sheep are indeed military weapons.

It is noteworthy, nevertheless, that the messianic figure in 1 *En.* 90:37-38 is exempt from this bloody eschatological battle where even the sheep engage in combat wielding the sword the Lord gave them. This messianic figure does not lead the flock into this carnage, for he arrives *only after* the sheep have lay down their sword (90:34). This procedure, not necessarily the detailed contents, fits the pattern in Ezekiel 34, 37 where the Davidic Shepherd arrives only after the eschatological shepherd fights to rescue the flock from the mouths of the evil shepherds.

2.1.4 The Eschatological White Bull and the Final Transformation

Whether or not there is any messianic expectation in *An. Apoc.* has been the subject of inquiry. Collins states that "there is little role for a messiah in this apocalypse."[44] This statement may or may not be correct, and demands explanation. The reason why Collins sees the messianic figure this way is that he cannot find much of the warrior image, viz., military activities, ruling and judging, in the eschatological white bull in 1 *En.* 90:37. As we noted above, the warrior role is assigned to the Lord of the sheep and to the sheep themselves. Then it is no wonder why we see this messianic figure exempt from any engagement in war or involvement in any bloodbath. Yet this is not to discredit it as non-messianic.

The vision of *An. Apoc* 89-90 does conform nicely to the Ezekiel pattern in chapter 34 and also in chapter 37 where the vision of chapter 34 is recapitulated but with a different angle. In *An. Apoc.*'s vision, the Lord of the sheep rescues the sheep from the nations (90:18; cf. Ezek 34:11-16) and judges the evil shepherds (90:25; cf. Ezek 34:1-10). The house is rebuilt as the Lord of the

---

[44] Collins, *Scepter*, 34; likewise, R. H. Charles, *The Book of Enoch or 1 Enoch* (Mokelumne Hill, Calif.: Health Research, 1964), 215, claimed that "he [the end-time white bull in 90:37-38] is not really the prophetic Messiah; for he has absolutely no function to perform .... and the Messiah-hope must be regarded as practically dead at this period."

sheep dwells in it (90:29, 34; cf. Ezek 34:30).⁴⁵ The sheep return with their eyes opened (90:29, 35; cf. Ezek 34:11-13, 16), and then the eschatological white bull arrives to transform Israel and nations alike (90:37; cf. Ezek 34:23-24). As Nickelsburg properly points out, the seer's description of the sheep in exile described in 90:2-4 (cf. 89:75) alludes to Ezek 37: "and I saw until those sheep were devoured by the dogs and by the eagles and by the kites; and they left them neither flesh nor skin nor sinew, until only their bones remained; and their bones fell on the earth, and the sheep became few" (v. 4).⁴⁶ As the vision of the dry bones in Ezek 37:1-14 is concluded by the reference to the one Davidic Shepherd in Ezek 37:24-25, so is the gruesome sketch of the exilic situation in 90:2-4 followed by the arrival of the messianic figure, the snow-white bull in 90:37-39.

Certainly, the characteristics of the messianic figure in both visions are quite similar. The Davidic Shepherd is referred to as מֶלֶךְ 'over' the flock (Ezek 37:24; cf. 34:23); in a similar vein, the end-time white bull in *An. Apoc.* presents as a figure of authority in that 'all the beasts' and 'all the birds of the sky' fear and make petition to him (1 *En.* 90:37). Neither Ezekiel's Davidic Shepherd/נָשִׂיא, nor *An. Apoc.*'s eschatological white bull wage war, but are nevertheless imposing figures vested with authority.⁴⁷

More significantly, they perform another aspect of the eschatological salvation eventually extended to the nations. After the eschatological battle, the next step in Ezekiel's vision is to appoint the Davidic Shepherd over the restored flock to rule and to lead them according to the laws and decrees of YHWH (Ezek 34:23; 36:27; 37:24). In contrast, the messianic white bull transforms both sheep and beasts into the white bulls. This transformation is in keeping with the Davidic Shepherd, that is, the role 'to sanctify' the rescued flock according to the laws and decrees of the Lord – a sanctifying salvation eventually extended to the nations (Mic 5:1-5a). Thus the color 'white' of the transformed sheep and beasts symbolizes the sanctified new humanity in the new age.

Concerning the identity of this eschatological white bull, the Ethiopic word *nagar* ['word, thing'] in 1 *En.* 90:38, referring to what the white bull was or was to become [Eth. *kona*], has long been recognized as a textual corruption.⁴⁸ While

---

⁴⁵ 1 *Enoch* 90:34 suggests that God dwells 'in that house' where no traditional cult is necessary due to God's presence and the full and permanent purification of the sins of the human race. See Nickelsburg, *Commentary*, 405. Thus here it differs from the vision of Ezek 40-48, which calls for the cult and the mandatory sacrificial offerings.

⁴⁶ Nickelsburg, *Commentary*, 359, 371.

⁴⁷ Nickelsburg, *Commentary*, 407-408.

⁴⁸ Tiller, *1 Enoch*, 385, reads it as "the first one became *a thing* among them." He further notes that with various MSS in view it is not clear where the phrase 'the first' belongs, though most likely it is with the following clause as a reference to the white bull in v. 37.

some read *nagar* to mean a 'thing,'[49] various emendations are supported by 'word' which appears more intelligible in the text. If this is correct, the Ethiopic word may be the rendition of the following possible Hebrew terms: (i) the Hebrew ראם ['buffalo'/'Stier'] transliterated into Greek as ρημ which is again read by the Ethiopic translators as ῥῆμα;[50] (ii) the Hebrew טלה ['lamb'] misread by the Greek translators as מלה ['word'];[51] (iii) the Aramaic אמר ['lamb'] read by Greek translators as אמר ['word,' 'speech'] and translated as λόγος which is translated literally into Ethiopic;[52] and (iv) the Aramaic דבר ['leader'] read in Hebraizing form as 'word.'[53]

The major weakness of the first option is the lack of a compelling reason for the transliteration of ראם ['buffalo'] into Greek ρημ rather than a simple translation.[54] The second option seems unlikely because it assumes the more obvious and expected word, i.e., טלה ['lamb'], taken to be the less expected one, מלה ['word'], twice in the same sentence.[55] The third option falters because of the weak assumption that the white bull turns into a lamb that represents an Israelite, particularly at that eschatological moment when both Israel and the nations are transformed and transferred back to the *Urzeit* before history, to a time when "there is no longer any Israel."[56]

According to Barnabas Lindars, the Aramaic term אמר ['lamb'] was intentionally rendered by the Greek translator(s) as λόγος to make the vision of a prophecy of Christ, and he argues that this may be the reason why the Jewish apocalyptic books were valued by Christians.[57] Lindars even suggests 'an allusion to the Davidic tradition' seeing that David is a lamb (1 *En.* lxxxix. 45f., Gr. text ἄρνα)."[58] Yet as we have already pointed out, David is pictured as a sheep/ram and not as a shepherd in *An. Apoc*. David being a ram or even a lamb alone in *An. Apoc*. does not necessarily mean this is an allusion to the Davidic tradition.

Rather, as we have argued, the fact that David is appointed 'ruler and *leader*' (ἄρχων καὶ ἡγούμενος) in 1 *En.* 89:46 suggests a possible connotation of

---

[49] E. Issac, *OTP* 1:71 renders it 'something'; Tiller, *1 Enoch*, 383.
[50] Sigbert Uhlig, "Das äthiopische Henochbuch," in *Apokalypsen* (*JSHRZ* 5.6; Gütersloher Verlagshaus Gerd Mohn, 1984), 704, likewise renders *nagar* as "Stier."
[51] L. Goldschmidt, *Das Buch Henoch* (Berlin, 1892), 90-92 informed by Barnabas Lindas, "A Bull, a Lamb and a Word: 1 Enoch XC. 38," *NTS* 22 (1976): 484; also, later R. H. Charles, *1 Enoch* (1912), 216.
[52] Lindars, "Bull," 484.
[53] Nickelsburg, *Commentary*, 403 (misprinted as דרב; Refer to דבר ["to lead"] in 1 *En.* 89:14).
[54] Tiller, *1 Enoch*, 386.
[55] Lindars, 484.
[56] Tiller, 387.
[57] Lindars, "Bull," 485-486. Illinois
[58] Ibid., 485.

Davidic expectation (cf. 11QPs³ 28:11). If this internal evidence within the Vision is taken into consideration, Nickelsburg's emendation of the Aramaic רבר ['leader'] may bear more weight than Lindars' option. Nickelsburg's option handily satisfies the sense of the emendation. Tiller doubts that the Aramaic רבר is not well attested,[59] but Nickelsburg supplies some evidence such as 4QEnᵉ 4 2:16; *b. Sanh.* 8a. in his recent commentary. Furthermore, Nickelsburg argues that the suggested textual emendation in verse 38 points to Ezek 34:24: My servant David will be prince *among them* (נשיא בתוכם ועבדי דוד); the wording approximates verse 38, "became <leader> *among them.*"[60] A few more attestations of the Aramaic רבר ['leader'] would leave no other reason but to accept this option as conclusive to the matter.

That the end-time white bull is called the 'leader in the midst of them,' certainly verifies the relevance of the Davidic Shepherd tradition to the Vision of *An. Apoc.* Not only the structure of the end-time events, but also the characteristics of the figure of the Lord's appointee are found to be similar; both Davidic Shepherd and white bull are *the 'leaders' in the midst of the restored sheep.* The phrase, בתוכם ['in their midst'] defines the kind of leader in the context of Ezekiel 34 and 37. This is the prince-type leader who leads among but not necessarily over them. In Ezekiel 34-37, the Davidic Shepherd is appointed to lead the rescued sheep to follow the laws and decrees of YHWH – that is, the process of sanctification! Likewise, the end-time white bull initiates the process of transforming the rescued sheep into white bulls in order to lead a new sanctified humanity free of any further defilement.

If compared to Ezekiel's Davidic Shepherd, the only difference worthy of note is the absence of commentary on the Mosaic law related to the role of the end-time white bull. For instance, in 1 *En.*, while the law is mentioned, it is hardly identified or associated with the Mosaic law (5:4; 63:12; 79:1-2; 99:2; 104:10); the term Torah is nearly absent.[61] Instead, in *An. Apoc.*, the blindness and openness of the eyes of the sheep are the prime indicators relaying to the readers the nature of their relationship with the Lord. For instance, their eyes were opened to see the Lord leading their Exodus before the Mosaic law was given (89:28).

Along with *An. Apoc.*'s decision to refrain from calling any Israelite leader a shepherd, the seer seems to consider the history of Israel as beyond repair, for obedience to the laws and decrees alone could not suffice. The Mosaic ideal and hope seem to have been discarded and replaced by some supernatural and

---

[59] Tiller, *1 Enoch*, 388.
[60] Nickelsburg, *Commentary*, 406 (italics mine).
[61] Eckhard J. Schnabel, *Law and Wisdom from Ben Sire to Paul: A Tradition Historical Enquiry into the Relation of Law, Wisdom and Ethics* (WUNT 16; Tübingen: Mohr Siebeck, 1985). 107, 112, notes "the law per se occurs rather seldom and signifies first of all the comprehensive order of creation."

mystical interpretation of Israel's history, coupled with an expectation of the Lord's intervention. The mystical revelatory vision is much more important for the sanctification of the flock.[62] Perhaps this fits that Enoch, born before Moses, is the chosen seer. The ideal presented in the Vision is not only to restore the Deuteronomic blessings, but even more significantly, signals a return to the primordial righteousness and perfection that looks back before the age of Moses to the age of Adam.[63]

The symbolism of the white bull indicates also that this figure belongs to those seen as white bulls and not as sheep/rams, such as Abram or Isaac (89:10, 11), Noah or Shem (89:1, 9), Seth (85:9), and Adam (85:3). Thus one can regard the end-time white bull as an eschatological counterpart to one of these specific figures, though the role of transforming entire peoples and not exclusively Israelites means that eventually this white bull can be seen as a new Adam. Yet, even in the OT Davidic Shepherd tradition, the Davidic Shepherd figure is in touch with this universal dimension of his reign. The rulership of the shepherd will not be restricted within Israel, but will extend even to עַד־אַפְסֵי־אָרֶץ ("to the end of the earth," Mic 5:4b), and as it reaches to the ends of the earth, there will come שָׁלוֹם (v. 5; v. 4b LXX; cf. Zech 9:10; 12:6-9; 14:7).

Particularly at the end of Ezekiel's description of the age, when the Davidic Shepherd arrives in the midst of the restored flock, the prophet records in 34:31: "You are my flock, flock of my pasture, people [הָאָדָם] you" (cf. Jer 23:1); here הָאָדָם may demonstrate that the vision does indeed go beyond Israel if הָאָדָם represents mankind.[64] Both in the OT Davidic Shepherd tradition and *An. Apoc.*, the pen of the flock of the messianic figure is to expand to include the nations.

2.1.5 Summary

The *Animal Apocalypse*, probably triggered by the Maccabean revolt, is a result of the seer's creative interactions with Zechariah 9-11, Ezekiel 34-37, and Micah 2-5. In this respect, Bryan's analysis, based primarily on the kosher legislation found in Leviticus and Deuteronomy, overlooks *An. Apoc.*'s intense interactions with the Davidic Shepherd tradition as he claims that the antitheti-

---

[62] Markus N. A. Bockmuehl, *Revelation and Mystery: In Ancient Judaism and Pauline Christianity* (Grand Rapids Mich.: Eerdmans, 1997), 35-36, names Enoch's revelation as 'eschatological mysteries' which underscores not their hiddenness but their belonging to the heavenly world as God's pre-existent salvific design for the eschaton, thus, 'soteriological mysteries.'

[63] Nickelsburg, *Commentary*, 407, compares the the analogy to the Apostle Paul's theology of the two Adams.

[64] For the reading of הָאָדָם limited to Israel, see *B. Yebad m.* 61a; *b. B. Mes.* 114b; *b. Ker.* 6b. See W. Zimmerli, *Ezekiel* 2:221.

cal relation of Israel (clean) with the nations (unclean) is resolved by the return to the past separation not the eschatological union between them.[65]

The continuation of the Davidic Shepherd tradition in the Vision is apparent especially in view of the overall structure as to the history of Israel up to the *eschaton*. The sins of both sheep and shepherds lead to exile, and contribute to the deteriorating conditions. The sovereign Lord is the eschatological shepherd of Israel, who will visit the earth to rescue the sheep from the hands of the evil shepherds and from the mouths of the nations. Following the carnage of the end-time battle is the arrival of the Lord's appointee as the end-time messianic figure to save them from this age and to guide them into a new epoch of peace that extends to the nations. His role is characteristic of sanctification of the rescued. Thus the primordial age of hostile tension between Israel and the nations is resolved forever. The nations are finally taken into the pen of the Lord of the sheep [of Israel] at the end. These themes are common both in *An. Apoc.* and the OT Davidic Shepherd tradition.

On the other hand, unique features are added by *An. Apoc.* to this age-old tradition. One feature that stands out in *An. Apoc.* is the fact that the shepherd in the Vision does not refer to the Lord nor any of Israel's leaders. While the functional descriptions of the shepherd are attributed both to the Lord and such leading figures as Moses, David, and Solomon, the term shepherd is squarely reserved for the *angelic* powers operating behind the historical scenes of the tragic exile.

This approach is innovative. The shepherds to be judged by the Lord of the sheep are actually powers in the heavenly realm. Since both sheep and beasts are controlled by these angelic powers, the blame for the tragedy is ultimately laid upon them. One may conclude that the afflicted shepherd imagery of Zechariah 9-11 finds its resolution in *An. Apoc.* The judgment of the shepherds required for the restoration in Zechariah 9-11, is finally executed as Enoch envisions the angelic shepherds being thrown into the abyss. Yet in *An. Apoc.*, the shepherds are not judged because of 'the detest of the sheep against the good shepherd' as in Zechariah 9-11, but are judged because of their disobedience to the Lord who sets the limits of destruction. The mystical elements in the affliction of the shepherd figure in Zechariah 9-11 evaporate with the placement of angelic shepherds on the seat of the accused in the eschatological court of the Lord.

Another unique feature of *An. Apoc.* is more thoroughgoing in its inclusion of the nations. The salvation to be wrought by the eschatological white bull will lead Israel beyond its own history. In this sense, the white bull plays the role of a new Adam rather than a new David. Although the Davidic hope is significantly modified in Ezekiel's vision from a military figure to a prince who would lead the people to obey the Lord, the Davidic Shepherd takes upon himself the

---

[65] Bryan, *Kosher Mentality*, 168-185 ("The Animal Apocalypse Read in the Light of the Kosher Mentality").

role of the sanctification of primarily the rescued Israel (Ezek 34:23-24), but only implicitly the nations as well (Mic 5:1-4). The *Animal Apocalypse* 90:18 preserves the military connotations of the staff/rod of the shepherd but fails to consider the later development of the non-violent usage of the staff-metaphor in the Davidic Shepherd tradition (Isa 11:1-4; Ezek 36:15-23; Zech 10:4-17). Rather, in *An. Apoc.* 90:19, the sheep clearly participate in the battle with the great sword, which is unclear in the OT Davidic Shepherd tradition (Mic 12:12-13, Mic 4:11-13 [cf. Zech 12:7-9]). Yet the bloody battle belongs to the Lord and his sheep, and not to the messiah. The white bull transforms or sanctifies both sheep and beasts into white bulls, the eschatological holy people of God.

Strikingly, there is no mention of Mosaic law to be fulfilled, and it appears that Kosher legislation will be done away with in the coming new age, indicating that both the clean and the unclean will be transformed as holy unto the Lord. What is only implicit in the extended pacifistic reign of the OT Davidic Shepherd is explicit in the role of the white bull. Yet the text of 1 *En.* 90:38 retains the echo of Ezekiel's vision of the new David, the prince who is found in the midst of restored Israel.

Lastly, the sin of the sheep is defined in terms of their ability or inability to gaze at the Lord. Eyesight or vision seems to be the more critical element by which to judge their apostasy than their disobedience to the laws and decrees of YHWH through Moses. The *Animal Apocalypse* rarely displays a harsh attitude toward the sins of Israel committed against the Law unlike the stereo-type rebuke of the people's sins exemplified in 4 Ezra 1:7-8: "Pull out the hair of your [Ezra] head and hurl all evils upon them [Israel]; for they have not obeyed my law; they are a rebellious people." The most recognized cause of the suffering in *An. Apoc.* is not the blindness of the sheep but the uncontrollable angelic powers, the shepherds. In this regard, *An. Apoc.* is shortsighted in its understanding of Israel's plight, sins, and hatred for the shepherd YHWH, the Lord of the sheep, particularly emphasized in Zechariah's interaction with Ezekiel's vision.

As an apocalypse, *An. Apoc.* presents a solution for Israel's exile and theodicy in view of the Lord's sovereignty, and this is accomplished by the use of in-depth interaction with shepherd imagery from the OT Davidic Shepherd tradition. In contrast to *An. Apoc.*, we encounter another resource for the modification of the OT Davidic Shepherd tradition: *Psalms of Solomon*. In this text, we find a hope deeply rooted in the "historical" rather than apocalyptic hope of Israel's restoration.

2.2 The *Psalms of Solomon* 17:1-18:12

Three integral themes of the OT Davidic Shepherd tradition, namely, Davidic expectation, shepherd imagery, and the restoration of Israel intertwine and interplay with one another in *Ps. Sol.* 17-18. With regard to Davidic expectation, the author(s) of *Ps. Sol* 17 recalls Davidic expectation (2 Sam 7:14) in v. 4 as he weaves this Davidic motif with various shepherd imageries such as the 'iron rod' motif [with the 'word of his mouth'] (17: 24, 35; 18:7; cf. Ps 2:7-9; Isa 11:2-5) while presenting the shepherd's 'compassion' as the fundamental motivation for Israel's promised future (17:45; 18:1,3,9; cf. 17:3). The hope of Israel's restoration is the immediate context of the author who modifies Davidic expectation by means of shepherd imageries. It is likely that these 'communal laments' or 'complaints' expressed in *Ps. Sol.* 17-18 have as their chief concern how to reestablish leadership over the restored community.[66]

The echo of accusations leveled at the Hasmonean dynasty as an illegitimate non-Davidic leadership cannot be missed (17:4),[67] though more than its illegitimacy is in sight. What might be more properly understood is the nature of the Maccabean regime: it falls miserably short of the expected promised time of the Davidic Shepherd, that is, "in the time known to you [Lord]" (17:21); "in those days" (17:44; 18:6) or "in the days of mercy" (18:9).

According to *Ps. Sol.* 17-18, two distinguishing characteristics of the time of the Davidic Shepherd's regime most likely are: (i) its righteousness, i.e., more truth- than power-encounter evidenced by extensive modifications of the rod [of iron] metaphor; and (ii) a universal appeal that involves the submission of the nations under the reign of the Davidic Shepherd.

---

[66] K. E. Pomykala, *The Davidic Dynasty Tradition in Early Judaism: Its History and Significance for Messianism* (SBL Early Judaism and Its Literature 7; Atlanta: Scholars Press, 1995),160.

[67] The majority of scholars agree that *Psalms of Solomon,* presumably composed by Pharisaic writer(s), reflects such historical events as the rise of the Hasmoneans, Pompey's seizure of Jerusalem in 63 B.C., Aritstobulus II's exile to Rome, the puppet kingdom of Hyrcanus II, and the death of Pompey in Egypt in 48 B.C. See Collins, *Scepter*, 49-51; likewise, Svend Holm-Nielsen, "Die Psalmen Salomos," in *Poetische Schriften (JSHRZ* 4.2; Gütersloher Verlagshaus, 1977), 58-60; G. L. Davenport, "The 'Anointed of the Lord' in Psalms of Solomon 17," in *Ideal Figures in Ancient Judaism* (ed. J. J. Collins and G. W. Nickelsburg; Chico, Calif.: Scholars, 1980), 71; Gerbern S. Oegema, *The Anointed and his People* (JSPSup 27; Sheffield: Sheffield Academic Press, 1998), 103-105. Recently K. R. Atkinson, "On the Use of Scripture in the Development of Militant Davidic Messianism at Qumran: New Light from *Psalm of Solomon* 17," in *The Interpretations of Scripture in Early Judaism and Christianity: Studies in Language and Tradition* (ed. Craig A. Evans; Sheffield: Sheffield Academic Press, 2000), 106-123, which is also incorporated in his book, *An Intertextual Study of the Psalms of Solomon* (Studies in the Bible and Early Christianity 49; Lewiston: The Edwin Mellen Press, 2001), 361-368, claims that Herod's seizure of Jerusalem in 37 B.C. was recounted in *Ps. Sol.* 17:11-20, and the rise of Herod the Great is the 'first-hand' effect of the psalmist's community that believed "the Davidic messiah would rise up to overthrow Herod in battle " (377). As Pomykala, *Davidic Dynasty*, 159, points out, however, the psalmist makes no reference to Alexander's revolt that occurred in 57 B.C. Cf. Joseph L. Trafton, "The Psalms of Solomon in Recent Research," *JSP* 12 (1994): 3-19.

On both of these points, the Maccabeans utterly fail in the eyes of the author responsible for the composition of *Ps. Sol.* 17-18. If compared to *An. Apoc.* in which the blame falls upon the angelic shepherds acting behind the scenes of the historical realm, what we have here, in contrast, is significantly down-to-earth, politically minded author who announces the Lord's rebuke to and invokes judgment unto his contemporary political/religious shepherds of Israel.

2.2.1 The Davidic Shepherd as Leader and Teacher

The psalmist recalls the Lord's promise of the Davidic throne (2 Sam 7:12-16) to present the alternative to his contemporary acclaimed shepherds "over Israel" (17:4). The illegitimate throne that "set up a monarchy because of their arrogance" (17:6) is justly overthrown by "a man alien to our race" (17:7) whom God raised by himself; Jerusalem with its temple is defiled by ὁ ἄνομος ("the lawless one," 17:11-14), and its people are scattered "over the whole earth" (17:18a).[68] As a consequence of the defiled temple and the Lord's absence the land becomes barren, infertile (17:18b-19a).[69] Particularly, *Ps. Sol.* 17:19 searches for the ultimate cause of the famine according to the Deuteronomic framework: "For there was no one among them who practiced righteousness or justice."[70] This is but the prelude to the arrival of the promised son of David (17:20-46).

This sequence typifies the address of Israel's plights made by Ezekiel concerning the restoration process in Ezekiel 34, 36 and 37. The sequence and language here echo Ezek 34:6b: "there is no one searching; there is no one seeking." Unlike Ezek 34:1-10 that highlights the faults of Israel's shepherds, here the psalmist delivers a guilty verdict – similar to Zechariah's intense reflections on the sins of the flock – aimed at all levels of Israelites: "For their leaders to the commonest of the people, [they were] in every kind of sin; the king was a criminal and the judge disobedient; [and] the people sinners" (17:19b-20). There is no hope.

Nevertheless, the psalmist refrains from turning to an apocalyptic vision beyond the historical scene as does the seer of *An. Apoc* in 1 *En.* 83-90. In reply to the imminent catastrophe, the down-to-earth politically minded psalmist turns

---

[68] Pompey, "a man alien to our race" (17:7b), his invasion (63 B.C.) and his later entrance into the holy of the holies (17:14) may be in view. Cf. Oegema, *The Anointed*, 104; Collins, *Scepter*, 49-51.

[69] Atkinson, *Intertextuality*, 366, identifies the famine described in *Ps. Sol.* 17:18-19 as the sabbatical year during Herod and Sosius's siege of Jerusalem (366; cf. Josephus, *Ant.*, 14.475).

[70] The *Psalms of Solomon* are preserved in both Greek and Syriac versions, but none of the *Ps. Sol.*'s surviving manuscripts date earlier than the tenth century A.D. Refer to Wright, *OTP* 2:640. The Greek text is taken from Atkinson's reconstruction, *Intertextuality*, 331; Atkinson often emends Rahlfs' edition of the LXX and Gebhardt's critical texts based on intertextual paralells (4); for further details as to the manuscripts, see Svend Holm-Nielsen, "Die Psalmen Salomos," 53-55.

to the OT promise of the eschatological though historical Davidic son: Ἰδέ κύριε καὶ ἀνάστησον αὐτοῖς τὸν βασιλέια αὐτῶν υἱὸν Δαυίδ εἰς τὸν καιρὸν ὃν ἴδες σὺ ὁ θεὸς τοῦ βασιλεῦσαι ἐπὶ Ἰσραὴλ παῖδα σοῦ ("See, Lord, and raise up for them their king, the son of David, in time known to you of God, to rule over your servant Israel," 17:21). Significantly, it is well known that this is "the only time in early Jewish literature that the phrase 'Son of David'(17:21) occurs with reference to a messianic figure."[71]

More important, the figure in the psalmist' mind approximates that of the OT Davidic *shepherd* who is "to smash the arrogance of sinners like a potter's jar; to shatter all their substance with an iron rod [ἐν ῥάβδῳ σιδηρᾷ]; to destroy the unlawful nations with the word of his mouth [ἐν λόγῳ στόματος αὐτοῦ]" (17:23b-24; cf. Ps 2:7-9; Isa 11:1-5). The people of Israel are scattered over the whole earth (17:18), and it is the figure who has the traits of the eschatological Shepherd, who, according to the psalmist, συνάξει λαὸν ἅγιον ("will gather a holy people," 17:26).

In Ezekiel's vison, the role of gathering is assigned to YHWH himself (34: 4, 12) while the Davidic Shepherd arrives only to lead the restored people to follow the laws and the decrees of the Lord (34:23-24), thus completing the return of the Lord's presence among them (37:26-28) with the restoration of the land and the blessings (34:25-31). In contrast, we find in *Ps. Sol.* 17 that the Son of David vested with the characteristics of the Shepherd executes both roles, though more emphasis is placed on the leading of rather than the rescuing of the flock.

Allusions to Ezekiel's Davidic Shepherd abound.[72] Particularly evident and dramatic is the echo of Ezek 34:16b-24 in *Ps. Sol.* 17:40b-42. The Davidic Shepherd will be faithfully and righteously *shepherding the Lord's flock* [ποιμαίνων τὸ ποίμνιον κυρίου] (*Ps. Sol.* 17:40b). The figure that is implied by these phrases, as Svend Holm-Nielsen suggests, "Messia als dem guten Hirten"; the relevance of Ezekiel 34 and 37:24 is explicit (cf. also, Mic 5:4).[73] This implies that his righteous judgment will be exercised among the flock as in Ezek 34:16b: "I will shepherd the flock with justice." As the psalmist describes, under the Shepherd's leadership "no sheep will stumble in their pasture" (v. 40b) because "no arrogance is found among them" and "any should be oppressed" (v. 41). This recalls the Davidic Shepherd who will practice justice within the rescued community in Ezek 34:20, 22: "See, I, myself will judge between the fat sheep and lean sheep"; "they will no longer be plundered."

---

[71] Pomykala, *Davidic Dynasty*, 162, cites D. C. Dulling's "The Therapeutic Son of David: An Element of Matthew's Christological Apologetic," *NTS* 24 (1978): 407-408.

[72] Atkinson, *Intertextuality*, 356-358, lists Ezek 34:23; 34:25; 17:24.

[73] Svend Holm-Nielsen, "Die Psalmen Salomos," 105, n. 40 (b).

In short, the psalmist sums up the role of the Davidic Shepherd: "he will *lead* [ἄξει] *all in righteousness/equity*" (17:41).[74] As the eschatological shepherd, the Son of David in *Ps. Sol* 17 is to 'lead them' – the eschatological community – 'to holiness'/'with equity' (ἐν ἰσότητι πάντας αὐτοὺς ἄξει, 17:42),[75] the similar task assigned to Ezekiel's Davidic Shepherd is to lead the restored flock "to follow my [YHWH's] laws and be careful to keep my decrees" (37:24). The role of the Shepherd in *Ps. Sol.* 17:42 parallels that of the eschatological white bull in *An. Apoc.* 90:17-18, who will transform both Israel and nations into one holy people. The Shepherd in *Ps. Sol.*17:42, however, is appointed primarily as a 'king over [the restored] Israel' as the Davidic Shepherd of Ezekiel's vision is placed 'over them [the restored Israel]' as 'one shepherd' (34:23); a 'king over them' (37:24) as again, 'one shepherd' (37:24).

Nevertheless, the roles of the eschatological Shepherds, YHWH and his Davidic Appointee in Ezekiel 34 and 37, are not clearly separated in *Ps. Sol.* 17. This may raise the question of the character of the Davidic Shepherd: Is he more a militant warrior or more a pacifistic messiah? Collins sees this Davidic Shepherd as 'undeniably violent' and considers him to be 'a warlike figure'[76] Likewise K. Atkinson supports the militant Davidic messiah in *Ps. Sol. 17*.[77] However, many do not think the psalmist's Davidic figure is militant. Rather they see him as an extensively modified peaceful righteous leader.[78]

Pomykala points out that throughout the plea (esp. 17:22-25) royal power is implemented not with military force but with the king's words (vv. 33, 35, 36b, 43); thus, he says, "while it is correct to view the Davidic messiah in political

---

[74] Schnabel, *Law and Wisdom*, 112-119, aptly states,"both Davidic King and the Messiah (in *Ps. Sol*.) are described as enemies of lawlessness, as aiming at establishing righteousness, and as possessing wisdom" (119): the figure in view does not bring "new law" (118), but stands as new Teacher.

[75] 'Holiness' is rendered by Wright, *OTP* 2:668; Atkinson, *Intertextuality*, emends it ἰσότης ['equity'] to parallel Ps 9:8; 67:4; 98:9. Yet these texts have nothing to do with the eschatological ἄξει. Furthermore, the LXX retains εὐθύτης ['uprightness'] in those Psalm texts in stead of ἰσότης; similarly, Svend Holm-Nielsen, "Die Psalmen Salomos," 105, renders it as "Ohne Unterschied"; the different renditions do not lessen the significance of the righteous character, goal, and effect of the Davidic Shepherd's eschatological leadership.

[76] Collins, *Scepter*, 55, assumes it without arguing for the case.

[77] Atkinson,"Militant Davidic Messianism," 106-123; also in *Intertextuality*, 333-378.

[78] The traditional view, predating Collins and Atkinson, has predominantly focused on the non-violent character of the Davidic messiah in *Ps. Sol*.17-18: J. Klausner, *The Messianic Idea in Israel* (London: Bradford and Dickens, 1956) 323; M. Delcor, "Psaumes de Salomon," in *Supplément au Dictionnaire de la Bible*, fasc. 48 (1973) 214-245; J. H. Charlesworth, "From Messianology to Christology: Problems and Prospects," in *The Messiah: Developments in Earliest Judaism and Christianity* (ed. J. Charlesworth with J. Brownson, M. T. Davis, S. J. Kraftchick, and A. F. Segal; Minneapolis: Fortress Press, 1992), 20; Chester, "Jewish Messianic Expectations," 28; J. D. Crossan, *The Historical Jesus* (San Francisco: Harper, 1991) 108: "And this messianic leader does not use violence, neither the actual violence of normal warfare nor the transcendental violence of angelic destruction."

terms, it is mistaken to understand him in military terms."⁷⁹ Davenport would agree: "The king will depend upon God, not upon human weapons and forces, and that his victories are to be through wisdom and the overwhelming authority of his word indicate that although God does indeed at times use violence, his lasting victories are won not by violence but by wisdom and manifestation of his glory."⁸⁰

The roles of the Davidic Shepherd laid out in *Ps. Sol.* 17-18 underlines his semi-divine qualities as the eschatological leader. To rule a great people, the psalmist says, the Davidic Shepherd will be καθαρὸς ἀπὸ ἁμαρτίας ("[will be] free from sin," 17:36), a quality necessary for carrying out his eschatological task of the sanctification of the eschatological community. The Lord will raise him over the house of Israel, indeed, "to *discipline* it" (παιδεῦσαι, 17:42). Atkinson maintains that the Davidic messiah is "a violent warrior" who yet "[will be] pure from sin,"⁸¹ which sounds obviously contradictory; how do we reconcile these two opposite natures of the Davidic figure? Atkinson advocates that the Davidic messiah is free from all blame since the messiah acted only on behalf of God who appoints him as his legal agent. The Davidic messiah is therefore "legally innocent" since he only follows God's orders "just like soldiers in today's military are considered innocent of their actions when they follow the orders of their commanders."⁸²

However, this proposal falters as one examines the context in which the phrase, καθαρὸς ἀπὸ ἁμαρτίας (17:36) occurs. Its immediate context, *Ps. Sol.* 17: 34b-37 explicitly displays a collection of the inner qualities in praise of the righteous Davidic messiah who "will not rely on horse and rider and bow" (v. 33): "compassionate" (ἐλεήσει, v. 34b), "wisdom and happiness" (v. 35b), "powerful in the holy spirit," "wise in the counseling of understanding", and "strength and righteousness" (v. 37b). The idea of 'legal innocence' hardly coheres with the view of the psalmist in this context. In fact, Collins notes that this is "a striking feature," and even refers to Ezek 36:25 as "the closest example" to this sin-free [Davidic] Shepherd qualified to lead the restored people! Collins, yet still is somewhat vague here: "It may be that the sinlessness of the king is required by the purity of the people in the eschatological age."⁸³

The role of the Davidic Shepherd in *Ps. Sol.* 17:23, if compared to Ezekiel's vision, overlaps with that of YHWH when the battle ensues to rescue the flock before he appoints over them a new Davidic Shepherd. Perhaps this is where the confusion starts. In the beginning the son of David has been portrayed as the Davidic king, but later he becomes the 'Leader and Teacher' not only of Israel

---

⁷⁹ Pomykala, *Davidic Dynasty*, 162.
⁸⁰ Davenport, "The Anointed," 83.
⁸¹ Atkinson, *Intertextuality*, 349.
⁸² Atkinson, *Intertextuality*, 349-350.
⁸³ Collins, *Scepter*, 55; Davenport, "The Anointed," 80, points out the link between the 'sinless messiah' and Ezekiel's vision, but not specifically Ezek 36:25.

but also of the nations.[84] A close examination of the use of the rod metaphor in *Ps. Sol* 17, especially v. 23, may help clarify the peaceful and didactic nature of the Davidic Shepherd.

2.2.2 The [Iron] Rod and the New Davidic Regime

The extended reign of the Davidic Shepherd over the nations in *Ps. Sol.* 17:26-46 is characterized by the psalmist's use of the 'rod' imagery of the shepherd-king. In *Ps. Sol.* 17 this imagery appears for the first time in this psalm in 23b-24. The 'iron' rod is often thought of as a physical weapon that suggests the militant nature of the son of David.[85] The expression found in 17:23b-24a, "to smash the arrogance of sinners *like a potter's jar* (ὡς σκεύη κεραμέως); to shatter all the substance *with an iron rod* (ἐν ῥάβδῳ σιδηρᾷ)," is one of the representative shepherd imageries for the Lord 's Anointed who will dash the nations like pottery "with the iron scepter" in Ps. 2:9 (בְּשֵׁבֶט בַּרְזֶל; LXX, ἐν ῥάβδῳ σιδηρᾷ; cf. Hos 6:5).[86]

In Isa 11:4b, we notice a tendency to modify the symbolism of this 'iron rod' metaphor; here the rod of the Anointed becomes the rod of 'his mouth' (פִּיו שֵׁבֶט). The same metaphor of 'the rod' in Ps 2:9 reoccurs in Isa 11:4b but with a critical modification. The LXX's rendition of Isa 11:4b, the 'word' (λόγος) of his (Davidic branch's) mouth, clarifies the intended modification by Isaiah's view of the 'rod' in Ps 2:9. Isa 11:4b has a formative influence on the conception of the Son of David in *Ps. Sol.* 17:21-25.[87] The psalmist here takes a step further by incorporating Isa 11:4b as it reads Ps 2:9 in the light of the LXX's rendition of Isa 11:4b, particularly in the parallelism between 17:23 (ἐν ῥάβδῳ σιδηρᾷ, Ps 2:9) and 17:24 (ἐν λόγῳ στόματος αὐτοῦ, Isa 11:4b [LXX]).

This is legitimate and natural when considering the development of the metaphor in the OT Davidic Shepherd tradition. In fact, the symbolism of the term שֵׁבֶט ('rod, staff, tribe') evokes the imagery of a shepherd holding the tribes of Israel in his hand in some texts, such as Ezek 37:15-28 and Zech 11:4-14. On the other hand, the 'rod' or 'staff' are able to represent certain principles of the shepherd's leadership as seen in Ps 23:4, 6 in which the Lord's rod and staff are paralleled to his 'goodness' (טוֹב) and 'faithfulness' (חֶסֶד). Similarly, the picture of shattering the nations like a potter's jar shattered with an iron rod in *Ps.Sol.*17:23b is modified in the immediate context of 17:23-24.

---

[84] Oegema, *The Anointed*, 105-106.

[85] Atkinson, *Intertextual*, 349.

[86] Pomykala, *Davidic Dynasty*, 163, notes, "In Ps 2:9, the nations are broken by the iron rod; here, the sinners [errant Jews] are its victims."

[87] B. L. Mack, "Wisdom Makes a Difference," in *Judaisms and Their Messiahs at the Turn of the Christian Era* (ed. J. Neusner, W. S. Green, and E. S. Frerichs; Cambridge: Cambridge University Press, 1987), 168.

The Davidic Shepherd will destroy the unrighteous gentile rulers ἐν σοφίᾳ ἐν δικαιοσύνῃ ("in wisdom and in righteousness," v. 23a), which means he destroys the unlawful nations using ἐν λόγῳ στόματος αὐτου (v. 24). Davenport rightly comments that here the power of the "word of his mouth" is a power which, though not to be equated with wisdom, is so thoroughly characterized by wisdom that "any separation of the two is out of the question."[88] Thus, sinners will be condemned by "the thoughts of their hearts" (v. 25) and not by the sword. As the text unfolds the vision of this Shepherd's reign over the nations in 17:26-46, the means of his dominion are more clearly established as his 'words.'

The peaceful character of the Davidic Shepherd is explicitly spelled out in *Ps. Sol.* 17:33, "he will not rely on horse and rider and bow, nor will he collect gold and silver for war," because he will drive out sinners (17: 36; cf. v. 23) "by the power of his word" (ἐν ἰσχύει λόγου, v. 36). This evidences a clear modification of the traditional Davidic expectation. Here again, his staff is consistent with the word of his mouth. The characteristics suitable for the warrior-messiah, if any, are either absent or modified. As Pomykala notes, the emphasis lies on righteousness, holiness and wisdom.[89]

This future Davidic king's refusal to trust in the instruments of warfare, to amass the wealth needed to purchase mercenaries or to maintain a standing army, are seen as growing out of the king's dependence upon the strength that issues from God, "upon his determination to show mercy rather than to threaten with violence, and upon the sheer weight of his words."[90] The *Psalms of Solomon* 17:35 in particular yields a significant parallel to verse 23b: "He will strike the earth with the word of his mouth (τῷ λόγῳ τοῦ στόματος αὐτοῦ, cf. Isa 11:4 [LXX]) forever; he will bless the Lord's people with wisdom and happiness." The same image of smashing with the rod in verse 23b is present, but note that it is clearly accomplished by the word of his mouth, i.e., the "iron rod" in verse 23b. The kingdom is established "not by the force of arms, but by the words issuing from the mouth of the Davidic king" (17:36).[91]

The predominant picture of the Son of David in *Ps. Sol.*17-18 is a shepherd vested with the wisdom of the Lord rather than a warrior riding on a horse (17:33). Hence, the psalmist praises this king whose "words will be purer than the finest gold, the best" (τὰ ῥήματα αὐτοῦ πεπυρωμένα ὑπὲρ χρυσίον τὸ πρῶτον τίμιον, 17:43) and whose "words [will?] be as the words of the holy angels in the midst of the sanctified people" (οἱ λόγοι αὐτοῦ ὡς λόγοι ἁγίων ἐν μέσῳ λαῶν ἡγιασμένων, 17:43b), as he exclaims: "This is the beauty of the king of Israel *which God knew to raise him over the house of Israel* [ἣν ἔγνω

---

[88] Davenport, "The Anointed," 73.
[89] Pomykala, *Davidic Dynasty*, 165.
[90] Davenport, "The Anointed," 77.
[91] Pomykala, *Davidic Dynasty*, 168.

ὁ θεὸς ἀναστήσαι αὐτὸν ἐπ' οἶκον Ἰσραήλ] to discipline it" (17:42). These numerous passages affirm that his reign will not be characterized by military might and carnage. The rod of this Davidic Shepherd is particularly "the rod of discipline of the Lord Messiah [ὑπὸ ῥάβδον παιδείας χριστοῦ κυρίου]" (18:7).

In this respect, the shepherd imagery plays several distinct roles. Atkinson finds more parallels in the Qumran texts (4Q252, 4Q174, 4Q161, 4Q285, 4Q246) for a militant Davidic messiah across sectarian lines as he assumes it is indicated in *Ps. Sol* 17.[92] But his equation between the Prince (נשיא) of the Congregation and the Davidic Branch in 4Q161, 4Q285, originally suggested by Pomykala,[93] as we shall present later, would not support a militant Davidic messiah. Nor is the prince hardly a warrior figure in the eschatological vision of Ezekiel's Davidic Shepherd (34:23-24; 37:24-25). In fact, the Son of David has been modified by the shepherd imagery in the OT Davidic Shepherd tradition, and is to arrive only after YHWH's eschatological confrontation with the nations and with the wicked shepherds of Israel. The psalmist's depiction of the Davidic messiah in *Ps. Sol.* 17 does not seem to widely deviate from this tradition. Yet, for the psalmist, the role of YHWH as the eschatological rescuer of the flock (17:23) overlaps with the Davidic Shepherd as Leader and Teacher for the community at the *eschaton*. But, for the remainder of the psalms, the picture coheres with the Davidic Shepherd at the end-time whose role is primarily to sanctify the eschatological people as set out in Ezekiel 34-37 (cf. *An. Apoc.* 90:17-18).

2.2.3 The Compassion of the One Shepherd

The utter hopelessness of the Israelites from the kings down to the commonest of the people is spelled out in *Ps. Sol.* 17:20. Yet a new hope springs up in the next verse: "See, Lord, and raise up for them their king" (v. 21). The promise of the perpetuity of David's throne is recalled by the psalmist already in 17:4: "Lord, you chose David to be king over Israel." Facing the utter failure on the part of Israel, the psalmist appeals to the covenant (2 Sam 7:14), but more precisely, to the mercy of the Lord: "for the strength of our God is forever with mercy" (17:3). Thus, as Zechariah cries out for the Lord's compassion toward Israel in 10: 6 (also, Zech 12:10; cf. Mic 7:18; Isa 40:9-11; 61:1-14), so the psalmist characterizes the eschatological shepherd by his compassion as in 18:2: "Your eyes [are] watching over them and none of them will be in need" (cf. Ps 23:1).

Compassion is a primary characteristic of the Davidic Shepherd in *Ps. Sol.* 17-18 (17:34b, 41, 45; 18:3, 5, 9). It serves as the most profound motivation for the Davidic Shepherd as God to "dispatch his mercy to Israel" (17:45); thus will

---

[92] Atkinson, "Militant Davidic Messianism," 112-121.
[93] Pomykala, *Davidic Dynasty*, 198-199, 205-212, 243.

Israel be cleansed in "the day of mercy in blessing" (εἰς ἡμέραν ἐλέους ἐν εὐλογίᾳ, 18:5). Davidic expectation in these psalms is definitely modified by means of the abundance of shepherd imageries since the compassionate Son of David, the eschatological shepherd of the psalmist, reigns with the rod of his words, the rod of discipline in wisdom and righteousness.

Interestingly, the Lord Messiah's compassion is directed toward all the nations and is not limited to Israel. The Davidic Shepherd will be compassionate also to all of the nations (17:34b), since his compassionate judgments [are] over the whole world (18:3) to those living in the fear of God "in the days of mercy" (18:9). The idea that the Davidic Shepherd will be the Lord's appointed king at the *eschaton* is clear in *Ps. Sol.* 17:19-46, though he is to be placed not only over Israel (17:19-25), but also over the nations (17:26-46; 18:1-12) since he will "have gentile nations serving under his yoke" (17:30).[94] His yoke is not harsh since he will be "compassionate" to all the nations (17:34b) and will lead them [all nations as well as Jerusal] into holiness (17:41).[95]

Davenport comments on *Ps. Sol.* 17:29-32 that "the Gentiles will become instruments in the reconstruction of the nation; the entire Gentile world shall become holy!"[96] Particularly the phrase in 17:31, "for nations to come from the ends of the earth (ἐπ' ἄκρου τῆς γῆς) to see his glory" recalls Mic 5:4 which foresees the extension of the Davidic Shepherd's reign. Thus, the third person plural pronoun in "he will be a righteous king over *them* (ἐπ' αὐτούς)" in 17:32 refers not only to the Jews but to the nations as well (cf. Ezek 34:23).[97] The expression, "in their midst" (ἐν μέσῳ αὐτῶν) in verse 32, "there will be no unrighteousness in their midst," indicates that this king is the eschatological Teacher who would lead them into holiness reminiscent of the task performed by the prince in Ezek 34:23-24; 37:24-25.

The new David will be *one shepherd over Jerusalem and the nations* in *Ps. Sol.* 17, and not merely over Israel and Judah as indicated in Ezekiel 34 and 37. The concept of the singleness of the Davidic appointee at the end-time seems to be modified as the psalmist looks beyond the restoration of Israel to the extended reign of the eschatological Davidic Shepherd over the nations.

This modification is also evidenced in the picture of the white bull in *An. Apoc.* 90:18, i.e., the eschatological leader among both Israel and the nations to transform them to be holy. The difference is that the Davidic Shepherd in *Ps. Sol.* 17:19-46 would be more a political/historical figure than the apocalyptic white bull. The ideal of a united nation in terms of the imagery of bringing together the two staffs and the succeeding arrival of the "one shepherd" over

---

[94] Cf. Matt 11:28-30.
[95] But the psalmist sees that the alien and foreigner will not participate in the distribution of the restored land (*Ps. Sol* 17:28); this differs from the vision of Ezek 45:8; 47:13, 21.
[96] Davenport, "The Anointed," 76.
[97] Pomykala, *Davidic Dynasty*, 164.

them in Ezekiel 36 is modified once again in the broken staffs of the covenant and the mysterious figure of the afflicted shepherd(s) in Zechariah 9-11. If this line of argument is valid as well for the author of *Ps. Sol.* 17 as well as for the seer of the *Animal Apocalypse*, it is already presupposed that the Davidic Shepherd at the *eschaton* will be the leader beyond Israel who will reign over Israel and the nations (cf. Mic 5:3-4). In this sense, he will be the One Shepherd over the eschatological community for the renewed Israel (*Ps. Sol.* 17:21-46) and the incorporated nations (18:1-12) as well. At the end, mercy and righteousness will characterize the reign of this Davidic Shepherd as One Shepherd of all nations as well as of the restored Israel.

2.2.4 Summary

The Son of David in *Ps. Sol.* 17:21 appears to involve military conflict. Both aspects of the eschatological shepherd – warrior and Davidic Leader/Teacher – may overlap. Nevertheless, the author of *Psalms of Solomon* predominantly treats this figure as the assimilation of the Davidic Shepherd as Leader and Teacher of the restored community at the time of Israel's restoration. Contrary to Atkinson's argument that the psalmist's use of the Hebrew Scripture to portray the messiah as a king and a warrior who would destroy God's enemies,[98] the modifications made by the psalmist using the Davidic Shepherd tradition suggests that the renewed expectation of the Son of David in *Ps. Sol.* 17-18 approximates the traditional picture of the Davidic Shepherd, though with additional new features.

Regarding the intention and historical backdrop of Psalms of Solomon, Atkinson proposes that the Davidic messiah would defeat Herod's regime through violent means, which possibly echoes the atmosphere of the time of the second Jewish revolt led by Bar Kokbar (A.D. 132-135).[99] Nevertheless, the Son of David tradition is extensively modified in *Ps. Sol.*17-18. This can be explained by looking at Oegema's proposal which adopts Schüpphaus' form-critical hypothesis that distinguishes the 'older psalm' – presenting a violent Davidic messiah who emerges during the Hasmonean regime – from the 'younger psalms' that depict the Teacher figure following Pompeis's attack on Jerusalem in *Ps. Sol.*17-18.[100] However, *Psalms of Solomon* 17-18 appears to point to a different paradigm to a regime profoundly different from the Hasmonean dynasty that lacks divine endorsement, namely, the legitimate Davidic lineage (cf. Matt 1:17-21).

Acknowledging the dearth of evidence for the surviving expectation of the Davidic messiah during 500-163 B.C., Pomykala may be correct when he states, "The davidic dynasty tradition did not generate disappointment with the

---

[98] Atkinson, *Intertextuality*, 347.
[99] Atkinson, *Intertextuality*, 373-377.
[100] J. Schüpphaus, *Die Psalmen Solomos* (Leiden: Brill, 1965), 140-151.

126   Chapter 2. The Davidic Shepherd Tradition In Second Temple Judaism

Hasmoneans; rather, disappointment with the Hasmoneans generated this appropriation of the davidic dynasty."[101] Passing harsh judgment on the shepherds of the Maccabeans by labeling them 'sinners' destined to be dashed like pottery (17:23), the coming son of David will be "all the things the Hasmonean kings were not" and "Israel would be all the things that the Maccabean kings never could be."[102] Charlesworth puts it this way: "Hence the corruption of the Hasmonea 'kings' apparently stimulated a messianology that portrayed a Messiah who was *not* a king."[103]

Presenting Davidic expectation as the promised hope for the future at the defining moment of crisis in Israel's national history, the psalmist extensively modifies the militant Davidic figure. Here the psalmist interacts with the OT Davidic Shepherd tradition in order to make the Davidic hope alive and adoptable for the changing face of his contemporary history. Now the Davidic Shepherd is hoped to be one shepherd not just over unified Israel (Ezek 37; Zech 9-11), but he reigns over both restored Israel and the nations coming from the ends of the earth. Nevertheless, the motivation for the Lord to restore all remains the same. It is the Lord's compassion by which he fulfills, despite all, the age-old promise given to David (Zech 10: 6; 12:10; cf. Mic 7:18; Isa 40:9-11; 61:1-14 10; Ezek 34:6, 10). This will all come true "in the days of mercy" (*Ps. Sol.*18:9) now fully extended to the ends of the earth (*Ps. Sol.* 17:31; cf. Mic 5:1-5a).

2.3 Qumran Writings

In order to conduct a close examination, only selected passages from the vast array of texts in the Dead Sea Scroll library will be dealt with in this section. The criterion for this selection is the presence of either Davidic expectation or shepherd imagery.[104] Johannes Zimmerman classifies 4Q161 (4QpIsa^a), 4Q285 (4QSefer ha-Milhamah), 4Q174 (4QMidrEschat^a), 4Q252 (4QQpGen^a), and CD-A 7:18-21 (CD-B 19:8-13) as the texts that entail the features of "Fürst der [ganzen] Gemeinde und ‚Sproß Davids'" under the rubric of ",königliche'

---

[101] Pomykala, *Davidic Dynasty*, 167.

[102] Pomykala, *Davidic Dynasty*, 166.

[103] Charlesworth, "From Messianology," 22 (italics his); see also, M. Hengel, J.H. Charlesworth, D. Mendels, "The Polemical Character of 'On Kingship' in the Temple Scroll: An Attempt at Dating 11QTemple," *JJS* 37 (1986): 28-38.

[104] The Ezekiel texts in 4QEzek^a (4Q73) and 4QEzek^b (4Q74) do not include any of those themes from Ezek 34-37; cf. John Lust, "Ezekiel Manuscripts in Qumran: A Preliminary Edition of 4QEzek a and b," in *Ezekiel and His Book: Textual and Literary Criticism and Their Interrelation* (ed. J. Lust; Leuven: Leuven University Press, 1986), 90-100; P. W. Flint *The Bible at Qumran: Text, Shape, and Interpretation* (Grand Rapids: Eerdmans, 2001), 61.

2.3 Qumran Writings

Gesalbtenvorstellungen."[105] Similarly, Atkinson finds the Davidic dynasty tradition or Davidic expectation in texts such as 4Q161, 4Q285, 4Q174, 4Q252, and additionally 4Q246, while he does not list CD-A 7:18-21.[106]

To include 4Q246 as a Davidic text would stir up controversy. Zimmermann defines 4Q246 as a 'Sohn Gottes' text and treats it separately from the other texts involving Davidic connotations in terms of 'Fürst der [ganzen] Gemeinde und ‚Sproß Davids'.' Pomykala, likewise, suggests that 4Q246 as 'Son of God' text does not belong to the Davidic tradition.[107]

Nevertheless, Davidic connotations are not entirely absent in the text. Although 4Q246 as 'a stitch-like construed [early Jewish] apocalyptic Text' displays a close relationship with Dan 7 and presents the Danielic Son of Man as "der ‚Sohn Gottes,'" Zimmermann argues that the Son of God in this pre-Christian tradition can be readily associated with OT Davidic messianism (cf. Ps 2:7; 89:27; 2 Sam 7:14).[108] More specifically, Zimmerann suggests that if בא in 4Q246 col.1, 1:9 indicates [בר מלכא ר], then 'the great King could recall a David.'[109] Nevertheless, the Davidic connotations in 4Q246 are only implicit and, at best, marginal; further, as Danielic [heavenly, high-exalted] Son of Man is recognized as the primary figure in the text, shepherd imagery is virtually absent.

Disqualifying 4Q246 from our examination, we shall add 1Q28b (1QSb) 5:20-29 to the list of the Qumran texts vested with Davidic connotations. Collins observes regarding this text that the figure of the Prince of the Congregation is merged with the Branch of David.[110] While Atkinson omits this text, Zimmermann places it under the heading of "Fürst der [ganzen] Gemeinde und ‚Sproß Davids'"[111]

Another indispensable text for our examination in view of Davidic expectation is 4Q521 (4QMessianic Apocalypse), which also contains shepherd language.[112] García Martínez sorts 4Q521 under the rubric of Davidic

---

[105] J. Zimmermann, *Messianische Texte aus Qumran: Königliche, priesterliche und prophetische Messiasvorstellungen in den Schriftfunden von Qumran* (WUNT 2; Tübingen: Mohr Siebeck, 1998), 49-127; he also presents (ix-xvi) under the same rubric other categories such as "Priesterliche Gesalbtenvorstellungen" (4Q375; 4Q376; 4Q541; 1QSb 3:1-6, 4:22-28; 4Q491) and "Prophetische Gesalbtenvorstellungen" (CD 2:12; 1QM 11:7; 4Q377; 5Q521; 11QMelch; 4Q558; 4Q175).

[106] Atkinson, "Davidic Messianism," 112-113, misses CD-A 7; cf. Pomykala, *Davidic Dynasty*, 171.

[107] Pomykala, *Davidic Dynasty*, 172-180.

[108] Zimmermann, *Messianische Texte*, 137-138, 157, 169-170; cf. Collins, "A Pre-Christian 'Son of God' Among the Dead Sea Scrolls," *BR* 9 (1993): 34-38, 57.

[109] Zimmermann, *Messianische Texte*, 143, n. 295.

[110] Collins, *Scepter*, 62.

[111] Zimmermann, *Messianische Texte*, 53-59.

[112] Atkinson omits 4Q521 in "Davidic Messianism"

128   *Chapter 2. The Davidic Shepherd Tradition in Second Temple Judaism*

messianism.[113] While considering 4Q521 as a messianic text and yet subsuming it under the category of 'prophetische Gesalbtenvorstellungen,' Zimmermann argues that in 4Q521 the kingly, and priestly aspects of the messianic figure are fused with prophetic connotations (cf. 11QDavComp); further, he detects Ezekiel 34 and 37 as a significant tradition standing behind the texts of 4Q521 2, 2:5, 6-7, 12.[114] As we investigate these supposedly Davidic texts, the focus, if any, will be on how shepherd imagery interacts with Davidic expectation in the context of Israel's restoration.

On the other hand, the texts from the Dead Sea Scrolls selected for their relevance to shepherd imagery include CD-A 13:7-12 as well as CD-B 19:8-12 (CD-A 7:10-21), known primarily for its Davidic connotations.[115] With regard to shepherd imagery, besides CD-A 13:7-12, 4Q504 (4QDibHam^a) must not be neglected. Although Atkinson dismisses this text based on its disputable association with Davidic expectation,[116] 4Q504 is worthy of a detailed examination not only for its possible Davidic expectation, but for its rich language related to shepherd imagery. As we shall demonstrate in the section examining these primarily 'shepherd' texts, we will test how the shepherd imagery interacts with any Davidic expectation in view of the restoration of Israel.

2.3.1 4Q504 (4QDibHam^a)

The fragments of 4QDibHam^a represent a collection of prayers of supplication recited on each day of the week.[117] The texts germane to our study are 4Q504 fragments 1-2, columns 2-4, that is, the prayers for the fifth or sixth day. These sections along with columns 5-6 are filled with supplications for the restoration (esp. 6:1-17) as the prayers recall God's completion of his creation on the sixth day. The text that particularly deserves our attention is 4Q504 1-2, 4:5-14:

For you loved (l.5) Israel more than all the peoples; and you chose the tribe of (6) Judah, and established *your covenant you established with David* (so he would) be (7) *as a shepherd, a prince over your people* [כרעי נגיד על עמכה]. He would sit in front of you on the throne of Israel (8) for ever; and all the nations saw your glory, (9) [by] which you were honored as holy *in the midst of your people* Israel; and to your (10) *great Name* they will carry their offerings: silver,

---

[113] García Martínez, "Messianic Hopes," 168-170; Atkinson omits 4Q521.

[114] Zimmermann, *Messianische Texte*, 343-388 (esp. 354-356, 364, 373, 382).

[115] Köstenberger, "Good Shepherd," 94, lists CD-A 7:13-21; 13:9. Vancil, *Symbolism*, 265-266, mentions CD-A19:2-3, 10 from the Dead Sea Scrolls.

[116] Atkinson, 112; Zimmermann omits 4Q504 as well.

[117] Thus, *Dibre* ('words') *Hamme'orot* 'the luminaries' used as a term for the day). Three copies of *Dibre Hamme'orot* (4Q504-506) have survived at Qumran, all in Hebrew. 4Q504 is the oldest and most complete manuscript and dates paleographically to the middle of the second century B.C. See Esther G. Charon, "Dibrê Hamme'orot: Prayer for the Sixth Day (4Q504 1-2 v-vi)," *Prayer from Alexander to Constantine: A Critical Anthology* (ed. Mark Kiley et al.; London: Routledge, 1997), 23-27.

gold, precious stone(s), (11) with all the treasure(s) of their land in order to glorify your people and (12) Zion, your holy city and your wonderful house; There is no adversary (13) or evil attack, but peace and blessing ... [...] (14) And they a[t]e, were replete, and became fat [...]¹¹⁸

Pomykala gives three reasons why 4Q504 1-2, 4:6-8 does not involve any expectation of a Davidic messiah. First, the grammatical possibilities lend themselves to read verses 6-8 as reminiscent of the historical king David of the past: the temporal horizon of the Davidic reign designated by כול הימים in verse 8 may refer only to 'all the days of David's own rule' i.e., 'lifetime,' or 'a long time,' though it can be translated 'forever.'¹¹⁹ Second, the literary structure of 4Q504 1-2, 4:2-5:18, Pomykala says, resembles Neh 9:6-37 as a prayer of supplication; the lines that speak of the Davidic covenant in 4Q504 1-2, 4:6-8 are recited merely as evidence of God's gracious acts toward Israel 'in the past.'¹²⁰ Third, the text of 4Q504 1-2, 5:6-14 speaks about the restoration of Israel, but in association with adherence to the Sinai covenant; Pomykala contends there is no mention of a restored Davidic covenant or a call for God to bring forth a new seed of David.¹²¹

Nevertheless, as Pomykala himself acknowledges, it is not impossible to see David in verse 6 in terms of 'David's corporate personality' or as 'a typologically new David' as in Hos 3:5; Jer 30:9; Ezek 34:23-24; 37:24-25 with כול הימים meaning 'all the days' or 'forever' in verse 8.¹²² In fact, this passage has been recognized by J. J. Collins and D. Dimant as evidence of the expectation of 'a future Davidic messiah' within the Qumran sect.¹²³ Pomykala acknowledges that the Davidic covenant is clearly meant to earn its perpetuity through the offspring of David in texts such as 2 Sam 7:11-16; Ps 89:3-4, 28-37, but in contrast, in 4Q504 God establishes his covenant with David so that he – the historical David – may be a shepherd prince over Israel.¹²⁴ However, if it is grammatically possible to read the text as open to the future and not limited to the past, the impression of the text's emphasis on David as an individual can be

---

¹¹⁸ Dennis T. Olson, *The Dead Sea Scrolls: Hebrew, Aramaic, and Greek Texts with English Translations*, (vol. 4, ed. J. H. Charlesworth; Tübingen: Mohr Siebeck, 1994), 131 (italics mine); the imperfect both in וישב in l.7 and ויביאו in l. 10 is read futuristic as in the rendition by Martínez and Tigchelaar, *Scrolls*, 1015.

¹¹⁹ Pomykala, *Davidic Dynasty*, 174-176.

¹²⁰ Ibid., 176-178.

¹²¹ Ibid., 178-179.

¹²² Ibid., 175, n.16.

¹²³ Collins, "Messianism in the Maccabean Period" in *Judaisms and Their Messiahs at the Turn of the Christian Era* (ed. J. Neusner, W. S. Green, and E. S. Frerichs; Cambridge: Cambridge University Press, 1987), 105; D. Dimant, "Qumran Sectarian Literature," in *Jewish Writings of the Second Temple Period* (CRINT 2:2; Assen: Van Gorcum/Philadelphia: Fortress, 1984), 539, n. 265.

¹²⁴ Pomykala, *Davidic Dynasty*, 174, gains this insight from M. Ballet, "Un recueil liturgique de Qumrân Grotte 4: 'Les paroles des luminaires'," *RB* 68 (1961): 222.

used against his own conclusion, that is, David in verse 6 as 'a typologically new David.'

The plausibility of this futuristic option increases as one finds that Neh 9:6-37, the Levites' prayer of repentance concerning Israel's sins in the past, makes no mention at all of the Davidic covenant. Although the patterns and literary practices of Neh 9:6-37 and 4Q504 might be similar, the paucity of references to the Davidic covenant in Neh 9:6-37 weakens Atkinson's point of the comparison itself. In fact, the prayer for the sixth day in 4Q504 1-2, 4 demonstrates a close interaction with Ezekiel's vision. For instance, we find all three significant designations of the Lord's appointed for the time of the restoration – *David, shepherd, and prince* (l. 7).

While the cluster of these designations could be simply a summary of the historical David as to his position, nevertheless, the way it is summed up indicates a possible allusion to Ezekiel's view of the Davidic Shepherd. Significantly, the texts in the Hebrew Bible where these three designations are clustered together occur only in Ezek 34:23-24 and 37:24-25.[125] Indeed, the term for 'prince' in l.7 is נגיד not נשיא. But the use of נגיד instead of נשיא does not create a serious problem as to the Davidic texts in the Second Temple period as illustrated in *An. Apoc.* 89:46 and 11QPs[a] 28:11.

There is no shortage of evidence for the allusion of 4Q504 1-2, 4 to Ezek 34-37. Expressions such as God's glory 'in the midst of' Israel in l.9 (cf. Ezek 34:24; 37:26-27), 'to your great Name' in view of the nations in l.10 (cf. Ezek 36:22-23) are characteristic to Ezekiel's vision of the shepherd in chapters 34-37. The terminology in particular enhances the case for the allusion. Again, the shepherd imagery of 'feeding' the flock recurs in l.14: "and they a[t]e, were replete, and became fat" (cf. Ezek 34:10, 14; Zech 11:16). The implied plight that calls for supplication can be identified as "the absence of the shepherd" (l. 14) who is like David (l. 6-7), similar to what is found in the prologue of the vision of the future restoration in Ezek 34:1-16.

There is more. The prayer for the sixth day in 4Q504 1-2, 5 is to be recited as reminiscent of the golden days of King David.[126] Yet this motif is preceded by earnest supplication for the Lord's 'healing' of their 'madness, blindness and confusion [of heart]' (1-2, 2:14) and 'illness, famine, thirst, plague' (1-2, 3:8). The prayers illuminate the cause of these disasters: It is their failure to follow the laws of the Lord, "for your name has been called out over us; with all [our] heart and with all [our] soul and to implant your law in our heart" (1-2, 3:13). This Deuteronomistic sin-exile-return pattern is the framework into which the prayers are embedded, and draws heavily from the divine predictions and promises set out in Lev 26, which is also distinctively Ezekielian (cf. Ezek 20;

---

[125] The tripartite designations are not even found in 2 Sam 7 (cf. 1 Sam 17:20).
[126] Charon, "Prayer for the Sixth Day," 23-27.

34:25-31; Lev 26:4-13).[127] The context certainly indicates Israel's ongoing exile.[128]

In this context, David is conjured up, highlighting the absence of the shepherd/prince who would lead them into the righteousness of the laws of the Lord. Most likely 4Q504 is a collection of 'communal laments' as J. C. R. de Roo suggests, or a composition of 'community prayers of supplications' dating back to 150 B.C., and was used 'for a very long time' among the Qumran sect.[129] Considering the grammatical possibility of a futuristic reading and the allusions to Ezekiel's shepherd imageries, this historical background only amplifies the Davidic expectation in the sixth day prayers found in 4Q504 1-2, 4.[130] Thus when recalling David, what is seen behind the prayers is the figure that approximates the Davidic Shepherd for the community.

As in Ezekiel's vision, the Davidic Shepherd is expected to come after the Lord's eschatological shepherding the flock, especially healing the sick and restoring their heart to follow the law. The Davidic Shepherd as the Prince reigns them in view of the nations for the sake of God's glory. The sixth day prayers in 4Q504 depicts this eschatological state in terms of the pastoral image, i.e, the sheep fed by their Shepherd.

2.3.2 4Q252 (4QpGen)

The pivotal blessing of Judah by Jacob in Gen 49:8-12 is interpreted 'in a clear messianic sense' within the Qumran community.[131] Along with *Ps. Sol.* 17, Zimmermann asserts that 4Q252 indicates a critical link between OT expectation of the kings from the house of David and the later Christian and Judaic expectations of בן דויד משיח.[132]

---

[127] Wacholder, "Ezekielianism," 187; Charon, "Prayer for the Sixth Day," 24.

[128] Likewise, Craig A. Evans, "Aspects of Exile and Restoration in the Proclamation of Jesus and the Gospels" in *Exile: Old Testament, Jewish, and Christian Conceptions* (ed. James M. Scott; Leiden: Brill, 1997), 308-309, views 4Q504 as replete with an abundance of exile/restoration motifs.

[129] Esther G. Charon, "4QDIBHAM: Liturgy or Literature?," *RevQ* 15 (1991): 447-455; idem,"Is *DIVREI HA-ME'OROT* a Sectarian Prayer?" in *The Dead Sea Scrolls: Forty Years of Research* (ed. D. Dimant and U. Rappaport; Leiden: Brill, 1992), 3-17; J. C. R. de Roo, "David's Deeds in the Dead Sea Scrolls," *DSD* 6 (1999): 44-65; also, Nitzan, *Qumran Prayer and Religious Poetry* (STDJ 12; Leiden: Brill, 1994), 89-116.

[130] Roo, "David's Deeds," 65, links the remembrance of David's deeds with God's preservation of the remnant.

[131] F. G. Martínez, *The People of the Dead Sea Scrolls* (Leiden: Brill, 1995), 161; J. T. Milik, *Ten Years of Discovery in the Wilderness of Judea* (SBT 26; London: SCM/Naperville, Ill.: Allenson, 1959), 96 n.1, dates 4QpGenª to the first century A.D.; N. Avigad, "The Paleography of the Dead Sea Scrolls and Related Documents," in *Aspects of the Dead Sea Scrolls* (2d ed.; ScrHier 4; Jerusalem: Magnes Press, 1965), 72, sets the range from 50 B.C. to A.D. 70.

[132] Zimmermann, *Messianische Texte*, 124-125.

4Q252 5:1-7 defines the blessing of Judah (l. 1-3a) in terms of 'the covenant of kingship' (ברית המלכות, l. 2b) for everlasting generations (l. 4), and articulates the promised one explicitly as 'the anointed' of righteousness (משיח הצדק, l. 3). Eventually the text identifies him with the 'Branch of David' (צמח דויד, l. 3-4) who obeys the Law with his people of the community (l. 5-7).

As the Davidic connotations are apparent, our focus as we examine this passage will be its possible link with shepherd imageries in view of the restoration of Israel. First, while 4Q252 5:1 cites Gen 49:10, לֹא־יָסוּר שֵׁבֶט מִיהוּדָה ('the scepter will not depart from Judah'),[133] it interprets the referent of MT's שבט ('scepter') as שליט ('ruler, dominion'),[134] which occurs primarily in the book of Daniel (2:15; 4:14, 22, 29; 5:21, 29); thus it reads, "the *ruler* will not depart from ...."[135]

At the same time, the Qumran text alters the phrase 'from Judah' to 'from the tribe (משבט) of Judah.' Here, the term for 'tribe' in 4Q252 5:1 is שבט, referring either to scepter or tribe. The changes appear casual. The scribe might have attempted to delineate the meaning of MT's scepter to mean 'dominion,' and thereby simply added שבט in the later phrase by bringing in its other meaning, i.e., 'tribe,' though this time with the intent to intensify the significance of the tribe of Judah in its association with the imagery of the scepter. Interestingly, this word play with the different meanings of שבט possibly reflects the Davidic Shepherd tradition such as we find in Ezek 37:15-23, Zech 11:4-14 (cf. Ps 78:68) where the texts intentionally play with staff and tribe in association with the Davidic Shepherd imagery.[136]

Second, more importantly, the text identifies the one for whom the throne of David is promised – not only as the anointed but also as צמח דויד הצדיק ('the Branch of David,' l. 3-4), and whose title occurs in Jer 23:25 and 33:15.[137] Pomykala suggests that the text is influenced by Jer 23:5-6 (cf. Jer 33:15-17) with the hope that the re-establishment of the Davidic kingship would occur when Israel achieved dominion in the eschatological war against the Sons of

---

[133] To establish the genre of 4Q252 as 'a form of biblical commentary,' see G. J. Brooke, "4Q252 as Early Jewish Commentary," *RevQ* 17 (1996): 385-401.

[134] E. Tov (ed.), DJD 17, 205, reads שליט as 'scepter'; Zimmermann, *Messianische Texte*, 114, "Machthaber" (cf. Gen 42:6; Qoh 7:19; 8:8; 10:5; Sir 9:3).

[135] García Martínez translates the term 'a sovereign' in the first edition, but 'a staff' in the study edition.

[136] Ps 78:68 confirms YHWH choosing 'the tribe (שֵׁבֶט; LXX, φυλή) of Judah'; Zimmermann, *Messianische Texte*, 115, notes this 'Wortspiel' as a step for the following messianic interpretation.

[137] Brooke, "Commentary," 392; Collins, *Scepter*, 252; Pomykala, *Davidic Dynasty*, 188.

Darkness for the perpetual kingship that would follow.¹³⁸ Further, Zimmermann notes that 4Q252 affirms the presence of the expectation of a Davidide as משיח in pre-70 Judaism and thus underscores the messianic interpretation "nicht ersetzen sondern erklären" of Gen 49:10 with the help of other texts such as Jer 33:17, 2 Sam 7:14 (ברית המלכות, l. 1b) and Isa 11:1-5 (משיח הצדק, 1.3).¹³⁹

In fact, as a precursor of the shepherd imagery in Ezek 34:1-16, the context of Jer 23:1-6 (cf. Jer 33:15-17) also depicts the eschatological Shepherd confronting the wicked shepherds and appoints the Davidic Shepherd who will lead the restored people with righteousness. This fits well the intent of 4Q252's interpretation of the Davidic expectation that likely criticizes and points to the illegitimacy of the contemporary kingly figures.¹⁴⁰ On the other hand, Isa 11:1-6 also makes a suitable backdrop for 4Q252 5:1-7 since both texts involve such shepherd imageries as the staff or scepter, the ideal and eschatological David, and the righteousness that characterizes David and his reign.

Third, besides the ambivalent use of the term שבט, there is another description of the Davidic messiah in 4Q252 5:1-7, namely, his righteousness (משיח הצדק, 1.3). The eschatological Davidic Shepherd is singularly marked by his righteousness, and thus his capability to lead his eschatological people to obeying the law in the new age (4Q252 5:5-6) similar to what we find in Ezek 34:23-24; 37:24-25; *Ps. Sol.* 17-18; *An. Apoc.* 90:17-18. In these texts, the eschatological Shepherd arrives and gathers his scattered flock (cf. Jer 23:1-4) before he appoints his Davidic Shepherd for peaceful administration and legitimate leadership similar to the eschatological Teacher figure (cf. Jer 23:5-6). To execute righteous leadership, the Branch of David must be righeous/holy, evident in *Ps. Sol.* 17-18; Ezekiel 34, 37; *An. Apoc.* 90:17-18.

In this respect, Atkinson's thesis regarding a militant messiah in 4Q252 suffers, which lacks a due explanation for the righteous character of this Anointed, the Branch of David.¹⁴¹ The eschatological Davidic messiah would be called upon not for his militant power and prowess, but for his ability to lead people in righteousness. A substantial part of the eschatological paradigm assigned to the Davidic messiah revolves around his role as Teacher of the Law.¹⁴²

---

¹³⁸ Pomykala, *Davidic Dynasty*, 188, dates 4Q252 around 30 B.C. to A.D. 70 on the basis of the historical counterpart of the Branch of David being Herodian kings or the king of the Kittim, the Roman emperor.
¹³⁹ Zimmermann, *Messianische Texte*, 114-118.
¹⁴⁰ Atkinson, "Davidic Messianism," 115 says the Herodian kings are in view.
¹⁴¹ Atkinson, *Intertextual*, 368-378.
¹⁴² G. J. Brooke, "The Thematic Context of 4Q252," *JQR* 85 (1994): 54-55, suggests that the Interpreter of Law is featured here in 5:5 as in close proximity with other community texts such as CD 6:2-11 in which Brooke finds considerable thematic correlation with 4Q252, i.e., 'the promise of the land'; Collins, *Scepter*, 62, makes the same connection.

Although quite fragmentary, 4Q252's interpretation of Gen 49:10 in column 5 exhibits both Davidic expectation and shepherd imagery combined with it, plausibly around the time of the Herodian regime (30 B.C.- A.D.70), implying a continuation of the earlier OT (Jer 23:1-6; Isa 11:1-5) and contemporary (*Ps. Sol.* 17-18) Davidic Shepherd tradition. Although the shepherd imagery is not explicit, the text of 4Q252 affirms the figure of the non-militant Davidic messiah in pre-70 Judaism, which fits well with the pastoral and the righteous figure of the Davidic Shepherd in the Davidic Shepherd tradition.

2.3.3 4Q174 (4QFlorilegium)

The genre of 4Q174 is, generally, referred to as 'thematic pesher,' i.e., an interpretation of a collection of biblical verses from various books linked by a common theme.[143] Since the common theme in the case of 4Q174 may be the restoration of Israel at the eschaton, it can be called, particularly, 'Midrasch zur Eschatologie' concerning the 'Ende der Tage' (אחרית הימימ).[144] The first block of 4Q174 1-2, 1:1-5 envisions the return of YHWH's presence and his reign centered around the eschatological sanctuary (מקדש) built by YHWH himself "where there shall never more enter [... for] and the Ammonite and the Moabite, and bastard and alien and sojourner, forever, for his holy ones are there; Y[HW]H [shall reign for]ever" (l. 3-5).[145]

This vision is generically similar to the concluding scenes of Israel's future restoration found in the salvation oracles of Ezekiel 34-37, 40-48 and Zechariah 9-14, which end with YHWH's return to the temple where unholy ones ['Canaanite' or 'traders'/'merchants'] are no longer found (Ezek 44:9; 48:35; Zech 14:21). In fact, the author of 4Q174 explicitly cites from Ezek 44:10 later in l. 16-17: "it is written in the book of Ezekiel the prophet, ['They shall] no[t defile themselves any more with] their [i]do[l]s'."

4Q174 1-2, 1:6 then briefly indicates that the ruin of the past sanctuary occurred 'because of their [Israel's] sins,' the ultimate cause of the destruction of the sanctuary by the nations. The quest for restoration thus requires two major tasks, i.e., Israel obeying the Law and the rebuilding of the sanctuary, yet this time, 'the sanctuary of man' (מקדש אדמ, l. 6b) after the destruction of 'the sanc[tuary of I]srael' (l. 6a) because of their sins.

This is a critical connection with which 4Q174 recalls YHWH's promise of 'building a house [בית]' for David (2 Sam 7:12-14) in l. 10. For David, YHWH 'shall obtain rest from all your [David's] enemies' (l. 7; cf. Ps 2). Further, line 11 identifies the seed of David of l. 10 with the "Branch of David" found in Jer

---

[143] Pomykala, *Davidic Dynasty*, 192.
[144] Zimmermann, *Messianische Texte*, 105, 112.
[145] John M. Allegro, *Qumrân Cave 4: I (4Q158-4Q186)* (DJD 5; Oxford: Clarendon, 1968), 53-54; Martínez and Tigchelaar, *Scrolls*, 353, read גר as 'proselyt' instead of 'sojourner.'

23:5; 33:15 and Isa 11:1-5 who, according to 4Q174, will arise with the Interpreter of the law (דורש התורה, l. 11).

There is more. In the middle of the cadenza of the OT texts recalling Davidic expectations (l. 11-12), the author of 4Q174 cites from Amos 9:11 and assigns the task of 'David,' 'the Branch of David,' 'to save' Israel: "[will rise up] in Zi[on in the l]ast days; as it is written, 'And I shall rise up the tabernacle (סוכת) of David that is fallen'; That is 'tabernacle of David that is fal][len' is he] who will arise to save (להושיע) Israel."[146] We find the interesting common theme of building a house, either of David by YHWH or of the people by David [i.e., the sanctuary]; the double meaning of בית in 2 Sam 7:5-14 is equally valid to the relevant texts of 4Q174.[147]

Likewise, the ultimate goal of the tasks of the Branch David is tightly related to the restoration of YHWH's sanctuary (l. 3-4, 16) as YHWH raises up the fallen house of David. For this task, the Davidic promise of 2 Samuel 7 is recalled while being interpreted through texts such as Isa 6 and Amos 9. The task of the new David is defined in terms of 'saving' (ישע) Israel, while the Teacher/Interpreter of the Law is to arise along with him. Passing the citation from Ezek 44:10 in l. 16, 4Q174's allusion to Davidic expectation becomes more patent. In the context of the vision of the eschatological fight with the kings of the earth, 4Q174 interprets 'his Anointed' with a collective reference to "the elect ones of Israel in the last days" (l. 19, cf. Ps 2; cf. Mic 4:11-13).

The Branch's extended reign toward the nations is already hinted at in the phrase, 'the sanctuary of man' (מקדש אדם, l. 6b). This mention of מקדש אדם after the destruction of the temple of Israel because of her sin may well echo YHWH's claim for his own flock, eventually, 'people (הָאָדָם) you!' in Ezekiel's vision of restoration (Ezek 34:31) as well as in *An. Apoc.*'s vision for the eschatological sanctification of both Israel and the nations through the messianic white bull after the destruction of the temple because of the blindness of the sheep (*An. Apoc.* 90:17-18).

4Q174 certainly interacts with the OT Davidic Shepherd tradition. The paradigm for the future restoration laid out in 4Q174 is identical with the basic pattern found in Ezekiel 34-37, 40-48, Micah 2-5, and Zechariah 9-14: The return of YHWH's presence – the theme absent in 2 Sam 7 – precedes the appearance of the Davidic Shepherd over and among his restored flock who will face the final battle. Here Davidic expectation is tied closely with the vision of the restoration of Israel, particularly centered around the rebuilding of YHWH's sanctuary.

Contrary to Pomykala's observation that the overall characterization of the Davidic messiah in 4Q174 is sparse and even colorless, the characteristics and

---

[146] Allegro, DJD 5, 53-54; Martínez and Tigchelaar, *Scrolls*, 353, renders סוכת as 'hut.'
[147] Zimmermann, *Messianische Texte*, 108.

role of the new [eschatological] David are clearly laid down.[148] Of import as well is that the promised seed of David in 2 Samuel 7 is identified as the 'Branch of David' in Jeremiah 23 and Isaiah 11. This Branch of David, associated with shepherd imageries, is the eschatological figure who will appear in 'the end of days' (באחרית הימים, l. 12,19).[149]

In Ezekiel 34-37, the Davidic Shepherd himself is distinctively the eschatological Leader/Teacher, the 'prince' among the renewed people in terms of his role to lead them to follow the law of YHWH.[150] Likewise, 'righteousness' as the main characteristic of the Branch of David in Jeremiah 23 and Isaiah 11 fits the picture of the Son of David in *Ps. Sol.* 17-18. In 4Q174, however, the Branch of David is a separate royal figure from the priestly 'Interpreter of the law' (l. 11) who is said to appear alongside with him.

Atkinson contests that 4Q174's author underlines this Davidic messiah's function as a warrior. 4Q174 1:13b claims the function of the Branch as God's agent in a battle in "which (the hut of David) he (YHWH) will rise up to save (להושיע) Israel." Atkinson argues that the same verb יָשַׁע is used in 1QM 10-11 to describe God's deliverance of the righteous from their enemies in battle: "By the hands of our kings, besides, you (YHWH) saved us many times" (1QM 10:3).[151]

However, יָשַׁע alone does not delineate the (militants or other) means by which the Branch was expected to save Israel. In the OT, the use of יָשַׁע is often differentiated from נָצַל; while the latter definitely involves rescue from apparent external threats, the former, in comparison, frequently involves moral troubles.[152] For instance, Israel is held captive "not by human enemies but by their own uncleanness" in Ezek 36:29 where the verb יָשַׁע describes YHWH's deliverance.[153]

Similarly, 4Q174 1-2, 1, 1:6-9 refers to YHWH destroying the enemies of Israel in order to provide David with promised rest. In the following section (l. 10-17), however, the main concern shifts to YHWH saving Israel from 'the path of the wicked' (l. 14, 15; Ps 1:1) through the means of an Interpreter of the law with whom the Branch of David will arise (l. 11). It is in this context that YHWH will raise up the house of David 'to save' (להושיע) Israel (l. 13b), especially in view of restoring the Lord's sanctuary, and particularly against the

---

[148] Pomykala, *Davidic Dynasty*, 197.

[149] A. Steudel, "אחרית הימימ in the Texts from Qumran," *RevQ* 16 (1993): 225-246.

[150] Collins, *Scepter*, 61, identifies the 'Branch of David' in 4Q285 and 4QpIs^a with the Prince of the Congregation seen in light of Isa 11 in 1QSb.

[151] Atkinson, "Davidic Messianism," 115.

[152] Robert L. Hubbard Jr., *NIDOTTE* 2:556, says, "In general, the root *yš* 'implies bringing help to people in the midst of their trouble rather than in rescuing them from it." *BDB*, 447, includes Ezek 37:23 as a case for the verb conveying the sense of saving 'from moral troubles.'

[153] Block, *Ezekiel 25-48*, 357.

backdrop of the destruction of the temple as a vivid reminder of Israel's sin (l. 6a); the role of the Branch David is to save them particularly from their sin.

The Ezekiel pattern of YHWH rescuing Israel from external enemies precedes YHWH's appointment of the Davidic Shepherd who will 'save' Israel from their own uncleanness, does fit into the flow of the hopes expressed in 4Q174. Further, the eschatological conflict between the nations and 'the elect ones of Israel' – not merely 'the anointed' but 'the righteous' depicted as a collective (cf. Ps 2:1-9) in the following section in 4Q174 1:18-19 – reflects less of militant nature as to the last battle. Above all, the ultimate goal is identical as expressed in 4Q174 in terms of Ezekiel's vision: the return of YHWH's presence into the sanctuary among the eschatologically sanctified people.

To summarize, while the shepherd imagery is indirect, Davidic expectation, and various themes and motifs of Israel's restoration are certainly expressed. 4Q174 does not display any significant difference from the modifications of the Davidic expectation already evidenced in the OT Davidic Shepherd tradition, except for the separate roles of 'saving' and 'teaching' between the Branch David and the Teacher of the Law.

2.3.4 4Q161 (4QIsaiah Pesher^a)

4Q161 is a continuous *pesher* in which the consecutive biblical text, particularly Isa 10:21-11:5, is cited and interpreted. In fragments 2-6, we find the description 'Prince of the Congregation' used in a context where 4Q161's author interprets Isa 10:22-27. In fragments 8-10, Isaiah 11's 'shoot of the stump of Jesse' (l. 18) is clearly identified as a Davidic figure, the 'Branch of *David*' (דויד צמח, 4Q161 8-10:18); Isa 11:1 says 'a Branch' (חֹטֶר).[154] Undoubtedly, what we have here is the combination of the eschatological Davidic expectation and the restoration of Israel 'in the last days' (4Q161 2-6:12, 22; 8-10:18). The question remains: Do we find any shepherd imagery or any type of modification of Davidic expectation as a result of interacting with the OT Davidic Shepherd tradition?

The pesher of Isa 10:28-32 in 4Q161 5-6 is found in l. 2-3: "[...] when they returned from the Desert of Peo[ple]s [...] [...] the Prince of the Congregation (נשיה העדה), and afterward [...] will depart from [*them*]."[155] The messianic figure, נשיה העדה or just נשיה obviously does not appear in Isa 10:28-32. The OT precursor of the title נשיה in this eschatological context seemingly appears only in Ezek 34:23-24; 37:24-25 and chapters 40-48; the title נשיה is particularly Ezekielian. The reasons for mentioning the Prince by the author of 4Q161 may be explained in two ways. First, in the context of 4Q161 5-6, the Qumran

---

[154] Pomykala, *Davidic Dynasty*, 200; Zimmermann, *Messianische Texte*, 71; Atkinson, "Davidic Messianism," 117.

[155] Allegro, DJD 5, 13.

138   *Chapter 2. The Davidic Shepherd Tradition in Second Temple Judaism*

exegete seems to extend the range of Israel's final return; it is the gathering not merely from Assyria but from מדבר העמים ('the wilderness of peoples,' l. 14) which is certainly Ezekielian vocabulary.[156] The author of 4Q161 sees this as another redemptive event on par with the first exodus as YHWH will lift 'his rod' against the sea in the fashion of Egypt (l. 11). It is not certain whether or not the Ezekiel pattern was in the mind of the author of 4Q161 when he introduced the figure of נשיה at this juncture. Since the texts related to this figure are lost in the lacuna, there is not much to say about the identity and function of this נשיה.

Nevertheless, the remaining sections describe the time when this Prince will appear. The Prince will appear "when they [the remnant] returned from the Desert of peoples" (5-6:2) and "after it [yoke] will be removed from them" (5-6:3). Ezek 34 exhibits a similar pattern: YHWH rescuing the scattered flock followed (Ezek 34:1-22) by the introduction of נשיה (34:23-24). Likewise, 4Q161 presents the eschatological נשיה as YHWH's appointee at the time of the restoration.

Second, the title נשיה identifies its bearer as the Davidic messiah: "my servant David will be their prince forever" in Ezek 34:24; 37:25.[157] Similarly, 4Q161 preserves both the Prince and the Shoot of the stump of Jesse, and identifies them as the Branch of David in the same eschatological context as in fragments 5-6 and 8-10. Collins makes a significant statement as to the influence of Ezekiel's pattern on these figures:

Another modification of the kingship had even more far reaching implications in the Second Temple period. Ezekiel prophesies the restoration of the Davidic line on a number of occasions. Typically, however, he refers to the Davidide as a 'prince' (נשיא, Ezek 34:23-24; 37:25), the title used in the Priestly source to refer to the lay leader of the tribes (David is also called 'king' in Ezek 37:24). In the vision of a new order in Ezekiel 40-48, the role of the נשיה is reduced to providing for the cult. He becomes an apolitical messiah, subordinate in importance to the Zadokite priesthood.[158]

4Q161 is a case in point that illustrates the validity of these statements. In Ezekiel 34 and 37, YHWH's Davidic Shepherd for the restored community is the Davidic king 'over' them (Ezek 34:23; 37:24) and the prince 'in their midst' (Ezek 34:24; 37:25) at the same time. Characteristically, while the 'Branch of David' will 'rule over' all [בידו ובכול] the peoples (8-10:21), the Prince of 4Q161

---

[156] Wacholder, "Ezekielianism," 188.
[157] García Martínez, *People*, 258, n. 249.
[158] Collins, "The Nature of Messianism in the Light of the Dead Sea Scrolls" in *The Dead Sea Scrolls in Their Historical Context* (ed. Timothy H. Lim, Edinburgh: T&T Clark, 2000), 206; see also, Émile Puech, "Messianism, Resurrection, and Eschatology at Qumran and in the New Testament" in *The Community of the Renewed Covenant: The Notre Dame Sympium on the Dead Sea Scrolls* (ed. Eugene Ulrich and James Vanderkam; Notre Dame: University of Notre Dame Press, 1994), 238.

is the Prince of 'the Congregation' (נשיא העדה, 2-6:15).[159] The roles of these two messianic figures, however, seem to be differentiated in 4Q161.

The Prince is the leader of the restored community while the Branch of David is a kingly figure who rules not only over the renewed community but over 'all the peoples' [ובכול הגואים] including Magog (cf. Ezek 38-39).[160] Here, it is explicit that the Branch of David will rule over all the peoples, while it is implicit that Ezekiel's Davidic Shepherd will rule over the united nation of Israel (Ezek 37:15-23), even though his reign is to extend to the ends of the earth (cf. Mic 5:1-4). We have already noticed how this universal reign of the Davidic Shepherd figure is highlighted in *Ps. Sol.* 17-18 and *An. Apoc.* 90:17-18. Yet the Prince and the Branch of David of 4Q161 could be seen identical with each other as depicted in Ezekiel 34 and 37.[161]

Despite the militant aspect of the Branch of David, especially possibly attested by 4Q161's insertion of the phrase 'his sword' (חרבו) in 8-10:22, the term חרבו should be understood in light of the Branch's righteous judgement in l. 22-23. Further, the Branch shares judicial authority under the supervision of and in consultation with the priests. The roles between the Branch and the Prince seem thus to be differentiated but overlapped in part in 4Q161.

To conclude, 4Q161 displays significant continuity with the OT Davidic Shepherd tradition, particularly Ezekiel's vision, in terms of the major messianic figures, the ideal David, the Prince, and the overall pattern for the eschatological drama.[162] As demonstrated in 4Q174 as well (cf. 2 Sam 7; Isa 11; Ezek 37, 40-48), the figure and role of the Prince and the Branch in 4Q161 are envisioned in the matrix of the Ezekiel pattern for the future restoration of Israel.

2.3.5 4Q285 (4QSefer ha-Milhamah)

Many assert that the Prince of the Congregation is identified with the Branch of David in 4Q285 5:4: [והמיתו נשיא העדה צמ]ח דויד] ("and the Prince of the Congregation, the Bra[nch of David] will kill him").[163] Here, 'the Branch of David' is set in apposition to 'the Prince of the Congregation.' Given that the *hiphil* form of והמיתו can be vocalized as third person plural ('they will kill') or

---

[159] Allegro, DJD 5, 13.

[160] Atkinson, "Davidic Messianism," 117, thus assumes that the Kittim symbolized the Romans and their Herodian allies; Pomykala, *Davidic Dynasty*, 198, dates 4Q161 between 30 B.C. and A.D. 20.

[161] Pomykala, *Davidic Dynasty*, 203, argues for the functional identity of the Branch with the roles of other messianic figures in Qumran such as the Messiah of Aaron and Israel (CD 12:23-13:1; 20:1) and the Prince of the Congregation (CD 7:20).

[162] Zimmermann, *Messianische Texte*, 70, aptly notes that the preserved text of 4Q161 is the case in point for that "der qumranische Ausleger in Jer 10-11 einen ähnlichen apokalyptischen Gesamtplan wie in Ez 37- 48 sah."

[163] Pomykala, *Davidic Dynasty*, 206; Collins, "Nature," 215; García Martínez, "Messianic Hopes," 167; The translation is Pomykala's, *Davidic Dynasty*, 205.

as third person singular with a suffix ('he will kill him'), the reading of this line quickly becomes complicated.¹⁶⁴ If the former is the case, those two figures—the Prince and the Branch – can be the plural subject of 'they will kill' (l. 4), yet the absence of any conjunctive between these two figures in l. 4 points toward the case of the apposition.

If the plural reading holds, and if neither the Prince nor the Branch is the plural subject of והמיתו, this could imply that the Prince/Branch is killed by 'they,' i.e., the Kittim in 5:2 and 6+4:5 where we find a description of the Kittim confronting the Prince in the eschatological war. The widespread idea of 'the slain messiah,' however, has been refuted by many; the descriptions of the triumphant Prince/Branch in the broad context of the Qumran testimonies as well as in the immediate context of 4Q285 hardly support this thesis.¹⁶⁵ M. Bockmuehl proposes that 'they' refers to both the Prince of the Congregation and 'all Israel' mentioned in 4Q285 6+4:2 and third person singular in the suffix of והמיתו, 'him,' indicates the chief of the Kittim in 6+4:6: "[... and] they ('all Israel' in ln. 2) shall bring 'him' before the Prince [of the Congregation ...]." Then 5:4 reads, "the Prince of the Congregation (who is none other than) the Branch of David, will kill him (the chief of the Kittim)."¹⁶⁶

What we have in 4Q285 is a triumphant Davidic messiah. In Ezekiel 34-37, the Davidic Shepherd is identified with the Prince against the background of YHWH's eschatological and victorious confrontation with the wicked shepherds.¹⁶⁷ Perhaps due to the fragmentary nature of the text, one can find in 4Q285 neither the nature of the confrontation nor the characteristics of the Prince and the Branch, the Davidic messiah. Yet, the militant language attributed to the Prince/Branch in 4Q285 – "go into the battle; will kill him; and with wounds" (4Q285 5:3-4) – coincide with the absence of the shepherd imageries related to the Davidic messiah in this Qumran text.

Pomykala rightly contrasts the picture of the Davidic messiah of 4Q285 with the righteous Davidic Shepherd of *Ps. Sol.* 17: "therefore, the Davidic messiah (in 4Q285) has a militant, even violent role attributed to him – a role that stands in contrast to the less martial characterization of the davidic figure in *Ps. Sol.* 17."¹⁶⁸ If this is correct, then 4Q285's Prince/Branch also stands in contrast to

---

¹⁶⁴ García Martínez, "Messianic Hopes," 166-168.

¹⁶⁵ Philp R. Davies, "Judaism in the Dead Sea Scrolls: The Case of the Messiah" in *Dead Sea Scrolls in Their Historical Context* (ed. Timothy H. Lim; Edinburgh: T. & T. Clark, 2000), 237; Collins, "Nature," 215; Pomykala, *Davidic Dynasty*, 207; cf. G. Vermes, T. H. Lim and R. P. Gordon, "The Oxford Forum for Qumran Research Seminar of the Rule of War from Cave 4 (4Q285)," *JJS* 43 (1992): 85-90.

¹⁶⁶ Likewise, Zimmermann, *Messianische Texte*, 84, 86-87; similarly, García Martínez, *Scrolls*, 643, reads: "and the Prince of the Congregation will kill him, the bu[d of David ...]."

¹⁶⁷ M.G. Abegg, "Messianic Hope and 4Q285: A Reassessment," *JBL* 113 (1994): 81, dates the fragments to the early herodian periods; Atkinson, "Davidic Messianism," 118, thus proposes the herodian rulers as the counterpart of the envisioned Davidic messiah in 4Q285.

¹⁶⁸ Pomykala, *Davidic Dynasty*, 210.

the Davidic Shepherd/prince in Ezekiel 34 and 37 where shepherd imagery abounds as in *Ps. Sol.* 17-18.

2.3.6 *1Q28b/1QSb (The Scroll of Blessings)*

1Q28b 5:20-29, called the 'Blessing of the Prince of the Congregation,'[169] renders the roles of the Prince of the Congregation primarily in light of the descriptions attributed to the Branch of David in Isa 11:1-5. Scholars vary in their judgments on whether or not this is sufficient evidence to equate 1Q28b's Prince with the Branch of 4Q285 and 4Q174. According to Atkinson, "such passages as Isa 11 are used apart from connotations of Davidic status (in 1Q28b), with no clear allusion to a Davidic Messiah."[170]

On the other hand, García Martínez has no problem identifying 1Q28b's Prince with the Branch of Isaiah 11. Although the technical term is not used, García Martínez states that the figure of the Prince is described as the instrument chosen by God to "raise up [establish] the kingdom of his people for eve[r ...]" (l. 21) and that this shows clearly that he is a traditional Messiah-king.[171] For Collins, the clear application of Isaiah 11 to the Prince of the Congregation in 1Q28b 5:24-26 itself leaves no doubt about its identification with the 'Branch of David' in 4Q285 and 4Q161.[172]

More methodology than content seems to be at stake here as we pose the question of how to identify the messianic figure. Is it determined by explicit titles or from the context or profile of the figure? If one accepts the non-titular method, then it is no real problem here that 1Q28b supports identifying the Prince with the Davidic Messiah. No description is provided about the Prince in1Q28b that would prevent readers from seeing this figure as messianic.

Our concern, however, is centered on 1Q28b's possible interaction with the Davidic Shepherd tradition. 1Q28b column 5 presents a unique interpretive structure that weaves together various OT Davidic Shepherd texts. From lines 20-23, the major figure is the Prince of the Congregation of the new covenant community: "For the Master, to bless, the Prince of the Congregation who [...]; his [strength] and the covenant of the Community he [i.e., God] shall renew for him [i.e., the Prince], so as to raise up [establish] the kingdom of his people

---

[169] J. Charlesworth and L. T. Stuckenbruck, *The Dead Sea Scrolls* (vol 1, ed. J. H. Charlesworth; Tübingen: Mohr Siebeck, 1994), 119; cf. Milik, DJD 1, 120.

[170] Atkinson, "Davidic Messianism," 120-121; cf. Pomykala, *Davidic Dynasty*, 240-241.

[171] García Martínez, "Messianic Hopes," 165.

[172] J. J. Collins, "Messiahs in Context: Method in the Study of Messianism in the Dead Sea Scrolls" in *Methods of Investigation of the Dead Sea Scrolls and the Khirbet Qumran Site: Present Realities and Future Proposals* (ed. Michael O. Wise, Norman Gold, J. J. Collins and Dennis G. Pardee; New York: The New York Academy of Science, 1994), 217.

for ever" (l. 21).¹⁷³ Again, outside of Ezekiel 34, 37 and 40-48, nowhere else do we find נשיא with such expressive messianic connotations.

Further, here the Davidic promise (2 Sam 7:14-15) is renewed. YHWH's appointment of the Davidic Shepherd as the prince for the renewed community in Ezek 34:23-24; 37:24-25 is well suited to the background of 1Q28b 5:20-23. In fact, the Prince's role described in l. 22 amounts to the figure of the Leader/Teacher of the law: "reprove with fair[ness the h]umble of the [l]and (cf. Isa 11:4); and to walk before him perfectly in all the way[s of God]" (l. 22).¹⁷⁴ This is the role YHWH entrusted to the Davidic Shepherd in Ezekiel 34 and 37, that is, the role of sanctifying/saving the already-rescued people: "They will follow my laws and be careful to keep my decrees" (Ezek 37:24; cf. 34:23) so that "they will no longer defile themselves" (37:23). The affinity continues. 1Q28b l. 24-26 is replete with imageries of the righteous Branch of Isa 11:1-5 (cf. Jer 23:5; 33:15):

May you be [...] with the might of your [mouth], with your scepter [שבט] may you devastate the land, and by the breath of your lips may you kill the wicked [ones ...] (l. 24-25a; Isa 11:4); (May he give) [you a spirit of coun]sel and everlasting strength, a spirit [...] knowledge, and of fear of God (l. 25b; Isa 11:2); May righteousness (be) the loincloth of [your loins, and loyalt]y the belt of your hips (l. 26a; Isa 11:5).¹⁷⁵

It seems natural for 1Q28b's author to move from the Prince's role of sanctification within the new covenant community to the righteous quality of the Prince who uses the language of the righteous Branch of Isaiah 11 since only the "righteous"/blameless leader can lead the eschatological community to the path of righteousness. We have already observed how the authors of *Ps. Sol* 17-18 and *An. Apoc.* 90 carefully elaborate on the righteous character of the Davidic Shepherd/leader figure: the Davidic Shepherd who is 'free from sin' (*Ps. Sol.* 17:36; cf. Ezek 36:25) and the eschatological 'white bull' who 'transforms' both Jews and Gentiles into one holy people (*An. Apoc.* 90:38). The last section of 1Q28b 5:26b-29 deals with the expansion of the Prince/Branch's 'mighty' reign over the nations. The particular, interacting texts may be Micah 4-5:

May he make your horns of iron and your hoofs of bronze (l. 26b; Mic 4:13); For God has raised you to a scepter for the rulers be[fore you...all the na]tions will serve you, and he will make you strong by his holy Name (l. 27-28; cf. Mic 5:1-5).

In this section, the invincibility of the Prince/Branch is emphasized rather than his 'righteous' reign extended over the nations as characterized similarly by the term 'peace' in Mic 5:5. Victory, yet not peace, is in the spotlight. This is not to suggest, however, that the eschatological battle with the nations through which the Prince/Branch is supposed to extend his reign constitutes the major thrust

---

¹⁷³ Charlesworth and Stuckenbruck, *Scrolls* 1,129-131.
¹⁷⁴ Ibid., 129.
¹⁷⁵ Ibid., 130-131.

of restoration. Rather, once the restoration process is achieved, the victory is only seen as part of the expansion program of the restored community.

Thus, it seems that the Prince is characteristically 'of the Congregation,' while it is the Branch who is in charge of the next phase. Nevertheless, either one can easily be identified or confused with the other. The text of 1Q28b exhibits that these figures can be seen separated or identified; nevertheless, the text affirms both motifs of the Prince/Davidic Branch's righteousness and his (or their) extended reign over the nations, which are integral to the Davidic Shepherd tradition.

2.3.7 4Q521 (4QMessianic Apocalypse)

In 4Q521 there are several shepherd imageries that possibly convey Davidic messianic connotations. As we shall see in the followings, the משיחו ('his anointed') in fragment 2, 2:1 assumes the role of the eschatological shepherd as healer in the context of Israel's restoration.[176] What we have here, it appears, is shepherd imagery used in the restoration context. Yet is this figure specifically Davidic?

'His anointed one' in 4Q521 is imbued with the authority over the heavens and the earth (כי הש[מ]ים והארץ ישמעו למשיחו) so they might listen to YHWH's appointee who will lead them by means of YHWH's precepts (l. 1; cf. Ezek 34:23; 37:25); he (YHWH) will bring about "the eternal kingdom" to the pious (l. 7; cf. 2 Sam 7:14-15), especially freeing prisoners, giving sight to the blind, straightening out the twisted (l. 7; cf. Isa 35:4-6; 61:1-2); in his mercy (ובחסדו, 'dans son amour il')...the Lord will perform marvelous, unprecedented acts (l. 9b-11); for he will heal (ירפא, 'il guérira') the badly wounded and will make the dead live and proclaim good news to the poor (l. 12; cf. Ezek 37:1-4; Isa 61:1-2); and he will satisfy the [poo]r (ודלים ישי, 'et les [pauvre]s il comblera'), lead the outcast (נתושים ינהל, 'les expulsés il conduira'), and feed the hungry/invite to the banquet (ורעבים יעשר, 'et les affamés Il enrichira/invitera au banquet,' l. 13; cf. Ezek 34:10, 14, 18)."[177]

In short, the anointed himself assumes the restorative roles of Ezekiel's eschatological Shepherd. With regard to l. 13, Émil Puech's emendation and rendition, compared to those by Zimmermann and Martínez/Tigchelaar,[178] fully reflects the characteristics of the Ezekielian eschatological Shepherd. The profile of the משיהו in 4Q521 presents distinctively pastoral tasks, i.e., healing

---

[176] Here, משיחו may not be plural, "his anointed ones"; the plural form should be משיחיה as in frg.8, l. 9. Furthermore, the third person singular suffix refers to God as in 2:6: "his spirit; he will renew ..." See García Martínez, "Messianic Hopes," 168-169.

[177] Émil Puech, Qumrân Grotte 4 (DJD 25; Oxford: Clarendon Press, 1998), 10-11.

[178] Zimmermann, Messianische Texte, 344-345, reads l. 13, "[... Vertrie]bene(?) führen und Hunger[nde(?)] reich machen(?)"; Martínez/Tigchelaar, Scrolls, 1044-1045, "and [...] ... [...] he will lead the [...] ... and enrich the hungry."

144   *Chapter 2. The Davidic Shepherd Tradition in Second Temple Judaism*

the sick, feeding the hungry, even raising the dead, and particularly as Peuch makes clear: "il conduira les expulsés." As 'his anointed' comes to search and bring the outcast back to the community, all those shepherding activities coincide (cf. Matt 8-9).

Yet we are left with the question whether 4Q521 associates this 'anointed' as the eschatological Shepherd with Davidic connotations as well? M. O. Wise and J. D. Tabor think it does. To prove it, they reconstruct the very fragmentary section, 4Q252 2, 2:10-14 to be read messianically:

A[nd in His] go[odness forever. His] holy [Messiah] will not be slow [in coming] (l. 11); And as for the wonders that were not the work of the Lord, when he (i.e. the Messiah) [come]s (l. 12) then he will heal the sick, resurrect the dead, and to the poor announce glad tidings...(l. 14) He (i.e. the Messiah) will lead the [Hol]y Ones, he will shepherd [th]em.[179]

Nevertheless, it still remains uncertain as to the subject in l. 12-14 – whether it be the messiah or not. Davidic messiahship applied to 'his anointed' in 4Q521 may need not depend on these shaky reconstructions from the extremely fragmentary texts. García Martínez bypasses this problem, namely, the question of the subject of l. 10-13 as he classifies 4Q521 fragment 2, column 2 under the category of 'Davidic messianism' in the Qumran writings along with 4Q252, 4Q161, 1Q28b and 4Q285, though for different reasons.[180] Besides the fragmentary reference to the scep[ter] ([...] שׁבט) in fragment 2, 3:6, which could conceivably point to the 'royal Messiah,'

García Martínez asserts that "the horizon of eschatological salvation which the Lord achieves during his age seems to be limited to the eschatological congregation, the assembly of the faithful in the last times."[181] García Martínez does not find it problematic to identify 'the anointed' with 'the Prince' whose roles are particularly associated with the rescued community to be further restored. The Prince is indeed identified with the Branch of David in other Qumran texts, especially 1QSb 5:20-29. Thus 'the anointed' in 4Q521 2, 2:1, he argues, can legitimately seem a Davidic messiah.

Dismissing Collins' proposal which argues that "it is likely that God acts through the agency of a prophetic messiah in l. 12"[182] as an attempt to throw away the problem itself, Zimmermann underscores that it makes sense only when one holds God as the subject that "ab Z. 5 nur vom Handeln Gottes die Rede ist und das ‚Hinzukommen' des Gesalbten."[183] That is, the sayings in l. 5-13 are altogether more likely assigned to God rather than to 'his anointed' The descriptions with regard to משיחו in 4Q521 2, 2 takes a pause at the end of l. 2 which is taken over by the speech to the people in the next line: "Strengthen

---

[179] M. O. Wise and J. D. Tabor, "The Messiah at Qumran," *BAR* 18 (1992): 61-62.
[180] García Martínez, "Messianic Hopes," 168-170.
[181] Ibid., 169.
[182] Collins, *The Work of the Messiah* (*DSD* 1; Leiden: Brill, 1994), 100.
[183] Zimmermann, *Messianische Texte*, 364.

yourselves, you who are looking for the Lord, in his service."[184] As Zimmermann rightly points out, the Lord's anointed is 'to follow' YHWH's eschatological shepherding activities for his flock.

Importantly that 4Q521 2, 2:5-13 assigns those tasks of eschatological shepherding in l. 5-12 to YHWH himself, if seen in conjunction with the vision of Ezekiel 34-37, makes much more sense. The messianic figure in l. 1 is the one who will be appointed at the end to reign over the restored community invested with authority over the heavens and the earth. But the structure of 4Q521 2, 2 indicates that his being anointed will be a major consequence of the Lord's eschatological shepherding of his own faithful ones; thus, we can understand why the pious are exhorted to remain in hope in l. 2-4. Besides the messiah's universal reign (cf. Mic 5:1-5a), the general structure of 4Q521 2, 2:1-14 certainly reflects the Ezekiel pattern – YHWH's eschatological shepherding precedes his appointment of the Davidic prince/shepherd. The allusion to YHWH raising the dead (Ezek 37:1-15) is evident, thereby signifying the promised restoration of Israel in l. 12. The restoration language of Ezekiel 34-37 is fully presented.

Yet, García Martínez's identification of 'the messiah' in 4Q521 2, 2:1 with the Prince of the Congregation presumably described in l. 5-14 is not altogether invalid since the tasks of the eschatological Shepherd of Ezekiel 34-37 are often ascribed to the messianic figures in other Qumran texts such as 1QSb 5:20-29, CD-A 13:7-12. While it is clearly YHWH who fulfills the tasks of eschatological shepherding – including healing the sick both in Ezekiel 34 and 4Q521 2, 2:12 – what we see in Ezekiel 34-37, however, is the new David who is *both* Shepherd-king and Prince. This poses the complex question of whether we should identify the Prince with the Branch of David in the Qumran writings, especially as we consider other messianic figures in the Qumran, such as 'Teacher of the Law,' 'messiah(s) of Aaron and Israel,' etc.

There appears to be a tendency that 'the Branch of David' plays a role more akin to Ezekiel's Davidic Shepherd over the community and eventually over the nations, while the Prince 'of the Congregation' is more akin to Ezekiel's Davidic prince who tends to the internal issues of the community within the community. But, as we have examined earlier, 4Q252 (4QpGen) fragment 5 identifies 'the anointed' of righteousness (משיח הצדק) also with the 'Branch of David' (צמה דויד, l. 3-4) in the context where the Branch is said to obey the Law with his people of the community (l. 5-7). There are complicated cross-references among the anointed, the Prince, and the Brand David within various Qumran texts.

Although Davidic expectation is not explicit, 4Q521 belongs to one of the variations of the OT Davidic Shepherd tradition, especially those that stress the shepherd's role in the restoration of the community. Here, the roles performed by Ezekiel's eschatological Shepherd, particularly YHWH's intervention, are

---

[184] Peuch, DJD 25, 11.

closely related to 'the anointed' in view of the restoration of Israel, though the text does not specifically name him either the Prince or the Branch in 4Q521 2, 2. Yet surely this משיח is to be appointed [by YHWH] with the authority of heaven and earth (l. 1-2), but only after the shepherding activities of the eschatological Shepherd are accomplished among his own troubled flock (l. 5-14); thus the readers are exhorted to remain strengthened and hopeful, continually seeking the Lord (l. 3-4).

2.3.8 Damascus Document

2.3.8.1 CD-A 13:7-12

Now we turn our attention to CD texts, CD-A 13:7-12 and CD-B 19:8-13 (CD-A 7:10-21).[185] As we shall see, both involve shepherd imagery. CD-A 13:7-12 in particular reflects the imageries used of the eschatological Shepherd in Ezek 34:1-16, and CD-B 19:8-13 reflects the smitten shepherd of Zech 13:7. Since these are explicit, we will concentrate on how these shepherd imageries work with Davidic expectation – if indeed they do – in the context of Israel's restoration. First, CD-A 13:7-12 exhibits a fascinating gallery of the themes of the Davidic Shepherd tradition:

(l. 7) And this is the rule of the Inspector (Examiner, המבקר) of the camp. Let him instruct the Many about the works of (8) God, and shall teach them his mighty marvels, and recount to them the eternal events with their explanations, (9) He *shall have pity on them* like a father on his sons, and will *heal* (וישקה) *all the afflicted among them like a shepherd his flock.* (10) He will *undo all the chains which bind them, so that there will be neither harassed nor oppressed in his congregation.* (11) And whoever joins his congregation, he should examine, concerning his actions, his intelligence, his strength, his courage and his wealth; (12) and they shall inscribe him in his place according to his inheritance in the lot of light.[186]

In the same column (13), line 22 indicates that these are the regulations for the Instructor [משכיל, 'Master']ardestruction[187] who will walk in the prescribed way(s) in the appointed time when God visits the earth; while the community does not know the day nor the hour, the belief that God would ultimately intervene to put an

---

[185] The two medieval manuscripts of CD date from the tenth century A.D., yet the CD fragments were found in Qumran Caves 4, 5, and 6 (4QD, 5QD, 6QD). Charlotte Hempel, *The Damascus Texts* (Sheffield: Sheffield Academic Press, 2000), 23, dates the final composition of the Damascus Document towards the end of the second century B.C., at least before its earliest copy 4QDª was produced in the fist half the last century B.C.; cf. J. H. Charlesworth, *The Dead Sea Scrolls* 2 (Tübingen: Mohr Siebeck, 1994), 6-7.

[186] J. M. Baumgarten and D. R. Schwarz, *Scrolls* (ed. J. H. Charlesworth; Tübingen: Mohr Siebeck, 1994) 2:55, renders שקד as וישקוד ("to watch over"; "to show concern for," l. 9); García Martínez and Tigchelaar, *Scrolls*, 573, emend it to וישקה as in the translation in the text.

[187] Baumgarten and Schwartz, *Scrolls* 2, 55.

end to wickedness was a definite part of the community's world-view.¹⁸⁸ This 'age of visitation' is attested as to the time when the unfaithful will be put to the sword (CD-B 19:10), and most likely the Damascus Document expected the coming of the messiah(s) when God visits the earth 'at the end of the days.'¹⁸⁹ The 'Instructor' is to be understood in light of God's eschatological visitation in order to judge the wicked and to save the remnant.

CD-A 13:7-12 depicts God's eschatological visit through his agent for the sake of his flock; the agent is המבקר who "shall instruct Many in the deeds [works] of God" (l. 7b-8a). Charotte Hempel summarizes the roles of this Inspector [Examiner] that emerge in the passage in terms of "pastoral oversight of the camp, responsibilities assumed concerning the admission of new members, and matters of trade."¹⁹⁰

Among these roles, 'pastoral oversight' in l. 9-10 is elaborated upon in explicit shepherd images. In his exegesis of Ezek 34:1-16, W. Zimmerli notes that the official title המבקר in CD-A 13:7 might have some connection with the role of YHWH as the eschatological Shepherd who will 'seek' (בָּקַר) the lost and wounded flock: "I [YHWH] myself will search for my sheep" (vv. 11, 12); "I will search for the lost and bring back the strays" (v. 16a).¹⁹¹ Both play the identical role of the compassionate Shepherd for the sake of the harassed and down-trodden flock (cf. Matt 9:10-13; 35-36).

The parallel between CD-A 13:9-10 and Ezek 34:11-12, 16 as G. J. Brooke rightly observes, is unmistakable.¹⁹² The Inspector "shall have pity on them," that is, the congregation (l. 9a). The compassion motif is not verbalized in Ezek 34:11-16 but nonetheless is present as one of the major motivations for YHWH as the eschatological Shepherd to seek his harassed and downtrodden flock. This is crystallized in Davidic Shepherd texts such as Zech 10:6; Mic 7:18; Isa 40:9-11; 61:1-14: "I will restore them because I have compassion on them" (Zech 12:10). Israel's restoration and YHWH's compassion are intrinsically bound together.

Further, out of this compassion for the harassed flock, the Inspector is said to "heal (שקד) all of the afflicted among them" like a shepherd his flock (l. 9b). The rendition of שקד as וישקה by García Martínez and Tigchelaar appears the

---

¹⁸⁸ Collins,"The Expectation of the End in the Dead Sea Scrolls" in *Eschatology, Messianism, and the Dead Sea Scrolls* (ed. C. A. Evans and P. W. Flint; Grand Rapids: Eerdmans, 1997), 90.

¹⁸⁹ A. Steudel, "אחרית הימים," 238.

¹⁹⁰ Hempel, *Damascus Texts*, 40.

¹⁹¹ Zimmerli, *Ezekiel* 2:215-216.

¹⁹² Brooke, "Ezekiel in Some Qumran and New Testament Texts," in *The Madrid Qumran Congress* (ed. J. T. Barrera and L. V. Montaner; Leiden: Brill, 1992), 1:332 (cf. Brooks also notes Matt 9:36, 18:12). Moreover, C. Hempel, *The Laws of the Damascus Document: Sources, Tradition and Redaction* (Leiden: Brill, 1998), lists the relevant OT texts to CD-A 13:9-10: Ps 103:13; Ezek 34:12; Isa 58:6; Hos 5:11.

better choice than וישקוד by Baumgarten and Schwarz since the former rendition does not require the insertion of the second waw (ו); further, the latter bears mostly negative connotations in the OT (Ps 127:1; Jer 1:11, 12; 5:6; 44:27), while וישקה carries the idea of relief by giving water to the needy in other Qumran texts (1Q20 17:12; 4Q530 ii+6-12:7).[193]

Yet, if וישקוד likely implies the positive caring for those in trouble, both renditions, in their intention in the context, would not differ much from each other.[194] More significantly, it is worthwhile to recall at this juncture that one of the major tasks of the eschatological Shepherd for the flock is identified with healing the afflicted in Ezekiel's vision, despite of the absence of the verb רפה: "I will bind up the injured and strengthen the weak" (Ezek 34:16b).

Thus CD-A 13:9 may be one of the few Second Temple texts that attests to the healing aspect of the eschatological shepherd's tasks (cf. 4Q504 1-2, 2:14; 4Q 521 2, 2:12; *Apocryphon of Ezekiel* 5). The activities of the Inspector – the 'Seeker' in CD-A 13 – are focused on restoration and the exercise of justice within the community as members enter the community. Line 10 describes the Inspector's restoration of justice within the community; "there will be neither harassed nor oppressed [עשוק ורצוץ] in his congregation." Those images triggered by the expression עשוק ורצוץ echo the detailed situation of the community to be restored in Ezek 34:16-22: The rams and goats trample the pasture with their feet (vv.17-18); the fat sheep shove with their shoulders, with their horns butting the weak sheep until they are driven away, plundering them (vv. 19-21).

The *inclusio* of Ezek 34:16-22 – "I will shepherd the flock with justice" (v. 16)/ "I will judge between one sheep and another" (v. 22) – confirms that YHWH's eschatological visitation as the true shepherd of Israel aims to restore justice starting from the rescued community. Further, YHWH appoints in vv. 23-24 the Davidic Shepherd/Prince, presumably to accomplish this job for the community. In short, the Instructor of the Damascus Document appears to take up the role assigned to the Davidic Shepherd/Prince in Ezek 34-37.

Who is המבקר? The Inspector or the Inspector of Many (cf. 1QS 6:12, 20) appears to be the dominant authority of the camp in CD-A 13:7-12. In this section the figures of the priest or Levite encountered in CD 12:22b-13:7a are absent. More questions emerge. What is the relationship between the Inspector and the Prince, the Branch, or the Messiah(s) of Aaron and Israel? Is this Inspector another messianic figure? If so, he certainly does not appear to be a militant figure. The role of המבקר, i.e., to establish justice or righteousness within the community, is nearly identical with the task assigned to the Prince or also possibly to the Branch of David found elsewhere in Qumran texts like 4Q521 2, 2:3-14 and 4Q252, 5:5-7. The characterization of המבקר by means of

---

[193] Martin G. Abegg, *The Dead Sea Scrolls Concordance* (Leiden/Boston: Brill, 2003), 1:938.
[194] Especially with 'those afflicted,' see Baumgarten and Schwartz, *Scrolls* 2, n. 199.

the shepherd imageries in CD 13:9-10 thus indicates that the Instructor can be seen as a messianic though not militant figure in the context of Davidic expectation at the end of days. Above all, being vested with the Davidic connotations, he is presented as the eschatological healer as well.

2.3.8.2 CD-B 19:7-13 (CD-A 7:10-21)

Redactional analyses of the relationship between CD-B 19:7-13 and CD-A 7:10-21 can be at best speculative, as P. R. Davies summarizes: "A and B are prior ... differences are deliberate, and they are accidental."[195] While CD-A 7:10-21 cites Isa 7:17, Amos 5:26-27 and then interprets Num 24:7 with the *star* (הכוכב) read as the 'Interpreter of the Torah' (דורש התורה) and the *staff/scepter* (השבט) the 'Prince of all the Congregation,' CD-B 19:8-13 – frequently referred to as the *Zechariah-Ezekiel midrash/pesher* – cites Zech 13:7 and Ezek 9:4 against the background of the coming of the 'messiah(s) of Aaron and Israel.'[196]

Based on the method of 'a network of interlocking references,'[197] Collins identifies the *star* as the Interpreter of the Law with the messiah of Aaron, and the *scepter* as the Prince with the messiah of Israel, particularly in CD-B 19:7-13 (CD-A 7:10-21).[198] The expression 'messiah of Aaron and Israel' is illusive in terms of its numeric reference to the messianic figure,[199] but to match any of these messianic figures with any other among them is a presumptuous task. The texts themselves simply do not present consistent relationships among these figures.[200]

Nevertheless, CD-B 19:10-11 is yet another passage in which the arrival of the messianic figures defines an eschatological moment.[201] The imagery of the shepherd smitten by the sword (CD-B 19:7b-9a) is obviously drawn from Zech 13:7. Then, the remnant is designated as 'the little ones' (הצוערים) and 'the poor

---

[195] Davies, "Judaism," 225-226. For instance, J. Murphy-O'Connor, "The Damascus Document Revisited," *RB* 92 (1987): 225-245, takes MS A as original; M. A. Knibb, "The Interpretation of Damascus Document VII, 9b-VIII, 2a and XIX, 5b-14," *RevQ* 15 (1991): 247, Amos-Numbers midrash as an insertion and Zechariah-Ezekiel midrash as original; yet, S. A. White, "A Comparison of the 'A' and 'B' Manuscripts of the Damascus Document," *RevQ* 48 (1987): 537-553, says that A and B coincidentally suffered haplography; also, see Jonathan G. Campbell, *The Use of Scripture in the Damascus Document 1-8, 19-20* (Berlin: Walter de Gruyter, 1995), 153-158.

[196] Baumgarten and Schwartz, *Scrolls* 2, 27. Cf. P. R. Davies, *The Damascus Covenant* (JSOTSupS 25; Sheffield: JSOT Press, 1982), 146.

[197] Timothy H. Lim, *Pesharim* (London: Sheffield Academic Press, 2002), 48, calls it a "thematic pesharim."

[198] Collins, *Scepter*, 82.

[199] For its singular reference, see CD 12:23;14:19;19:10; 20:1, for the plural, 1QS 9:11.

[200] Davies, "Judaisms," 228, suggests '*a single teacher-messiah*' as the most consistent representation of the Damascus community (italics his).

[201] J. Vanderkam, "Messianism in the Scrolls" in *The Community of the Renewed Covenant* (ed. E. Ulrich and J. Vanderkam; Notre Dame: University of Notre Dame Press, 1994), 214.

ones of the flock' (cf. Zech 11:11) who will escape in the age of the visitation, while "those who escape at the time of the visitation; but those who remain will be handed over to the sword when the Messiah of Aaron and Israel comes" (CD-B 19: 9b-11a).[202] Does the 'Messiah of Aaron and Israel' have anything to do with the sword that will strike the shepherd of YHWH? The text can be read in preference to either of these opposing interpretations. It could be YHWH himself or the 'messiah of Aaron and Israel' who will strike the shepherd; furthermore, it is far from certain whether the 'Messiah(s) of Aaron and Israel' refers to one or two messiahs.

Despite the ambiguity of the referents, we can safely conclude the following: (i) The Damascus document understands the smitten shepherd as the one set to oppose the coming messiah(s); and (ii) the death of the shepherd and the coming of the messiah(s) will coincide with the eschatological restoration of 'the poor ones of the flock.'[203] Perhaps the messiah(s) in CD-B 19:10-11 assume the role of the smitten shepherd. As the smitten shepherd is judged and removed, the messiah(s) of Aaron and Israel will lead the flock, fulfilling the tasks to be performed by the eschatological shepherd as illustrated by the eschatological shepherding activities of the Inspector among the community (CD-A 17: 9-10).

The context of Zech 13:7 in Zechariah 9-13 finds the people rejecting YHWH as their shepherd, followed by YHWH's consequential maltreatment of the flock and their shepherds. The context of CD-B 19 is similar to Zechariah 9-13 with the citation of Zech 13:7. Yet, other important motifs are omitted, such as the mysterious intimacy of YHWH with the smitten/pierced shepherd(s), the lament of the people over the pierced shepherd, and the theme of the forgiveness of sins. The remainder of the text in CD-B 19:10-35, however, is replete with lamentations over the group in which each one "has chosen the stubbornness of his own heart" and "did not keep apart from their sins" (l. 20; cf. l. 33).[204] Quoting Deut 32:33, the author of CD-B 19 continues to complain that Israel, rebelling with insolence, walks in the path of the wicked ones (l. 21).

In other words, CD-B 19 also shares with Ezekiel and Zechariah the realization that they cannot expect YHWH to fulfill his promise because of Israel's injustice and crooked heart: "You are going to possess the nations ... but because he loved your fathers and keeps the oath" (l. 28). Moreover, as the day of gathering arrives (l. 35), only the faithful will join the new community. In this context, 'the smitten shepherd' is clearly one of the wicked leaders of Israel, or one of the kings of the peoples (l. 23) who will be killed by the sword of the messiah(s) of Aaron and Israel (l. 10-11a).

It is noteworthy that this Qumran text weaves together texts from Zechariah and Ezekiel thus utilizing the Davidic Shepherd tradition, though it is not

---

[202] Baumgarten and Schwartz, *Scrolls* 2, 30-31.
[203] Cf. Vanderkam, "Messianism," 230-231.
[204] García Martínez and Tigchelaar, *Scrolls*, 577-578.

strictly limited to the Ezekiel 34-37 passage. While the identity of the messiah(s) of Aaron and Israel remains ambiguous, CD-B 19:7-13 still clearly exhibits Zechariah's Davidic Shepherd tradition. While 'the poor ones of the flock' will constitute the remnant at this eschatological time, the sins of the flock are attested, along with the consequential judgment upon the shepherd to be smitten, who is apparently the representative of the wicked leaders of Israel, set in contrast to the 'messiah(s) of Aaron and Israel.'

2.3.9 Summary and Conclusion

Qumran texts involving Davidic expectation have undeniably been influenced by the OT Davidic Shepherd tradition. Besides Isaiah 11 and Numbers 24, passages like Ezekiel 34-37, Zechariah 13, and Micah 5 can be verified as traditions with which Qumran interpreters engaged. While the case of Zechariah shows rather minimal influences (CD-B 19), we see that especially Ezekiel's eschatological pattern of the restoration and its shepherd imageries leave clear marks on those Qumran texts examined thus far. Explicit citations as well as implicit allusions abound. If David is remembered, he is often remembered through the lens of Ezekiel's modified vision of the Davidic Shepherd (4Q504).

Particularly, the variety of shepherd imageries found in Ezekiel's vision helps us recognize the modifications maintained by the Qumran community with regard to Davidic expectation. Where the plethora of shepherd imageries are associated with the messianic figure, the militant nature of the figure tends to be suppressed or altered (4Q174) in a fashion similar to what is found in *Ps. Sol.* 17-18. Ezek 34:23-24 and 37:24-25 are key texts to unlock the precise relationship between the 'Prince of Congregation' and the 'Branch of David' in Qumran texts such as 4Q161 and 1Q28b.

Even the figure of the Inspector in CD-A 13 closely resembles the Prince in terms of his shepherding activities including, significantly, "healing" (l. 9). It is not altogether unlikely to think that the 'anointed' one in 4Q521 2 is closely interrelated with the roles of the eschatological Shepherd who will 'heal the badly wounded' as he will 'lead the outcast' (l. 12; cf. Ezek 34:1-16). Conversely in those texts involving the Davidic messiah figure, but lacking in shepherd imageries often treat this figure with militant connotations (4Q285).

Summarizing the nature of messianism found in the Dead Sea Scrolls, J. J. Collins uses key terms like 'the limited kingship,' meaning *'the Davidic messiahship qualified by priestly authority,'* especially in the case of the 'dual messiahship' of Aaron and Israel.[205] Collins' assessment of the Qumran's understanding of the Davidic messiah as 'limited kingship' may well be correct. Likewise, this qualified Davidic kingship is the key feature of Ezekiel's identification of 'My Servant David' (עַבְדִּי דָוִיד) as YHWH's appointed

---

[205] Collins, "Nature," 217; idem, "Messianism," 227; idem, *Scepter*, 112.

Shepherd/King and Prince (נָשִׂיא) in 34:23-24; 37:24-25 as well. Importantly, *it is the set of shepherd imageries* drawn from the OT Davidic Shepherd tradition, including Jeremiah 23, Isaiah 11, and Psalm 2, that accomplishes these significant modifications concerning YHWH's promise for the perpetuity of the Davidic monarchy in those Qumran texts we have examined thus far.

Nevertheless, what we have observed is Qumran's particular understanding of the messianic figure(s) at the time of Israel's restoration. Notably it is both sketchy and complex. The case of 4Q521 demonstrates how Davidic expectation (2 Sam 7) is read in light of the later Davidic Shepherd tradition (Isa 6, 11; Ezek 34, 37, and 44; Mic 5; cf. Amos 9), particularly in view of the rebuilding of YHWH's sanctuary, that is, the return of the divine presence in the midst of the eschatological Israel.

Yet in the Qumran texts we encounter neither singular nor dual but plural Davidic messianic figures: the Prince of the Congregation, the Branch of David, the Interpreter of the Law, the Teacher of Righteousness, and even the Instructor. How these messianic figures should be paired in view of the title, 'messiah(s) of Aaron and Israel,' can be a significant task that goes beyond the scope of this chapter.[206]

What we can point out, however, is that all of these messianic figures seem to share the multiple roles of the Shepherd(s) from the OT Davidic Shepherd tradition. Particularly, the Davidic Shepherd in the relevant OT texts attests that the 'new David,' appointed by YHWH at the end of days, will not be simply a militant warrior like those kings before the exile, nor the Hasmoneans, or even the herodian kings. Rather, the eschatological David would assume all of the roles necessary for the sake of 'saving' YHWH's 'rescued' eschatological community as their Davidic Shepherd-king, Prince, Leader, and yes of course, as their Teacher of righteousness.[207] It is also likely that righteousness is a major characteristic of this coming Davidic Shepherd, which would be true of various ideal messianic figures presented by *An. Apoc.* 90, *Ps. Sol.* 17-18 as well as the Qumran texts examined in this section.

Further, those various Davidic messianic figures found in the Qumran texts mainly focus on the restoration of the eschatological community against the backdrop of the outer enemies, containing only minimal hints of a universal peaceful reign of the righteous Davidic messiah over the nations. The phrases that typically indicate the universal reign of the Davidic messiah – "to the ends of the earth" (*Ps. Sol.* 17:31; Mic 5:4), 'you people, man' (*An. Apoc.* 90:18;

---

[206] Cf. Davies, "Judaisms," 219-232.

[207] For instance, J. J. Collins, "Teacher and Messiah? The One Who Will Teach Righteousness at the End of Days," in *The Community of the Renewed Covenant: The Notre Dame Symposium on the Dead Sea Scrolls* (ed. E. Ulrich and J. Vanderkam; Notre Dame, Ind.; University of Notre Dame Press, 1994), 193-210, views this figure yet as a new Moses rather than a Davidic.

Ezek 34:31) – are rather scant in the Qumran Davidic messianic texts (cf. 'sanctuary of man,' in 4Q174 1-2, 1:6).

The Lord will visit the earth (1Q28b 5:24-28; CD-A 13:22; CD-B 19:10) and the Davidic 'Branch' will conquer all the peoples, the enemies of the chosen (4Q161 8-10:18); most likely 'his anointed' will receive authority over the heavens and earth (4Q521 2, 2:1), though this Davidic Shepherd/Prince figure is often identical with the messiah 'of the Congregation' (4Q521 2 2:4-13). The universal reign of the Davidic Shepherd beyond the national interests of Israel is already implied in Ezekiel, especially in terms of God's concerns for his holy Name exalted among the nations as well in other OT texts (Mic 5:4; Isa 11:1-5).

In the Qumran sectarian writings, however, the Davidic Branch plays only a part of the eschatological vision, and the extended pacifistic Davidic Shepherd-kingship envisioned in the tradition seems to be suppressed as well, while the Ezekiel pattern for YHWH's eschatological shepherding, followed by the arrival of 'his anointed,' seems to be maintained.

2.4 The *Targum of Ezekiel*

Samson H. Levey thinks the Targum of Ezekiel is unique because unlike the Targum of other prophetic books, 'the outright designation messiah' is nowhere to be found. Even though there is ample opportunity for messianic interpretation, Levey points out that "the absence of the Messianic in the Targumic exegesis of Ezekiel is clear and unmistakable."[208] While Levey proposes that *Tg. Ezek* substitutes Merkabah Mysticism for messianic activism in the wake of the catastrophe of A.D. 70, his argument for *Tg. Ezek*'s 'suppressed messianism' relies on the fact that the title משיחא is simply absent.[209] It is true that the Hebrew text of Ezekiel 34-37 does not contain the designation, משיחא.

While recognizing that the Targum Onkelos to the Pentateuch – a more literal Aramaic version than *Tg. Ezek* – gives an outright messianic interpretation to Gen 49:10 and Num 24:17, the absence of any designation in *Tg. Ezek* appears somewhat puzzling indeed.[210] Nevertheless, a titular approach alone does not prove either the absence or the presence of messianism in the text.

As we shall see, *Tg. Ezek* strips away the shepherd imageries from the Hebrew text of Ezekiel 34-37 through a literal translation, most likely in an attempt to make Scripture more intelligible to Aramaic-speaking Jews. Consequently, the complex nexus of shepherd imagery and Davidic expectation

---

[208] S. H. Levey, *The Targum of Ezekiel: Translated, with a Critical Introduction, Apparatus, and Notes* (The Aramaic Bible 13; Wilmington, Del.: Michael Glazier, 1987), 4-5.

[209] S. H. Levey, *The Messiah: An Aramaic Interpretation: The Messianic Exegesis of the Targum* (Cincinnati: Hebrew Union College Jewish Institute of Religion, 1974), 79.

[210] Ibid., 86-87.

in view of Israel's restoration dissolves. With shepherd imagery dismantled, Davidic expectation seems as if it might be unmodified or replaced by the activities of God's surrogates such as '*Shekinah*' (36:5,20; 37:27) and '*Memra*' (34:24, 30; 36:6, 9, 36, 37); and the vision for the eschatological restoration is kept strictly nationalistic.

The *Targum Ezekiel* takes the entire chapter of Ezekiel 34 with its poetic pastoral image as metaphor and translates it in terms of applied meaning. The *Tg. Ezek* renders 'shepherds' (רועי) as 'providers' (פרנסי) and the 'sheep' as 'people' throughout its translation of Ezekiel 34; likewise, 'one shepherd' is rendered 'a provider' (פרנס, v. 5).[211] Further, the various types of assaulted sheep in Ezekiel 34 are translated into rather more historically sensitive expressions. For instance, 'the strays' YHWH would search out with the lost in MT can be seen as 'those that have been exiled' and YHWH 'strengthening' the weak would imply that he will 'sustain' the weak 'with hope' in *Tg. Ezek* 34:16.[212] Thus the hope of restoration remains alive and readers are encouraged to endure and hold on to that hope.

On the other hand, 'the fat the strong' are identified with 'the transgressors and the sinners' (חטאיא וית חייביא) in *Tg. Ezek* 34:17. These people are the fat and the strong, and are seen as law-breakers in *Tg. Ezek*. Eventually they are identified with 'the rich' in 34:20: "I [YHWH] will judge between 'the rich man and the poor man' (גבר עתיר ובין גבר מסכין), ['the fat and the lean sheep,' וּבֵין שֶׂה רָזֶה בֵּין־שֶׂה בְרִיָה in MT])."[213] This line of reading shares the Similitudes' (1 *En.* 37-71) view of the wicked – the powerful and rich ones, often called 'landowners' and eventually identified with 'money-lovers' (63:10) – and it is these who 'oppressed with wickedness and force' in *Tg. Ezek* 34:21.

Clearly YHWH's oppressed flock in *Tg. Ezek* are the poor. Accordingly, the shepherd/leader's prime task is to provide. The *Tg. Ezek* translates MT's 'to shepherd' primarily as 'to provide' (34:3, 8, 10, 13, 14, 16), not just 'to lead'; thus YHWH will reveal himself (*Tg. Ezek* 34: 11, 20; cf. 36:9) first and foremost as the eschatological פרנס ("provider")! *Targum Ezekiel* points out how the leaders (MT, 'shepherds') exploit the people as they (just as their 'servants' copy them) and fail to provide them with their basic necessities, such as food and drink: "And My people must eat the food left over by your servants, and must drink the drink left over by your servants" (*Tg. Ezek* 34:20). This may indicate, as Levey suggests, that 'the fat and the strong' refers to the priests in the Temple since the mistreatment of the populace by the 'servants' of the

---

[211] Alexander Sperber ed., *The Bible in Aramaic* 3: *The Latter Prophet according to Targum Jonathan* (Leiden: Brill, 1962), 347-348; cf. Moshe Eisemann, *The Book of Ezekiel: A New Translation with a Commentary Anthologized from Talmudic, Midrash, and Rabbinic Sources* (Brooklyn/Israel: Mesorah, 1980), 2:372; cf. Levey, *Targum*, renders פרנסי as "leaders" in 34:2, likewise in v. 5.

[212] Following Levey's rendition in *Targum*, 97.

[213] Sperber, *Bible in Aramaic* 3:349.

priests is mentioned in Josephus (*Antiquities* 9:2, 206) and recorded elsewhere (*b. Pes.* 57a).[214] If this is right, the MT's shepherd(s) are interpreted to be the priestly leaders in the *Tg. Ezek* (cf. 34:23). Before we move to the Targum's view of Davidic expectation reflected in its translation, one more phenomenon deserves our attention, that is, the Targum's rendering of עֵץ and שֵׁבֶט in Ezek 37:16-19. The chart below summarizes and compares the various choices of the translations available among MT, LXX and the Targum:

	MT	LXX	Targum
Ezek 37:16	עֵץ [2x]	ῥάβδος [2x]	לוּח ('tablet,' [1x])
	Joseph	Joseph	שבט of Joseph
	עֵץ Ephraim	ῥάβδος Ephraim	שבט of Ephraim
37:17	עֵץ	ῥάβδος	לוּח
37:19	עֵץ	φυλή	שבט of Joseph
	hand of Epharim	hand of Ephraim	שבט of Ephraim
	שֵׁבֶט of Judah	φυλή	שבט of Israel
	עֵץ	φυλή	שבט of Judah
	one עֵץ	one ῥάβδος	one עַם ('people')

In Ezek 37:16-19 (MT), both עֵץ and שֵׁבֶט can mean 'tribe,' but in the case of Judah in verse 19 the word for staff is not עֵץ but שֵׁבֶט, though עֵץ refers to a staff in all of the instances in verses 16-19 (MT). The fact that MT reserves שֵׁבֶט exclusively for the tribe of Judah (v. 19) may reveal MT's sensitivity about Judah's significance in the passage's messianic connotations (Gen 49:10; cf. 4Q252 5:1). More important, in the context of Ezek 37:15-25, the united staff/tribe eventually is entrusted to the Davidic Shepherd in verses 24-25.

The MT's use of שֵׁבֶט for Judah and the interchangeability between its two meanings, i.e., staff and tribe, contributes to the dramatic build up of the shepherd imagery that points to the future David in Ezek 34:23-24 as the Davidic Shepherd who would hold the united scepter – the united eschatological people – in his hand.[215]

In the OT Davidic tradition, Ezekiel's word play of this kind may well be linked, on a semantic level, to 'the iron staff' that 'the Lord's anointed' holds in his hand as he shatters the nations at the eschaton (cf. Ps 2:9). The Greek translation barely preserves the richness of the metaphor, which basically serves to untangle the complexity. Where עֵץ clearly means 'tribe,' the LXX renders it so (φυλή, v. 19 [2x]); otherwise, it is translated simply as ῥάβδος. But as for *Tg. Ezek*, we notice that the Targum paraphrases those parabolic expressions, whether they mean staff or tribe to uniformly mean 'tribe' (שבט), except that it replaces 'the staff' with 'the tablet' in a few places. Given that *Tg. Ezek* tends to

---

[214] Levey, *Targum*, 97.
[215] Perhaps, *Tg. Ezek*'s rendition, one עַם in v. 19, may help more clearly bring out this point.

156   *Chapter 2. The Davidic Shepherd Tradition in Second Temple Judaism*

render various figurative expressions into literal language and that particularly *Tg. Ezek* 37:16-19 retains שבט, not only for Judah but also for Joseph, Ephraim, and Israel, Levey's rendition of the term meaning consistently 'tribe' is probably correct.[216] This shows that *Tg. Ezek* is not fully or as much as aware of the significance of the metaphor of שבט in the context of Ezekiel's vision for the Davidic Shepherd and Israel's restoration. What happens when the Targum renders shepherd as 'provider/leader' is similar to what occurs when the Targum paraphrases the metaphoric matrix among the various images of 'staff,' 'tribe,' and 'hand' in Ezek 37:15-23. The ambivalent yet rich semantic matrix of the term simply evaporates.

The *Targum Ezekiel*'s rendition displays, in this respect, a certain discontinuation of this biblical shepherd imagery. Devoid of shepherd imagery, Davidic expectation seems to have been set aside in *Tg. Ezek*'s translation. The Davidic 'shepherd' in *Tg. Ezek* 34:23 is *distinctly a 'provider'*(פרנס) set in opposition against the wicked leaders who exploit the people: "And I will raise up one provider (פרנס חד) who shall provide (ויפרנים) for them, My servant David; he shall provide for them and he shall be their provider."[217] The explanation inserted in the Targum indicates that its emphasis is on future provision.

The Targum's tendency of flattening the shepherd metaphor becomes evident once again as it renders both MT's נָשִׂיא (Ezek 34:24; 37:25) and מֶלֶךְ (37:24) uniformly into 'king': 'My servant David shall be מלכא among them' (same as in 37:25).[218] The expression, 'king *among* (ב, not על) them' indicates its unawareness of the subtle distinction between the roles of Prince and the [Davidic] King in Ezekiel 34 and 37 and, likely, also which exists between the Prince and the Branch David in Qumran texts.

Even if this is seen against the OT usage for 'the prince among (them)' as approaching a customary expression (Ezek 12:12; 34:24; 46:10; cf. Gen 23:6; 49:26; Deut 33:16; 2 Chr 11:22), the Targum's 'king *among* them' appears as an anomaly. One can only speculate why *Tg. Ezek* replaces MT's 'prince' with 'king,' but it is noticeable that removing the shepherd imagery coincides with the disappearance of the term 'prince,' thus leaving the Davidic expectation unmodified and vague.

Last, the vision of Israel's restoration consciously limits itself to national interests, and avoids making any possible offence to the foreign nations. For instance, as Levey rightly points out, *Tg. Ezek* retains the reference to 'wild beasts' literally in the Hebrew found in 34:25, "in spite of the fact that the normal Targumic exegesis would find in this a good opportunity to interpret it

---

[216] Levey, *Targum*, 104.

[217] Sperber, *Bible in Aramaic* 3:348; Levey, *Targum*, 104, renders פרנס as "leader" this time (cf. vv. 2-16).

[218] Levey, *Targum*, 98, does not note the change here in his translation while he does in 37:25 (104).

as the yoke of the foreigner."²¹⁹ In this respect, *Tg. Ezek* 34:31 intrigues the readers: "And you My people, the people *over whom My name is called*, you are *the House of Israel* (בית ישראל; cf. MT, 'people [הָאָדָם] you')."

It is noteworthy that the term הָאָדָם, YHWH's flock, is open to be read in terms of the general humanity in Ezekiel, but definitely restricted to the house of Israel in the Targum. Levey notes that "the Targum equates the Masoretic *'adam* with the house of Israel, perhaps implying that Israel personifies Adam, of humanity *par excellence*."²²⁰ The phrase, 'the house of Israel,' indeed appears frequently in the context of the restoration in Ezekiel 34-37 (34:30, 31; 35:15; 36:10, 21, 22, 32, 37; 37:11). Yet the intended ambiguity of הָאָדָם at the end of Ezek 34, which comes after YHWH's appointment of the Davidic Shepherd over his united eschatological people in verses 23-24, is simply flattened out by the Targum's paraphrase. Its interpretive rendition implies the Targum's restricted vision for Israel's restoration. The translation of *Tg. Ezek* decodes the various shepherd metaphors thus making interpretive choices.

Davidic expectation in *Tg. Ezek*, as compared to the normative Targumic exegesis, is rather deflated. Even the hope of the restoration expected in the Targum is downplayed in such a way that it neither threatens nor attracts the nations. Truly, at the expense of the richness and complexity of Ezekiel's shepherd imagery (MT) – especially its specific function to modify Davidic expectation – the Targum, rather, addresses its own readers and attempts to meet their needs: the primary role of the Leader/Shepherd is to provide for and to feed the heavily exploited flock.

2.5 The Fourth Book of Ezra

The theme of 'theodicy' lies at the heart of this piece of apocalyptic literature.²²¹ Through the mouth of Ezra, the author of the book raises questions about God's

---

²¹⁹ Levey, *Messiah*, 82.

²²⁰ Levey, *Targum*, 99, n. 19, refers to R. Simeon b. Yohai who cites this verse to demonstrate that Israel is designated by God as Adam in contradiction to the other nations of the world (*b. Yeb.* 60b-61a).

²²¹ J. Schreiner, "Das 4 Buch Esra," *JSHRZ* 5.4 (Güntersloher Verlagshaus Gerd Mohn, 1984), 301-303, dates the book after A.D. 70 during the last decade(s) of the first century; Tom W. Willett, *Eschatology in the Theodicies of 2 Baruch and 4 Ezra* (JSPSup 4; Sheffield: Sheffield Academic Press, 1989), 51, date 4 Ezra soon after the fall of Jerusalem in A.D. 70, so does M. P. Knowles, "Moses, the Law, and the Unity of 4 Ezra," *NovT* 31 (1989): 273; B. M. Mezger, *OTP* 1:520, dates the original composition around A.D. 100; likewise, Egon Brandenburger, *Die Verborgenheit Gottes im Weltgeschehen: Das literarische und theologische Problem des 4.Esrabuches* (Zürich: Theologischer Verlag, 1981), 13; for the later date around A.D. 150, see Frank Zimmermann, "The Language, the Date, and the Portrayal of the Messiah in IV Ezra," *HS* 26 (1985): 203-218, who suggests the vantage point of the author as Babylon (215).

justice over Israel and the nations (3:4-27; 4:4; 7:21-24; cf. 14:27-36).[222] He complains about the inequity of God's dealings with Israel; despite the apparent sins of the nations, he argues that God deals too harshly with Israel, and inflicts excessive punishment: "Are the deeds of those who inhabit Babylon any better? Is that why she gained dominion over Zion?" (4 Ezra 3:28, 31).

This motif of the excessive suffering of Israel is apparent in *An. Apoc.*, but 4 Ezra raises and pushes to the limit the question of theodicy. Ezra realizes that not only Israel but eventually all humanity was unable to do right because of the evilness of the human heart: "Yet you did not take away from them their evil heart, so that your Law might bring forth fruit in them" (4 Ezra 3:20, 26; 6:38-59; 7:1-16; 8:35).[223] This may recall the plight of Israel presupposed in the promise of 'a new heart' and 'a new spirit' set forth in Ezek 36:24-32. Thus, here, we witness the universalization of what Zechariah brought up as the fundamental issue of YHWH's flock, i.e., their sin.

Whether the charge be inequity or unfairness, sin is the root problem for the preservation of evil found in 4 Ezra. Eventually, 4 Ezra presents the eschatological settlement for these problems of theodicy. This poses the question: How could the root cause of evil – the people's sin – be overcome? Tom W. Willet explains this: "4 Ezra found the resolution of the sin problem in the mercy of God."[224] The plight is the sin of the people, and the ultimate motivation for God's faithfulness is grounded solely in his mercy. The thought-world of 4 Ezra thus echoes the plights and resolutions imbedded in the Davidic Shepherd tradition of both the OT and Second Temple periods, particularly that of Zechariah.

All three elements of the Davidic Shepherd tradition are discovered in 4 Ezra: Davidic expectation, shepherd imagery, and the restoration of Israel. In the first two chapters we read of the hope of Israel's restoration and note the shepherd imagery: "Await your *shepherd*; he will give you everlasting rest, because he who will come at the end of the age is close at hand" (2:34).[225] On the other hand, Davidic expectation is clearly envisioned in 12:32: "This is the Messiah whom the Most High has kept until the end of days, who will arise from the posterity of David, and will come and speak to them." This coming Redeemer figure is called 'Messiah' in 7:26 and the 'servant' in 7:29; 13:32, 37 and 14:9 (cf. both 'David' and 'Servant' in Ezek 34:23-24; 37:24-25).

---

[222] Brandenburger, *Verborgenheit*, 165-169, presents the problem of 'the way of God.'

[223] See Alden L. Thompson, *Responsibility for Evil in the Theodicy of IV Ezra: A Study Illustrating the Significance of Form and Structure for the Meaning of the Book* (SBLDS 29; Missoula, Mont.: Scholars Press, 1977), 332-339; also, Brandenburg, *Verborgenheit*, 169-176 ("Das Problem des 'bösen Herzens'").

[224] T. W. Willet, *Eschatology in the Theodicies of 2 Baruch and 4 Ezra* (JSPSup 4; Sheffield: Sheffield Academic Press, 1989), 75.

[225] The shepherd imagery in 5:18 is applied to Ezra and appears irrelevant to the Davidic hope.

2.5 The Fourth Book of Ezra

From the composition of the book, however, it becomes clear that Davidic expectation (chs. 7, 12) is rather isolated from the shepherd imagery that accompanies the hope of Israel's restoration (chs. 1-2). It is generally considered that the first two and the last two chapters of the book (chs. 1-2; 15-16) were added later by Christian author(s).[226] This is a way of explaining how the occurrence of the shepherd symbol in chapters 1-2 came to reflect New Testament thought and Christian theology regarding Israel.

For this reason, the shepherd figure in chapter 2 can be seen as a Christian revision of the future David in chapter 12, 'a warlike Davidic messiah' who will destroy the ungodly and wicked ones (12:33-34).[227] Though he is viewed as the Redeemer, M. E. Stone notes that this Redeemer figure is nowhere spoken of in the language of kingship: "He makes the survivors rejoice (7:28; 12:34), delivers them (12:34; cf. 13:26), and defends (13:49) or directs them (13:26); he never rules over them."[228] Davidic expectation is alive, but this is not exactly a revival of the monarchy.

The eschatological shepherd in 2:34, apart from this militant Davidic expectation, is the one who will give 'everlasting *rest*' (2:34b; cf. Matt 11:28-29) in the context of the predominantly restorative language found in the preceding section of 2:15-32. Here, the text echoes some of the Ezekielian shepherd tradition, most notably the raising of the dead (v. 16: cf. Ezek 37:1-14), the concern for his name (v. 16: cf. Ezek 36:22-23), and God's mercy as the foundational motivation for the eschatological restoration (4 Ezra 2:31-32). In addition, the exhortation to do good works in 2:15-32 lists several tasks to be performed by the eschatological shepherd: "Care for the injured and the weak, do not ridicule a lame man, protect the maimed, and let the blind man have a vision of my glory" (2:21).

In this context, we observe that the shepherd imagery, though it can be construed as messianic, tends to be associated more with justice and peace, and is adopted to modify a militant messianic figure. The isolated militant Davidic messianic figure in 12:32 – from the shepherd imagery in chapters 1-2 – suggests that if the figure of Davidic messiah is devoid of shepherd imagery, he tends to be described as more likely a militant figure similar to what is illustrated in 4Q285 and *2Baruch*.[229]

---

[226] The main reason is that a number of surviving oriental translations comprise with those chapters. Refer to J. Schreiner, "Das 4. Buch Esra," 233; Otto Eissfeldt, *The Old Testament, An Introduction* (trans. Peter R. Ackroyd; New York: Harper & Row, 1972), 27; B. M. Metzger, *OTP* 1:520; Geer Hallbäck, "The Fall of Zion and the Revelation of the Law: an Interpretation of 4 Ezra," SJOT 6 (1992): 265.

[227] Collins, *Scepter*, 68.

[228] M. E. Stone, *Fourth Ezra: A Commentary on the Book of Fourth Ezra* (ed. F. M. Cross; Minneapolis: Augsburg Fortress, 1990), 213.

[229] The shepherd imagery is closely linked to the concept of the Law in *2 Baruch* 72:1-16, while the Davidic expectation occurs apart from the prophet-like shepherd figure (40:1; 72:2).

The identity of the shepherd in the phrase 'await your shepherd' (4 Ezra 2:34) is the shepherd for whom 'the nations' should wait. That the Lord will turn to the nations because of Israel's refusal is already alluded to at the onset: "What shall I do to you, O Jacob? You would not obey me, O Judah, I will turn to other nations and will give them my name, that they may keep my statues" (1:24). The eschatological people of God includes anyone from the nations "who have fulfilled the law of the Lord" (2:40), that is, "a great multitude on Mount Zion" whom Ezra saw (2:42). Once again, the shepherd's association with the Law is attested, especially his leadership in showing the people the way to keep the law of God in the new age.

Significantly, the shepherd imagery for the messianic figure appears at the juncture of the nations joining in the eschatological salvation. The concept of 'one' shepherd in Ezek 37:15-24 has already been revised in *An. Apoc.* 90: 17-18 and *Ps. Sol* 17-18, to become 'one shepherd' over both Israel and the nations. Indeed, 4 Ezra's use of the shepherd imagery in the context of the Lord extending eschatological rest to the nations (2:34) reflects the Davidic Shepherd's eventual sovereign reign over the nations, as envisioned in the OT/Second Temple Davidic tradition.

What is innovative in 4 Ezra is the idea of Israel's rejection of the shepherd in the context of the shepherd becoming one ruler over the new people. The concept of Israel's rejection of the shepherd – a significant role in Zechariah's reinterpretation of the tradition – is implicitly evidenced in 4 Ezra. This text displays another non-militant trait of the shepherd imagery that points toward the Davidic messiah. It is worth noting that this Christian reinterpretation of Davidic expectation in 4 Ezra is mediated through shepherd imagery.

2.6 The Hellenistic Monarchy and the Shepherd Imagery

The Hellenistic idea of kingship in the period extending from Alexander the Great (356-323 B.C.) to the rise of Christianity can be set up as a foil to any serious reinterpretation of the Jewish Davidic Shepherd tradition.[230] In the first chapter of our study, we saw the militancy of the shepherd imagery in Greece, as captured in the Homeric phrase 'shepherd of the host.' In this section, we will attempt to present a more nuanced description of the shepherd imagery, especially in its association with the Hellenistic idea of kingship.

---

[230] E. E. Peters, *The Harvest of Hellenism* (New York: Simon and Schuster, 1970), 410-431; F. W. Walbank, "Monarchies and Monarchic Ideas," in *The Cambridge Ancient History*, vol. 7, 1 (Cambridge: Cambridge University Press, 1982), 100, notes that while the Senate of Rome considered the Roman consul or proconsul more than a match for any king, the legacy of Hellenistic kingship persisted throughout the Roman Empire.

2.6 The Hellenistic Monarchy and the Shepherd Imagery

Before discussing the characteristics of the Hellenistic monarchy, it may be helpful to understand how the Greeks came to accept the monarchic style of regime. Initially, 'for the Greek,' the monarchy was "intolerable ... whereas other peoples cannot live without a rule of that sort" (Isocrates, *Phil.*, 107).[231] Isocrates's influential proposal, *Panegyricus* (*Paneg.*, 380 B.C.), though it was not practiced by the leaders of his city, effectively represents the anti-monarchic Greek mentality. He argues that an Athenian leadership of Greece should be based on culture no less than on feats of arms as in the Persian monarchy. For Isocrates, the monarchy was a foreign institution with bizarre customs, a point on which he elaborates:

But those held in highest honor among them have never lived by standards of equality or fellow-feeling or patriotism, but live in a permanent attitude of arrogance towards some and servility toward others, which is calculated to be very damaging to human character. They overindulge their bodies because they are wealthy, but they humble their minds and cringe in fear because they have a single overlord, presenting themselves for inspection at the very gates of the palace, grovelling, and in every way practicing a lowly attitude of mind, abasing themselves to a mere mortal man and addressing him as a god, thereby showing less respect for gods than for men (*Paneg.* 115).[232]

Thus the king is not a god. Isocrates warns against elevating a king to a divine level since he, like any other human, is fragile and vulnerable to corruption from the misuse of power and wealth, and in particular is subject to unlawfulness and vain glory. This sort of criticism shapes the facets of the 'Hellenistic' monarchy as Greek cities eventually adopt and modify the eastern monarchic style in various realms of the Hellenistic regime.

The momentum of the Greek reception of the monarchy occurs within twenty years of Alexander's death when his empire split into separate states ruled by the *Diadochi*. These new kings relied on their armies because they mostly ruled in lands where monarchy was traditional.[233] In a surge of anti-democratic thought, monarchy was heralded and advocated as the most stable regime at the time. For instance, by 280 B.C. in Egypt, where monarchy already acquired status, the new ruler Ptolemy II made a public – and international – declaration of the divinization of his dead mother and father as 'savior gods' (*theoi soteres*).

According to E. Peters, this practice was gradually and widely imitated throughout the Hellenistic monarchies and was later practiced among the Romans, where "the act of canonization was simultaneously a gesture of filial piety" and a cementing of the dynasty's claim to rule.[234] Antiochus I saw his

---

[231] Walbank, "Monarchies," 76; Deirdre J. Good, *Jesus the Meek King* (Harrisburg: Trinity Press International, 1999), 43.

[232] S. Usher ed. and trans., *Isocrates: Panegyricus and To Nicocles* (Greek Orators vol. 3; Warminster, England: Aris & Phillips, 1990), 96-97.

[233] Walbank, "Monarchies," 62-63.

[234] Peters, *Harvest*, 156.

kingship as the sublime combination of the best of Hellenistic and Persian traditions. In a similar vein, he instituted worship of himself along with an assortment of an assembly of mixed Greco-Asian gods in sanctuaries scattered throughout his kingdom, thereby exhibiting his 'theocratic program.' One of the inscriptions testifies about his self-exalted king: "The Great King Antiochus, God Just and Manifest, Friend of the Romans, Friend of the Greeks."[235]

Yet it was different in Macedonia. It seems that the divinization of rulers was and remained an alien custom, for the rule of the king was grounded in the popular will; and the Macedonian army had the right to elect their king, and afterwards, the king had to rule according to the guidance of Stoic sages.[236] Nevertheless, it would be an inaccurate generalization to set the contrast between the Macedonian style as 'a national monarchy' and other styles as 'individual monarchies.'[237] Walbank is cautious and is probably right in his assessment that there was a gradual process of assimilation, which in time led the various monarchies to resemble each other more and more and to adopt similar institutions and conventions affecting their interstate relations.[238] If this is true, then we can legitimately discuss the general characteristics of Hellenistic kingship.

As we consider what those are, we must bear in mind Greek inventions and additions to earlier Eastern/Persian monarchies. Most of the Hellenistic kings differed from the Eastern type in that they granted to their 'friends' (*philoi*), under kingly supervision and scrutiny, unlimited power.[239] That is one of the ways Greeks modified the monarchy. The new monarchies presented the Greeks with an ideological problem, however. The dilemma was how to establish a monarchy according to their espoused political theory without discarding their traditional commitment to freedom and democracy.

Not surprisingly, this elicited new theories that justified monarchy for the Greek, but with modifications. Walbank's research on 'monarchies and monarchic ideas' examines the sources of the Hellenistic concept of the ideal king in this period.[240] Among the literary sources for the Hellenistic concept of the ideal king, Walbank selects Isocrates's *Ad Nicoclem, Nicocles* as the most

---

[235] R. R. Smith, *Hellenistic Sculpture* (London: Thames and Hudson, 1991), 227.

[236] Peters, *Harvest*, 91.

[237] For example, when Aristotle refers to monarchies with limited powers, he mentions Sparta and the Molossians, but makes no reference to Macedonia (Aristotle's Pol. V.II.2.1313a). Walbank, "Monarchies," 64-65, thinks the expression "king of *Macedonians*" is only intended for special occasions, and not as a title.

[238] Walbank, "Monarchies," 65.

[239] Erich S. Gruen, "Hellenistic Kingship: Puzzles, Problems, and Possibilities," in *Aspects of Hellenistic Kingship* (ed. Per Bilde, Troels Engberg-Pedersen, Lise Hannestad and Jan Zahle; Oxford: Aarhus University Press, 1996), 116.

[240] Walbank, "Monarchies," 62-100.

2.6 The Hellenistic Monarchy and the Shepherd Imagery 163

influential work of the period on this subject, and the *Epistle of Aristeas* as the best surviving source on the topic.[241]

Nicocles was a son of Euagoras, and at one time was Isocrates's pupil. At the beginning of his reign over Cyprus (374/3 B.C.), Isocrates offered Nicocles advice on the immediate problems facing a king in his own country.[242] Discussing the ideal leadership, Isocarates began a careful portrayal of a ruler who avoids all the vices of tyranny (τυραννίς: monarchy) as they had come to be recognized and abhorred by his time (*Ad Nic.* 1-8).

The style of *Ad Nic.* closely resembles a collection of proverbs that could also be regarded as a treatise of ideal kingship, taking pains to avoid unclear metaphoric expressions, not to mention that of shepherd and sheep. Nevertheless, Isocrates's modifications of the monarchy approximates some of the actual effects that the shepherd imageries would generate in the Davidic Shepherd tradition.

Isocrates did not openly require Nicocles to submit himself to the same law laid down for his subjects, but underlines various means by which the king can avoid excess of power, vice, and mistakes.[243] The king's self-control is emphasized (*Ad Nic.* 10-11, 29): "Govern yourself no less than your subjects" (29). As a quasi-formal institution, the king's *friends* not only were to advice him, they were also expected to correct him: "Make friends (Φίλους κτῶ) ... with those who will help you govern the state best" (27; cf. 53); "those who criticize your mistakes" (28).

There are also cautions against the vanity of wealth and its inferiority to the virtues (19, 32, 33). Besides the need for self-control, Isocrates states, "it is also necessary for a king to become a 'lover of mankind' and a 'lover of [his] country' for his subjects" (15, πρὸς δὲ τούτοις φιλάνθρωπον εἶναι καὶ φιλόπολιν). Kindness, gentleness, or meekness (πρᾶος) is likewise expected from a king as well as dignity (34), especially by "making punishments less severe than the offences" (23).

The motif of mercy/love in the term of φιλανθρωπία might be compared to the compassion motif in the Davidic Shepherd tradition, which later became an expected attribute of Hellenistic monarchs (*Paneg.* 29; *Phil.* 5.114, 116). Conjoined with names like Savior and Benefactor,[244] φιλανθρωπία is used

---

[241] According to Walbank, "Monarchies," 62-81, the full list includes other sources that are fragmental by nature: Xenophon's *Cyr.* and *Hier.*, Isocrates' *Paneg.*, pseudo-Plutarch's *Apoph. lac.*, Plato's *Pol.* and *Leg.*, Aristotle's *Pol.* and *On Kingship* (or *Rhetoric to Alexander*), Hecataeus of Abdera's *On the Egyptians* in the first book of Diodorus's *History*, *Ep. Arist.*, and pseudo-Archytas' *On Law and Justice*. Cf. Good, *Meek King*, adds no other significant sources.

[242] Usher, *Isocrates*, 117.

[243] Isocrates' request may only be implicit in *Ad Nic.* 38: "Feel obliged to follow yourself any advice you would give your own children."

[244] For the titles, see A. Deissman, *Light from the Ancient East* (trans. L. R. M. Strachan; New York: G. H. Doran Co, 1927), 362-364.

occasionally to strengthen the popularity of the monarchs and even to encourage king-worship.[245]

Besides love for the subject, justice, along with truth (22), is another essential principle of the Greek idea of kingship (16). The Greek emphases on the democratic ideal, justice, and compassion for the subjects indicates that the presence of the motif in the Jewish Davidic Shepherd tradition during this period finds, at least, its parallel in its Greek surroundings.

The *Epistle of Aristeas* takes a similar direction. Aristeas, a Jew from Alexandria who presumably participated in the mission of producing the Septuagint, wrote a letter to his brother Philocrates sometime between 250 B.C. and A.D. 100).[246] In *Ep. Arist.* 187-294, Aristeas describes seven days of banqueting during which time the Egyptian king Ptolemy II put forth seventy questions to each of the translators, who in turn offered their replies. All of the questions concerned kingship and centered around a king, though a host of questions did not explicitly involve this vocabulary. The replies of the seventy translators may represent the outlook of Diaspora Judaism regarding kingship, seeing how they gathered in Alexandria where Judaism and Hellenism seemed to live side by side in comparative harmony.

If a king were to remain undefeated in war, one of the seventy advised, the king should refrain from placing "his confidence in his numbers and his forces" but rather should ask God to direct his enterprises aright (*Ep. Arist.* 193). The most notable and perhaps noble characteristics of kingship is striving to keep oneself incorruptible, practicing moderation, and respecting justice because God himself loves justice (209). In numerous responses, the translators present God as the author and model for the king and the kingship.

One of them, whose comments represented the sentiments of the entire group of seventy, defined kingship as that which exhibits "real self-mastery, not being carried away by wealth and glamour...God likewise does not want anything and yet is merciful" (211). A king's self-control recurs consistently throughout the responses, yet admittedly self-control is "impossible to achieve unless God disposes the heart and mind toward it" (237). The essential possession a king must have is "love and affection" for his subjects as God fulfills these aims according to his will (265).

Regarding the laws, *Ep. Arist.* speaks clearly that the laws are the guidance a king must follow (279). Thus with peace and justice will a king be able to establish his subjects, which is regarded as "the most important feature in a kingdom" (292). Suffice it to say, a Hellenistic monarch must be someone of highest moral stature. In *Ep. Arist.*, the Jewish translators link highest moral

---

[245] Usher, *Isocrates*, 207; Cf. Tromp de Ruyter, "Φιλανθρωπία," *Mnemosyne* 59 (1937): 271.

[246] R. J. H. Shutt, "Letter of Aristeas" in *OTP* 2, 7-8.

stature to God, acknowledging Him as the ultimate source and model for the monarchs' integrity.

Turning to the Jewish realm during the Hellenistic period, the Hasmoneans adopted in stages a package of practices associated with Hellenistic kings: erecting monuments, minting coinage in their name, hiring mercenaries, displaying their achievements on *stelai*. Even Aristobulus, who conducted an aggressive policy of converting conquered Gentiles to Judaism, adopted diadem and labeled himself '*philhellene*' (Josephus, *A.J.* 13. 301, 318).[247]

From its inception, an emphasis on military achievement was the hallmark of the Hasmonean dynasty; furthermore, "a remarkable new conception of a warrior high priest" emerged.[248] The high priesthood under Hasmonean rule was not a monarchy, even though the role had a political and cultic dimensions. Therefore, the Hasmoneans come under the rubric of contemporary visions of rulers.[249]

Against this background, reading from the texts of the Davidic Shepherd tradition is illuminating. The shepherd imagery for God and his Davidic appointee could be seen as the Jewish counterpart of the Greek modifications of monarchy. Indeed, the biblical idea of kingship and the desiderata for a Jewish king had already been spelled out in uncompromising terms in Deut 17:14-20. Highly exalting one pre-eminent individual and placing untold wealth into his hands was prohibited by the Torah.

The subordination of the king/shepherd to the Torah was another key point made in the biblical modifications of monarchy.[250] David's faithfulness to YHWH's laws and decrees was a major reason for depicting the future messianic figure as Davidic. David is remembered and idealized in many places, including *Sir* 47:4-7, as the one with whom God could and would share his shepherd-rulership over Israel and the nations.

T. Rajak reflects on the contractual nature of the kingship in biblical tradition; the two way contract was made with God, as expressed in Deut 17:14-20: "then he and his sons will reign long over his kingdom in Israel."[251] For king David, however, the contract was unconditional (2 Sam 7:11-14), though the future ideal David is expected to be like him by being faithful to the law and leading the new people according to the law.

The Hasmoneans in their attempt to restore the monarchy certainly failed, because of their violence as well as their illegitimate status, to meet the standards of these earlier modifications for the Davidic Shepherd to be invested with wisdom, Spirit, and righteousness. This is also illustrated in *Ps. Sol.* 17-18,

---

[247] Gruen, "Hellenistic kingship," 124.
[248] Tessa Rajak, "Hasmonean Kingship and the Invention of Tradition," in *Aspects of Hellenistic Kingship* (Oxford: Aarhus University Press, 1996), 112-113.
[249] Rajak, "Hasmonean," 102.
[250] Ibid., 99-100.
[251] Rajak, "Hasmonean," 99.

where the intent of the monarchy modifications is tightly associated with shepherd imagery. Having examined Jewish understanding, we now turn our attention to shepherd imagery to see whether or not it contributes to the Greek understanding of kingship.

Shepherd and sheep were common in everyday Greek life. For instance, a tombstone dedicated to 'Shepherd Papias Klexos' (Πάπιας Κλέξος ποίμην ἥρως χρηστός παροδείταις χαίρεν) is found in Laodicea circa A.D. 2.[252] The shepherd in this source is also called a 'benevolent hero' (ἥρως χρηστός) who 'greets passengers' for reasons unknown. The term 'benevolent' recalls a high status figure, nevertheless, the identity of the shepherd remains unclear.

Further, direct links between shepherd imagery and a kingly figure/ kingship in the Greek world are not as conventional as we might find in the ANE and in the Judaism of the Second Temple period. No indisputable shepherd image is detected in Walbank's listing of the major literary sources that highlight the Greek understanding of kingship mentioned earlier.

The *Thesaurus Linguae Graecae* brings up a scant seventeen texts that contain both of the Greek roots ποιμ- and βασιλ- within the period from the third B.C. to the first century A.D. An interesting text is found in Plutarch's *An seni* 792. B. 5: εἰ δύναται παρὰ τῷ Φιλοποίμενι βασιλεύς ("if [the] king had any influence with Philopoemen"). Philopoemen was an Arcardian of Megalopolis born about 252 B.C. He was a general of Archadia, well known and respected throughout Greece: "The memory of this Philopoemen is most carefully cherished by the Greeks, both for the wisdom he showed and for his many brave achievements" (Pausanias, *Descr.* 49.1-2). Φιλοποίμην literally means "loving-the-flock [of sheep]."[253]

Granting the fact that Φιλοποίμην was remembered as a legendary commander for the Greeks, the shepherd/sheep metaphor in his name draws special attention. This example can support the Greek use of the shepherd metaphor, yet mainly in a military context. If we consider that the kings are often called φιλανθρωπία (*Ep. Arist.*15; cf. *philhellene* in Josephus, *A.J.* 13. 301, 318), Φιλοποίμην suggests that the sheep metaphor might have been applied only to the Greek army.

One can conjecture that in the eastern monarchies it was not difficult to view their subjects as sheep; on the other hand, the Greeks might be uncomfortable adopting a sheep image, which for free citizens in their cities could easily be construed as demeaning. Yet the common prefix, φιλ-, indicates that the Greeks

---

[252] T. Corsten, *Die Inschriften von Laodikein am Lykos. Teil I* (IK 49.1; Bonn: Habelt, 1997). Also, see the papyrus (BC 160) about 'Lost sheep' described in S.R. Llyewelyn, *A Review of the Greek Inscriptions and Papyri,* vol 9 (Grand Rapids: Eerdmanns, 2002), 54-55. We are indebted to E. Schnabel for these data.

[253] *LSJ*, 384.

recognized the significance of compassion or affection conveyed in the idea of leadership, be it for soldiers or civilians.

Similarly, Philo of Alexandria (20 B.C.- A.D. 50), in *De Agricultura* 39-66, presents an extended and highly intriguing argument concerning a shepherd's task, and expands the analogy to include the idea of kingship. The purpose of this treatise is to supplement his explanation for the difference between a real husbandman and a mere 'tiller of the ground' by comparing 'a shepherd' with a mere 'keeper of sheep' (*Agr.* 67). Philo defines the 'keepers of sheep' as those who treat the sheep negligently without prudence. On the other hand, real 'shepherds' refer only to those who really know how to benefit the flock, and who are skilled at shepherding, especially exerting wisdom to control the sheep as well as themselves (39).[254]

The key to the analogy is the ability to control unbridled passions and irrational impulses. This ability constitutes the essence of true mastery whether it is applied to agriculture, shepherding, or the rulership of a nation – 'any herdsman of any kind' (48). Thus shepherds can represent kings: "so that the race of poets has been *accustomed to call kings the shepherds of the people* [τοὺς βασιλέας ποιμένας λαῶν εἴωθω καλεῖν]; but the lawgiver gives this title to the wise, who are the only real kings, for he represents them as rulers of all men of irrational passions, as of a flock of sheep." (*Agr.* 41).

Philo's use of shepherd imagery in this context of dealing with various types of rulership is unique when compared to the OT Davidic Shepherd tradition. As Philo allegorically reinterprets the biblical uses of the shepherd imagery applied to Jacob (Gen 30:36, "for he is the shepherd of Laban's sheep"), Moses (Exod 3:1, "Moses was the shepherd of the sheep of Jethro"), and David (Ps 23:1, "the Lord is my shepherd, I shall not be in want"), exhibits a reflection of the Greek idea of rulership characteristic of the emphasis on self-control and virtue, though resorting to Jewish images and stories.[255] Interestingly, Philo mentions 'anarchy' and 'tyranny' (monarchy in Greek) as different sides of the same coin, i.e., lack of the right herdsman-ship (46). The τυραννίς, a ruler of cities, is like the hostile and unbridled soul of a body.

Yet surprisingly, Philo criticizes the gentleness of the ruler or governor, seeing it as a lack of discipline, and considering it injurious to both rulers and subjects. In the end, Philo concludes: "Our mind should govern all our conduct, like a goatherd, or a cowherd, or a shepherd, or, in short, like any herdsman of any kind; choosing in preference to what is pleasant that what is for the advantage both of himself and his flock" (48).

---

[254] Philo eventually calls God "a blameless and in all respects a good shepherd" (*Agr.* 49; cf. 44).

[255] The name Laban means "whitening," and by such colors the sheep are easily deceived; thus Jacob is appointed to be a true shepherd over Laban's sheep (42); Likewise, Jethro means "superfluous" (43).

The monarchy is modified by the general (Greek) principle of the superiority of rationality over passion. In Philo' usage of the shepherd imagery for the ideal kingship, the OT prophetic Davidic Shepherd tradition is completely missing. There is *no eschatological expectation* in Philo's use of shepherd imagery associated with kingship. Despite the biblical imageries, the idea of leadership remains predominantly Greek. Philo's treatment of shepherd imagery in his search for ideal leadership displays an early attempt to interact with the biblical tradition in a milieu where Hellenistic ideals flourished.

Turning to Greek papyri, *P. Oxy* 1611 presents an interesting case for the image of the scepter. The relevant text reads: καὶ οὗτός ἐστιν ὡς ἀληθῶς ὁ τῷ σκήπτρῳ βασιλεύων, οὐ τῷ δόρατι καθάπερ ὁ Καινεύς (lines 42-46, "and this is the king who really rules *by his scepter*, not *by his spear* like Caeneus)."[256] Justice is a classic virtue in the Greek idea of kingship; the same holds true for the biblical tradition. The text is a quotation from Theophrastus's second book *Concerning Kingship*, which did not survive. Theophrastus had in mind that king Caeneus failed because he relied on military force rather than on justice. What is more significant is the contrast set between τῷ σκήπτρῳ and τῷ δόρατι. Here the Greek metaphoric use of scepter is illuminating. For a king to rule successfully, he is to rule 'by the scepter' and not 'by the spear.' As one of the primary shepherd imageries, the scepter plays a crucial role in the modified understanding of a militant monarchy.

To sum up, regardless of the requests reflected in the letters of these literary sources, the Hellenistic kings themselves often violated justice, and deemed themselves 'free to legislate as they wished.' Thus it is a Greek idea of kingship or Greek hope, if you will; they resorted to ways to justify theoretically the rising up of monarchs in the expanded Hellenistic world following the death of Alexander the Great. Even at this theoretical level, however, we notice similar rather than totally antithetical tendencies in both the Hellenistic and Jewish views of the ideal monarchy.

2.7 Conclusion: The Profile of the Shepherd Revisited

Vancil concluded his research on the shepherd imagery in the Second Temple period by saying: "the shepherd image in the intertestamental literature amounts to a continuation of the ideals presented in the Old Testament; eventually, the authors who used the shepherd image followed the pattern of the Old Testament."[257] Our observations of the imagery associated with the idea of kingship

---

[256] Bernard P. Grenfell and Arthur S. Hunt, eds., *The Oxyrhychus Papyri* (London, Egypt Exploration Fund: Greco-Roman Branch, 1919), 13:139.

[257] Vancil, "Symbolism," 251.

2.7 Conclusion: The Profile of the Shepherd Revisited

reveal, not only the continuity, but a more diverse and complex development of the OT Davidic Shepherd tradition in this period.

Even in Greek usage, there is evidence of this imagery associated with monarchs and kingship but not limited to the military context. In most instances, it is not the shepherd imagery that modifies the Hellenistic idea of kingship. Still, Hellenistic modifications of the shepherd imagery may not be labeled as the 'antithetic' Jewish idea of kingship. Both share an emphasis on the kingly figure's subjection to the superior authority, whether it be the law, God/gods, or virtues. Likewise both elaborate on the king's devoted interest in his subjects' protection, peace, and prosperity, often characterized by compassion.

Nonetheless it is also true that it is difficult to find the rich texts in terms of shepherd imagery related to the idea of kingship in the Greco-Roman sources as we find in Ezekiel 34-37, *An. Apoc.* 89-90, *Ps. Sol.* 17-18, or in a sampling of Qumran texts such as 4Q504, 4Q174, 1Q28b, 4Q521 and CD-A 13. Above all, the eschatological feature is a distinctive of Jewish traditions – a feature entirely absent in Hellenistic sources.

While the theocratic type of monarchy is constrained by the Greek pursuit of virtues to be exhibited by a human king, the hope for theocracy is preserved, modified, and even extensively developed in Jewish sources. Comparing Hellenistic with Jewish traditions, shepherd imagery is more vivid and alive in its association with kingship in Jewish sources. The biblical and Jewish sources might even appear striking when set against Greek descriptions, in terms of its frequency and sophistication.

At the end of the first chapter, we outlined a profile of the eschatological/Davidic Shepherds based on the OT Davidic Shepherd tradition. We will now revisit that profile in order to interact with it based on the observations of this chapter. The profile involves the following characteristics: (i) hope for YHWH's theocratic intervention as the eschatological shepherd; (ii) linguistic clues pointing to the shepherd's various rescuing/restoring activities; (iii) renewed eschatological obedience to the law under the leadership of the Davidic Shepherd appointee; (iv) the coming of the Davidic appointee: the shepherd-king and the prince-leader; (v) YHWH's compassion and his concerns for his holy name; the plight, cause, and solution of the exile; (vi) the extension of the Davidic Shepherd's reign over the nations; (vii) the staff/scepter metaphor; (viii) the Ezekiel pattern in terms of the shepherd's absence, return, and perpetual presence.

We have already mentioned the characteristic Jewish expectation for theocracy in contrast to the Greek denunciation of a theocratic monarchy. The hope for God's intervention was kindled and often intensified in various Jewish visions. The *Animal Apocalypse* 89-90 is a remarkable text that testifies to the continuation of the OT Davidic Shepherd tradition in this period. Yet, the seer of the Vision radically departs from the tradition and how it seeks a solution for

Israel's plight of sin when he includes all of humanity in the Vision, and goes beyond the confines of Israel's history. The mystery of Israel's suffering – considered as excessive beyond the punishment due – calls for a redefinition of the plight itself; the wicked shepherds to be blamed and destroyed are now angelic powers. Besides *An. Apoc*.89-90, *Ps. Sol.* 17-18 and countless Qumran texts evince the fact that the Hasmoneans, and later the Herodian kings, might have been the catalyst for the resurgence of various Davidic expectations in view of God's promised intervention. Moreover, it is no challenge to find the shepherd imagery used in these texts.

The eschatological activities of the Shepherd in Ezekiel 34-37 set out the criteria for the shepherd imageries appearing in those Jewish texts. The restorative tasks performed by the Shepherd – seeking, rescuing, gathering, healing, feeding the flock, are far from the Greek idea of kingship. Nevertheless, these activities are confirmed in a significant number of Qumran texts. Importantly, the healing activity of the eschatological shepherd in Ezekiel 34 is attested by select sources such as CD-A 13:9; 4Q521 2, 2:12; 4Q504 1-2, 2:14; *Apoc. Ezek.* 5.

Most often, David is remembered for his faithfulness to the laws and decrees of the Lord (Ezek 34-37; 4Q504). In this respect, *An. Apoc.* 89-90 is an exception, for the Mosaic law is surpassed by the revelatory experience of the vision of the Lord's glory. Still, the law of God remains at the center of Israel's plight and exile, and is at the core of the restoration solution. The new David will shepherd and be a prince among the restored people. As Leader and Teacher of the eschatologically renewed community, he will be expected to usher in the community justice and righteousness (4Q174; 4Ezra 2:40-42). The nature of his role is more salvific in that he will save the flock from injustice and unrighteousness rather than rescue the people from the jaws of the wild beasts, that is, the nations as well as Israel's evil leaders.

Concerning the characteristics and roles of the Davidic Shepherd figure, the modifications made during this period challenge us with their complex descriptions. The Prince (נשיא) as an alternative title for Ezekiel's Davidic Shepherd (Ezek 34:23-24; 37:24-25) gains much more prominence in the Qumran texts, more so than in any other sources.

In Ezekiel, these figures are identical, but this is not necessarily the case in the Qumran materials. Ezekiel's נשיא is the Prince of the Congregation, and this figure may be seen as involving matters within the community, whereas the Branch of David takes charge, and extends the Lord's eschatological reign over the nations. This distinction remains unclear, however, for both figures may be seen as identical (4Q285; 4Q252) or separate (4Q161; 1Q28b).

Yet, the function of the shepherd imagery in its association with either of these figures indicates that once modified by shepherd imagery, neither could be seen as a militant warrior (*Ps. Sol.* 17-18; 4Q285; CD-A 13; 4Q521; 4 Ezra; *Tg. Ezek*; even in *P. Oxy* 1611).

2.7 Conclusion: The Profile of the Shepherd Revisited

As the eschatological Leader and Teacher, and due to the nature of this Davidic Shepherd, he is characterized as holy (*An. Apoc.* 90:17-18), free from sin (*Ps. Sol* 17:36), often invested with wisdom, the Spirit, righteousness according to the description found in Isa 11:1-5. Furthermore, the two distinct roles – YHWH's rescuing task as the eschatological Shepherd, and the saving task of the Davidic Shepherd – are maintained certainly in 4Q521, 4Q174 and *An. Apoc.* 90; yet at the same time, the tasks of YHWH the eschatological Shepherd are seen to be possibly taken up by the Son of David/the Branch/the Davidic Shepherd in *Ps. Sol* 17-18.

The Qumran texts affirm this tendency. In CD-A 13, the 'Inspector of the camp' is the one who assumes the tasks of Ezekiel's eschatological Shepherd; likewise, it is possible to see 'his anointed' in 4Q521, as one whose role is not specifically defined besides his the incomparable authority given to him, in fulfilling the similar tasks for the community.

The Zechariah tradition is also attested in this period. The *Animal Apocalypse* 89:59-90:39 adopts the idea that the Lord handed Israel over to the wicked shepherds, seen here as angelic shepherds. CD-B 19:7-13 is a rare text that preserves the image of the smitten shepherd, but as the representative of Israel, and probably not a messianic figure. Compassion and mercy – the fundamental motivations for YHWH to restore Israel – are, of course, recognized in numerous places, particularly in *Ps. Sol* 17-18, *An. Apoc.* 89-90, and also in 4Q521 and 4Ezra 2.

Interestingly, Ezekiel's vision of the eschatological unification of the divided nation of Israel now begins to be interpreted as the unification of Israel and the nations under one Davidic Shepherd. The expansion of the future Davidic Shepherd's rule is envisioned in the OT texts such as Ezekiel, Micah, and Zechariah, but the feature of the unification of Judah and Israel ceases to be an issue for the authors who interacted with the OT Davidic Shepherd tradition at this time. Particularly in *An. Apoc.* 90 and *Ps. Sol.* 17-18, the one shepherd over the unified Israel in Ezekiel 34 and 37 is seen now as the one shepherd over both Israel and the nations.

In the OT shepherd tradition, the scepter metaphor is used interchangeably with the concept of the tribe of Israel (Ezek 37; Zech 11; cf. *Tg. Ezek.*). The Davidic Shepherd becomes one shepherd over the unified tribes (scepters) of Israel in Ezekiel 37. What he holds in his hand at the end time will be thus none other than the new Israel. On the other hand, the scepter metaphor can represent the principles of the shepherd's rulership as hinted at in the LXX's translation 'word.'

Likewise, *Ps. Sol* 17-18 shows the scepter metaphor underlining the truth-encounter of the Davidic Shepherd with the nations. As an exception, *An. Apoc.* 90:18 might convey a sense of militancy. Yet, generally speaking, the scepter metaphor represents the shepherd's rule of justice and truth in this period as well (4Q252; CD-B 19; *P. Oxy* 1611).

Ezekiel 34-37 (cf. Ezek 40-48), however, remains the richest text regarding the Davidic Shepherd tradition in the OT and probably in the Second Temple period as well. Ezekiel's vivid and rich shepherd imagery revises Israel's idea of monarchy.

Moreover, Ezekiel's programmatic vision that the returning presence of the eschatological Shepherd is followed by YHWH's appointing the Davidic Shepherd left an indelible mark on various revisions of the tradition. One of them, I suggest, is the First Gospel.

Chapter 3

Matthew's Textual Interaction with the Davidic Shepherd Tradition

This chapter intends to demonstrate how the text of the First Gospel interacts with the Davidic Shepherd tradition. Of the synoptic gospels, only Matthew interacts with the text of Mic 5:2 (5:1, MT/LXX; Matt 2:6). Another occurrence is Matt 26:31, which quotes Zech 13:7, along with the phrase 'thirty silver coins' of Zech 11:11 alluded to in Matt 26:15, which is Matthean *Sondergut*. The Evangelist places Mic 5:2 at the beginning of the narrative and enhances his interaction with Zechariah by adding Zech 11:11 to the passion narrative.

Ezekiel's vision of the eschatological and Davidic Shepherds (chs. 34-37; cf. chs. 40-48) occupies a significant place in the development of the OT Davidic Shepherd tradition. Yet the First Gospel, while explicitly quoting and re-interpreting the texts of Micah and Zechariah, surprisingly does not cite even a single verse from this portion of the OT tradition. It is our assumption, however, that Matthew extensively echoes Ezekiel's vision of YHWH as the eschatological Shepherd and the Davidic Shepherd-Appointee, particularly in the central section of his narrative. The allusions to the tradition, placed between Matt 2:6 (Mic 5:2) and Matt 26:31 (Zech 13:7), include Matt 9:36 ('sheep without a shepherd' cf. Num 27:17; Ezek 34:5), Matt 10:6, 16; 15:24 ('Jesus sent to the house of Israel', cf. Ezek 34:30; 36:37; 37:11, 16), and Matt 25:31-46 ('sheep and goats', cf. Ezek 34:17, 20). If Matthew indeed interacts with Ezekiel's vision of Israel's restoration, it is our proposal that many of Jesus' messianic activities are authenticated in terms of Ezekiel's descriptions of the promised eschatological and Davidic Shepherds to come at the time of Israel's restoration.

Besides the aforementioned allusions, a plethora of shepherd/sheep images echo the tradition in the Gospel: Matt 7:15 ('wolf and sheep'); Matt 12:9-14 ('saving a sheep on the Sabbath'); and Matt 18:10-14 ('the lost sheep'). While these Matthean shepherd/sheep images echo the tradition, the Gospel's quotations from Micah and Zechariah, along with the allusions to Ezekiel, indicate that Matthew is conversant with the OT Davidic Shepherd tradition from the vantage point of the Jesus event. That is, the Gospel presents him in terms of the shepherd images of the tradition. In the following sections, we will examine Matthew's quotations and allusions associated with the Davidic Shepherd tradition, followed by an overview of a sampling of shepherd/sheep images that also echo the tradition.

3.1 Quotations

3.1.1 Matthew 2:6 (Micah 5:1-4)

Does Matthew clearly interact here with the Davidic Shepherd tradition? The reference in Matt 2:6, introduced by the formula, οὕτως γὰρ γέγραπται διὰ τοῦ προφήτου (Matt 2:5b), meets the criterion for 'citation' as discussed above.¹ Yet, regarding the exact sources of the content in Matt 2:6, some have questioned whether Matt 2:6 really is or is not a quotation.² The options vary among 'an *ad hoc* interpretation of the MT's consonantal text' similar to a *midrash*,'³ 'a cumulative exegesis' of Gen 49:10, 2 Sam 5:2, Mic 5:2,⁴ or 'a theological and messianic reinterpretation of Mic 5:2.'⁵

These suggestions indicate that Matthew's exegesis of Mic 5:2 may not quite fit any methods of citing the text, either the first century or the modern age. Yet the question really concerns 'how' Matthew reads Mic 5:2, and leads us to consider the point of his particular reading of the tradition. Matthew's own work can be seen in the shape of the quotation;⁶ yet, the differences are significant for Matthew's purpose of presenting Jesus in his narrative. Here are the variants of Mic 5:2 (5:1, MT/LXX):

MT: וְאַתָּה בֵּית־לֶחֶם אֶפְרָתָה צָעִיר לִהְיוֹת בְּאַלְפֵי יְהוּדָה
מִמְּךָ לִי יֵצֵא לִהְיוֹת מוֹשֵׁל בְּיִשְׂרָאֵל מִקֶּדֶם מִימֵי עוֹלָם

LXX: καὶ σύ Βηθλέεμ οἶκος τοῦ Ἐφραθα ὀλιγοστὸς εἶ τοῦ εἶναι ἐν χιλιάσιν
Ἰούδα ἐκ σοῦ μοι ἐξελεύσεται τοῦ εἶναι εἰς ἄρχοντα ἐν τῷ Ἰσραήλ

---

¹ We followed the stricter definitions suggested by Thompson (*Clothed with Christ*, 30) and Stanley (*Paul and the Language of Scripture*, 37); refer to Introduction, 3. "Methodology."

² M. Silva, "Ned. B. Stonehouse and Redaction Criticism. Part I: The Witness of the Synoptic Evangelists to Christ," *WTJ* 40 (1977): 77-88; idem, "Part II: The Historicity of the Synoptic Tradition," *WTJ* 41 (1978): 281-303, argues that the modern definition of 'quotation' does not quite fit Matthew's use of Mic 5:2; Davies and Allison, *Matthew*, 1:242, see it as a quotation yet close to an interpretation; Robert H. Gundry, *Matthew: A Commentary on His Literary and Theological Art* (Grand Rapids: Eerdmans, 1982), 29, finds no serious difficulty to call it a quotation; Ulrich Luz, *Das Evangelum nach Matthäus* (vol. 1: Matt. 1-7; vol. 2: Matt. 8-20; vol. 3: Matt. 21-25; vol. 4: Matt. 26-28; Evangelisch-Katholischer Kommentar zum Neuen Testament; Benziger/Neukirchener: Düsseldorf und Zürich, 1987, 1990, 1997, 2002), 1:119, sees it a 'Zitat' with serious redactions.

³ K. Stendahl, *The School of St. Matthew and Its Use of the Old Testament* (Lund: Gleerup, 1968), 166.

⁴ Homer Heater, "Matthew 2:6 and Its Old Testament Sources," *JETS* 26 (1983): 395-397.

⁵ A. J. Petrotta, "A Closer Look at Matt 2:6 and Its Old Testament Sources," *JETS* 28 (1985): 47-52.

⁶ Donald A. Hagner, *Matthew* (WBC 33a and 33b; Dallas: Word, 1993), 1:29, notes that the changes are minor (the most significant one, Hagner says, is "by no means the least "), and the last line of the quotation is similar to Mic 5:3 (LXX).

Matt: καὶ σὺ Βηθλέεμ, γῆ Ἰούδα, οὐδαμῶς ἐλαχίστη εἶ ἐν τοῖς ἡγεμόσιν Ἰούδα· ἐκ σοῦ γὰρ ἐξελεύσεται ἡγούμενος, ὅστις ποιμανεῖ τὸν λαόν μου τὸν Ἰσραήλ

Luz views vv. 5b-6 as part of the tradition handed down independently rather than the Evangelist's addition inserted for the specific purpose of his narrative strategy, stating: "Obviously Matthew was unwilling to attribute the fulfillment formula to the hostile high priests and scribes."[7]

On the other hand, while reconstructing the history of Matt 1:18-23, some propose that the text in vv. 5b-6 was added by the Evangelist to the pre-Matthean narrative; Davies and Allison think that it is an expansion of the Mosaic narrative (1:18-21, 24-25; 2:13-15, and 19-21) in the interest of a Davidic Christology (cf. 'the legend of the magi and the star' in 2:1-2, 9b-11).[8] This analysis of the layers of the text history according to Davies and Allison is based on their preference for a Mosaic typology. Martin C. Alble thinks, instead, that Matthew drew directly on a written *Testimonia* source compiled to prove that the messiah would be born in Bethlehem.[9]

Brown suggests that the text of Mic 5:1 came to Matthew in a form already fixed by Christian usage, and thinks Matthew himself would have us believe that Mic 5:1 was accepted by Jews as a reference to the birthplace of the Messiah (Bethlehem).[10] It is difficult to establish when the citation was brought into narrative, but the Davidic connotation, whether seen as part of the tradition or as the Evangelist's addition, is evident (cf. messianic interpretations of Mic 5:2 in the targum on Micah and Tg. Ps.-J. On Gen 35:21).[11]

Matthew does not quote the last clause of Mic 5:2, וּמוֹצָאֹתָיו מִקֶּדֶם מִימֵי עוֹלָם ("whose origin is from of old, from ancient days"). It may be, as Davies and Allison suggest, that readers are supposed to fill in the clause for themselves.[12] Yet the ruler's ancient origin, predating the kings of the Davidic dynasty in the context of Micah 5,[13] is already evident in the changes Matthew makes in his reading of Mic 5:2. Both the MT and the LXX retain 'Judah' in the phrase, "among the clans/chiefs of Judah," while Matthew uses γῆ Ἰούδα instead of the MT's archaic אֶפְרָתָה (LXX, οἶκος τοῦ Ἐφραθα).

---

[7] U. Luz, *Matthew 1-7: A Commentary* (trans. W. C. Linss; Continental Commentaries; Minneapolis: Augsburg, 1989), 130.

[8] Raymond E. Brown, *Birth of the Messiah: A Commentary on the Infancy Narratives in Matthew and Luke* (New York: Doubleday, 1977), 184; Davies and Allison, *Matthew*, 1:193-195.

[9] Martin C. Albl, *"And Scripture Cannot Be Broken": The Form and Function of Early Christian* Testimonia *Collections* (Leiden: Brill, 1999), 184.

[10] Brown, *Birth of the Messiah*,184.

[11] Davies and Allison, *Matthew,*1:242.

[12] Ibid., 1:244.

[13] May, *Micah*, 115-116, sees 'ancient days' as the time of David, viewed as an era of the distant past.

Further, by using 'Judah' twice in the same verse, the Evangelist evokes the significance of Judah in view of YHWH's promise of Israel's ruler, the שֵׁבֶט in Gen 49:10: "The scepter (שֵׁבֶט) will not depart from Judah, nor the lawgiver/ruler's staff (וּמְחֹקֵק) from between his feet, until he comes to whom it belongs and the obedience of the nations is his" (cf. Matt 1:2).[14] Besides Matthew's omission of the last clause of Mic 5:2, the discrepancies found in Matthew's rendition of Mic 5:2 raise three major exegetical issues to be dealt with in the following sections: (i) Matthew's reading of Micah's מוֹשֵׁל; (ii) the meaning of οὐδαμῶς in Matthew's version; and (iii) the context of Micah 2-5 for understanding Matt 2:6 in its narrative context.

3.1.1.1 Matthew's Reading of מוֹשֵׁל in Mic 5:2

While many detect the wording of 2 Sam 5:2 in Matthews rendition of Mic 5:1 in Matt 2:6,[15] Homer Heater, more specifically, views Matt 2:6 as a 'cumulative exegesis' of Gen 49:10, 2 Sam 5:2 and Mic 5:2, and argues that the LXX has personified the scepter of Gen 49:10 with ἄρχων, and the ruler's staff with ἡγούμενος, and that the LXX calls David ἡγούμενος, the prince (נָגִיד) in 2 Sam 5:2. Heater suggests that Matthew's choice of ἡγούμενος for מוֹשֵׁל in Mic 5:2 reflects the ἡγούμενος who is promised to come from Judah in Gen 49:10, and the Shepherd 'Greater' than David who is also called ἡγούμενος in 2 Sam 5:2.[16] Heater's search for the solution launches with his suggestion that all those texts, Gen 49:10, 2 Sam 5:2, and Mic 5:2, involve the term ἡγούμενος as well as in Matt 2:6.

The foremost difficulty we encounter with this proposal is the fact that the LXX renders מוֹשֵׁל of Mic 5:2 as ἄρχων, not as ἡγούμενος. Further, in Gen:49:10, the LXX renders שבט ('scepter') as ἄρχων and מחקק ('lawgiver/ruler's staff') as ἡγούμενος. If Matthew were following the LXX, the Evangelist would have had ἄρχων instead of ἡγούμενος. On the other hand, Heater argues that Matthew picks up ἡγούμενος from 2 Sam 5:2 as a designation for David as a shepherd, while Luz thinks λαός is the catchword taken from 2 Sam 5:2 since "it contains the idea of the people of God, which is important to Matthew," especially "in an anti-Jewish point."[17]

Obviously, Matthew's rendition of Mic 5:1 is reminiscent of 2 Sam 5:2, and in fact, the oracle in Mic 5:1-2 is often interpreted as a revision of Nathan's

---

[14] Davies and Allison, *Matthew* 1:242.
[15] Luz, *Matthew 1-7*, 130; Davies and Allison, *Matthew*, 1:243; Joachim Gnilka, *Das Matthäusevangelium* (Freiburg/Basel/Wein: Herder, vol. 1, 1988; vol. 2, 1992), 1:39; Brown, *The Birth of the Messiah*, 184.
[16] Heater, "Matthew 2:6," 396.
[17] Luz, *Matthew*, 1:130, 136.

oracle (2 Sam 7:6-7), namely, Davidic expectation.[18] If 2 Sam 5:2 were the main text from which Matthew takes the wording in verse 6b, then λαός would refer to the people of Israel,[19] easily confirming Luz's emphasis on the anti-Jewish stance of the Evangelist's view of the people of God. In Mic 5:1-4 (MT), however, the Davidic Shepherd envisions the extension of his reign beyond Israel and over the nations, a point underscored by the phrase, עַד־אַפְסֵי־אָרֶץ ("to the end of the earth," v. 4b).

Likely, the λαός in Matt 2:6 refers primarily to Israel (cf. 2 Sam 5:2) but still remains open to the nations (Mic 5:3-4). In this respect, it is noticeable that even if Matthew takes both images of ἡγούμενος and the shepherd figure from 2 Sam 5:2, the rendition in Matt 2:6 displays the reversed order; that is, unlike in 2 Sam 5:2, the designation ἡγούμενος precedes the shepherd image as is found in Mic 5:1-2 and 3-4. Matthew's rendition in 2:6, although reminiscent of 2 Sam 5:2, likely reflects Micah's vision of the Davidic Shepherd.

This possibility is enhanced as we observe that the Evangelist associates the designation ἡγούμενος with the first phrase of Mic 5:1. Yet he inserts the significant phrase, γῆ 'Ιούδα, which is absent both in Mic 5:1 (MT) and 2 Sam 5:2, but, as Heater proposes, this attests the influence of Gen 49:10. In Genesis 49:10 (LXX), the ἄρχων (שֵׁבֶט, MT) is said not to depart from Judah, while ἡγούμενος (מְחֹקֵק, MT) is said to come from "between his [Judah's] feet" (רַגְלָיו מִבֵּין).

Whether the feet (רֶגֶל) in this obscure phrase refers to the 'feet' of a dignitary, in between which he rests his staff,[20] or is a euphemism for one's private parts (e.g., Judg 3:24; 1 Sam 24:3; Isa 7:20) – that is, 'descendants' (cf. Deut 28:57)[21] – the subtle distinction between the scepter/Judah and the staff/his feet should not be neglected. If the metaphor of 'feet' represents the negligible or small parts of the body, i.e., Judah, then Matthew's insertion of γῆ 'Ιούδα matches well with the designation ἡγούμενος. This designation refers not to the ἄρχων (שֵׁבֶט, MT) from Judah *per se* but to מְחֹקֵק who is to come from a negligible part of Judah, a place like Bethlehem as found in Mic 5:1.[22]

---

[18] Shaw, *Speeches*, 146; cf. Wolff, *Micah*, 144; Mays, *Micah*, 115.

[19] H. Frankemölle, *Jahwebund und Kirche Christi: Studien zur Form und Traditionsgeschichte des Evangeliums nach Matthäus* (NTAbh 10; Münster: Aschendorff, 1974), 199-200, thinks the referent of the λαός is exclusively Israel; likewise, Jean Miller, *Les Citations d'accomplissenment dans l'évangile de Matthieu* (Rome: Editrice Pontificio Istituto Biblico, 1999), 41, suggests that the phrase τὸν 'Ισραήλ comes for the precision to clarify what τὸν λαόν μου means.

[20] E. A. Speiser, *Genesis* (AB; New York: Doubleday, 1964), 365.

[21] Gordon Wenham, *Genesis* 16-50 (WBC 2; Nashville: Nelson, 1994), 447.

[22] Matthew's view of Bethlehem as one of ἐν τοῖς ἡγεμόσιν backs up this interpretation.

Noticeably the LXX renders מוֹשֵׁל of Mic 5:2 ἄρχων, and not ἡγούμενος. This could be seen as a minor change.[23] Also, it is possible that Matthew reads מוֹשֵׁל of Mic 5:2 as ἡγούμενος independent of the LXX.[24] As we will see from other variants in Matthew's reading of Mic 5:2 in following sections, Matthew does not exactly follow the LXX. Further, the LXX renders מוֹשֵׁל in several ways,[25] and Matthew could have chosen ἡγούμενος as an available option reminiscent of David who is referred to as ἡγούμενος in 2 Sam 5:2.

Nevertheless, the Hebrew term for ἡγούμενος in 2 Sam 5:2 is נגיד and not מוֹשֵׁל (Mic 5:2). The importance of this difference can be illuminated by the case of 11QPs³ 28:11-12 since this text involves these two designations, yet presenting a subtle distinction between them. In this passage, מושל is rendered 'ruler' while נגיד refers to 'leader' or 'prince.'[26] As we have argued in the previous chapter,[27] the full text of 11QPs³ 28:11-12a reveals the double titles of David, one in relation to people, and the other in relation to God: "and [YHWH] made me prince of his people (נגיד לעמו) and ruler over the sons of his covenant (בבני בריתו ומושל)."

If it is correct that this subtle distinction is also valid in Gen 49:10 between שבת (ἄρχων, LXX) and מחקק (ἡγούμενος), then we may have uncovered a clue to Matthew's choice of ἡγούμενος over ἄρχων for מוֹשֵׁל of Mic 5:2. In fact, Mic 5:2 does not present two distinctive designations as in Gen 49:10 (cf. Num 24:17) but only one – מוֹשֵׁל. Yet this figure in Mic 5:2 is described by two different sets of modifications: (i) the ruler figure is said to come from Bethlehem, from "a small/insignificant clan of Judah"; and (ii) he is to rule "over Israel" (בְּיִשְׂרָאֵל).

Turning to Matt 2:6, the Evangelist identifies ἡγούμενος as the one who is to come from Bethlehem, by no means the least of Judah (οὐδαμῶς ἐλαχίστη εἶ ἐν τοῖς ἡγεμόσιν Ἰούδα). Of greater importance is the phrase τὸν λαόν μου τὸν Ἰσραήλ ([ruler over] my people Israel; בְּיִשְׂרָאֵל in Mic 5:2) in Matt 2:6, which as the objective clause, is tied to the phrase, ὅστις ποιμανεῖ, rather than to ἡγούμενος! To be precise, the equivalent of Micah's מוֹשֵׁל (5:2) might not be

---

[23] Hagner, *Matthew*, 1:29, suggests that Matthew takes ἡγούμενος as a synonym for ἄρχων.

[24] Johan Lust, "Mic 5, 1-3 in Qumran and in The New Testament and Messianism in the Septuagint," in *The Scriptures in the Gospels* (BETL 131; ed. C. M. Tuckett; Leuven: Leuven University Press, 1997), 79, 82.

[25] For instance, for the MT's מוֹשֵׁל, the LXX retains, besides ἡγούμενος (2 Chr 7:18, 9:26; Pro 29:26), also ἄρχων (1 Ch 29:12; Mic 5:1[2, MT]), and the various terms of authority and power such as βασιλεία (2 Chr 20:6); βασιλεύς (1 Kgs 5:1); δυνάστης (Prov 23:1); κράτος (Ps 89:10), and ἐξουσιαζόντων (Eccl 9:17).

[26] J. A. Sanders, *The Dead Sea Scrolls v. 4a: Pseudepigraphic and Non-Masoretic Psalms and Prayers* (ed. J. H. Charlesworth; Tübingen: Mohr Siebeck, 1994), 164-165, renders נגיד 'leader' as do García Martines and Tigchelaar, *Scrolls*, 1179; for 'prince,' see Nicklsburg, *Commentary*, 383, n. 77.

[27] Chapter 2, 1.1.1.

ἡγούμενος but ὅστις ποιμανεῖ in Matt 2:6, i.e., the eschatological Shepherd whom[ever] he may be.[28]

In this respect, the change of the person of the verb after the indefinite relative pronoun ὅστις appears to reinforce the probability of our proposal.[29] Likely, Matthew inserts ἡγούμενος for Micah's מוֹשֵׁל since ἡγούμενος would come from "the least part [feet] of Judah," and renders מוֹשֵׁל primarily as ὅστις ποιμανεῖ on the basis of מוֹשֵׁל being the one who is "to rule over Israel [and even the nations]" as the Evangelist picks up this [Davidic] shepherd motif from Mic 5:4. The implication of this exegesis is that the key thread by which Matthew pulls out the OT Davidic Shepherd tradition lies neither in ἡγούμενος (Heater) nor in τὸν λαόν (Luz) but rather in ὅστις ποιμανεῖ.

Consequently, it is not 2 Sam 5:2 but the shepherd image that is reserved for מוֹשֵׁל primarily in Mic 5:2-5, consisting of the climactic part of Micah's vision for the Davidic Shepherd in view of Israel's restoration (Mic 2-5). It is here where the restoration context in view of the nations enters the picture. In this respect, Davies and Allison's argument – "*the switch from Micah to Samuel was probably motivated by a desire to underline Jesus' status as the Son of David*" – is unnecessary.[30]

The point of Jesus having been adopted as the Davidic royal lineage was the main issue for the Evangelist in 1:18-22. In 2:6, however, the narrative looks forward to Jesus' mission as the Shepherd among the people and far more, to the marvelous salvation of God beheld in the eyes of men. Matthew's point of the rendition of Micah's מוֹשֵׁל is found indeed in the figure of the Davidic Shepherd of Mic 5:4 who is said "to stand and shepherd [his flock]" for the sake of YHWH's great Name, with his reign to be extended to the nations.

The indefinite relative pronoun in ὅστις ποιμανεῖ is grammatically defined by ἡγούμενος; these figures are identical. It is probable that Matthew's insertion of ἡγούμενος, however, reflects the distinction as to the messianic figure(s) illustrated in Gen 49:10, i.e., between ἄρχων (שֵׁבֶט, MT) and ἡγούμενος (מְחֹקֵק, MT).[31] Then it is no minor change explicable in terms of 'an *ad hoc* interpretation of the MT's consonantal text similar to a *midrash*'[32] or 'a cumulative exegesis'[33] of Gen 49:10, 2 Sam 5:2, Mic 5:2. Perhaps Matthew interacts with the tradition behind the subtle distinction. To examine this feature more closely,

---

[28] In this respect, Matthew's use of the indefinite relative pronoun ὅστις seems to match with the similar tenet shown in his omission of the last clause of Mic 5:2, מִקֶּדֶם מִימֵי עוֹלָם וּמוֹצָאֹתָיו.

[29] J. Lust, "Mic 5,1-3 in Qumran," 82, makes a note of this subtle change.

[30] Davies and Allison, *Matthew*, 1:243 (italics mine).

[31] Gnilka, *Matthäusevangelium*, 1:39, notices the christological relevance for the introduction of ἡγεμόσιν / ἡγούμενος in v. 6.

[32] Stendahl, *School of Matthew*, 166.

[33] Heater, "Matthew 2:6," 395-397.

the following chart (examples are not comprehensive) offers examples of how these distinctions are maintained in various revisions, not only in the OT but in the Second Temple period as well:

Ruler Figure in Relation to God		*Leader Figure in Relation to People*
(a) Num 24:17	כּוֹכָב [star] from Jacob	שֵׁבֶט [scepter] from Israel
(b) Gen 49:10	שבט from Judah	מחקק [lawgiver] from his feet
(c) (LXX, Gen 49:10)	ἄρχων [ruler] from Judah	ἡγούμενος [leader] from his feet
(d) Mic 5:2	מוֹשֵׁל [ruler] over Israel	
(e) (LXX, Mic 5:1)	εἰς ἄρχοντα over Israel	
(f) *An. Apoc.* 89:46[34]	mekwennan (ἄρχων)	wamārāhe (ἡγούμενος, David)
(g) 11QPs[a] 28:11-12[35]	מוֹשֵׁל [ruler]	נגיד [leader/prince]
(h) CD-A 7:18-22[36]	והכוכב הוא דורש התורה [star / Interpreter of Law]	השבט הוא נשיא כל העדה (Num 24:17) [scepter/ Prince of all Congregation]
(i) Ezek 34:23-24	רֹעֶה [shepherd] over the flock	נָשִׂיא [prince] in their midst
(j) (LXX Ezek 34:23f)	ποιμήν	ἄρχων
(k) Ezek 37:24-25	רוֹעֶה אֶחָד [king]/מֶלֶךְ [one shepherd]	נָשִׂיא [prince]
(l) (LXX Ezek 37:24f)	ἄρχων	ἄρχων
(m) Matt 2:6	(ὅστις ποιμανεῖ)	ἡγούμενος

Before we examine how Matt 2:6 echoes the distinction between the 'ruler' figure defined in relation to God and the 'leader' figure defined in his relation to people, a word of caution is due. As illustrated in several cases such as שבט, both in Num 24:17 and Gen 49:10, and ἄρχων in Ezek 34, 37 (LXX), the distinction is never clear cut in terms of linguistics in these diverse texts. While we can only maintain a general tenet, the distinction itself is still far from negligible.

Indeed, the question of the identity of these figures poses a very complex issue. For instance, CD-A 7:10-21 (cf. CD-B 19:7-13) interprets הכוכב ('star') of Num 27:17 as דורש התורה (the 'Interpreter of the Torah') and השבט ('staff/scepter') as נשיא כל העדה ('Prince of all the Congregation'). John J. Collins identifies הכוכב as the Interpreter of the Law with the messiah of Aaron, and השבט as the Prince with the messiah of Israel, however.[37] The expression 'messiah of Aaron and Israel' is elusive in terms of its numeric reference to the messianic figure.

---

[34] Isaac, *OTP* 1:67, renders it 'a judge and a leader'; Tiller, *1 Enoch*, 304, 'administrator and leader.'

[35] J. A. Sanders, *Scrolls 4a,* 164-165 renders נגיד as 'leader' as so do García Martínes and Tigchelaar, *Scrolls,* 1179; for 'prince,' see Nicklsburg, *Commentary,* 383, n. 77.

[36] Baumgarten and Schwartz, *Scrolls* 2:26-27.

[37] Collins, *Scepter,* 82.

Thus matching up any one of these messianic figures remains an unsettled issue.[38] In the case of Matt 2:6, it is singularly Jesus, in view of its narrative context, who is said to be ἡγούμενος and ὅστις ποιμανεῖ for Micah's singular מוֹשֵׁל; thus, these two separate figures or functions are found in one person, Jesus.

Yet the distinction seems to be carefully maintained in the tradition. From the above chart, we note that מוֹשֵׁל is set in contrast with נגיד as in 11QPs^a 28:11-12, and while מוֹשֵׁל is rendered ἄρχων (Mic 5:2), נגיד is another term for נָשִׂיא meaning both, 'prince'/'leader.' The case in *An. Apoc.* 89:46 presents the same distinction between two designations for David: ἄρχων and ἡγούμενος. What we suggest is that the relationship of three different functional titles for 'My Servant David,' מלך ('king'), רעה ('shepherd') and נָשִׂיא ('prince') in Ezek 34:23-24 and 37:24-25 might shed some light on Matthew's rendition of מוֹשֵׁל as ἡγούμενος [and] ὅστις ποιμανεῖ.

In Ezekiel's vision of the Davidic Shepherd in chapters 34 and 37, YHWH's Shepherd-Appointee at the time of Israel's restoration is said to be the king 'over' the rescued flock, and the prince 'in their midst'; not the other way around in terms of the prepositions. This distinction is carefully maintained while the term רעה ('shepherd') in its connotations sides primarily with מלך, yet it maintains an intimacy with the people characteristic of the role of נָשִׂיא within the eschatological community.[39]

Obviously both 'king' and 'shepherd' are interchangeable and convey *the notion of authority granted by God* himself in these texts, while the 'prince' or 'leader' refers to the figure who characteristically *ministers 'in the midst of' the community*. Figures such as (primarily) 'the Prince of all the Congregation' and (sometimes) 'the Branch of David' from the Qumran writings may approximate the Prince figure in Ezekiel's vision in this respect (cf. 4Q161; 4Q285; 4Q174; 4Q28b; and 4Q521).

Particularly, 4Q161 illustrates this point. While the Branch of David will 'rule over' all [בידו ובכול] the peoples (8-10:21), the Prince of 4Q161 is the Prince of 'the Congregation' (נשיה העדה, 2-6:15).[40] The roles of these two messianic figures seem to be differentiated in 4Q161. In other words, Matthew's choice of ἡγούμενος – absent in Mic 5:2 (LXX) – and his identification of מוֹשֵׁל as ὅστις ποιμανεῖ appear to reflect this subtle distinction, while evidently relying on the Davidic Shepherd tradition of the OT.

If this is correct, then Matthew's ὅστις ποιμανεῖ can be understood as the designation applied to Jesus, particularly invested with authority from God, and

---

[38] For its singular reference, see CD 12:23;14:19;19:10; 20:1, for the plural, 1QS 9:11; cf. Davies, "Judaisms," 228, suggests '*a single teacher-messiah*' as the most consistent representation of the Damascus community (italics his).

[39] Refer to Chapter 1.4.6 "The Shepherd in Ezek 34-37."

[40] Allegro, DJD 5, 13.

characterized by his intimacy with τὸν λαόν μου τὸν Ἰσραήλ. In short, the phrase ὅστις ποιμανεῖ in Matt 2:6 as the Evangelist's rendition of מוֹשֵׁל emphasizes the representative authority of Jesus as the Shepherd-Appointee at the eschaton, while the designation ἡγούμενος stresses Jesus' leadership as the Shepherd particularly in his relationship with the people.

In the immediate context of Matt 2:1-12, Matthew's careful rendition of Micah's מוֹשֵׁל primarily as ὅστις ποιμανεῖ and secondarily as ἡγούμενος in verse 6 indicates Matthew's narrative strategy in terms of the characteristics of Jesus as the eschatological leader of the people Israel. Clearly one of the central issues of 1:18-2:23 is kingship.[41] In the whole of Matthew 2, the term βασιλεύς occurs three times: first applied to Herod in v. 2, again in v. 3, and to the anointed in v. 2 (βασιλεὺς τῶν Ἰουδαίων). Herod the Great is designated in terms of βασιλεύς at the time when Jesus was born 'in Bethlehem of Judah' (vv. 1, 3), but he fears that his kingship will be challenged by the coming βασιλεὺς τῶν Ἰουδαίων (v. 2). Second, we see, however, that the Evangelist chooses terms such as ὅστις ποιμανεῖ and ἡγούμενος over of βασιλεύς for the eschatological figure of authority in his rendition of Mic 5:2.[42]

It is noteworthy that the avoidance of the designation βασιλεύς, both in Ezekiel and the aforementioned Qumran texts may perhaps indicate the tradition of the varied revisions of the hope frustrated by Israel's failed monarchy. Matthew's choice of ὅστις ποιμανεῖ and ἡγούμενος for the OT expectation of the Davidic ruler in Mic 5:1-4 thus affirms a plausible, resurgent expectation of the age-old yet persistent, vibrant Davidic expectation.

The shepherd image of YHWH applied to the Davidic ruler makes the difference unique from the failed monarchy, its Hasmonean revival, or even Herodian rule (cf. *Ps. Sol.* 17-18). The shepherd image, which is stressed explicitly by Matthew's rendition of ὅστις ποιμανεῖ and implicitly by his choice of ἡγούμενος in verse 6, likely represents well the varied, yet shared characteristics of the future Davidic 'Shepherd' in the texts of Ezekiel 34 and 37 as supported by the Qumran texts.

In Matt 2:6 at this point in the narrative of the Gospel, however, it is unclear how and in what sense Jesus will *'shepherd'* YHWH's flock of Israel; it is only implicit that Jesus' kingly rulership will be contrasted with that of Herod, whose leadership of Israel is in Jerusalem. We argue that the attuned readers might expect that Jesus is the expected Shepherd-Appointee who will *shepherd/lead* his people in a manner through which the *authority* of YHWH himself would be revealed. Nevertheless, at this opening stage of the narrative, the text in Matt

---

[41] Brown, *The Birth of the Messiah*, 193.

[42] Cf. Warren Carter, *Matthew and Empire: Initial Explorations* (Harrisburg, Penn.: Trinity Press International, 2001), 57-69, reconstructing "Matthew's Presentation of Jesus," presents Jesus as βασιλεύς who proclaims and enacts the empire (βασιλεία) and underlines the motif of the conflict between Jesus and Herod.

2:6 does not provide any further information. The detailed answer yet awaits the unfolding of the narrative. Now we turn our attention to the Evangelist's use of οὐδαμῶς in 2:6.

3.1.1.2 The Meaning of οὐδαμῶς in Matt 2:6

Matthew replaces בְּאַלְפֵי (LXX, ἐν χιλιάσιν) of Mic 5:2 with ἐν τοῖς ἡγεμόσιν; also, this choice of 'ruler' over 'clan' may touch upon the rational used for another change Matthew made: οὐδαμῶς. Perhaps the Evangelist personifies the clans,[43] or emends the Hebrew by having it refer to 'leaders/princes'[44] In fact, the interchangeability of 'ruler' and 'clan ' was not a foreign idea to Matthew. As shown in Gen 49:10, while the word שֵׁבֶט ("scepter") can be read as 'ruler' (LXX), it can also mean 'tribe.' What seems more useful is to ask about 'the point' of comparison, and not whether this is contradictory or not compared to what is meant by Micah.[45]

The emphasis could be Jesus' superiority over the predecessors in the Davidic dynasty (Matt 1:6-11),[46] or setting the aside of Jerusalem,[47] or as Carter argues, the impact Jesus' birth made on the Roman political powers.[48] Yet, the point of Matthew's change introduced by οὐδαμῶς may well be YHWH's 'confrontation' with the leaders of Israel,[49] regardless of the identity of the counterpart, whether it be the past kings of the Davidic dynasty, or the contemporary Roman political leaders who settled in Jerusalem, including Herod. Bethlehem never pales in significance because it is linked to the legitimacy of Jesus' kingship.

This becomes particularly clear in that Matt 1:18-23 underscores Jesus' Davidic royal lineage, that is, the way to legitimate his claim that he is the Son of David according to the promises given to the Jews (cf. Deut 17:15; 2 Sam 23:5). Zimmermann underscores this point stating, "Auch wenn es sich bei den

---

[43] J. Lust, "Mic 5, 1-3 in Qumran," 78; Gundry, *Matthew*, 28.

[44] Luz, *Matthew*,1-7, ascribes it to a different pointing of the Hebrew term of אלפי; W. Carter, *Matthew and the Margins* (JSNTSS 204; Sheffield: Sheffield University Press, 2000), 79; A. Schlatter, *Der Evangelist Matthäus* (Stuttgart: Calwer Verlag, 1963), 35; Hagner, *Matthew*, 1:29.

[45] Gnilka, *Matthäusevangelium* 1:39, the term enhances "die Bedeutungslosigkeit Bethlehems." Similarly, Carson, *Matthew 1-12* (Grand Rapids: Zondervan, 1995), 87, concludes that the phrase οὐδαμῶς is added in Matthew formally, but a wholistic reading of the verses shows the contradiction to be merely formal; likewise, Brown, *The Birth of the Messiah*,184-187, notes that although both the MT and the LXX state the insignificance of Bethlehem, this insignificance enhances the greatness to be.

[46] Gundry, *Matthew*, 28.

[47] Hill, *The Gospel of Matthew* (London: Marshall, Morgan and Scott, 1972), 83.

[48] Carter, *Margins*, 79.

[49] Carter, *Empire*, 60-74, underscores the issue of sovereignty between Roman Empire and God through their agencies, Herod and Jesus: "whose world is it?"

Hasmonäern nicht um Ausländer handelt, könnte sich hier die Ablehnung bereits auf ihre nichtdavidische Dynastie beziehen – erst recht gilt das für Herodes und römische Herrscher."[50] Both Hasmoneans and Herod, although not 'foreigners' (Deut 17:15), fail to meet the criterion 'from the house of David' (2 Sam 7:12-14; 23:5; דָּוִד עַבְדִּי, Ezek 23:23-24; 37:24-25).

By making meaningful changes, Matthew, echoing the tradition, announces that YHWH's Shepherd-Appointee has finally arrived; and this Shepherd-Appointee confronts Israel's past and present leaders. Matthew's omission of the MT's לִי ('to me,' that is, to YHWH) may be understood in this respect, since the quote is delivered by the leaders of Jerusalem. Matthew sees Jesus as the coming ruler siding with YHWH, and not with them.

Eventually Matthew's interpretive insertion, οὐδαμῶς, can be best understood as YHWH's eschatological means of confronting and reversing Israel's leadership; this demands a closer look at the context of Mic 2-5 for its relevance to Matt 2:6 in its immediate narrative context.

3.1.1.3 Matthew 2:6 in Light of Micah 2-5

This last motif of confrontation implied in Matthew's οὐδαμῶς leads us to consider the function of Matt 2:6 in view of the narrative flow. While many commentators discuss the variations of Matthew's reading of Mic 5:2, few have paid attention to the role of Matthew's quote in view of his narrative as a whole.[51]

To unearth a clue, we may need to examine Mic 5:2 in its context.[52] Micah 5:1-4 is part of Micah's tripartite salvation oracle set out in Mic 2-5: 2:12-13, 4:11-13 and 5:1-4. The context of the first oracle Mic 2:12-13 unfolds with YHWH confronting false prophets (2:6-11; 3:1-12). YHWH's raising up the new shepherd for his flock includes indictments and judgments heaped upon the false shepherds of Israel.

The situation appears similar to what is behind a statement such as Matt 2:6. From a historical point of view, texts such as *Ps. Sol.* 17:11-20, 4Q252, 4Q161 and possibly 4Q285 may testify that the resurgence of the expectations of the Davidic Shepherd, or the Branch of David, coincides with the Hasmonean Dynasty and the rise of Herod the Great and his successors. In similar surroundings, Matt 2:6 explicitly proclaims that the Davidic expectation is fulfilled in Jesus the Messiah, the Son of David (Matt 1:1).

---

[50] Zimmermann, *Messianische Texte*, 127.

[51] Noticeable exceptions are Carson's *Matthew 1-12*, 87-88; Carter, *Matthew*, 79.

[52] For the significance of the entire Old Testament context from which the New Testament writers take the quotations, besides Hays' *Echoes of Scripture*, see C. H. Dodd, *According to the Scriptures: the Sub-Structure of New Testament Theology* (London: SCM, 1952), 126, 133. Specifically on this text, Petrotta, "A Closer Look at Matt 2:6," 47-52, says Matthew draws upon a whole tradition of texts and interpretations for the purpose of proclaiming the gospel.

Further, the integrating theme of Mic 5:1-4 – the portion from which Matthew cites in 2:6 – is the coming Davidic Shepherd, while the focus of Mic 2:12-13 and 4:11-13 is on the tasks of the Shepherd (2:12; 4:6-8), the 'Breaker' (הפֹרֵץ, 2:13; 4:8), and the gathered flock (4:13) at the time of the restoration. In Micah 5:1-4, YHWH's choice of ruler from a small, insignificant, and even despised rank (cf. צָעִיר in Ps 119:41) among the clans of Judah, points to the divine mystery that marvels and provokes human expectation.[53] What is most unlikely finally happens.

This indicates that it can be no other than YHWH himself who intervenes to raise up the new ruler – and surely a unique, different kind of ruler; he is the Davidic Shepherd-Appointee. It is he who 'will stand and shepherd his flock' (Mic 5:4; LXX, καὶ στήσεται καὶ ὄψεται καὶ ποιμανεῖ τὸ ποίμνιον αὐτοῦ), and he envisions the extension of his reign beyond Israel and over the nations: עַד־אַפְסֵי־אָרֶץ ("to the end of the earth," v. 4b).

Likewise, the immediate context of Matthew's quotation of Mic 5:2 presents the scene of the coming of the magi from the east (Matt 2:1-12), which can be considered proleptic, that is, as the first-fruits of the Gentiles, if viewed in the light of Mic 5:4b (cf. Matt 28:19).[54] One might expect πρόβατα instead of λαός in Matt 2:6, recalling *Tg. Ezk*'s rendition of 'sheep' as 'people' (עַם, ch.34). Perhaps in Matthew's context of 2:6, λαός instead of πρόβατα may look beyond the constriction of Jesus' shepherd mission to Israel; the Davidic Shepherd in Mic 5:3-4 indeed envisions the extension of his reign beyond Israel and over the nations.

In this regard, some assert that Matthew's Davidic Shepherd in 2:6 is already the messiah for both Jews and Gentiles.[55] Nevertheless, unlike what we discover in Mic 5:1-4, the Davidic Shepherd in Matt 2:6 is worshiped by the magi, the Gentiles from afar. For Ezekiel, the one Shepherd rules over unified Israel (34:23; 37:24), and not explicitly over Israel and the nations. Only later does the vision of the one Shepherd over both Israel and the nations begins to crystallize as attested in *An. Apoc.* 90 and *Ps. Sol.* 17-18.

Likewise, in Mic 5:1-4, the vision of one Davidic Shepherd for both Israel and the nations remains in the future; he is expected to bring about the restoration of Israel (the twelve) first, and afterwards his reign will be extended to the nations.

---

[53] Wolff, *Micah*, 145. Cf. Ch.1.3.3 in our study: "The Davidic Shepherd in Mic 5:1-4 [4:14-5:3]."

[54] Brown, *The Birth of the Messiah*, 181, although not fully developing, notes that in 2:1-12, Matthew turns his attention to the ramifications of Jesus as son of Abraham by showing that the first to pay homage to the newborn King of the Jews were Gentiles from the East (cf. Matt 8:11).

[55] Carter, *Margins*, 79, who underlines 'the universal significance of Jesus' set against Herod; Luz, *Matthew*, 1:136-139, also underlines the motif of rejection of the Christ by Jerusalem.

Another illuminating aspect of Mic 5:1-4 is the pattern that presents the motifs of Davidic expectation (v. 2), the shepherd image (v. 4), and the restoration of Israel with the nations in peace (v. 4b-5a). Micah 5:1-2 is known as a revision of Nathan's oracle (2 Sam 7:6-7), but Micah expects a new David, the Shepherd at the end, who will restore Israel and fulfill her mission to the nations.

This pattern is a cue to understanding the narrative structure of Matthew's story up to its second chapter. The First Gospel unfolds in the same pattern we recognize in Micah's vision: Davidic expectation (Matt 1:1-17); the coming of the promised Shepherd (2:6); the end of the exile (2:18);[56] and the Davidic Shepherd's extended rule over the nations in terms of the proleptic account of the magi's worship of the infant Jesus (2:1-2; cf. 9-12; 28:18-20).

Furthermore, the appearance of the star that led the gentile magi might attest "the universal aspect" of this Davidic ruler in Matt 2:6; this supernatural phenomenon can be seen as a token of God's sovereignty over nature at the time of the birth of this promised ruler "as the world-king whom all (both Jews and the Gentiles) await."[57] Josephus attests that his contemporaries attached significance to signs just as the apostle Paul attested, "Jews seek for signs" (1 Cor 1:22); particularly Josephus' description of a star may illuminate the Jews' attachment to signs (*J. W.* 6.5.3).

At this juncture, the prophecy of Num 24:17 may well be relevant. It is often argued that Balaam's prophecy may have influenced Matthew's account of the leading star in Matthew 2: "A star [הכוכב] shall come forth out of Jacob and a scepter [השבת] shall rise out of Israel" (Num 24:17).[58] This seems more fitting to Matthew's world of reference as we have examined the influence of Gen 49:10 upon Matthew's rendering of Mic 5:2. That is, the appearance of the star in Matt 2:7-12 – reminiscent of הכוכב Num 24:17 – presents itself in the narrative as a perfect match with the appearance of Micah's מוֹשֵׁל whom Matthew reads as ὅστις ποιμανεῖ, the Davidic royal figure rather than ἡγούμενος who is close to the figure of the Prince (cf. Ezek 34:23-24; 37:24-25; *An. Apoc.* 89:46).

In other words, it seems more likely that Matthew's readers would have taken it as a sign with messianic overtones rather strictly assign a universal significance related to God's sovereign rule over the creation.

---

[56] The context of Jer 31:15 is the hope of the end of the exile culminating with the new covenant in 31:31-34. See J. A. Thompson, *The Book of Jeremiah* (Grand Rapids: Eerdmans, 1980), 551-552.

[57] Frederik Dale Bruner, *The Christbook: A Historical/Theological Commentary* (Waco: Word Books, 1987), 45-47; idem, *Matthew: A Commentary* (rev. and exp. ed.; vol. 1: The Christbook, Matthew 1-12; vol. 2: The Churchbook, Matthew 13-28; Grand Rapids: Eerdmans, 2004), 1:58-60.

[58] Craig A. Evans, "Exile," 319-320; Margaret Davies, *Matthew* (Sheffield: JSOT Press, 1993), 36; also, see Luz, *Matthew*, 1:131, who lists CD 7:18-21; 4QTest 11-13; 1QM 11:6f; *T. Levi* 18:3; *T. Judah* 24:1; Rev 22:16.

Thus Matthew, in his reading of Mic 5:2, associates the promised מוֹשֵׁל with the shepherd image in the context of the coming restoration of Israel (2:18), while the shepherd image is expected to play a crucial role explicating *how* the new ruler will lead his flock. Rereading Mic 5:2, the Evangelist announces the fulfilment of YHWH's promised *theocratic* rule over his lost flock through the Davidic Shepherd, who is the Shepherd-Appointee by YHWH at the time of restoration as envisioned in Micah. This is how God will be with them as Jesus is called Ἐμμανουήλ (Matt 1:23), in the sense that Jesus is portrayed as the one in whom God is with his people.[59]

Matthew's story of Ἰησοῦ Χριστοῦ υἱοῦ Δαυὶδ υἱοῦ Ἀβραάμ (1:1) up to Matt 2:6, places the title Χριστός within the sphere of the Davidic expectation in view of the Gentiles. The title Χριστός occupies a prominent place in each major block of the genealogical table in 1:1-17 (vv. 1, 17), the story of the virginal conception in 1:18-25 (v. 18), and the infancy narrative in 2:1-5 (v. 4). The messianic idea of the title Χριστός is thus narrated in these stories of Jesus,[60] a motif which the first verse of the Gospel succinctly summarizes.[61] Both sections of 1:1-17 and 18-25 begin with the title Χριστός (vv. 1, 18), and present stories that place it in the context charged with the Davidic expectation. In the genealogical table (1:1-17), Jesus as the son of David is presented as the consummation of the story (vv. 1, 17).

Moreover, the history of God's salvation of his people connected exclusively with Israel is thus brought to its culmination, just as Jesus' saying about his being sent only to the lost sheep of Israel (15:24; cf. 10:1-6) underscores this point.[62] In the story of the virginal conception (1:18-25), Verseput is probably right as he states, "when Matthew explains the irregularities in the 'origin' of Jesus Christ (1:18-25), he does so primarily to elucidate the crucial point of the adoption of this divinely conceived child into the Davidic line through Joseph"[63]

Further, the infancy account in Matthew 2 continues this distinctive interest in royal Davidic messiahship. The reference to ὁ χριστὸς in 2:4 is situated in the

---

[59] Mogens Müller, "The Theological Interpretation of the Figure of Jesus in the Gospel of Matthew: Some Principal Features in Matthean Christology," *NTS* 45 (1999): 166.

[60] For this approach of reading 'Christology in the whole story', refer to U. Luz, "Eine thetische Skizze der matthäschen Christologie" in *Anfänge der Christologie* (ed. C. Breytenbach and H. Paulsen; F. Hahn; Göttingen: Vandenhoeck & Ruprecht, 1991), 221-235.

[61] Morna D Hooker, "The Beginning of the Gospel," in *The Future of Christology* (ed. Abraham J. Malherbe and Wayne A. Meeks; Minneapolis: Fortress, 1993), 21, 28; Cf. J. M. Gibbs, "Mark 1, 1-15, Matthew 1, 1-4, 16, Luke 1, 1-4, 30, John 1, 1-51: The Gospel Prologues and their Function" in *SE* 6 (ed. Elizabeth A. Livingstone; Berlin: Akademie Verlag, 1973), 154-188.

[62] Müller, "Theological Interpretation," 165.

[63] Donald J. Verseput, "The Role and Meaning of the 'Son of God' Title in Matthew's Gospel," *NTS* 33 (1987): 533; yet, J. D. Kingsbury, "The Title, 'Son of David' in Matthew's Gospel," *JBL* 95 (1976): 591-592, argues, "though it is not said in 1:1, Jesus Messiah is above all the 'Son of God'"

context of the juxtaposition of the two kings, Herod (v. 1) and Jesus (v. 2). The hostility of and the plot devised by Herod the Great, the chief priests, and the scribes of the people, 'all Jerusalem' (vv. 3-4) testify 'little reticence' in Matthew associating Jesus' Davidic right with Herod's earthly political agenda.[64] Yet, Matthew's understanding of Jesus' Davidic messiahship may neither be squarely identified with that 'earthly political agenda' nor be replaced 'in favor of a Son-of-God christology.'[65]

It is certain that the messianic idea evoked by the title Χριστός is readily associated with Israel's Davidic expectation in the infant story in 2:1-5, but the Evangelist's reading of Mic 5:2 introduces the figure of the Shepherd (2:6). This implies that Matthew interacts with the revisions of the Davidic expectation in the OT Davidic Shepherd tradition. The Davidic christology in Matt 1:1-2:6 is not forsaken but is presented in terms of its fulfillment (1:17) and revision (2:6), which is traditional as well.[66] The story that follows 2:6 illustrates how Jesus is the Christ, the Son of David among the lost sheep of Israel, the promised Shepherd who will tend his flock by himself at the eschaton; the promised eschatological theocracy has finally arrived in the midst of YHWH's flock (cf. Ezek 34).

At this juncture in Matt 2:6, Matthew focuses his presentation of the Davidic Shepherd primarily on the authority to be granted by God himself in the near future. Noticeably his compassion for the downtrodden among the flock (9:36) is not yet in the picture. Of greater importance is Matthew's rendition of מושל as ὅστις ποιμανεῖ (cf. Mic 5:3-4), indicating that the Evangelist envisions not only the characteristic of Jesus' eschatological mission but also YHWH's immanent appointment of or the raising up of the Davidic ruler over his eschatological community (cf. 28:16-20).

In this regard, motifs such as 'the rejection of the Christ' and 'an anti-Jewish point' are not clear as to the identity of the λαός in verse 6b.[67] Yet it is implied that this Davidic Shepherd is the one who will be raised 'by God' as YHWH's Appointee at the end – an utter marvel for human eyes to behold. His figure, however, also overlaps with the expectation of what it would be like if YHWH the Shepherd of Israel would tend his own flock, by himself, among them (cf. Matt 1:21; 8-9). Certainly, the marvelous and mysterious rule of YHWH the

---

[64] D. J. Verseput, "The Davidic Messiah and Matthew's Jewish Christianity," *SBLSP* (1995): 107-108.

[65] J. D. Kingsbury, "The Title 'Son of David' in Matthew's Gospel," *JBL* 95 (1976): 595, states, "In chap. 2, Matthew does not develop his portrait of the earthly Messiah by means of a Son-of-David christology but forsakes this in favor of a Son-of-God christology"

[66] Kingsbury, "Son of David," 594-596, in this respect, toils to explain the association of Jesus' healing ministry with the title, Son of David, especially recognizing the significance of healing ministry during Jesus' public ministry as summarized in the critical passages such as 4:23; 9:35 for his literary analysis of the Gospel.

[67] Luz, *Matthew*, 1:136, 139.

Shepherd of Israel continues, yet assumes different forms, as Matthew cites in Matt 26:31, and articulates in yet another text extracted from the Davidic Shepherd tradition set out in Zechariah 9-14.

3.1.2 Matthew 26:31 (Zechariah 13:7)

Not only Zech 13:7 but the entire vision of Zechariah 9-14 – clearly one of the most important sections of the OT Davidic Shepherd tradition – greatly influences Matthew's passion narrative.[68] Four explicit or implicit citations occur in the passion narrative of the First Gospel: 'The humble king on a donkey' in Matt 21:4-5 (Zech 9:9), the 'thirty silver coins' in Matt 26:15 (Zech 11:12), the 'striking [of] the shepherd' in Matt 26:31 (Zech 13:7), and the 'throwing [of] the money into the temple and [the] buying [of] the potter's field' in Matt 27:3-10 (Zech 11:13).[69]

Among all of these, Matthew shares only one quotation (Zech 13:7) with Mark (14:27-28). This may indicate Matthew's unique interest in Zechariah's reading of the OT Davidic Shepherd tradition. As Matthew reaches the conclusion of his narrative, he continues – more intentionally than other synoptic writers – the shepherd image applied explicitly to Jesus in Matt 2:6 (Mic 5:2).

3.1.2.1 Matthew's Passion Narrative and Zechariah 9-14

Beginning with Jesus as the humble king on a donkey as pictured in Zech 9:9, Matthew carefully weaves the passion narrative with the threads of Zechariah's vision of the shepherd-king, and through whose suffering YHWH would restore Israel in view of his sovereign rule over the nations. In the context of Zechariah 9-14, Zechariah's vision of Israel's restoration opens with Zech 8:6 announcing, vis-a-vis the use of the shepherd image, that YHWH bringing together scattered Israel will seem 'marvelous to the remnant at that day' (cf. 8:1-5).

Douglas J. Moo regards the individual figures in those four passages – 'the humble king' in Zech 9:9, 'the good shepherd sold for thirty silver coins' in 11:12-13, 'the pierced one' in 12:10, and 'the smitten shepherd who is YHWH's associate' in 13:7 – as different descriptions of a single figure, suggesting the

---

[68] Raymond E. Brown, *The Death of the Messiah: A Commentary on the Passion Narratives in the Four Gospels* (ABRL; New York: Doubleday, 1994), 126-133, notes the influence of Zechariah 9-14 upon the Gospel descriptions of Jesus' last days; Davies and Allison, *Matthew*, 3:483-484; also, see Douglas J. Moo, *The Old Testament in the Gospel Passion Narratives* (Sheffield: The Almond Press, 1983), 173; likewise, Dodd, *According to the Scriptures*, 64-67.

[69] Introductory formulae are found only with Zech 9:9 (Matt 21:4-5) and Zech 13:7 (Matt 26:31). Other two cases can be regarded as citations according to more loose definition, for instance, by Porter's: "the focus would be upon formal correspondence with actual words found in antecedent texts" ("The Use of the Old Testament," 95).

theme of suffering and/or rejection for this leadership figure in Zechariah 9-14.[70] The identity of the one who is to suffer remains uncertain.

Yet whether it is the good shepherd, the representative of YHWH (9:9; 11:12-13; 13:7), the mysterious pierced one (12:10), or the worthless shepherd(s) (11:14-17; cf. 10:3), the unmistakable theme is the suffering of the shepherd figure.[71] Yet it is YHWH's marvelous restoration of Israel presented in the later part of the vision in chapters 13-14 that overshadows this gruesome picture of the suffering shepherd.

Compared to Ezekiel 34-37, Zechariah's contribution to the Davidic Shepherd tradition is characterized by its exposition of the conflict between the shepherd and the flock that detests their good shepherd, YHWH. Particularly, the context of Zech 11:4-7 is illuminating as the passage engages in the reception and interpretation of the tradition of Ezekiel's two-staff analogy in Ezek 37:15-28.

The conflict between YHWH the shepherd and the flock of Israel finally erupts; it becomes clear that YHWH is despised by the flock (11:3-4), and consequently, the covenant of Favor and Union is broken at last (v. 10). It is likely that Zech 11:10-11 demonstrates the chosen flock's unfitness for the task of mediating YHWH's covenantal Favor to all the peoples. The thirty pieces of silver – the wage for YHWH as their shepherd (11:12-13), and the selling price of a slave – are thrown into the temple, which is at the center of the whole community.[72] Then the smitten shepherd in Zech 13:7 can best be seen as a consequence of this conflict between YHWH and his flock as a whole.

This sequence of Zechariah's vision is rearranged in Matthew's passion narrative. The Evangelist places the scene of YHWH smiting his associate shepherd of Zech 13:7 (cf. Matt 26:31) against the background of the story of Zech 11:4-17. In this scenario, the flock detests the good shepherd, pay the price of a slave for him (cf. Matt 26:15), and throw the coins into the temple (cf. Matt 27:3-10). Also, Jesus' establishment of the new covenant story (Matt 26:26-29) in Matthew's narrative context, as in Mark's (14:22-25), tightly interlocks with the prophecy Jesus quotes from Zech 13:7. The broken covenant lies at the center of Zech 11:4-16, a backdrop against which Matthew places Jesus' reference to Zech 13:7 in his Gospel. Likewise, the shepherd being smitten in Matt 26:31 has significant bearing on Jesus' establishment of the new covenant (cf. Matt 26:26-29).

---

[70] Moo, *Passion*, 173-174.

[71] S. L. Cook, "The Metamorphosis of a Shepherd," *CBQ* 55 (1993): 453-466, suggests that particularly Zech 13:7-9 envisages the martyrdom of an eschatological, Davidic king.

[72] Larkin, *Eschatology*, 130, sums up what the conflict in Zechariah 11 really meant: "it is possible that just as the wages are supposed to pay the prophet for his mission to the whole community (Zech 11:4b-6), so his rejection of them symbolizes his rejection of the whole community, which however is centered upon the temple."

Just as the flock detested YHWH as their shepherd, and paid for him with a trivial thirty silver coins in Zech 11:4-16, likewise Jesus in Matt 26:17-35 is mistreated by Judas, and is abandoned by all of his disciples, as well as by the leaders of Israel, whose chief priests cut a deal in advance with Judas (Matt 26:15). Matthew's particular arrangement of Zechariah's sequence thus indicates that he understood the smitten shepherd of Zech 13:7-9 as part of the same vision of the good shepherd vision in Zech 11:4-17. Furthermore, each appearance of the motif of YHWH's judgment upon the shepherd and his flock in Zech 11:14-17, 12:10-14 and 13:7-9 is followed by the respective announcement of YHWH's salvation of Israel in Zech 12:1-9, 13:1-6 and 14:1-21.[73]

This judgment-salvation pattern likely applies to Matthew's narrative strategy. The judgment upon the smitten shepherd is presented as the solution for the conflict between YHWH and his flock. In Zechariah 11, YHWH discharges the good shepherd from Israel and breaks up the covenant, coinciding with his judgment of the worthless shepherd(s). Yet the consequence is the return of YHWH's blessings upon restored Israel.[74] Matthew likely adopts this pattern in the passion narrative as shown in Zechariah's vision.

Eventually, as a modification of Ezekiel's vision (esp. Ezek 37:15-28), Zechariah 11-13 expects Ezekiel's vision of the Davidic Shepherd to be reactivated (cf. Matt 27:51-53; 28:16-20). This end of Zechariah's modification, i.e., fulfilling Ezekiel's vision of the Davidic Shepherd by introducing the smitten shepherd (thus resolving the ultimate conflict between the shepherd and the flock), appears to shape Matthew's narrative strategy of Jesus' death, resurrection, and the hereafter. We will investigate this possibility in the next chapters. For now, by turning to the micro and surveying the details, we hope to effectively examine the characteristics of Matthew's version of Zech 13:7.

3.1.2.2 Characteristics of Matthew's Version of Zech 13:7

Jesus' reference to Zech 13:7, according to Matthew's report, is abbreviated and only represents the middle part of the verse: "Strike (הַךְ) the shepherd, that the flock (הַצֹּאן) may be scattered" The first part of the verse – "Awake, O sword, against my shepherd (עַל־רֹעִי) against the man who is my associate (גֶּבֶר עֲמִיתִי וְעַל)" – and the last part – "yet I will bring back (וַהֲשִׁבֹתִי) my hand upon the little ones (עַל־הַצֹּעֲרִים)" – are not mentioned in Jesus' saying in Matt 26:31. The designation, 'my associate,' refers to the smitten shepherd in Zech 13:7. Thus as we discuss the identity of the smitten shepherd in this section, we will cover the omitted parts as we consider Zechariah's context of Jesus' saying in Matt

---

[73] For instance, D. L. Petersen, *Zechariah 9-14*, 129.

[74] Roy P. Schroeder, "The 'Worthless' Shepherd," *CTM* 2 (1975): 344, aptly notes on Mark 14:27, "Just as time of salvation for the remnant of Zechariah began with the destruction of the 'worthless shepherd,' so the time of salvation for the remnant of God's people began with Jesus' Passion."

26:31. The last part of Zech 13:7, YHWH's promise for the little ones, likewise omitted in Matt 26:31 will be dealt with in the next section as we examine Matt 26:32.

We will begin by focusing on Jesus' saying in Matt 26:31, "I will strike the shepherd, then the flock of the sheep will be scattered" (πατάξω τὸν ποιμένα, καὶ διασκορπισθήσονται τὰ πρόβατα τῆς ποίμνης). This presents two major changes: (i) the addition of the phrase τῆς ποίμνης; and (ii) the question as to the subject in πατάξω as to the identity of the smitten shepherd.

As to sources, the literary connection between Matt 26:30-35 and Mark 14:26-31 is quite evident. Thus, Davies and Allison argue for some slight evidence for Markan priority, saying that additional phrases such as ὑμεῖς and ἐν τῇ νυκτὶ ταύτῃ in Matt's version appear to be elaborations. Luz finds that the second line, καὶ διασκορπισθήσονται τὰ πρόβατα τῆς ποίμνης corresponds literally with LXX A, arguing that Matthew lifts the entire citation from Mark and LXX A.[75]

Yet even Mark's case indicates the earlier origin of the saying and the authenticity of Jesus' saying in Matt 26:31 has been firmly established without much difficulty.[76] Particularly, the place of the citation in the Gospel of Mark is distinctive since it is the only formal, explicit citation of Scripture in the whole Markan Passion narrative, and the citation is presented as a statement of Jesus.[77] This quotation is probably part of the Jesus tradition.

(1) The Implication of τῆς ποίμνης in Matt 26:31: If compared to the MT's version, 'the flock' in Zech 13:7, is altered to 'the sheep *of the flock*' in Matthew's version.[78] Further, Matthew retains πάντες ὑμεῖς instead of Mark's πάντες, while the phrase, ἐν ἐμοὶ ἐν τῇ νυκτὶ ταύτῃ does not appear in Mark's record.[79] While it appears insignificant, the changes in Matthew's version might be taking in the entirety of the scope of YHWH's judgment upon the flock, the whole people of Israel as God's own covenantal community. Likewise, with regard to the added phrase τῆς ποίμνης ('of the flock') in Matt 26:31, Gundry rightly argues that it is reminiscent of Ezek 34:31, particularly "to underscore the antithesis between the shepherd and sheep" while stressing the motif of "the dispersal of the flock."[80]

---

[75] Luz, *Matthäus*, 4:123-124.

[76] Refer to J. Jeremias, *New Testament Theology I: The Proclamation of Jesus* (trans. John Bowden; London: SCM, 1971), 297-298; also, Davies and Allison, *Matthew*, 3:485. Luz, *Matthäus*, 4:124, n.7, underlines the lack of conformity of the Jesus's saying in Matt 26:31 to the standard LXX; Wright, *Victory*, 554-556.

[77] Brown, *Death of Jesus*, 128.

[78] Only LXX A has "the flock *of the sheep*" as in Matt 26:31. Refer to Moo, *Passion*, 183-184, for the hypothesis of Mark/Matt's alleged dependence on "a pre-Christian Palestinian Jewish Greek recension" that might have influenced LXX A, Q.

[79] Is this related to the motif of YHWH's visitation "on a day of clouds and darkness" (Ezek 34:12)?

[80] R. Gundry, *The Use of the Old Testament in St. Matthew's Gospel* (Leiden: Brill, 1967), 25-28.

This interpretation fits Matthew's intention of utilizing Zechariah's pattern (chs. 9-14) in his passion narrative, as we argued in the preceding section. The unique feature of Zechariah's interaction with the Davidic Shepherd tradition is his emphasis on the irrevocable conflict between the flock and YHWH their shepherd. The additional phrase τῆς ποίμνης in Matthew's version has the effect of indicating the whole community of Judah as the one who throws the thirty silver coins into the temple (Matt 26:15; cf. Zech 11:12-13). In Jesus, who takes up the role of the smitten shepherd in Matt 26:31, the whole flock of Israel undergoes YHWH's judgment, though it will be followed by the promised restoration envisioned both in Ezekiel 34-37 and Zechariah 9-14.

That Jesus assumes the role of Israel's shepherd – whether good or wicked in terms of Zechariah 9-13 – is very likely in view in Matt 26:31.[81] With little difficulty we can associate the shepherd image of Matt 26:31 with 9:36 and 10:5-6, especially when one considers that the flock of the sheep in 26:31 refers to the entire house of Israel. Likewise, Luz aptly points out, "Beim 'Hirten' [ποιμήν in 26:31] werden die Leser/innen nach 9,36 selbstverständlich an Jesus denken, bei den 'Schafen der Herde' an die Jünger, aber nicht nur an sie, sondern auch an das ganze Volk, dessen Hirte Jesus ist (vgl. 10.5f)."[82]

Yet if Jesus is both the smitten shepherd in 26:31 and the compassionate shepherd in 9:36, what is the logic that runs through the narrative? Did Jesus as the compassionate shepherd in 9:36 suffer for the entire flock in 26:31 to be smitten by God? Also, if it is correct that τὰ πρόβατα τῆς ποίμνης in 26:31 refers to the house of Israel in 10:5, 6, what is happening to the house of Israel as Jesus becomes the smitten shepherd of Zech 13:7?

To answer these questions, we need to assemble other evidences from the following sections and chapters. Suffice it to say here that from the change of τῆς ποίμνης, we conclude that Jesus taking upon himself the role of the smitten shepherd of Zech 13:7 should effect the whole house of Israel and Jesus' mission to her. To probe the extent of this effect, we will attend to the other details in the changes we find in Matt 26:31.

(2) The subject in πατάξω and the Identity of the Smitten Shepherd: The subject in πατάξω in Matt 26:31 is God himself (πατάξω τὸν ποιμένα); while the implied subject of the verb in the MT, 'strike (הך),' is third person singular masculine, it is the second person plural in the LXX: πατάξατε τοὺς ποιμένας καὶ ἐκσπάσατε τὰ πρόβατα. Who then is to strike the shepherd of Zech 13:7, and why?

The text of CD-B 19:8-35 faithfully reflects Zechariah's identification of the plight of Israel's exile with her loathing for her Shepherd: "each one has chosen

---

[81] Schroeder, "The 'Worthless' Shepherd," 342-422, identifies Jesus in Matt 26:31 with the worthless shepherd of Zech 13:7; on the other hand, Brown, *The Death of the Messiah*, 129, sees Jesus in 26:31 as the good, compassionate shepherd (cf. John 10:1-18; 18:8-9; Mark 6:34).

[82] Luz, *Matthäus*, 4:125.

194 *Chapter 3. Matthew's Textual Interaction with the Davidic Shepherd Tradition*

the stubbornness of his own heart"; "they did not keep apart from their sins" (l. 20; cf. l. 33). The solution for this problem, as it was for Zechariah, is that the shepherd is to be smitten by the sword, likely that of the messiah(s) of Aaron and Israel, or possibly by YHWH (CD-B 19:7-9). Yet, 'the smitten shepherd' of Zech 13:7 is clearly distinguished from the messianic figure(s), and refers to one of the wicked leaders of Israel, or one of the kings of the peoples (l. 23), that is, the figure of the representative of the wicked leadership of Israel. The LXX's rendition of 'my shepherd' in Zech 13:7 as the plural 'shepherds' (ἐπὶ τοὺς ποιμένας μου) may enhance this possibility.

Eventually, the remnant – designated as 'the little ones' (הצוערים) and 'the poor ones of the flock' (עניי הצאן; cf. Zech 11:11) – will escape in the age of the visitation, while "those who remain will be handed over to the sword when the Messiah of Aaron and Israel comes" (CD-B 19: 9-11).[83]

Even though the identity of the messiah(s) of Aaron and Israel remains ambiguous, CD-B 19:7-13 still clearly exhibits the Zechariah Davidic Shepherd tradition, often called the *Zechariah-Ezekiel midrash/pesher*.[84] While 'the poor ones of the flock' will constitute the remnant at this eschatological momentum, the sins of the flock are nevertheless confirmed.

More important, the consequential judgment upon the shepherd to be smitten, who is apparently the representative of the wicked leaders of Israel, is set in contrast to the 'messiah(s) of Aaron and Israel.' The text of CD-B 19:8-13 cites Zech 13:7 and Ezek 9:4 against a backdrop of the coming of the 'messiah(s) of Aaron and Israel,'[85] the messianic figures who define eschatological moments.[86]

The ambiguity as to the subject of the striking – both in the LXX's version and CD-B 19:8-13 – is contrasted with Matt 26:31, which clearly sees it as God himself without any mention of other messianic figure(s). Nevertheless, Jesus hands down the sentence of YHWH's eschatological judgment – according to CD-B 19 it is reserved to fall upon the wicked shepherd(s) of Israel – to fall upon himself as the shepherd of the [whole] flock (cf. 9:36; 10:5-6). If it is correct that the emphasis of Matthew's version of Zech 13:7 in Matt 26:31 is on the conflict between all (πάντες) the sheep of the flock and their shepherd, then it is plausible that Matt 26:31-32 speaks to the renewal of Israel – to her shepherd and her flock, and the covenant established between them.

As we have examined earlier, the shared idea implied in both Matt 26:32 and the second half of Zech 13:7 is *the reversal* of the scattering. In particular, the scattering of the disciples in Matt 26:31 appears to recapitulate Israel's exile once again, as God's judgment is inflicted upon their shepherd, yet only to be

---

[83] Baumgarten and Schwartz, *Scrolls* 2, 30-31.
[84] Ibid., 27.
[85] P. R. Davies, *The Damascus Covenant* (JSOTSup 25; Sheffield: JSOT Press, 1982), 146.
[86] J. Vanderkam, "Messianism in the Scrolls" in *The Community of the Renewed Covenant* (ed. E. Ulrich and J. Vanderkam; Notre Dame: University of Notre Dame Press, 1994), 214.

reunited and renewed. Jesus going ahead of them into Galilee is not (exclusively) a geographical or chronological comment, but an indication of the reversal of the exile of the whole nation, Israel.[87]

The scattering of the flock in Matt 26:31 may represent the climax of the state of Israel's exile, signaled by God's judgment inflicted upon their shepherd, yet accompanied with the promise of the re-gathering of the remnant at the end (v. 32). The [twelve] disciples in Matt 26:31-32, including Judah who betrayed him earlier (Matt 26:17-35; cf. Zech 11:4-17), and Peter as the representative of all the others (Matt 26:69-75), recapitulates the moment of Israel's exile, followed by God's judgment upon the whole flock of Israel and the eventual renewal.

This reading presupposes the motif of exile as the backdrop of Matt 26:31, which is more than a possibility.[88] The OT expectation of the restoration of Israel had not yet been fulfilled.[89] One of the critical features of Jesus' activities that justifies exile theology and its significant role is Jesus' appointment of the twelve disciples. Most likely it is to be thought of as symbolizing the reconstitution of the twelve tribe of Israel.[90]

Later when we discuss Matt 9:36 and 10:1-8, we will see how Jesus assumes the role of the eschatological Shepherd for 'the sheep without a shepherd' (9:36), and calls the twelve disciples who would represent the reconstitution of Israel. Similarly, N.T. Wright argues that the very existence of the twelve speaks of the reconstitution of Israel, recalling the fact that Israel did not have twelve visible tribes since the Assyrian invasion. Thus, for Jesus to give twelve

---

[87] Brown, *The Death of the Messiah*, notes that Matthew's addition, "of the flock," underscores the nature of the scattering in v. 31 not as geographical, while seeing Jesus' going ahead in v. 32 as Jesus resuming the role of leading, "*a shepherding role that will reconstitute the flock*" (131, italics mine).

[88] For the textual evidence of the possibility, refer to Craig A. Evans, "Aspects of Exile and Restoration in the Proclamation of Jesus and the Gospels" in *Exile: Old Testament, Jewish, and Christian Conceptions* (ed. James M. Scott; Leiden: Brill, 1997), 305-315, who claims to find evidence that Israel is still in exile in the following texts: *Sirach* 36:6, 15-16; *Tobit* 13:3; *Baruch* 2:7-10; *2 Macc* 2:7; *1QM* 1:3; *1QpHab* 11:4-6; 4Q504; *1 Enoch* 90:20-42; *Test. Moses* 4:8-9; *2 Bar* 68:5-7, while finding expectations of restoration in *Tobit* 13:5; *Ps. Sol* 8:28; *2 Baruch* 78:7; *Targum Isaiah* 28:1-6; 53:8; *Jubilees* 23:18-31; idem, "Jesus and the Continuing Exile of Israel," in *Jesus and the Restoration of Israel* (ed. C. C. Newman; Downers Grove: Inter-Varsity Press, 1999), 78-86; cf. also see, N. T. Wright, *The New Testament and the People of God* (Minneapolis: Fortress, 1992), 268-272.

[89] Steven M. Bryan, *Jesus and Israel's Traditions of Judgment and Restoration* (Cambridge: Cambridge University Press, 2002), 12-20, distinguishes between the notion of 'the continuing exile' and that of 'the incomplete restoration,' while critiquing Wright's 'equation of the two' (16).

[90] Evans, "Restoration," 316-318, 326.

followers a place of prominence indicates that he was thinking in terms of the eschatological restoration of Israel (cf. Ezek 37:15-23).[91]

Besides CD-B 19:8-13, the expectation of the restoration of Israel, particularly related to Zech 13:7, is attested by *An. Apoc.* 89:59-90:39, the text which adopts the idea that the Lord handed Israel over to the wicked [seventy] shepherds (cf. Zech 11:15-16). In this instance, however, the shepherds are not historical leaders of Israel but angelic beings upon whom YHWH's judgment befalls. The *Animal Apocalypse* shares this sentiment with Jews of this period that Israel is still left in the wilderness exposed to the wild beasts without a shepherd (cf. Matt 9:36); the Lord of the sheep is in heaven and his entrusted angelic shepherds are excessive in their execution of the Lord's judgment upon the flock.

Steven M. Bryan contests that the case of *An. Apoc.* disapproves the notion of the continuing exile since (i) the seventy shepherds represent seventy years of exile (within God's control; thus not an ongoing exile); and (ii) the unseasonable revelation (opening of the eyes) prior to the end time implying the 'emergence of righteous Israel.'[92] Yet as we examined in the preceding chapter,[93] Bryan's suggestion is unlikely.

First, the seventy shepherds probably do not present God's control for seventy years, but the angelic shepherds whom the Lord of the sheep judges at the momentum when the messianic white bull arrives and relieves the misery of the exile (90:1-28). Second, the unseasonable revelation prior to the end time hardly repudiates (a) the horrible and desperate state of the sheep as prey exposed to the wild beasts (89:53-58; 90:3; cf. Matt 9:36), and (b) the (once-and-for-all) eschatological metamorphic change of both sheep (Israel) and beasts (nations) to holiness at the end of the vision in *An. Apoc* (90:17-18; 37-39), which presupposes the preceding era as the stage of exile.

Further, it is noteworthy that these wicked [angelic] shepherds (cf. Zech 9-13) are separate figures from the end-time messianic white bull in *An. Apoc.* 90:17-18. In Matt 26:31, however, we find that Jesus also, depicted as the compassionate shepherd for the sheep without a shepherd (9:36), takes up the role of Zechariah's smitten shepherd. Also, in contrast to the *An. Apoc.*'s

---

[91] N. T. Wright, *Jesus and the Victory of God* (Minneapolis: Fortress, 1996), 299-300, operating in terms of a "remnant theology, return-from-exile theology," suggests that "the call of the twelve ... this is where YHWH was at last restoring his people of Israel"; Wright, ibid., 318, also comments on disciples as shepherds: "the twelve disciples 'will sit on twelve thrones, judging the twelve tribes of Israel' (Matt 19:28//Luke 22:30); the twelve are to rule over Israel" (see Judg 3:10; 4:4; 10:2, 3; 12:7, 8, 9, 11, 13, 14; 15:20; 16:31; 1Sam 4:18; 7:6, 15); cf. Zimmermann, *Messianische Texte,* 126-127; Cf. E. P. Sanders, *Jesus and Judaism* (Philadelphia: Fortress, 1985), 98, while not approving the exile-theology, comments that "'twelve' would necessarily mean 'restoration.'"

[92] Bryan, *Restoration*, 20.

[93] Refer to Chapter 2.1 "The *Animal Apocalypse*."

apocalyptic solution of the Lord's judgment upon the angelic shepherds, Jesus' death and its gospel tradition attest to a 'historical' solution for the sins of the flock. In this regard, it is important for the seer(s) of the *An. Apoc*, that YHWH's judgment, accompanied by eschatological bliss, confronts the angelic shepherds alone and not necessarily the flock itself. In Matthew's Gospel, however, according to the pattern laid in Zechariah 9-11, YHWH's smiting the shepherd implies that his judgment is meted out 'upon the whole flock.' Jesus' rendition of Zech 13:7 in Matt 26:31 underscores that the judgment upon the shepherd will also be meted out upon the whole community (τῆς ποίμνης).

Concerning the identity of the smitten shepherd, the MT texts in Zechariah 9-13 remain ambiguous. The suffering shepherd figure could refer to the good (Zech 11:4-14), the wicked (11:14-17), YHWH's associate (13:7-9), or to YHWH himself (12:10?) in the context of Zechariah 11-13.[94] In Matthew 26:31, Jesus' citation of Zech 13:7 omits the phrase 'the man who is my associate (עֲמִיתִי וְעַל־גֶּבֶר)' found in Zech 13:7 (MT). With this omission of the phrase, most likely a positive nuance of the figure of 'the man,'[95] it is more probable that Jesus is taking up the destiny of the wicked shepherds in his reference to Zech 13:7 as we argued earlier in CD-B 19 and the LXX's 'shepherds' in Zech 13:7.

It is probable in Matt 26:31-32 that the scattering of the whole flock and the shepherd's re-gathering of the little ones according to Zech 13:7 can be seen as the eschatological reworking of YHWH's judgment upon his own flock, the breaking-up of the old covenant, the consequential end of exile, and the arrival of the promised restoration. The message of what really happens when the shepherd was smitten in Matt 26:31 can readily be seen as the momentum of YHWH putting an end to Israel's exile (Zechariah 9-14). Thus he fulfills the promise of the reconstitution of the twelve tribes, that is, he initiates the union of his flock, particularly according to Ezek 37:15-28.

To conclude, Matthew 26:31 is approached by commentators with a variety of explanations of what Jesus intended by quoting Zech 13:7, and how Matthew placed it in its current narrative context. To illustrate, Matthew's intention of this quote is often identified as 'divine sovereignty over Jesus as the shepherd being smitten';[96] God is in control behind Jesus' passion. Some find the theme

---

[94] Antti Laato, *David Redivivus*, 287, presents a list of the scholars such as B. Stade, J. Wellhausen, R. Mason, who regard the shepherd in 13:7-9 as a reference to the wicked shepherd in ch.11; yet Laato sees it as the good shepherd in 11:4-14; likewise, I. Duguid, "Messianic," 274.

[95] Cf. figures such as the Persian conqueror Cyrus who was called רֹעִי (Isa 44:28); the "Prince" in Ezek 34, 37. Refer to Chapter 1.4.6.

[96] David Garland, *Reading Matthew* (New York: Crossroad, 1995), 252; also, David Hill, *Matthew*, 340, stresses that the subject of "striking" is God himself in Matthew's version of Zech 13:7; Luz, *Matthäus*, 4:125, says, "Die Leser/innen der Passionsgeschichte wußten immer schon, daß er im Hintergrund des Passionsgeschehens steht-nun sagt es Jesus bzw. Gott durch die Schrift direkt."

of discipleship to be the main intention of the Evangelist's quote.[97] Yet others attempt to view Matt 26:31 more closely in light of the Zechariah context. For instance, Carter sees in the context of Zechariah 13 a purging of Jerusalem by YHWH, or a re-creation of Jerusalem by scattering and destroying two-thirds of the people, and restoring the remaining one-third who will eventually participate in the reestablished covenant.[98] Without excluding those motifs such as God's control and discipleship, we conclude that the emphasis of Matthew's version of Zech 13:7 is on the plight and the solution of Israel's exile, the conflict between the flock and their shepherd, and Jesus taking upon himself the role of the wicked shepherd of Israel as the representative of the whole flock.

It is important that the event of Jesus' death as the smitten shepherd would proceed the process of the promised restoration according to the pattern of the Davidic Shepherd tradition. The next expected steps according to the tradition would be YHWH's appointment of the Davidic shepherd over the renewed flock (Ezek 34-37; cf. Zech 12:6-9), and the extension of his reign through this Appointee over the nations, thus finally ushering in peace (cf. Mic 2-5; Zech 12-14). The smitten shepherd will be raised again to lead his little flock. This will be our immediate concern as we examine the last element of the shepherd image in Matt 26:32.

3.1.2.3 Προάγω in Matt 26:32 and the Shepherd Image

In Matt 26:32, Jesus says, μετὰ δὲ τὸ ἐγερθῆναί με, προάξω ὑμᾶς εἰς τὴν Γαλιλαίαν. Here the verb προάγω seems to echo Zech 13:7d: "I will bring back (ἐπάξω, LXX) my hand upon the little ones," thus implying that he continues to lead them, i.e., the disciples whom Jesus gathers again, the remnant, the restored Israel. While many detect the shepherd image of Matt 26:31 continuing in verse 32,[99] others disagree. Luz is reluctant to affirm the shepherd image in this verse, setting up two options: (i) Jesus moves forward to Galilee with his disciples as the shepherd goes before the sheep following him; or (ii) the risen Jesus goes ahead before the disciples get to Galilee and waits for them.[100]

The first option is refuted by Luz based on 'Palestinian shepherding skills': "Im übrigen gingen palästinische Hirten normalerweise hinter ihrer Herde und

---

[97] Daniel Patte, *The Gospel According to Matthew* (Philadelphia: Fortress, 1987), 365; Davies and Allen, *Matthew,* 3:484; Gundry, *Matthew,* 529-530; Keener, *A Commentary on the Gospel of Matthew* (Grand Rapids: Eerdmans, 1999), 633-635.
[98] Carter, *Margins,* 508; R. T. France, *Matthew* (Leicester: Inter-Vasity Press, 1985), 371, speaks in the same tone yet with more Christological emphasis.
[99] For instance, Davies and Allison, *Matthew* 3:486; Brown, *The Death of the Messiah,* 126-132; France, *Matthew,* 371; Carter, *Margins,* 508-509.
[100] Luz, *Matthäus,* 4:125-126..

trieben sie, nur metaphorische Hirten gehen ihrer Herde voran."[101] Thus, while pointing out that in Matt 28:10 the risen Jesus mentions Galilee, and tells his disciples to go there, Luz considers Jesus' mention of Galilee in Matt 26:32 'eine Doppelprophetie' with Matt 4:15 where Galilee is mentioned as a significant place because of Jesus' presence in terms of his bringing the light to the Gentiles.[102]

In this respect, many scholars suggest that Jesus' going ahead into Galilee (cf. Mark 16:7) refers to his *parousia* in Galilee, and not to his resurrection.[103] Brown aptly refutes this view by pointing out that the *parousia* of the Son of Man, however, surely involves a manifestation to the world, and "strange indeed would be a parousia selectively directed to the seeing of the risen Jesus by the disciples and Peter (as Matthew interprets it)."[104] Perhaps, Galilee as mentioned in Matt 26:32 is significant as it relates to the expected mission to the Gentiles in the Gospel.

Nevertheless, this is not incompatible with the presence of the shepherd image in the verse! (the shepherd image applied to Jesus is assumed by Luz to be restricted for Jesus in his mission to Israel only.) We argue, however, that the expected Gentile mission rather *necessitates* the continuation of the shepherd image in verse 32. The picture of the shepherd leading his flock 'in front of' them frequently occurs in the OT (Exod 13:21; Num 27:16-17; Mic 2:13; Zech 12:8).[105] Further, Jesus saying to his disciples in 28:10, 'go to Galilee,' obviously does not involve the verb προάγω with Jesus as the subject.

Last, the context of Matt 26:31-32 is highly charged with concerns about Israel's destiny, its covenant, its leadership, its members, and its future, namely, the critical matters to be settled as Israel's prerequisites for fulfilling the vision for the Gentiles. Davies and Allison affirm that the shepherd image continues in Matt 26:32, saying, "It [the shepherd image in v. 32] refers not to spatial ('go before') nor even chronological ('to arrive before') priority so much as to leadership."[106]

The idea entailed in the verb προάγω in Matt 26:32 indicates the continuation of the shepherd image of verse 31,[107] as Rudolf Pesch regards προάγω as

---

[101] Luz, *Matthäus*, 4:126, n.17.

[102] Luz, *Matthäus*, 4:136.

[103] So Lohmeyer, R.H.Lightfoot, Marxsen, Weeden, Kelber; cf. Brown, *The Death of the Messiah*, 132.

[104] Brown, *The Death of the Messiah*, 132.

[105] Luz, *Matthäus*, 4:126, n.18, points out the use of προάγω in Ex 12:42 for the possibly messianic implication suggested by Munoz León, but questions the relevance of its context to that of Matt 26:31-32.

[106] Davies and Allison, *Matthew*, 3:486.

[107] Gnilka, *Matthäusevangelium*, 2:407, notes Jesus' going ahead to Galilee after his resurrection, which indicates "die Erscheinungsgeschichte in 28, 16-20," echoes "das Hirtenmotif" anew; likewise, Stuhlmacher, "Matt 28:16-20 and the Course of Mission in the

'Terminus technicus der Hirtensprache.'¹⁰⁸ The ἄγω verb with various prefixes, particularly in its association with the shepherd image, often suggests YHWH's redemptive or restorative actions in the OT. Although the exact verb προάγω does not occur, Ezek 34:13 illustrates a case in point: "I (YHWH) will cause them to come out (וְהוֹצֵאתִים; LXX, ἐξάγω); I will gather them (וְקִבַּצְתִּים ; LXX, συνάγω) from the peoples and the countries; I will lead them into (וַהֲבִיאֹתִים; LXX, εἰσάγω) their own land; I will shepherd them (וּרְעִיתִים; LXX, βόσκω) on the mountains of Israel."

Moreover, the ἄγω verb linked to the shepherd image continues in Ezek 37:12: "I will open your graves; and I will raise (וְהַעֲלֵיתִי; LXX, ἀνάγω) you, from your graves, my people; and I will bring (וְהֵבֵאתִי; LXX, εἰσάγω) you into the land of Israel" (cf. Matt 27:51-53). The series of these various types of ἄγω verb suggest that the shepherd's primary task is to lead the flock. The verb προάγω may well indicate a kind of leadership the shepherd exercises for his flock in such a context where the shepherd image occurs.

Further, we note with interest that the MT's verb וַהֲשִׁבֹתִי in the sentence, "I will bring back my hand upon the little ones" (וַהֲשִׁבֹתִי יָדִי עַל־הַצֹּעֲרִים), is missing in Jesus' citation of Zech 13:7 in Matt 26:31. The LXX renders the verb ἐπάξω, which literally means "I will *turn/lead* (ἄγω) my hand *upon* (ἐπί) the flock." The meaning of YHWH turning his hand upon the flock can be read negatively, seeing how both the NRSV and NIV render עַל־הַצֹּעֲרִים *'against* the little ones,' often attaching the first phrase of Zech 13:8 (MT) 'through all the land.'

Yet, it is not so with the LXX's rendition, in which the first adverbial phrase, 'throughout all the land' in 8a, is attached clearly to the rest of verse 8: "And it will come to pass (καὶ ἔσται), in all the land, says YHWH the Almighty, two parts in it will be cut off, and they will perish; but the third will remain in it." In fact, the idea of 'the little ones' does not fit the wide range envisioned in the phrase, 'throughout the land.' The latter phrase is more in keeping with the scope introduced in the explanation of what would happen to the two-thirds and the one-third, that is, to the entire community of Israel. Then, the 'little flock' refers to 'the one-third,' and ἐπάξω [וַהֲשִׁבֹתִי, MT] may indicate YHWH's positive or redemptive leadership for his little flock.¹⁰⁹

This becomes more understandable if the group of 'the little ones' is likely identified with the poor ones of the flock (עֲנִיֵּי הַצֹּאן), those "who will escape at

---

Apostolic and Postapostolic Age," in *Mission of the Early Church to Jews and Gentiles* (ed. Jostein Ådna and Hans Kvalbein; Tübingen: Mohr Siebeck, 2000), 25-27, yet links the motif exclusively with Psalms (80, 87, 92).

¹⁰⁸ Pesch, *Das Markusevangelium* (Freiburg:Herder,1977), 2:381; cf. the verb προάγω used in Matt 2:9, 14:22 and 21:9 is not associated with the shepherd/sheep images.

¹⁰⁹ Likewise, Brown, *The Death of the Messiah,* 131, notes Mark 14:28/Matt 26:32 the positive promise, "however, after my resurrection I shall go before you into Galilee" may retain implicitly the majority LXX reading of Zech 13:7 ("draw out the sheep").

the time of the visitation" (CD-B 19:9). Indeed, Baumgarten and Schwartz render the phrase והשיבותי ידי על הצוערים as "and I will *turn* my hand *to* the little ones "[110] – thus producing motifs both of judgment and salvation by YHWH in each of verse 7 and verse 8, in step with Zechariah's pattern of both themes evidenced throughout Zechariah 9-14. The phrase 'the little ones' in the MT refers to the one-third, that is, the remnant whom YHWH will lead by his hand (ἐπάγω, LXX).

Therefore, Jesus' use of προάγω in Matt 26:32 echoes YHWH's ἐπάγω in the last part of Zech 13:7, whose middle part is quoted by Jesus in terms of YHWH's smiting the shepherd in the preceding verse in Matt 26:31. In Zech 13:7, we find a tripartite scene: (i) the shepherd will be smitten by the sword; (ii) the flock will be scattered; but (iii) YHWH will turn his hand upon [ἐπάγω] the little ones. Likewise, in Matt 26:31-32, we find the corresponding tripartite echoes: (i) YHWH will smite the shepherd (v. 31a); (ii) the sheep of the [whole] flock that is scattered (v. 31b); and (iii) Jesus will lead [προάγω] 'you' ahead (v. 32b).

Matthew 26:32 presents two significant discrepancies, however. First, the subject of leading the little flock in Matt 26:32 is not YHWH but surprisingly, it is Jesus, the once-smitten shepherd himself. Second, unlike the original tripartite actions in the eschatological scene of Zech 13:7-8, which coincides with the appearance of the messiah(s) of Aaron and Israel in the case of CD-B 19:7-11, Matt 26:32 involves another critical eschatological event, namely, μετὰ δὲ τὸ ἐγερθῆναί με.

The fundamental idea implied in ἐγερθῆναι, especially in this kind of text charged heavily with eschatological and messianic connotations, seems reminiscent of YHWH's promise of the eschatological Davidic Appointee. The particular association of the notion of YHWH's 'raising/establishing up' (קוּם) with various messianic figures is evidenced in numerous texts in the Davidic Shepherd tradition: (i) 'My Servant David,' who is identified with 'king,' 'shepherd,' and 'prince' in Ezek 34:23-24 (cf. 37:24-25); (ii) 'the Branch of David' in Jer 23:5 (cf. Isa 11:1-6; 4Q521 2, 2:3-14; 4Q161 8-10:18; 4Q252 5:5-7); (iii) 'the Son of David' in *Ps. Sol.* 17:20-21, 42; (iv) 'the house of David' in 4Q174 1-2, 1, 1:13; (v) 'the Prince of the whole Congregation' in 1Q28b 5.

Significantly, Matt 2:6 renders וְרָעָה of Mic 5:4 as ὅστις ποιμανεῖ but omits וְעָמַד ("he shall stand"), which can possibly be seen as a consequence of YHWH's raising up the figure of his Appointee. Upon arrival at Matt 26:32, the reader finds the notion of God's establishment of the Shepherd Appointee in terms of ἐγερθῆναι in conjunction with YHWH's eschatological leadership for his little flock entrusted to the risen Jesus.

---

[110] Baumgarten and Schwartz, *Scrolls* 2:30-31.

A close look reveals that the shepherd image in verse 31 is taken up in verse 32 in terms of both the notion of God's eschatological 'raising up' of the shepherd figure (v. 32a), and that of 'leading the flock ahead of them' reminiscent of Zech 13:7. If the smitten shepherd is understood as the representative of the wicked shepherds of Israel as in CD-B 19 and the LXX's reading of Zech 13:7, the shepherd image implied in the phrase προάγω in Matt 26:32 involves clearly another messianic figure besides the Zecharian smitten shepherd in verse 31 (cf. *Ps. Sol.* 18:5).[111]

In other words, the continuation of the shepherd's leadership by the risen Jesus in verse 32 likely points toward a further step within the eschatological process envisioned in the Davidic Shepherd tradition, namely, God's establishment of the Davidic Shepherd Appointee. In this respect, the ambiguity of the identity of the smitten shepherd in Zech 13:7 (MT) seems justified in light of the complex identity of Jesus revealed in Matt 26:31 as the smitten shepherd, and in verse 32 as the risen shepherd raised by God. Particularly in the context of the eschatological restoration of Israel, YHWH or the Davidic Shepherd figure, is often described as leading the flock, '*going ahead* of them.' Considering Zechariah 11-13 as the OT context of Matt 26:31-32, especially the vision of Zech 12:8 that coincides with the suffering shepherd(s) in Zech 11:4-16 and 13:7, this text presents the picture of the messenger/angel of the Lord. The immediate context of the prophecy regarding 'the house of David' that appears in Zech 12:6-9 is obviously the eschatological restoration of Israel 'on that day' (vv. 5, 8, 9).

According to Zech 12:8, on that day "the feeblest among Jerusalem's inhabitants will be like David, and the house of David will be like God, *like the angel of the Lord* (מַלְאַךְ יְהוָה) *[going] before them* (לִפְנֵיהֶם)." Considering the references to David and the leadership of the house of David (cf. 1Kgs 22:17; 2Chr 18:16), the phrase '[going] before them' can readily be seen as a semi-technical term for a shepherd figure developed particularly from the exodus tradition (cf. Ex 13:21). In Zech 12:6-9, it is the messenger who defeats the nations rising against Jerusalem. In eschatological contexts such as Zech 12:6-9, the eschatological shepherd going before the flock often indicates a decisive 'breakthrough' for the flock to achieve liberation, security, and abundance.

---

[111] The use of the verb ἄγω in *Ps. Sol.* 18:5 in this respect draws particular attention: καθαρίσαι ὁ θεὸς Ἰσραὴλ εἰς ἡμέραν ἐλέους ἐν εὐλογίᾳ εἰς ἡμμέραν ἐκλογῆς ἐν ἀνάξει χριστοῦ αὐτοῦ ("May God cleanse Israel for the day of mercy and blessing, for the day of election as he *brings forth/back* his anointed one"; italics mine). While Atkinson, *Intertextuality*, 386, renders ἀνάξει as "brings forth," Charlesworth, "From Messianology," 30, renders it as "brings back," stating, "the author was referring ..., as seems more probable, to the return of the Messiah, who is like the wonderful King David;" the function of this kind of eschatological 'leading' is "transferrable from God to Messiah and then back again to God."

As we shall see, Mic 2:12-13 is another case in point.[112] While Mic 2:12 clearly affirms the shepherd image – "I (YHWH) will surely gather; I will bring them together like sheep in a pen" – verse 13 says, "their king will go in advance before them (וַיַּעֲבֹר מַלְכָּם לִפְנֵיהֶם), the Lord at their head." The LXX's rendition of Mic 2:13 presents a variety of images of the shepherd figure going before the flock: διὰ τῆς διακοπῆς πρὸ προσώπου αὐτῶν διέκοψαν καὶ διῆλθον πύλην καὶ ἐξῆλθον δι᾽ αὐτῆς καὶ ἐξῆλθεν ὁ βασιλεὺς αὐτῶν πρὸ προσώπου αὐτῶν ὁ δὲ κύριος ἡγήσεται αὐτῶν (emphasis mine).

Once again, the link between the shepherd image and the phrase, 'going before the flock,' is affirmed. In Mic 2:13, the shepherd figure is closely intertwined with concepts such as 'gathering,' 'breaking through,' 'the king,' and 'the Lord' in the context of the eschatological restoration. The shepherd/king in Mic 2:13 bursts open the way, and goes before the flock who also break through 'the gate' and 'go out.'

In light of these shepherd images from the eschatological contexts of Zechariah 12 and Micah 2-5, what Jesus says in Matt 26:31-32 fits well the pattern of the Davidic shepherd's eschatological breakthrough. Jesus saying μετὰ δὲ τὸ ἐγερθῆναί με in verse 31 primarily indicates restoration from being smitten and his flock being scattered.

While the phrase, τὸ ἐγερθῆναί με, probably refers to Jesus' resurrection rather than *parousia*,[113] a more appropriate implication is YHWH's restoration of the shepherd as the head of the re-gathered flock. The Judgment will be executed by the smitten shepherd, and the shepherd's recovery would signal the re-gathering of the scattered community, which would be the foundational group of the churches of the Evangelist. Probably Matthew sees his churches in terms of this little flock.

The idea of Jesus' resurrection and the continuation of the shepherd image in verse 32 need not be seen as mutually exclusive, since OT Davidic shepherd texts such as Mic 2:12-13 and Zech 12:6-9 retain both ideas combined in the same context: the shepherd's eschatological breakthrough and the shepherd going before them to lead them. Similarly, in Matt 26:31-32, the shepherd is said to be raised again to re-gather his flock in order to lead them into the promised eschatological future. Yet, it is Jesus himself in both verse 31 and verse 32 in Matt 26 who combines the gruesome picture of the smitten shepherd of Zech 13:7 with that of the 'Breaker [הַפֹּרֵץ]/Leader,' 'the king,' and 'the Lord' of Mic 2:12-13 (cf. 'the figure like the angel of the Lord' in Zech 12:7-8).

As we can see in the broad context of Matt 26:14-27:10, the Evangelist is clearly aware of the OT Davidic Shepherd tradition. The overall theme is

---

[112] Davies and Allison, *Matthew*, 3:486, suggests Mic 2:12-13 is a reference to Matt 26:32.
[113] Luz, *Matthäus*, 4:123, 125; R. H. Stein, "A Short Note on Mark xiv.28 and xvi.7," *NTS* 20 (1974): 445-452; Davies and Allison, *Matthew*, 3:486; Don Senior, *Matthew* (Abingdon New Testament Commentaries; Nashville: Abingdon Press, 1998), 301; Carter, *Margins*, 509.

YHWH's judgment upon the whole community of his flock, symbolized by his action of smiting their shepherd. The conflict between YHWH and his own flock (eventually led to the break-up of the old covenant of Favor and Union; cf. Zech 11:4-14) is acted out by Judas at the beginning and at the closing of the story of the smitten/sold shepherd in Matt 26:14-27:10. Yet, the restoration of the remnant – that is, the new community – is promised as the shepherd is raised again, breaking through the final gate to be opened for the flock, although what is meant by 'to break through the final gate' (to liberate the flock) remains unclear at this juncture in the narrative.

What is certain is that Matt 26:31-32 is indeed a continuation of the picture of Jesus as the shepherd set out in the gospel beginning in 2:6. Similarly as various aspects of the shepherd image(s) are permeated throughout Zechariah 9-14 and Micah 2-5, Matthew presents different aspects of Jesus as the Davidic Shepherd in 2:6 (Mic 5:1-4) and 26:31-32 (Zech 13:7; Mic 2:12-13). In both of these two texts, Matthew explicitly cites from the OT Davidic shepherd tradition. In the former, Jesus is expected primarily to be established as the Davidic Shepherd Appointee over his people according to Micah's vision, and in the latter, Jesus fulfills the vision, yet this time, by assuming the role of the smitten shepherd according to Zechariah's vision.

3.1.3 Summary

Having assessed the results of our research of the Davidic Shepherd tradition in the OT and Judaism, in chapter 3 we investigated Matthew's textual interaction with the tradition. We selected the texts in the Gospel that reveal a certain degree of intertextual activity with the tradition. Clearly drawn from the tradition, Matt 2:6 (Mic 5:1) and Matt 26:31 (Zech 13:7) are treated as 'quotations.' Matthew's reading of מוֹשֵׁל in Mic 5:2 indicates his interest in the shepherd image in its association with Davidic expectation and the hope of the restoration of the nation (cf. Matt 1-4).

The Evangelist announces the coming of the Davidic Shepherd in Jesus while preparing readers for the pastoral activities of Jesus as the eschatological Shepherd after 2:6 (cf. 9:35-36; 10:1-6; 15:24). Jesus' reference to Zech 13:7 is part of Matthew's consistent use of Zechariah 9-14 in the passion narrative. As the smitten shepherd in 26:31, Jesus' use of προάγω in 26:32 continues to take up the shepherd role – this time the Shepherd as the Breaker (Matt 27:51b-53; cf. Mic 2:12-14), while later he is Davidic Shepherd-Appointee (Matt 28:16-20; cf. Ezek 34:23-24; 37:24-25).

These pictures of Jesus as the Davidic schatological Shepherd remain incomplete in view of the entire narrative of the Gospel. To hear the whole chorus of how Matthew presents Jesus as the eschatological Davidic Shepherd, besides these two explicit texts from the OT Davidic Shepherd tradition, we need to search for more missing voices.

In the following sections, we will examine Matthew's allusions to the OT Davidic Shepherd tradition and the rest of the shepherd/sheep images in the Gospel.

3.2 Allusions

Besides the citations examined above, a few passages can be labeled 'allusions' to the Davidic Shepherd tradition: Matt 9:36; 10:6, 16; 15:24 and 25:31-46. While less explicit than citations, allusions still manage to interact with the tradition. To be precise, by the term 'allusion' or 'allusive echo,' we seek to describe, eventually, 'the system of codes or conventions' rather than to find 'the genetic or causal explanations to specific texts.'[114] Thus in our attempt to interpret allusions, we hope to recover the underlying discourse that runs through those texts in the Gospel.

Although isolated in the narrative, once those allusions are blended together, they will contribute to our understanding of the theological scheme of the Gospel. The frame of reference with which those allusions in the Gospel interact, we propose, is found mainly in Ezekiel 34-37. Although we will not meticulously apply any set of criteria in order to find and evaluate allusions for each text, we will keep alert to their intended echoes of the tradition, and will focus on how Matthew is conversant with Ezekiel's vision of the Shepherd figure(s).

3.2.1 Matthew 9:36: The Shepherd with Compassion

Matthew depicts Jesus in 9:36 in terms of the compassionate shepherd for the crowd as the harassed and downtrodden flock without a shepherd: ἰδὼν δὲ τοὺς ὄχλους ἐσπλαγχνίσθη περὶ αὐτῶν, ὅτι ἦσαν ἐσκυλμένοι καὶ ἐρριμμένοι ὡσεὶ πρόβατα μὴ ἔχοντα ποιμένα. This characterization of Jesus, particularly followed by the Evangelist's summary of Jesus' ministry-teaching, preaching and healing (9:35; cf. 4:23), bears a crucial significance to understand Jesus' identity and mission.

Matthew's phrase ἦσαν ἐσκυλμένοι καὶ ἐρριμμένοι is without parallel in Mark 6:34, while Matthew and Mark share ὡσεὶ πρόβατα μὴ ἔχοντα ποιμένα. It is uncertain whether Matt relied upon Mark 6:34.[115] The latter phrase occurs three times in the LXX: Num 27:17; 2 Chr 18:16; and Jdt 11:19.[116] Characteristic of Matthew's use of the phrase is its location in his narrative; in Matthew, the phrase works in tandem with the imperative in 10:6: "Go rather to the lost sheep of the house of Israel." In this regard, texts such as 3 Βασ 22:17; Ezek 34:5-6; Josephus,

---

[114] For the discussions about method and definition adapted to this study, see introduction.
[115] Luz, *Matthew*, 2:61.
[116] Hagner, *Matthew*, 2:259, regards it as the citation of Num 27:17.

Ant. 8:404; and 2 Bar 77:13 ("the shepherds of Israel have perished") could be highlighted as the background.[117] However, the exact phrase ὡσεὶ πρόβατα μὴ ἔχοντα ποιμένα in Matt 9:36 is not found in any of the texts listed above. The shepherd/sheep images in Matt 9:36 do not deviate from the typical picture of historical Jesus.[118] Further, its placement in the narrative reflects the Evangelist's distinctive understanding of the occasion involved in the text.

The text depicts the 'harassed and downtrodden' (ἐσκυλμένοι καὶ ἐρριμμένοι) state as the condition of the people at the time of Jesus. Both the cause and the cure of this miserable state are implied in the shepherd image – they are 'sheep without a shepherd' (πρόβατα μὴ ἔχοντα ποιμένα). Moreover, Jesus' response is described in terms of compassion: ἐσπλαγχνίσθη περὶ αὐτῶν.

Thus, Jesus is depicted as the Shepherd with compassion for the leaderless people trapped in a devastating situation. For the description in 9:36, there arise three motifs to be examined in light of the Davidic Shepherd tradition: (i) the motif of compassion; (ii) the state of the sheep; and (iii) the absence of the shepherd.

(1) The Compassion Motif: The vivid verb σπλαγχνίζομαι is challenging to translate and perhaps indicates a Semitic influence; it is not the typical NT verb that means 'to feel compassionate,' i.e., οἰκτίρω (Rom 9:15; 2 Cor 1:3; Phil 2:1; Col 3:12; Jas 5:11). The OT background for the use of σπλαγχνίζομαι in the LXX is not certain either. The usual LXX equivalent for the use of רַחֲמִים (mercy/compassion) is οἰκτίρμοι not σπλάγχνα.[119]

Further, H. Köster observes that the combination of חֶסֶד and רַחֲמִים – common in OT Hebrew usage – corresponds to ἔλεος and σπλάγχνα in later Hebrew usage such as in *T. 12 Patr.* and the Dead Sea Scrolls (e.g., 1QS 1:22; 2:1); thus not any longer to ἔλεος and οἰκτίρμοι as is found in the LXX (e.g., 2:21).[120] The translation of רַחֲמִים by σπλάγχνα, often associated with the covenantal connotation of ἔλεος, thus likely sets the presupposition of NT usage.

R. T. France notes that in the NT the verb is exclusively used of Jesus himself, and like his mercy, it regularly issues forth in action to meet the need that evokes it.[121] Noticeably, the verb appears only in the Gospels: in Matt 9:36; 14:14; 15:32; 20:34 (cf. Matt 18:27) and their synoptic parallels.[122] It literally describes a tortured physical reaction in the intestines (σπλάγχνα) brought about by an irresistible pity one feels toward another; it is a compelling compassion.[123]

---

[117] Davies and Allison, *Matthew*, 2:147-148.
[118] Cf. J. Jeremias, *Jesus' Promise to the Nations* (2d ed.; London: SCM Press, 1967), 19-20, 28.
[119] H. Köster, *TDNT* 7:550.
[120] H. Köster, *TDNT* 7:552.
[121] France, *Matthew*, 175. Cf. Hagner, *Matthew*, 1:260.
[122] Here, the use of the verb in Matt 20:34 is Matthew's *Sondergut*.
[123] *LSJ*, 1628. Likewise, the term σπλάγχον frequently refers to the heart as the seat of emotions, which can be illustrated in the expression σπλάγχνα ἔλεος (a merciful heart),

In Matthew, with the exception of the parable of the unforgiving servant (18:24-35), the verb σπλαγχνίζομαι is used to describe Jesus' response to the crowds in desperate need of rescue (9:37-10:1; cf. 9:35), healing (9:37-10:1; 14:14; 20:34) or food (15:32; cf. 14:13-21). Suffice it to say, this particular verb is closely associated with Jesus' mission among the harassed and downtrodden flock in Matthew.

Against the background of the Greco-Roman world, Matthew's depiction of Jesus as the compassionate leader for the harassed flock deserves special attention. The Hellenistic kings were often linked with φιλανθρωπία (*Ep. Arist*.15; cf. *philhellene* in Josephus, *A.J.* 13. 301, 318).[124] It is hardly likely that any Hellenistic king figure was ever called 'shepherd,' as illustrated in the name such as Φιλοποίμην who is not a king, but a famous Greek general (Pausanias, *Descr.* 49:1-2). Neither is the compassion of a royal figure considered a prime virtue.

For instance, Philo of Alexandria applies the analogy of a shepherd's task to the idea of kingship. He notes that "the race of poets has been accustomed to call kings the shepherds of the people [τοὺς βασιλέας ποιμένας λαῶν εἴωθω καλεῖν]; but the lawgiver gives this title to the wise, who are the only real kings, for he represents them as rulers of all men of irrational passions, as of a flock of sheep" (*Agr.* 41). It is noteworthy that Philo also provides an explanation for the difference between a real husbandman and a mere 'tiller of the ground' by comparing 'a shepherd' with a mere 'keeper of sheep' (*Agr.* 67).

According to *De Agricultura* 39, the 'keepers of sheep' are defined as those who treat the sheep negligently and without prudence; yet, real 'shepherds' only refers to those who really know 'how to benefit the flock,' who are skilled at shepherding, and in particular, who exert wisdom to control the sheep as well as themselves.[125] The distinction between 'keepers of sheep' and 'a shepherd' greatly intrigues us, but the primary qualification for the [true] shepherd – as Philo argues and many other Greeks would agree – is self-control. Philo even critiques the quality of 'gentleness' in a ruler or governor as a lack of discipline (*Agr.* 48).

If seen against the backdrop of Philo's treatise, it appears that Matthew's description of Jesus in 9:36, ἰδὼν δὲ τοὺς ὄχλους ἐσπλαγχνίσθη περὶ αὐτῶν, ὅτι ἦσαν ἐσκυλμένοι καὶ ἐρριμμένοι ὡσεὶ πρόβατα μὴ ἔχοντα ποιμένα, presents Jesus as the true shepherd, and as an ideal kingly figure, who is indeed able to benefit the flock. From the immediate context of Matt 9:36, that is, at least in Matthew 8-9, Jesus as the true shepherd of Israel benefitting them is

---

BDAG, 938.

[124] This Greco-Roman background may illuminate *Tg. Ezek*'s rendition of 'sheep' as 'people' in 34:31.

[125] Philo eventually calls God "a blameless and in all respects a good shepherd" (*Agr.* 49; cf. 44).

illustrated in his activities of healing the sick (9:9-13; 9:27-31), casting out demons (8:28-34; 9:32-34), and raising the dead to life (9:18-26). As we shall demonstrate in the following chapters, these comprise the tasks of the promised eschatological Shepherd of Israel, and his 'compassion' is the critical sign that describes the ground motivation for the restoration.[126]

(2) The State of the Flock: Matthew's use of the image of the sheep/shepherd, and notably its specific association with the compassion motif warrants our careful attention. As J. R. C. Cousland aptly points out, 'sheep without a shepherd' is one of the most important descriptions of the crowds provided in the Gospel.[127] In both OT and Second Temple Davidic Shepherd traditions, YHWH's compassion toward His 'scattered' flock is often introduced as *a critical signal for the initiation of the restoration* (διεσπάρη, Ezek 34:6, 10; cf. Mic 7:18; Zech 10:6; 12:10; *PsSol* 17:3, 34, 45; 18:1, 3, 9; cf. Isa 40:1-11; 61:1-14). Along with YHWH's zeal for his holy name among the nations (Ezek 36:20, 21, 22; 39:7, 25; 43:7, 9; 1Q28b 5:27-28; cf. Matt 6:9; 28:19), YHWH's compassion for the lost sheep of Israel in exile constitutes the foundational motivation for Israel's undeserved and marvelous restoration.

Matthew's description of Israel's state in 9:36 thus suggests that the whole people of Israel are still in exile, as Jesus saw them as a plundered flock without a shepherd (cf. *An. Apoc.* 89:50-58; 90:3; Judith 11:19). First, it is the whole Israel that is in view in 9:36, just as Luz rightly notes, "Für Matthäus leidet offenbar das ganze Volk Not zu; die in Kap.8 und 9 erzählten Geschichten von Kranken sind für das ganze Volk repräsentativ."[128] The summary verse 9:35 also supports the totality of the description of the flock in verse 36, especially when seen as preparation for the imminent commissioning of the disciples to go to the whole house of Israel, indicated in 9:37-10:8, whereas the short-term mission of the disciples is focused on Galilee.[129]

---

[126] Cf. Gnilka, *Matthäusevangelium,* 2:352, says, Jesus' compassion here is merely "eine menschliche Rührung," while he states, as he comments on the shepherd figure in Matt 25:31-46, "In Jesus ist der göttliche Erbarmungswille erschienen (9:36)" (377); cf. Koester, *TDNT* 7:552, notes that the translation of רַחֲמִים by σπλάγχνα in the later usages retains especially "the eschatological element in the Hebrew word."

[127] J. R. C. Cousland, *The Crowds in the Gospel of Matthew* (Leiden: Brill, 2002), 86-88.

[128] Luz, *Matthäus*, 2:81; also, France, *Matthew*, 178.

[129] Cf. Alex von Dobbeler, "Die Restitution Israels und die Bekehrung der Heiden: Das Verhältnis von Mt 10,5b.6 und Mt 28,18-20 unter dem Aspekt der Komplementarität. Erwägungen zum Standort des Matthäusevangeliums," *ZNW* 91 (2000): 30, argues that the phrase, "sheep without a shepherd" does not represent the whole Israel but, according to the prophetic ruler-criticism (Jer 50:6; Isa 11; Ezek 34), only the people who suffer under their leadership; E. J. Schnabel, *Urchristliche Mission* (Wuppertal: R. Brockhaus Verlag, 2002), 293, notes, from a historical point of view, that the disciples in their short-term mission to Galilee might have met none of those ruler-class people who comprised of only 5% of the population.

Likely, the state of the people is reminiscent of the exile. The expression 'harassed and downtrodden' sheep evinces the severe injustice endured within the pen, that is, the community, exposing the sheep and leaving them vulnerable to attack by wild beasts ready to devour them, both within and outside the community.

The picture of Israel as the suffering sheep exposed to the wild beasts recalls Ezekiel 34 in the OT and *An. Apoc.* 89 (cf. *Apoc. Ezek.* 5; Judith 11:19), and both scenarios involve exile situations. As a later elaboration of the sheep/shepherd images of Ezekiel 34 and Zechariah 9-11, *An. Apoc* vividly captures the horrific condition of Israel's life experienced in the exile: the wild beasts "tear those sheep into pieces" (89:55); "eat them" (v. 56). They were "being devoured, swallowed, and snatched" (v. 58); so he (YHWH, the Lord of the sheep) "abandoned them to all of the wild beasts for food" (v. 58). Israel's desperate situation in exile finds a voice in 'Enoch' as the Lord abandons the sheep: "Then the sheep cried aloud because their flesh was being devoured by the birds; I too cried aloud and lamented in my sleep because of that shepherd who was pasturing the sheep" (90:3). The Lord of the sheep is not with them.

From the perspective of the seer(s) of *An. Apoc.*, the beasts are the nations, and not the shepherds of Israel. In contrast, from the perspective of Ezekiel, heading the list of beasts devouring the sheep is none other than the leaders of the flock itself. The relevance of Ezekiel 34 to Matt 9:35-36 is far more likely than what is found in *An. Apoc.* 89 in which the shepherds refer to the angelic beings with power.

In Ezek 34:4, 16 we find a description of the state of the plundered flock of sheep that employs similar language: they are the weak, the sick, the injured, the outcast, and the lost. Thus, YHWH – the ultimate shepherd of Israel – promises to come and shepherd his flock by himself, thereby seeking the lost, bringing back into the fold the outcast, binding up the injured, healing the sick, and strengthening the weak (Ezek 34:16). Matthew 9:35-36 presents the exact picture composed of these three images: the plundered sheep, the shepherd, and healing.[130]

Moreover, in Zech 10:2 we find an expression, "like a flock oppressed because of not having a shepherd" (כְמוֹ־צֹאן יַעֲנוּ כִּי־אֵין רֹעֶה), which the LXX translates, "like sheep and mistreated because there is no *healing*" (ὡς πρόβατα καὶ ἐκακώθησαν διότι οὐκ ἦν ἴασις). Here, the absence of the shepherd is epitomized by the absence of healing for the oppressed flock.

In the OT, the phrase, 'sheep without a shepherd,' occurs also in Num 27:17, 1 Kgs 22:17, 2 Chr 18:16, yet without the full context of the Davidic Shepherd and restoration of Israel. As many scholars would agree, Ezek 34:5-8 exceeds

---

[130] The link between the shepherd image and teaching will be discussed later (cf. *Sir.* 18:13).

all other parallels with reference to Matt 9:36: "because there is no shepherd" (LXX, διὰ τὸ μὴ εἶναι ποιμένας, Ezek 34: 5, 8).[131]

In Ezek 34:1-16, the deplorable condition of the scattered flock drives home the point of the passage, as YHWH cries out, וְאֵין דּוֹרֵשׁ וְאֵין מְבַקֵּשׁ ("there is no one searching; there is no one seeking," Ezek 34:6). Now Matt 9:35-36 proclaims that the eschatological Shepherd has finally arrived (cf. Matt 2:6). Thus, healing occurs for the harassed and downtrodden flock, and the shepherd even sets out to seek the lost in 10:1-6. The crux of Matthew's composition (that is, 9:36) lies in his identification of Jesus as the eschatological Shepherd who has finally arrived among his 'sheep without a shepherd.'

(3) The Absence of the Shepherd: In Ezek 34:5-8, the presence/absence of the true shepherd and his potential confrontation with the wicked shepherds of Israel are tightly interwoven, crucial themes. Ezekiel 34:1-16 presents YHWH's indictments aimed at the wicked shepherds of Israel, and the consequential justification of his theocratic visitation among the flock.

Similarly, in the case of Matthew 9:36, by means of the phrase, 'the sheep without a shepherd,' Israel's status is defined as leaderless, and the validity of the current leadership of the nation is implicitly denounced. Thus, the absence of the shepherd epitomizes their crisis, and the presentation of Jesus in this verse underscores his rightful motivation for restoring the flock – the compassion of YHWH upon his own flock.

This association of the themes of divine presence and divine confrontation with the false shepherds of Israel is attested by various motifs in the Gospel. The return and comforting presence of the eschatological Shepherd among the harassed and downtrodden sheep serves as a significant characterization of Jesus as Ἐμμανουήλ, God with us (Matt 1:21). Not only does this reveal the scheme of the Gospel's narrative at the macro-level, this also shows how well Matthew understands – indicated by fulfilment citations – Jesus' ministry on the continuum of God's dealing with Israel.[132]

In Matt 2:6, the announcement of the coming shepherd from Bethlehem threatened the throne of Herod the Great and startled 'all Jerusalem' with him (2:4-5). Likewise, the presence of the eschatological Shepherd among the 'sheep without a shepherd' is predictive of Jesus' presence among the crowds, necessarily implying the invalidation of Israel's current leaders.[133] When Herod

---

[131] W. Eichrodt, *Ezekiel*, 470; Carson, *Matthew* 1-12, 235; Carter, *Margins*, 230; France, *Matthew*, 175; Gundry, *Matthew*, 181.

[132] Similarly, Cousland, *Crowds*, 303, says that by identifying the people of Israel as 'sheep without a shepherd,' Matthew is "contextualizing or re-judazing" (to use his term) Jesus' ministry within the trajectory of God dealing with Israel, illustrating "the outworking of *Heilsgeschichte* within the context of Israel."

[133] Cf. Luz, *Matthäus* 2:81, states with caution, "Der Singular ποιμέν legt keine direkte Polemik gegen die jüdischen Führer nahe, am ehesten könnte man noch von 2,6 her vermuten, daß Matthäus beim Hirten an Jesus selbst denkt;" also, refer to Senior, *Matthew*, 113.

3.2 Allusions

asks the 'chief priest and scribes of *the people*' where the Messiah is to be born (2:4), they connect the birth of Jesus with the prophecy of a 'ruler' (מוֹשֵׁל) whose task will be that of 'shepherding [God's] people Israel.' The difference between Herod and Jesus as two rival 'kings' (2:1, 2) becomes explicit in the narrative where Herod's brutal power leads to the massacre in Bethlehem (2:16-17), the place where the Evangelist announces the coming of the Shepherd who will rule in peace (cf. Mic 5:4).

Similarly, the critical difference of Jesus' leadership from that of the Pharisees and scribes is witnessed by the crowd especially in terms of Jesus' unparalleled ἐξουσία in his teaching (7:29) and healing (9:6, 8). Also this ἐξουσία epitomizes Jesus' mission toward the lost sheep of Israel in 10:1-6, and his mission through his disciples underscores the arrival of the time for the restoration of Israel and her leadership. Jesus' activities among the crowd thus echo YHWH's indictments upon Israel's false shepherds who neither seek nor heal (Ezek 34:1-16).[134]

The result was that the crowd 'followed' Jesus as the flock follows its shepherd (4:25; 8:1; cf. 19:2; 20:29).[135] Jesus also warns the people against the false prophets using the sheep image in 7:15, whose identity possibly involves the Pharisees of Jesus' time.[136] Jesus takes the role of the true shepherd who benefits his own flock as he rescues and protects them.

In Matthew 8-9, the conflict with Israel's leadership appears to be implicit, but the presence of the eschatological Shepherd among his own flock is explicit. At this moment in Matt 9:36, all depends on how Jesus as the eschatological Shepherd would minister among the sheep, and how the sheep would respond to him. Although we shall closely examine Jesus' shepherding activities in the following chapters, the inclusion of 'healing' in 9:35 as one of the critical tasks of Jesus' mission of compassion in verse 36 likely recalls the association of YHWH as the eschatological Shepherd with his healing mission in Ezek 34:16 (cf. ἴασις in Zech 10:2 [LXX]).

Finally, in Matt 9:36, it becomes obvious that it is Jesus (and not the Jewish leaders) whose role is that of 'shepherding' the Jewish people.[137] Consequently, as Evert-Jan Vledder highlights, the differing reactions of the crowd and the

---

[134] Did the Pharisees or their disciples heal? Matt 12:27b, οἱ υἱοὶ ὑμῶν ἐν τίνι ἐκβάλλουσιν, indicates that they cast out demons not mentioning healing. See, Keener, *Matthew*, 363; Ben Witherington, *The Christology of Jesus* (Minneapolis: Augsburg Fortress Press, 1990), 164; for the common practice of Jewish exorcists, see J. P. Meier, *Matthew* (Wilmington, Del.: Michael Glazier, 1980), 134-135 with his notes on the differences as well.

[135] Cousland, *Crowds*, 146-148.

[136] Keener, *Matthew*, 251, n.243; Carson, *Matthew*, 190; cf. Otto Böcher, "Wölfe in Schafspelzen," *TZ* 6 (1968): 405-426; Hill, "False Prophets and Charismatics: Structure and Interpretation in Matthew 7:15-23," *Bib* 57 (1976): 327-348.

[137] Dorothy Jean Weaver, *Matthew's Missionary Discourse: A Literary Analysis* (JSNTSup 38; Sheffield: Sheffield Academy Press, 1990), 77.

Pharisees emerge at the conclusion of 'the miracle circle in Matt 8-9.'[138] Their reaction becomes more pronounced as the narrative continues: the Pharisees try to prevent Jesus from healing the crowds (9:32-34; 12:22-24); they balk as he seeks out the lost (9:11); they blaspheme (12:31-32); they also bar admission to those who would seek to enter the Kingdom of heaven (23:13), and they mislead their flock by 'teaching as doctrines the precepts of men' (15:9).[139]

The shepherd image applied to Jesus in Matt 2:6 (ὅστις ποιμανεῖ) involves clearly the figure of Davidic Shepherd whose reign is envisioned to extend over the nations (cf. Mic 5:3-4). Given that both shepherd images in Matt 2:6 and 9:36 involve the OT expectation of Israel's restoration, which one would bears more weight to identify Jesus remains unclear; the readers must await the narrative that follows in the Gospel.

3.2.2 Matthew 10:6, 16; 15:24: The Lost Sheep of Israel

Matthew 10:6 retains Jesus' command for his twelve disciples to "go rather to the lost sheep of the house of Israel": πορεύεσθε δὲ μᾶλλον πρὸς τὰ πρόβατα τὰ ἀπολωλότα οἴκου Ἰσραήλ. Here, the people of Israel are seen as the lost sheep and Jesus sends his disciples as his under-shepherds to seek and heal them. The sheep image reoccurs again in verse 16a, yet this time it depicts the disciples as sheep among the wolves: Ἰδοὺ ἐγὼ ἀποστέλλω ὑμᾶς ὡς πρόβατα ἐν μέσῳ λύκων. Jesus reasserts this particular understanding of his own mission limited within the house of Israel as he faces the request of the Canaanite woman in 15:24: ὁ δὲ ἀποκριθεὶς εἶπεν· οὐκ ἀπεστάλην εἰ μὴ εἰς τὰ πρόβατα τὰ ἀπολωλότα οἴκου Ἰσραήλ.

These shepherd/sheep images in these texts are tied closely to Jesus' understanding of his earthly mission. Both the use of the images and the motif of Jesus' mission to Israel indicate the authenticity of the texts regarding the historical Jesus:

Concerning the source of Matt 10:6 (v. 5b), some defend the authenticity of the saying,[140] while others ascribe it to Matthew's own creation in accordance with the shepherd/sheep image of 9:36.[141] In fact, not only is the image of πρόβατα μὴ ἔχοντα ποιμένα ('sheep without a shepherd') in 9:36 Matthew's *Sondergut*, but also the references to the mission of Jesus to τὰ

---

[138] Evert-Jan Vledder, *Conflict in the Miracle Stories: A Socio-Exegetical Study of Matthew 8 and 9* (JSNTSup 152; Sheffield: Sheffield Academic Press, 1997), 42; Cousland, *Crowds*, 304, reads Matthew's community as "irrevocably broken with the leadership but not with the people as a whole."

[139] Cousland, *Crowds*, 92.

[140] Maria Trautmann, *Zeichenhafte Handlungen Jesu: ein Beitrag zur Frage nach dem geschichtlichen Jesus* (FB 37; Würzburg: Echter, 1980), 221-225; J. Jeremias, *Jesus' Promise to the Nations* (2d ed.; London: SCM Press, 1967), 19-20, 28; D. Bosch, *Die Heidenmission in der Zukunftsschau Jesu* (ATANT 36; Zürich: Zwingli-Verlag, 1959), 76-80.

[141] F. W. Beare, "The Mission of the Disciples and the Mission Charge: Matthew 10 and Parallels," *JBL* 89 (1970): 9; Frankemölle, *Jahwebund*, 126-130.

3.2 Allusions

πρόβατα τὰ ἀπολωλότα οἴκου Ἰσραήλ ('the lost sheep of the house of Israel') in 10:5-6 and 15:24 are also unique to Matthew. Matthew's unique material, in principle, does not necessarily indicate that he created it. Indeed, opinions vary concerning the historicity of Matt 10:5-6 (cf. 15:4) in view of its relation to Jesus' command for the universal mission at the end of the Gospel (28:19-20).

Some argue that the saying goes back to Jesus since it does not reflect Matthew's own conviction of the mission to the Gentiles, while other argues that it was manufactured by Matthew's Jewish Christian community who opposed the Gentile mission.[142] An argument based on the hypothetical reconstructions of Matthew's community may be inadequate either to prove or to disprove the authenticity of the sayings in 10:5-6 and 15:24.

While attractive, the last option is highly unlikely as well: the Evangelist himself created the texts in order to advance his own understanding of salvation-history.[143] It is an attractive option that the saying might be Matthew's own interpretation of Jesus' particular attitude toward his earthly mission to Israel.[144] Yet it is stretch to think that Matthew, simply to prove his point, would put words in Jesus' mouth based on an unstable juxtaposition of 10:5-6 and 28:19-20. Thus it is not likely that if Matthew were to do that, he would not offer an explanation for the seemingly contradictory views on mission in his narrative, or fail to at least present the issue itself.[145] Instead, Matthew regarded the disciples' mission restricted to Israel as a thing of the past (10:5-6; 15:34).[146] Of greater import is this mission of the disciples 'to the lost sheep of the house of Israel' likely assumes the continuation of the historical mission of Jesus to seek the lost among Israel (cf. Matt 9:10-13; Matt 25:31-46).

Bound up with the question of the authenticity of 10:5-6 is the issue of its relationship to 15:24. If the Evangelist received from the Lord the commission to go to the Gentiles (28:19-20) and placed 10:5-6 as Jesus' historical mission to Israel, it is also likely that 15:24 is historical.[147] Once we argue for the authenticity of the saying in 10:5-6, no convincing reason is offered not to view the saying of 15:24 as authentic rather than Matthew's redactional version of the traditional login preserved in 10:5-6. Particularly, the controversial trait of the saying in the context of 15:21-28 likely affirms its historicity as well.

The narrative connection between 9:36 and 10:5-6 has been widely noticed.[148] Compared to Mark 3:13-19a and 6:7-13, it is characteristic of Matthew that Jesus' sending his disciples to the lost sheep of Israel (10:1-16) immediately

---

[142] Davies and Allison, *Matthew*, 2:168-169, list the scholars such as Kümmel and Bartinicki for the former view, and Wellhausen and Hahn for the latter.

[143] Hubert Frankenmölle, *Jahwebund und Kirche Christi: Studien zur Form und Traditiongeschichte des Evangeliums nach Matthäus* (Münster: Verlag Aschendorff, 1973), 123-143; S. Brown, "The Two-fold Representation of the Mission in Matthew's Gospel," *ST* (1977): 21-32.

[144] G. N. Stanton, "Matthew as Creative Interpreter of the Sayings of Jesus," in *Das Evangelium und die Evangelien: Vorträge vom Tübinger Symposium, 1982* (ed. P. Stuhlmacher; WUNT 28; Tübingen: Mohr Siebeck, 1983), 276-277.

[145] Davies and Allison, *Matthew*, 2:169.

[146] Luz, *Matthew*, 2:74, lists scholars, starting with Tertullian and Jerome, who support this interpretation, e.g., Strecker, G. Bornkamm and A. Vögtel.

[147] Cf. G. Strecker, *Der Weg der Gerechitigkeit: Untersuchingen zur Theologie des Matthäus* (Göttingen: Vandenhoeck und Ruprecht, 1966), 109; Luz, *Matthew*, 2:340.

[148] Luz, *Matthäus* 2:87; Carson, *Matthew*, 244; Keener, *Matthew*, 309.

follows the summary of the shepherd's ministry among the sheep without a shepherd (9:35-38). We recognize that the shepherd/sheep image constitutes the nexus of these two pericopes. Luz, while noting that the motif of harvest in 9:37 is associated with judgment in the OT and in Judaism, addresses the unsolved exegetical question: "How are the merciful shepherd and the Lord of judgment [Son of Man] to be understood together?"[149]

Yet, it is noteworthy that the setting for the shepherd's mission is restricted to Israel, so it is not universal; the judgment theme should be relevant in view of this restriction. Further, the shepherd/sheep images both in 9:35-36 and 10:1-6, 16 would be key to understanding the association of the shepherd image for Jesus and the theme of judgment. In the following, we will argue that the Ezekiel pattern in 34-37, and partially in 40-48 will provide the theological framework for Jesus' mission to Israel as represented using shepherd/sheep images in 10:1-6, 16 and 15:24.

Matthew's preference of the shepherd/sheep image is surely affirmed in the phrase ὡς πρόβατα ἐν μέσῳ λύκων in 10:16 – Mark omits it, while Luke uses ἄρνας ('lambs', 10:3) instead of πρόβατα. It is also likely that the 'sheep without a shepherd' in 9:36 are, in effect, the same group as 'the lost sheep of the house of Israel' in 10:5-6.[150] Furthermore, since the genitive in the phrase τὰ πρόβατα τὰ ἀπολωλότα οἴκου Ἰσραήλ is likely understood as epexegetical and not partitive,[151] both phrases refer to the entirety of Israel as a nation. This particularly Matthean construction raises a question about Jesus' identity and his ministry: Is Jesus the eschatological Shepherd?[152]

In fact, Matthew's presentation of Jesus as the compassionate shepherd for the sheep without a shepherd in 9:35-36, followed by Jesus' mission to the lost house of Israel in chapter 10, matches perfectly with the pattern embedded in Ezekiel 34, in which the absence of one shepherd necessitates YHWH's intervention to seek his own flock. Could this be how Jesus understood his own identity and ministry?

Risto Uro acknowledges that the possibility of its authenticity cannot be categorically denied based on the probability of dominical authority for the earliest Christian mission in Palestine. Yet as he also argues, "it seems that the prohibition of Matt 10:5b somehow indicates an ongoing Gentile mission and thus falls beyond the earthly life of Jesus."[153] One assumes an active 'ongoing Gentile mission' in Matthew's time, yet why would Matthew take a contradictory stance against a missionary thrust directed at Gentiles in his time? Unable

---

[149] Luz, *Matthew*, 2:65.

[150] Cousland, *Crowds*, 90; Gnilka, *Matthäusevangelium* 2:352; Davies and Allison, *Matthew* 2:167.

[151] For instances, France, *Matthew*, 178; Evans, "Restoration," 318.

[152] See Davies and Allison, *Matthew* 2:148, who suggest this interpretation as a possibility.

[153] R.Uro, *Sheep Among the Wolves: A Study of the Mission Instructions of Q* (Helsinki: Suomalainen Tiedeakatemia, 1987), 54.

3.2 Allusions

to explain the seemingly contradictory missions in Matt 10:5-6 and 28:19-20, Uro leaves the matter unsolved.[154]

Likewise, something similar Jesus said in 15:24 would create more difficulties if its authenticity were to be denied. Thus it is less problematic to insist, as does Gundry, that "Matthew's construction is based on Jesus speaking of himself as sent by God, out of OT phraseology for Israel, and out of the historical limitation of Jesus' mission to Israel."[155]

Jesus' compassion for the crowds described in Matt 9:36 effectively conveys Matthew's identification of Jesus as the promised Shepherd in view of Israel's restoration, as manifested in 2:6, echoing Mic 5:1-4. As to the phrase, τὰ πρόβατα τὰ ἀπολωλότα οἴκου Ἰσραήλ ("the lost sheep of the house of Israel"), besides the passing reference to the lost sheep found in Jer 50:6 (cf. Isa 53:6), no other compatible text exists in the OT except Ezekiel 34.

Indeed, many refer to Ezek 34:1-16 as the most plausible background for Matt 9:36, 10:5-6, and 15:24.[156] Yet we must ask what this intertexual connection implies for Jesus' ministry and Matthew's unique presentation of it. While some see in Matt 10:5-6 (cf. 15:24) and Matt 28:19-20 a contradiction seeking the solution mainly in 'the religio-social situation of the Matthean community,'[157] it is likely that the seemingly contradictory verses can best be understood in terms of subsequent stages in accordance to *Heilsgeschichte*.[158] Yet, it is rarely asked what specific pattern is implied in Matthew 10; and from where in the OT it is taken. Though Ezekiel 34 is often consulted as the background of the particularly Matthean shepherd/sheep images in 9:36 and 10:5-6, the pattern implied in Ezekiel's vision is scarcely considered for those concerns.

The most distinguished common element in this regard for both Matthew and Ezekiel is the critical issue facing Israel: the absence of its shepherd. In Matt 10:5-6 and 15:24, Jesus sends his twelve disciples to seek the lost sheep of the house of Israel (cf. Ezek 37:15-28). In Ezekiel 34, we find both the motif of the sheep without a shepherd, and the image of Israel as the lost sheep in the context of YHWH promising to come to, seek, and restore them. In the broader context of Ezekiel 34-37, the pattern becomes clear that after YHWH seeks the lost

---

[154] Risto Uro, *Wolves*, 55-56.

[155] Gundry, *Matthew*, 313.

[156] For instance, Davies and Allison, *Matthew* 2:167; Keener, *Matthew*, 309, 315; Luz, *Matthew*, 2:64; Heil, "Ezek 34," 702; Carter, *Margins*, 234; Hagner, *Matthew*, 1:260.

[157] S. Brown, "The Two-fold Representation, 21-32; idem, "The Mission to Israel in Matthew's Central Section," *ZNW* 69 (1978): 215-221; Y. Anno, "The Mission to Israel in Matthew: The Intention of Matthew 10:5b-6 Considered in the Light of the Religio-Political Background" (Th.D. diss.; Lutheran School of Theology, Chicago, 1984).

[158] J. P. Meier, *Law and History in Matthew's Gospel: A Redactional Study of Mt. 5:17-48* (Rome: Biblical Institute Press, 1976), 27-30; Carson, *Matthew*, 244; France, *Matthew*, 177-178. Cf. Axel von Dobbeler, "Die Restitution Israels," 18-44.

sheep of Israel and establishes justice, he secures his presence by appointing the Davidic shepherd who rules as King over the restored flock, and is also the Prince among them.

If Ezekiel's vision (Ezek 34:1-16) is relevant for the priority of Jesus' mission to Israel in Matthew 9-10, then Jesus' mission to 'the lost sheep without a shepherd' can be taken as a critical sign of the restoration of Israel. In other words, Jesus seeking the lost sheep of the house of Israel would be seen as preparatory for YHWH to appoint the Davidic Shepherd at the end of the process of gathering the scattered flock (cf. Matt 26:31-32) to reach out to the nations with peace (Matt 28:19-20; Ezek 34:23-24; 37:24-25; cf. Matt 2:6; Mic 5:1-4).

This end vision for the nations under the authority of the Davidic Shepherd, we suggest, is characteristic of Ezekiel 34 and 37 (cf. 40-48): "You are my flock, the flock of my pasture, people (הָאָדָם) you!" (Ezek 34:31; cf. Jer 23:1). In Ezekiel 34, the intended ambiguity of הָאָדָם at the end of Ezek 34:31 comes after (i) YHWH's eschatological shepherding for his own flock, the lost sheep of the house of Israel (34:1-23), and (ii) YHWH's consequential appointment of the Davidic Shepherd over his rescued flock in verses 23-24. Compared with the MT's version of Ezekiel, this open vision toward the nations is flatten out by the Targum's explicit paraphrase of the significant verse: "And you My people, the people over whom My name is called, you are *the house of Israel* (בית ישראל, *Tg. Ezek* 34:31).[159] For the translator(s) of *Targum Ezekiel*, the range of the mission of the Davidic Shepherd-Appointee is seen as restricted within the house of Israel.

Nevertheless, the notion of humanity in general still remains in the Targum's rendition of the MT's הָאָדָם as בית ישראל, for as Levey suggests, the Targum's rendition implies that "Israel personifies Adam, of humanity *par excellence*."[160] Compared with the *Ezekiel Targum*'s rather nationalistic view of the Davidic Shepherd's mission within the house of Israel, the First Gospel seems to follow the original vision and pattern of Ezekiel's vision of the Davidic Shepherd. Already in Matt 2:6, the Evangelist applies the shepherd image to Jesus as he renders the MT's מוֹשֵׁל as ὅστις ποιμανεῖ (cf. Mic 5:3-4) and places this citation in the context of Matthew 2 where the gentile magi come and worship Jesus, the Davidic [thus, legitimate] Shepherd (cf. Matt 1:1, 18-23).

According to Ezekiel's vision, YHWH's eschatological mission to the house of Israel as the lost sheep is to precede the Davidic Shepherd's extended reign over the nations. The phrase 'house of Israel' appears frequently in the context of the restoration in Ezekiel 34-37 (34:30, 31; 35:15; 36:10, 21, 22, 32, 37;

---

[159] Sperber, *Bible in Aramaic* 3:349.
[160] Levey, *Targum*, 99, n.19, refers to R. Simeon b.Yohai who cites this verse to demonstrate that Israel is designated by God as Adam in contradiction to the other nations of the world (*b. Yeb*. 60b-61a).

37:11). Jesus as the eschatological Shepherd in Matthew 10 undertakes first the mission to seek the lost sheep of the house of Israel, which is, intentionally, a very nationalistic mission; yet it follows exactly Ezekiel's pattern.

For the seer(s) of *An. Apoc.,* the sheep metaphor for Israel serves as the Vision's view of *the historical Israel* as particularly led by 'the Lord of the sheep' (89:16, 26, 29, 30) only until the time of the eschatological transformation (90:25-27). Yet, after the messianic white bull appears and transforms both the sheep and the wild beats into one people of the Lord's sanctified community, the characterization of Israel as "the sheep [of the house of Israel]" becomes invalid.

Likewise, Jesus' mission to 'the lost sheep of the house of Israel' in Matthew 10 as the mission of the eschatological Shepherd can be seen as the rescue mission of the harassed flock; yet, it is accomplished only when the new reign of the Davidic Shepherd-Appointee is established. Compared with Ezekiel's vision, Matthew's introduction of the Shepherd's mission to the lost sheep of the house of Israel exhibits certain differences as well. In Ezekiel 34, it is YHWH himself who seeks the lost sheep, whereas in Matthew, it is Jesus (cf. Matt 9:9-13) and his disciples sent by Jesus (cf. Matt 10:1-5), assuming YHWH's role for the task of Israel's restoration.

What we perceive in Jesus sending the disciples to the house of Israel is perhaps the picture of the eschatological community, presumably executing one of YHWH's roles at the time of Israel's restoration. In the broad vision of Ezekiel 34-37 and 40-48, the shepherd-leadership is transferred from YHWH himself (who mainly rescues the flock from the mouths of the beasts, Ezek 34:1-22) to the hand of the Davidic Shepherd (34:23-24; 37:24-25), and finally to the Princes whose leadership will be more cultic than political (40-48; esp. 45:8,9).[161] This pattern is attested in Micah 5 as well. The vision of the Shepherd's mission to feed the flock and establish the Davidic Shepherd in Mic 5:1-4 is followed by the vision of YHWH raising up 'the [under-]shepherds' in 5: 6-15 (רֹעִים, 5:6; v. 5 LXX) as YHWH extends his rulership among the nations through the Davidic Shepherd-Appointee.

Turning to Matthew 10, Jesus, appointing the twelve disciples as under-shepherds can be seen proleptic, in light of Ezekiel's pattern implied in 34-37 and 40-48. Jesus confers upon the disciples the role of seeking the lost, and healing the sick, implying that he considers them under-shepherds active in the restoration process.[162]

---

[161] Refer to Ch. 1.4.1, "The Structure of Ezekiel's Vision of Restoration in Chapters 34-48."

[162] Scot McKnight, "New Shepherds of Israel: An Historical and Critical Study of Matthew 9:35-11:1" (Ph.D. diss.; University of Nottingham, 1986), esp. 189-193, likewise calls Matthew's twelve apostles 'New Shepherds of Israel' confronting the Pharisees and their leadership.

In Matt 9:36-10:6, the twelve are designated apostles (10:2) whose role and work extends beyond Easter and beyond Matthew's time. Jesus' twelve disciples, sent as his under-shepherd yet as at the same time as 'the sheep among the wolves,' perhaps even symbolize the representative of the reconstituted Israel, similar to the seer(s) in *An. Apoc.* 89:12-14 that picture Israel as 'twelve sheep' who 'grew up in the midst of the wolves [Egyptians].'[163]

By assuming YHWH's role, Jesus gathers YHWH's flock as promised in the OT Davidic Shepherd tradition. Not only are the twelve disciples considered under-shepherds in the restoration process, they also may well be symbolically identified with the restored flock by themselves. Jesus is depicted as the eschatological Shepherd to gather 'das eschatologisch restituierte Zwölfstämmevolk,'[164] and the task is now entrusted in Matthew 10 to the twelve apostles. Likewise, the reason for the restriction of the mission within Galilee in Matthew 10 was, as Eckhard J. Schnabel notes, surely not even pragmatic but rather 'theologisch-heilgeschichtlich'; also, the number twelve is emphasized in Matt 10:1, 2, 5, which means that "the twelve represents Israel and are sent therefore to the Jews."[165]

In this context of Jesus taking charge of the role of the eschatological Shepherd in view of the restoration of Israel, the term ἐξουσία in Matt 10:1 is particularly engaging. What kind of authority is this? In the same verse, the following descriptions are only suggestive as to the nature and function of the authority mentioned in 10:1: ἐξουσίαν πνευμάτων ἀκαθάρτων ὥστε ἐκβάλλειν αὐτὰ καὶ θεραπεύειν πᾶσαν νόσον καὶ πᾶσαν μαλακίαν ("the authority [that enables] to cast out the unclean spirits and heal all sickness and every malady"). There is no clear explanation in the text about how Jesus received and how he could impart this authority to his disciples as the eschatological [under-]shepherds of Israel. Perhaps the authority given to the disciples is the authority described as conferred from YHWH himself as the eschatological Shepherd in Ezekiel 34.

According to Matt 10:1 — and drawing upon the language of Ezekiel 34 — this authority is imparted for the rescue of the sheep from the mouths of wild beasts, and for the purpose of restoring, and even healing them. It is not clear whether preaching (10:7) should be included or merely thought of as synony-

---

[163] The wolves in Matt 10:16 are highly symbolical, and are not to be identified with the lost sheep themselves, but with the leaders of the people (10:17-18). Against R. Uro, *Wolves*, 46, who contends that the image of the people as 'sheep without a shepherd' (9:36) drastically changes into 'wolves' (10:16). In Ezekiel 34, the failed shepherds play the role of wild beasts.

[164] Zimmermann, *Messianische Texte*, 126-127.

[165] Schnabel, *Urchristliche Mission*, 293. For the cases other passages, refer to Wright, *Victory*, 299-300, 318; Zimmermann, *Messianische Texte*, 126-127; cf. E. P. Sanders, *Jesus and Judaism*, 98.

mous with those type of activities. Notably, teaching is not yet included (cf. 28:19-20).[166]

What is important is those missionary tasks performed in the context of Israel's restoration seem to define the source and kind of authority Jesus assumes for himself (cf. 4Q521 1:1-13). That is, the context of the rescue-mission for the lost sheep of the house of Israel recommends YHWH himself, with his own authority, as the true Shepherd of Israel.

Is not Jesus' ἐξουσία, in view of this context, descriptive of not only his restricted mission but also his status? The ἐξουσία of healing the sick, casting out demons, raising the dead, and teaching, seems to belong squarely to the person of Jesus (cf. 4:23; 9:35). Jesus' imparting the ἐξουσία upon his disciples would not mean that they are expected to perform miracles independent of him. The absence of the teaching activity in Jesus' commission in 10:1-6 may also underlines that the teaching ἐξουσία remains exclusively bound to Jesus (cf. 28:19-20).[167]

Similarly, in Ezekiel 34-37, leading the rescued and restored flock to the way of righteousness according to the law – that is, teaching them – is to be initiated by YHWH's appointment of the Davidic Shepherd over them. It is likely that the ἐξουσία found in Matt 10:1 is particularly ascribed to YHWH himself. Now Jesus exercises and even imparts the authority to the twelve whom he appoints as the eschatological [under-]shepherds over and in the midst of the lost flock.

The language and context of Matthew 10 suggests that Jesus is the eschatological Shepherd who reconstitutes Israel, exercising YHWH's authority to snatch the lost sheep of Israel from the mouth of wild beasts. This implies judgment upon the leadership of Israel, while the poor of the flock are restored as envisioned in Ezekiel 34-37 (cf. Mic 5:1-4).

Nevertheless, the final harvest for both Israel and the nations appears to be deferred. Also, as an observation central to our thesis, it is important to note that the Evangelist depicts the scene of universal judgment with colorful images of the shepherd/sheep taken from the Davidic Shepherd tradition.

3.2.3 Matthew 25:31-46: The Shepherd as the Final Judge

Just as the shepherd/sheep images in 9:36, 10:6, 16 and 15:24 are unique to Matthew, the same can be said of Matt 25:31-46. In Mark and Luke we find no parallel of this impressive depiction of the son of Man as the final Shepherd-Judge who separates the sheep from the goats. The parable in Matt 25:31-46 is

---

[166] Refer to Heil, "Ezek 34," 702. Cf. Oscar S. Brooks, "Matthew xxviii 16-20 and the Design of the First Gospel," *JSNT* 10 (1981): 9, mistakenly combines Jesus' authority and teaching in Matthew 10.

[167] Samuel Byrskog, *Jesus the Only Teacher: Didactic Authority and Transmission in Ancient Israel, Ancient Judaism and the Matthean Community* (Almqvist & Wilsell International: Stockholm, 1994), 284.

indeed seen by many as a Matthean composition with many features characteristic to Matthew.[168]

Though this parable is unique to Matthew in our extant sources, this fails to prove that it should be attributed to Matthew's creation. Neither Matthew nor Luke uses every detail of Mark, and this warns us against rejecting material not documented more than once in extant sources. Some argue that a few linguistic features found present only in Matthew, such as ἔριφος (v. 32, 'goat'), ἐριφίον (v. 33, 'goats'), γυμνός (vv. 36, 38, 43, 44, 'naked'), ἐπισκέπτομαι (vv. 36, 43, 'visit'), καταράομαι (v. 41, 'curse') and κόλασις (v. 46, 'punishment'), can best be explained as the preservation of the oral tradition.[169] Moreover, as Keener notes, the parabolic form, the 'Son of Man,' the use of *Amēn* (25:40, 45), and similar features recall Jesus' own style.[170] In fact, nowhere else does Matthew enumerate corporal works of mercy or associate the *parousia* of the Son of Man with the title 'king.'[171]

These 'Matthean' features may signify that Matthew has more freely recast the oral tradition as some purportedly 'Matthean' features (like parallelism) that make just as much sense in Jesus' Palestine milieu.[172] Those who deny that Jesus could have planned the admission of the Gentiles – proffering a theological reason – naturally see this parable as inauthentic.[173] But even the notion of the inclusion of the Gentiles to the extended flock under one shepherd, as we shall argue in the following, may not be foreign to the Jesus tradition. As argued earlier, if Matt 26:31 is rightly to be ascribed to Jesus himself, it is not difficult to assume that Jesus regarded himself as the shepherd of the little flock in 25:31-46 as well.

The simile of the shepherd image found in 25:32c – "as a shepherd separates the sheep from the goats" – may be seen as a mere passing remark.[174] Nevertheless, the question of why the simile is necessary is worthy of further exploration. The obvious focus of the simile is the idea of separation, yet the idea of *gathering* all of the nations (πάντα τὰ ἔθνη) might indirectly support the image of the shepherd found in v. 32a: καὶ συναχθήσονται ἔμπροσθεν αὐτοῦ πάντα τὰ ἔθνη.[175] As Gnilka rightly affirms, the term συνάγω in v. 32a is "ein Term der

---

[168] Gnilka, *Matthäusevangelium* 2:367-370; Gundry, *Matthew*, 511. Cf. J. Jeremias, *Parables of Jesus* (rev. ed.; London: SCM, 1972), 206-209.

[169] Davies and Allison, *Matthew*, 3:417-418; T. W. Manson, *The Sayings of Jesus* (Grand Rapids: Eerdmans, 1979), 249, underlines the startling originality of the saying of Matt 25:31-46 that must be ascribed to Jesus himself; see also, J. Friedrich, "Gott im Bruder?," *CTM* 7 (1977), 9-45.

[170] Keener, *Matthew*, 602.

[171] Davies and Allison, *Matthew*, 3:418.

[172] Keener, *Matthew*, 602; cf. John A. T. Robinson, *Twelve New Testament Studies* (SBT 34; London: SCM Press, 1962), 88-89); Ingo Broer, "Das Gericht des Menschensohnes über die Völker: Auslegung von Mt 25,31-46," *BibLeb* 11 (1970): 273-295.

[173] For instance, E. P. Sanders, *Jesus and Judaism* (Philadelphia: Fortress Press, 1985), 111.

[174] Hill, *Matthew*, 330.

[175] The picture of Jesus gathering his lost flock is vivid in the gospel. Cousland, *Crowds*, 171, sees that the absence of the motif of the shepherd gathering the flock does in fact highlight that 'the crowd gather itself' to Jesus and 'follow him' as a consequence of the shepherd's ministry of seeking them out (cf. Matt 9:12-13).

Hirtensprache," which is part of the shepherd image in the context.[176] In order to separate his own flock, the shepherd must gather them first, which is a vital component of the shepherd image.

Morever, as the simile of v. 32c extends into the following verse, we discover that the shepherd in v. 33 is designated ὁ βασιλεύς who calls God 'my Father' in v. 34. Then, the shepherd-king as the Son of Man judges between the sheep and the goats in accordance with the acts of compassion – feeding, seeking, and healing – that ought to be administered to the little ones among his flock: ἑνὶ τούτων τῶν ἀδελφῶν μου τῶν ἐλαχίστων (v. 40; cf. ἀδελφῶν μου omitted in v. 45). Many affirm the relevance of Ezek 34:17-22 for the picture offered in Matt 25:31-46.[177] In fact, the image of the division of the sheep from the goats occurs nowhere else in the OT except in Ezekiel 34.[178] In Ezek 34:17-22, goats probably represent a group of people who enjoin the leaders of the flock to exploit the weaker ones in the community. YHWH confronts them as well as those apparently holding leadership positions: "My anger is hot against the shepherds [הָרֹעִים] and I will punish the he-goats [הָעַתּוּדִים]" (Ezek 34:2).

Thus, they may not be the shepherds themselves but the stronger ones in the community. This is apparent once again in 34:17 where YHWH proclaims: שֹׁפֵט בֵּין־שֶׂה לָשֶׂה לָאֵילִים וְלָעַתּוּדִים ("I will judge *between* one sheep and another, and between rams [male sheep] and goats"). The parallel structure in the verse likely indicates that Ezekiel mentions the issue of justice and mercy within the community where both the strong and the weak are found, which is the major theme of Matt 25:31-46.[179]

No other comparable text exists besides Ezek 34:17-22 that involves sheep and goats in the context of God's judgment to establish the eschatological community. The allusion is nearly irrefutable. Once we accept the possibility of an allusion between Matt 25:31-46 and Ezek 34:17-22, this proposal can be buttressed with additional, and rather striking correlations between two texts.

---

[176] Gnilka, *Matthäusevangelium* 2:371.

[177] W. Eichrodt, *Ezekiel*, 474; G. A. Cooke, *Ezekiel*, 377; Cousland, *Crowds*, 88; Gundry, *Matthew*, 512; Hill, *Matthew*, 330; Keener, *Matthew*, 603; Schlatter, *Matthäus,* 724; C. Landmesser, *Jüngerberufung und Zuwendung zu Gott: Ein exegetischer Beitrag zum Konzept der matthäischen Soteriologie im Anschluß an Mt 9,9-13* (WUNT 133; Tübingen: Mohr Siebeck, 2001), 108, particularly as the continuation of the Shepherd/Physician image of Jesus taken from Ezekiel 34, states, "Ausdrücklich bezieht sich der Verfasser des Matthäusevangeliums auch an anderer Stelle auf Ez 34,17; nach der nur von ihm aufgenommenen Schilderung des Weltgerichts in Mt 25,31-46 wird der Menschensohn, also Jesus, als Richter aller Völker auftreten und dabei die Schafe von den Böcken unterscheiden" (v.32f).

[178] Schnackenburg, *Matthew*, 257.

[179] K. Weber, "The Image of Sheep and Goats in Matthew 25:31-46," *CBQ* 59 (1997): 657-678, takes exception and doubts the relevance of Ezek 34:17-22 based on the reason mainly that Matthew's use of the goats applicable to the nations is an innovation which she thinks is not found in Ezek 34:17-22; thus it is Matthew's innovation.

In the followings, we will trace each of the following three lines of allusive echoes in order: (i) the identity of the Final Judge in Matt 25:31-46 as the eschatological Shepherd-Judge in Ezek 34:17-22 and the Davidic Shepherd-King in Ezek 34:23-24 (37:24-25); (ii) the criterion of the Judgment and the Shepherd's mission of compassion; and (iii) the identity of the least brothers and the One Shepherd.

3.2.3.1 The Identity of the Shepherd-Judge

The Shepherd-Judge in Matt 25:32 is designated as the Son of Man (v. 31) and also as 'the king' (v. 34). Further, this Shepherd-King calls God 'my father' (v. 34). The king (ὁ βασιλεύς) in v. 34 is not God himself, but is the Shepherd as the Judge invested with righteousness; he is the one who identifies himself with the least of his flock highlighted in the remainder of the account (vv. 35-46). Without identifying any tradition as background, Gnilka observes here a 'Querverbindungen über das christologische Königs- und Hirtenmotiv' and also underscores that 'the King-Judge' identifies himself with 'the least of his brothers' just as Jesus presents himself as 'the gentile [non-militant] King' (21:5) and 'the good Shepherd who has compassion for the sheep without a shepherd' (9:36).[180]

The image of the humble and merciful shepherd in these texts of 9:36 and 25:31-46 indeed echoes the same tradition (cf. Matt 21:5).[181] In Ezekiel 34-37, we find that YHWH the eschatological Shepherd, at the end of the restoration process, raises up a new David who will be the Shepherd *and* the King (מֶלֶךְ) over the flock (34:23-24; 37:24-25).[182]

Morever, YHWH defines his relationship with the Davidic Shepherd by calling him 'My Servant David' (עַבְדִּי דָוִיד), a term reminiscent of 2 Sam 7:12-16 (cf. Ps 2) in which YHWH refers to the son of David as his 'son' (2 Sam 7:14; Ps 2:7; cf. Matt 3:17).

Also, the eschatological Shepherd in Ezekiel 34 identifies himself with the harassed and downtrodden sheep by seeking, healing, feeding, and restoring them (vv. 4-16), and as we particularly note, by judging between the sheep and goats to restore justice within the community (vv. 17-22). Judging between the

---

[180] Gnilka, *Matthäusevangelium*, 2:377-378.

[181] Davies and Allison, *Matthew*, 3:424, notes, "ὁ βασιλεύς (v. 34) harks back to 2:2 and 21:5, recalls Jesus' status as David, and reinforces the irony which will come to expression in 27:11, 29, 37, and 42 (where Jesus' kingship is mocked or questioned)"; Keener, *Matthew*, 603, notes the shepherd motif, identifying the shepherd as God himself in Jewish tradition.

[182] David C. Sim, *Apcoalyptic Eschatology in the Gspel of Matthew* (SNTMS 88; Cambridge: Cambridge University Press, 1996), 125-126, regards "the unexpected presence of 'the king' in these verses (34, 40)" as an important Matthew's redaction in this passage, and notes that Matthew "dramatically altered" the synoptic parable tradition where the king normally represents God; but Sim fails to presents the rationale for this change, while paying no attention to the presence of the shepherd image in the passage.

members of the community is part-and-parcel of the restoration process that immediately precedes YHWH's appointment of the eschatological Davidic kingship over the rescued community. While the Davidic Shepherd is supposed to continue what YHWH does, he is especially appointed to lead the restored flock to the righteous path in accordance with the laws and decrees of YHWH.

In Matt 25:31-46, Jesus as the Shepherd/King/Judge, assuming the throne as the future king (cf. Ezek 34:23-24), warns his flock of the final judgment, though he actually intends to lead them to the righteous way to follow him, and to remain with his flock. Jesus appears to present himself as the Davidic Shepherd/King sitting on the royal throne (cf. Ezek 34:23-24; 37:24-25), and is expected to judge his own flock as does YHWH as the Shepherd-Judge judging his own flock near the end of the restoration process (cf. Ezek 34:17-22).

Therefore, as we consider this Ezekiel pattern as relevant to Matt 25:31-46, we find a mixture of pictures featuring YHWH/the eschatological Shepherd, and the Davidic Shepherd in Jesus as the Son of Man who, at the end, will separate the sheep from the goats. Both the final judgment and the designation of 'king' in the Gospel are ordinarily associated with God (5:35; 18:23; 22:2; cf. 'the Son of Man on the throne of the glory,' in *1 Enoch* 45:3; 51:3; 55:4; 61:8 etc).[183] Likewise, the role of judgment is applied to YHWH the eschatological Shepherd in Ezek 34:17-22, but in Matt 25:31-33, it is transferred to the Shepherd enthroned as the king as God's son.[184] 'King' as the referent to the Son of Man in 25:31-33 reflects "Matthew's emphasis on the Davidic lineage of Jesus and his role of Messiah" (1:16; 2:2, 4; 21:4-5, 9; 27:11, 29, 37, 42).[185]

Again, seen in the light of Ezekiel's vision in chapters 34-37, the Son of Man in Matt 25:31-33 can be identified primarily as the Shepherd, yet both are eschatological in terms of the tasks, and Davidic in terms of future status. The shepherd image functions successfully, projecting the role of YHWH on to the Davidic Shepherd as the appointed king for the futuristic eschatological community yet in the making.

In Matt 25:31-33, we find another modification of Ezekiel's vision of the eschatological Shepherd in chapters 34-37- the setting of the *parousia* of the Son of Man, but this is not to be understood strictly as the momentum of the historical restoration of the nation Israel. The judgment of the eschatological Shepherd is scheduled to fall upon the 'rescued' flock of Israel supposedly from exile in Ezek 34:17-22; the main purpose is to establish justice within the rescued community. In Matt 25:31-33, we see instead a background that is characteristically universal: καὶ συναχθήσονται ἔμπροσθεν αὐτοῦ πάντα τὰ

---

[183] Schnackenburg, *Matthew*, 256.
[184] Cf. France, *Matthew*, 359.
[185] Hill, *Matthew*, 331.

ἔθνη.¹⁸⁶ All the nations (πάντα τὰ ἔθνη) will be gathered before the throne of the Son of Man (v. 32). This gathered flock, without a doubt, will involve all humanity, Jews as well as Gentiles.¹⁸⁷

The vision of extending YHWH's flock to draw in peoples from all nations, as we have seen repeatedly often underscored, is characteristic of the OT Davidic Shepherd tradition (cf. הָאָדָם, Ezek 34:31). Particularly Mic 5:1-5 validates YHWH's concern to embrace the nations within his own flock. The Davidic Shepherd's reign will extend beyond Israel. Thus when "he (the Davidic Messiah) will stand and shepherd" (v. 4a), his reign "will extend to the ends of the earth" (v. 4b), and then will come peace (שָׁלוֹם, 5:5). The picture of Matt 25:31-46 may be seen as the far end vision within the restoration process through the ministry of the eschatological and Davidic Shepherd(s). After all of the flock is gathered before him (25:31-32), the Shepherd will then execute judgment upon them to settle the final restoration of his community.

The idea of 'one Shepherd over both Israel and the nations' may not be a theme entirely novel to first-century audiences. Ezekiel's vision of the Davidic Shepherd over Israel and Judah (37:15-24) is found modified in *An. Apoc.* 90:17-18, in which both the sheep and the wild beasts are transformed into the image of the messianic white bull, thereby creating the eschatological community composed of both Israel and the nations. Also, the *Psalms of Solomon* attest to the vision of the Davidic Shepherd figure whose leadership is envisioned to reach the nations (17:21-46; 18:1-12; cf. 4 Ezra 2:34) as well as Israel.¹⁸⁸

The Davidic Shepherd will be "faithfully and righteously shepherding the Lord's flock" (ποιμαίνων τὸ ποίμνιον κυρίου, *Ps. Sol.*17:40b), and the Lord's flock will include all nations, since this Davidic Shepherd's "compassionate judgments [are] over the whole world" (18:3) to those living in the fear of God "in the days of mercy" (18:9). It is also clearly expressed as well in *Ps. Sol.*17:19-46 that the Davidic Shepherd will be the appointed King by the Lord. But, he is to be placed over Israel (17:19-25) and also the nations (17:26-46;

---

[186] Sim, *Apocalyptic Eschatology*, 126, argues that the phrase "before him will be gathered all the nations," besides "the king" (v. 34), as another "important Matthew's redaction"; he suggests the language of Matthew here is reminiscent of Joel 3:11-12 (cf. Gundry, *Matthew* 511) lacking any discussion of the shepherd image in this regard.

[187] Davies and Allison, *Matthew*, 3:428-429, list various options but prefer "all humanity," thus agreeing with many others such as Jerome, Augustine, Bengel, Bornkamm, Gundry, Schnackenburg, Gnilka etc.; David R. Catchpole, "The Poor on Earth and the Son of Man in Heaven: a Re-Appraisal of Matthew xxv. 31-46," *BJRL* (1978): 389; Carter, *Matthew*, 493, refers to Matt 8:11-12; 12:41-42; 13:36-43 for similar use of the phrase, as does Cousland, *Crowds*, 281, yet only tentatively.

[188] These texts are missing in Andreas J. Köstenberger's article, "Jesus the Good Shepherd Who Will Also Bring Other Sheep (John 10:16): The Old Testament Background for the Familiar Metaphor," *BBR* 12 (2002): 77-96.

18:1-12) when he will "have gentile nations serving under his yoke" (17:30). An echo of Ezek 34:16b-24 is unmistakable.

In this respect, Kathleen Weber's doubt about the relevance of Ezek 37:17-22 to Matt 25:31-46 can be properly refuted.[189] The major objection Weber raised is that Matthew's use of the image of the goats – applicable to the nations for eternal condemnation in 25:31-46 – displays a discrepancy from the usage of the image in Ezek 34:17-22, in which the goats are conceived, along with the sheep, as members of the flock. Weber contests that Matthew's 'antithetical' use of the image of the goats is the Evangelist's own innovation.[190] It is likely that the image of goats is not explicitly associated with the nations in the OT, and might be normative during the Second Temple period. This can be illustrated in *An. Apoc.*; where as a metaphor, the goats often function as a symbol of arrogance, violence, and lawlessness (cf. Dan 8:1-12, 20-21; Ps 22:12; Prov 30:31; Isa 14:9), but probably not Gentiles.

Nevertheless, as we have examined the image of Jesus as the Shepherd-Judge of Matt 25:31-46 in light of Ezekiel's One Davidic Shepherd over both Israel and the nations, the goats as the members of the eschatological community can represent the nations (cf. Ezek 34:31; *An. Apoc.* 90:17-18). At the same time, the stronger [goats] mistreat the weaker [sheep] within one flock under the Shepherd (Ezek 34:17-22).

The use of the image of the goats applicable to the nations in Matt 25:31-46 thus may not be original. This use of the goats echoes the development of the concept of the one flock of Israel and the nations in the Davidic Shepherd tradition. In other words, Jesus' use of the image of the goats in Matt 25:31-46 reflects the merging of the traditional idea of YHWH the eschatological Shepherd's judgment within his own flock of Israel with the vision of the nations' joining as one flock under the Shepherd-Appointee at the eschaton.

Of great import is that this double image of the goats as representing both the nations and the sinners within the one eschatological community in Matt 25:31-46 coincides with the complexity of the identity of Jesus as the Davidic Shepherd-King over one flock, and as the eschatological Shepherd-Judge restoring justice through mercy within his own flock. Further, the criteria by which the Shepherd/King/Judge in Matt 25:31-46 will judge his own flock consisting of both Israel and the nations are characterized by mercy. In the following section, we will examine the criteria and the christology in Matt 25:31-46.

---

[189] K. Weber, "The Image of Sheep and Goats in Matthew 25:31-46," *CBQ* 59 (1997): 657-678.

[190] Weber, "Sheep and Goat," 677; as to the reason for this use in Matt 25:31-46, Weber proposes that Matthew uses the image of goats on two different levels: first on the level of 'discourse' where Matthew warns the nations about the final judgment, and secondly on the level of 'story' where Matthew warns against Christians against the possible complacency in view of the request for the better righteousness (656, 678).

3.2.3.2 The Criterion of Judgment and the Shepherd-Judge

The intention of the parable also reflects the particular use of the shepherd image in Matthew's narrative, especially in terms of the criterion of the judgment by the Shepherd-King. Jesus mentions six exemplary acts of mercy – feeding the hungry and giving water to the thirsty; inviting the stranger and clothing the naked; caring for the sick and visiting the prisoners (Matt 25:35-36; 37-39; 42-43, 44).[191]

Some say these are the typical acts of kindness identifiable in the tradition (Isa 58:6-10; Ezek 18:7, 16; Job 31:16-20, 31-32; Tobit 1:16-17; 4:16; Testament of Joseph 1:4-7; Testament of Benjamin 4:4; 2 Enoch 9:1; 10:5).[192] Yet this catalogue of six acts of kindness might only be representative, for it "covers the most basic needs of life in order to represent the meeting of human need of every kind."[193]

However, as we ponder the significance of these criteria within the context of Matthew's narrative, we find it irresistible to recall the mission of Jesus' disciples in Matthew 10, in which Jesus authorizes his apostles by identifying himself in the treatment they would receive from the people (vv. 40-42) as he grants them the right to be received with hospitality, food, and drink (vv. 8-13, 42).[194]

Furthermore, the need for clothing can be included as well since they are told not to bring an extra tunic (v. 10); caring for the sick and visiting the prisoners are in view since the disciples would readily face persecution and the imprisonment (vv. 17-19). The comprehensiveness of this list of needs Jesus' disciples depict in Matthew 10 suggests the link between the criteria in Matthew 25 and the mission of the disciples in Matthew 10.

The implications of the criteria tightly interconnect with the issue of the identity of the Shepherd-Judge as well as with the identity of the 'least brother' in Matt 25:31-46. If Jesus can be seen as the Davidic Shepherd-King, whose role at the end will be that of the Shepherd-Judge over his flock (cf. Ezek 34:17-22), the criteria may well be conceived in the context of Jesus' mission as the eschatological Shepherd in the Gospel. As we have argued previously, the characteristic of this mission of Jesus as the eschatological Shepherd is found

---

[191] Davies and Allison, *Matthew*, 3:425, notes that the chief distinguishing feature of Matthew's list (which might be a development of Isa 58:7) is its poetic quality, as vv. 37-39 reveal, of three pairs.

[192] Catchpole, "The Poor on Earth," 389-392; Hagner, *Matthew*, 2:743-744.

[193] Hagner, *Matthew*, 2:744; as G. Gray, *The Least of My Brothers, Matthew 25:31-46: A History of Interpretation* (SBLDS 114; Atlanta: Scholars, 1989), 353, notes that these are "only 'parabolic stageprops,' as it were, used to convey the primary meaning of the parable."

[194] Keener, *Matthew*, 605; Gundry, *Matthew*, 513. Cf. Catchpole, "The Poor on Earth," lists only Luke 10:25-37, Mark 9:33-37, 41, and Matt 7:12 (Golden Rule) among the synoptics.

in his 'compassion' toward the harassed and downtrodden sheep (9:36). Likewise, the criteria of the final judgment will be distinctively acts of mercy.[195]

If these acts of mercy are reminiscent of the mission of the disciples in Matthew 10, they also recall the figure of Jesus as the eschatological Shepherd who demonstrates his compassion in his earthly ministry among the troubled sheep by seeking the lost, healing the sick, feeding the hungry, and thus restoring the little flock. In other words, Jesus as the future Davidic Shepherd-Judge enthroned to judge God's own flock in Matt 25:31-46, in fact, demands a criteria of the acts of mercy that Jesus as the eschatological Shepherd demonstrated among the troubled flock. Thus we can trace the continuity in terms of the identity of Jesus as the Shepherd, and in terms of his mission of the Shepherd for the flock.

This also implies that the restoration process will be extended beyond Israel to reach out to the nations until the one flock is purged by the Shepherd-Appointee.[196] The mission of compassion is, Jesus ensures, to continue through his disciples as he identifies himself with them in their mission, and which reflects his own mission among the lost flock of the house of Israel. The Shepherd's ministry of compassion initiated by Jesus and once entrusted to his disciples among the lost sheep of Israel (Matt 10:1-6), is now envisioned to continue among the one flock of both Israel and the nations (Matt 25:31-46).

Yet, the criteria will remain the same even when the Davidic Shepherd-Appointee comes and purges his own flock at the end; it will be the mercy that the Shepherd-King himself lived out in his earthly ministry among the lost flock.

3.2.3.3 The Identity of "The Least of My Brothers"

The last exegetical issue facing us is the identity of ἑνὶ τούτων τῶν ἀδελφῶν μου τῶν ἐλαχίστων ("one of the least of these my brothers") in Matt 25:40, 45. The options vary from 'anyone in need' [whether Christian or not], 'Christians/disciples,' to 'exclusively Christian missionaries/leaders.'[197] The majority view in church history, held by contemporary NT scholars, is that the 'brothers' here are 'disciples,' whether missionaries in particular, or poor, fellow disciples

---

[195] Scot McKnight, *A New Vision for Israel: The Teachings of Jesus in National Context* (Grand Rapids: Eerdmans, 1999), 27, 145, 223-235; Hagner, *Matthew*, 2:744; cf. Davies and Allison, *Matthew*, 3:425; Keener, *Matthew*, 605.

[196] As Anthony J. Saldarini, *Matthew's Christian-Jewish Community* (Chicago: University of Chicago Press, 1994), proposes: "They (the saved/sheep) join the sheep of the house of Israel (10:6; 15:24).

[197] Davies and Allison, *Matthew*, 3:428-429, lists a list of the various options: (i) Every on in need; (ii) all Christians/disciples; (iii) Jewish Christians; (iv) Christian missionaries/leaders; (v) Christians who are not missionaries and leaders. For a comprehensive treatment of the various options, refer to G. Gray, *The Least of My Brothers, Matthew 25:31-46: A History of Interpretation* (SBLDS 114; Atlanta: Scholars, 1989).

in general.¹⁹⁸ This view, thus, eliminates the option for a non-Christian, the obscure Gentile people from the picture in Matt 25:40, 45.

On the other hand, there is a wide spectrum represented within this majority view. Those who argue for the close relevance of Matthew 10 to the identity of the least of the brothers with whom Jesus identified him with in Matt 25:31-46, tend to support that 'brothers' exclusively refers to Christian missionaries and leaders.¹⁹⁹ Others argue that 'these little brothers' can hardly denote an elite corps of Christian preachers in the church but to 'obscure people in the church' easily despised and prone to stray, but whose ecclesiastically low position leaders must themselves assume (Matt 18:1-14).²⁰⁰

Similarly, the group as Jesus' 'brothers,' i.e., his disciples (12:48-49; 28:10; cf. 23:8), whether missionaries or not, are charged with spreading the gospel, and do so in the face of hunger, thirst, illness, and imprisonment.²⁰¹ These options vary depending on one's interpretive emphasis, either on the relevance of the figure of 'these brothers' in Matt 25:31-46, to those disciples in Matthew 10 (e.g., Luz), or on the acts of mercy for those neglected ones (Gundry), or on both (Carson).

In the following, we will argue that 'the least of my brothers' primarily refers to the disciples encountering troubles and suffering oppression; yet at the same time, the Gentile nobodies are not to be entirely excluded. The acts of mercy as the criteria for the last judgment are not restricted within the Christian community. The flock in Matt 25:31-46 is to be the 'one' flock of the Davidic Shepherd, and this ever-widening 'one' flock envisions the people coming in from the nations as well as from Israel. This option may be supported by these reasons: (i) the parallel between Matthew 10 and Matthew 25 is likely, yet with some noteworthy differences; (ii) the focus of the final judgment between the sheep and goats is the justice within the community rather than the concerns for the missionaries; (iii) it is Jesus' identity in Matthew 25 that tends to determine the identity of the ones with whom he identifies himself.

First, the option for 'the Christian leaders and preachers' gains its support from the close relevance of Matthew 10 to the descriptions of the least ones and their situations in Matthew 25. This parallel is likely, but not entirely accurate; the case of Jesus' identification with the disciples/missionaries in Matthew 10 does not fully attest the case in Matthew 25. In particular, the merciful act of visiting prisoners as well as the act of caring for the sick in Matt 25:36, 38-39

---

[198] Keener, *Matthew*, 606.

[199] For instance, Luz, *Matthäus*, 3:539-542; Davies and Allison, *Matthew*, 3:429, lists others who support this view, such as Zahn, Colpe, Lambrecht, and Blomberg.

[200] Gundry, *Matthew*, 514.

[201] Carson, *Matthew 2*, 523.

3.2 Allusions

would not be expected in Matthew 10; it is not likely that the non-Christians would visit the disciples in prison and care for them in illness.[202]

Thus it is not the missionary context that is envisioned in Matthew 25. In the instance of Matthew 10, those who suffer are clearly the disciples of Jesus with their missionary tasks among the house of Israel. In the instance of Matthew 25, however, it is difficult to see that those who suffer refer only to the missionaries, and that Jesus judges the nations based on anyone's treatment of the Christian missionaries. Likely the warnings of Jesus in Matthew 25 are addressed to the members of his own flock. This weakens the case for the Christian missionaries with whom Jesus identifies in Matthew 25, unlike what we find in Matthew 10.

Second, the difference of the respective setting in Matthew 10 (mission) and in Matthew 25 (purging the community) is not without significance. The focus of the judgment in Matt 25 is justice among the flock, just as it echoes Ezek 34:17-22. The goats in this Ezekiel passage represent the stronger ones who trample and push the weak in the pen until they cast them out of the covenantal community. The main concern for YHWH in Ezek 34:17-22 is to establish justice within the community by judging the stronger, that is, the goats. This process of purging the community of his own coincides with YHWH raising up the Davidic Shepherd-Appointee over them (Ezek 34:23-24; 37:24-25); the new beginning of the eschatological [sanctified] community.

In Matt 25, similarly, the same images of the sheep and goats occur in the context of the Shepherd-King's judgment of the community. If this connection is likely, then 'one of the least of these my brothers' may not necessarily refer to the missionaries, but to those who are harassed and downtrodden by the other members of the eschatological community. Matthew 25 does not indicate who these outcasts are, or why they are subjected to and are enduring such great sufferings.

Furthermore, those with whom Jesus identifies in the story do not necessarily represent the missionaries as in Matt 10. The primary emphasis is on the merciful acts done in service to the vulnerable and the marginalized members within the community. This, then, is the emphasis, rather than addressing the responsibilities of what one – either Christian or non-Christian – ought to do in treating Christian missionaries.

Nevertheless, Jesus' identification with '*one of the least* of these my brothers' in Matthew 25 (ἑνὶ τούτων τῶν ἀδελφῶν μου τῶν ἐλαχίστων, v. 40; repeated without τῶν ἀδελφῶν μου, v. 45) is reminiscent of Jesus' earthly mission for seeking the lost sheep of Israel. Jesus' emphasis on the one lost sheep (cf. 18:12-14) fits the characteristic of his mission toward the lost sheep of the house of Israel in Matthew 10. This emphasis on 'one' of the least (not necessarily lost)

---

[202] Davies and Allison, *Matthew*, 3:429; Keener, *Matthew*, 607-608.

brothers is also maintained in the final judgment, but this time in the context of establishing justice within the community.

Perhaps both concepts – 'the needy' and 'the Gospel-bearer'– as the referents for the 'least of my brothers' in Matt 25, can readily be associated in Matthew's Gospel. The Gospel-bearers in Matthew are normally identified with those who suffer from persecution on the earth (5:3-10); they may also be easily identified with 'the little ones' within the community, among whom Jesus guarantees his presence (10:40, 42; 18:10, 14; cf. 26:31). Yet, the emphasis in Matt 25 is the care for the least within the community, not the lost ones outside the community (Matt 10).

Last, in this respect, there seems to be a continuity between Jesus' earthly mission to bring back the lost and outcast to the community (cf. Matt 10) and Jesus' identification with the little ones within the community (Matt 25); both emphasize the acts of mercy for the lost outside of the community, or least ones within it. Also, Jesus' identification with his disciples in Matt 10 in a mission context reflects the characteristic of his identification with the least one in Matt 25 in the sense that the disciples in Matthew 10 perhaps embody the devastated state of the flock themselves (cf. those instructions in 10:9-10).

It is interesting that Davies and Allison note that "what is new in Matthew is ... the identification of the needy with Jesus the son of man," and states as well, "this novel identification – another aspect of the messianic secret – is, however, left unexplained."[203] This 'novel identification' mentioned by Davies and Allison may be explained by Jesus' own identity, echoing the images of the Shepherds, of Ezekiel 34-37, i.e., that figures of YHWH as the eschatological Shepherd-Judge (Ezek 34:17-22) and the Davidic Shepherd over one flock of Israel and the nations (Ezek 34:23-24, 31; 37:24-25; cf. Mic 5:1-4, *An Apoc.* 89-90). While Jesus' identification with the disciples in Matthew 10 may reveal his identity as the eschatological Shepherd of Israel, his identification with the one of the least brothers in Matthew 25 reveals his identity as the Davidic Shepherd over one flock; yet it is the same Jesus.

The strength of this model is that it provides the continuity of Jesus' identification with the disciples in Matt 10 with that of the least ones in Matt 25; the continuity lies, in fact, with Jesus himself as both the eschatological Shepherd (who identifies himself with his disciples, i.e., his under-shepherds in his mission toward the lost sheep), and at the same time as the Davidic Shepherd (who identifies with the least ones in his flock).

The discontinuity in terms of Jesus' identification with his sheep in Matthew 10 and 25 is noteworthy as well. In Matthew 10, those with whom Jesus identifies himself are the missionaries (disciples), since the context is the Shepherd's mission to the lost sheep of Israel (cf. Ezek 34:1-16); but in Matt 25, they are primarily those neglected within the community, since the context is the

---

[203] Davies and Allison, *Matthew*, 3:430.

purging process of the community, both of Israel and the nations at the end, under the leadership of the Davidic Shepherd/King (cf. Ezek 34:17-22 and 23-24).

Thus, the confusion or the overlap with regards to the identity of the least ones in Matt 25:40, 45 between 'the needy' and the 'gospel-bearers,' coincides with the double identity of Jesus as the eschatological Shepherd of Israel (Matt 10) and the Davidic Shepherd over one flock of both Israel and the nations (Matt 25). Then it is understandable how the mission of the compassion of the eschatological Shepherd is reflected in the acts of mercy as the criteria of the Davidic Shepherd.

In other words, the criteria by which the future Shepherd-Judge will judge reflect Jesus' own mission of compassion in his earthly ministry. It is the compassion through which the Shepherd identifies himself with his flock (cf. Matt 9:9-13, 36), and with his under-shepherds/disciples (Matt 10). This mercy will be, Jesus reveals in Matt 25, the very criteria of mercy-acts by which the whole flock of both Israel and the nations will be judged before the throne of Jesus who will be their Shepherd-Judge.

Lastly, if our argument is valid, it is difficult to exclude completely the option of 'anyone' who is in need as 'the least of these my brothers' with whom Jesus identified in Matt 25:40, 45. For the readers of Jesus' parable as to the final judgment in Matt 25, it is mandatory to practice mercy following the pattern of Jesus' mission of compassion for the lost and outcast in his earthly ministry.

But one remaining problem concerning the issue of the identity of 'the least of these my brothers' is that as one practices mercy, following in the footsteps of Jesus' mission of compassion (just as the disciples did in Matt 10), it is hard to draw the line between 'the least among the brothers' and 'the lost and outcast' outside the community. The futuristic one flock of the Davidic Shepherd in Matthew 25 is to be ever-widening until the final judgment. The element of surprise at the end, on the part of the sheep as well as the goats in Jesus' saying of the final judgment in Matt 25:31-46, should remain challenging.

To conclude, if compared to the vision of the *An. Apoc.* 89-90, the scene of the Shepherd-King's judging his own flock at the end in Matt 25:31-46 is quite striking. The vision of YHWH purging his own flock (Ezek 34:17-22) is preserved in the act of the final judgment by the Shepherd-Appointee. It is the Lord of the sheep in *An. Apoc.* who takes the role of the Final Judge; further, it is the angelic shepherds and not the sheep whom the Lord of the sheep calls on the carpet, as it were, for the final judgment.

In Matthew's Gospel, it is the Davidic Shepherd-Judge, the King to be enthroned at the end, who takes the role of the Final Judge. He does not judge the angelic powers but his very own flock. According to Jesus' saying in Matt 25:31-46, the original vision of YHWH's judgment upon his own flock (Ezek

34:17-22) appears to have been deferred until the nations join in the one flock under the One Davidic Shepherd-Appointee.

A question arises in light of the OT Davidic Shepherd tradition: How will the establishment of the Shepherd-Appointee (Ezek 34:23-24; 37:24-25) come about without the purging-process within the flock of the house of Israel (Ezek 34:17-22)? Perhaps the narrative of the First Gospel, using various shepherd images echoing the tradition, provides an answer. As we have examined in the case of Matt 26:31-32, the renewal of the whole flock is perhaps to be achieved through Jesus' taking upon himself the role of Zechariah's smitten shepherd for the sake of the whole flock (Zech 13:7).

In view of the narrative structure, the presentation of Jesus as the Shepherd/King and as the Final Judge in 25:31-46 neatly sets up the seemingly disastrous scene of the smitten shepherd following in 26:31.[204] The implication designed in the narrative structure is that Matthew's Gospel announces that the very shepherd whom they are about to sell for thirty silver coins will be their Judge (26:17-35; cf. similarly in Zech 11:4-17). Thus, one's mistreatment of any of Jesus' followers who faces hardship and opposition will prove to be a costly mistake with no one more surprised than the offender.

The criterion of mercy, however, must not be novel to anyone since it is already manifested by the Judge/King himself as the eschatological Shepherd in the midst of the lost sheep of the house of Israel. Again, as the parable of the sheep and goats in Matt 25:31-46 is read in light of the Davidic Shepherd tradition, Jesus as the Shepherd-Appointee/King/Judge is envisioned as the "One" Shepherd over Israel and the nations who demands that his followers continue the restoration process to its full extent.

Seen at this momentum in the narrative context of the gospel, Jesus *was* the eschatological Shepherd full of compassion who sought the lost sheep of the house of Israel (cf. 9:36-10:16), and he *will be* the Shepherd/King who as Judge will judge his flock of both Israel and the nations (25:31-46), according to the criteria of the compassion which he lived out in the midst of the lost sheep of Israel. Further, the narrative prepares the readers that this future Shepherd yet now *is going to be* the shepherd to be smitten by God for the task of the reconstitution of the eschatological Israel (26:15-37).

3.2.4 Summary

In the Gospel, allusions to the tradition abound. We examined Matt 9:36; 10:6, 16; 15:24; 25:31-46, and discovered that Matt 9:36 suggests that Jesus' ministry summarized in 9:35 (4:23; cf. 10:1-6) is indeed the mission of the compassion of the eschatological Shepherd. The context of the miracle circle of chapters 8-9 underscores the significance of the summary verses in 9:35 (cf. 4:23). The

---

[204] Bauer, *Structure*, 70.

restricted mission of Jesus to the lost sheep of the house of Israel pointedly expressed in 10:1-6, 16 (15:24) is thus understood in light of the Ezekiel pattern: the coming of the eschatological Shepherd for the house of Israel is followed by his appointment of the Davidic Shepherd over the one flock of Israel and the nations.

Thus, the authority of Jesus in 10:1-6 is that of the eschatological Shepherd. Yet, the role of the judgment of the eschatological Shepherd over his own flock in Ezek 34:17-22 is transferred to the Davidic Shepherd/King as the Final Judge over the one flock of Israel and the nations in Matt 25:31-46. Further, the criteria of the judgment of the Davidic Shepherd at the end reflects the mission of the compassion of Jesus as the eschatological Shepherd in his earthly ministry. The flock of the Davidic Shepherd, which ought to conduct itself in light of the coming judgment within the community, ever increases, embracing all nations.

3.3 Shepherd/Sheep Images in Matthew 7:15; 12:11-12; 18:12-14

The shepherd/sheep images for Jesus and the disciples/crowds are evident throughout the narrative. Besides the explicit and implicit textual interactions, in terms of citations and allusions, within the Jewish Davidic Shepherd tradition, there occur several noticeable shepherd/sheep images in Matthew's gospel. Before we move on to thematic studies in our next chapter, a few brief comments on these images may prove useful.

The shepherd/sheep images in Matt 7:15; 12:11-12; 18:12-14, as we discussed in the Introduction, perhaps does not allude to the Davidic Shepherd tradition with explicit linguistic connections. Yet, the images in these texts should not be seen as isolated and irrelevant to the Gospel's particular interest in the tradition; they echo the tradition even though it is often with minimal linguistic connections. The shepherd/sheep images in these texts could be labeled metaphors, as distinguished from allusions, in that the latter clearly allude to certain texts in the tradition. Nevertheless, we would stress that the shepherd/sheep images in the Gospel might not simply be 'metaphorical.' Apart from literary definitions of metaphor,[205] these images in the Gospel likely are rooted in the tradition, and thus can be used as 'metalepsis.'[206] We might consider these images in these texts as echoes of the tradition, which resound

---

[205] For the definition of metaphor, refer to Hays, *Echoes*, 20; Dan O. Via, "Matthew's Dark Light," 354; also refer to the section for 'definition' in introduction.

[206] For the explanation of 'metalepsis,' refer to our discussion about the method (Introduction).The relevance of the shepherd/sheep metaphors in the Gospel to the Jewish tradition becomes more probable especially as we consider that the shepherd/sheep metaphors seem to have rarely been applied to the kingly figure in his/her relation with the subject in Greco-Roman surroundings during the second Temple period.

throughout the narrative as the discourse, thus can communicate Jesus as the shepherd figure in the Gospel.

Echoing particularly the tradition of Ezekiel 34, these texts likely support the thematic coherence of Matthew's interaction with the tradition.[207] It is noteworthy, in this respect, that among the Synoptics, only Matthew retains the shepherd/sheep image both in 7:15 ('wolves in sheep's clothing') and 12:11-12 ('sheep in a pit'); he shares the story of 'one lost sheep' found in 18:12-14 with Luke 15:4-5.

3.3.1 Matthew 7:15: False Prophets in Sheep's Clothing

Jesus says in Matthew 7:15: Προσέχετε ἀπὸ τῶν ψευδοπροφητῶν, οἵτινες ἔρχονται πρὸς ὑμᾶς ἐν ἐνδύμασιν προβάτων, ἔσωθεν δέ εἰσιν λύκοι ἅρπαγες ("Be aware of false prophets, who come to you in sheep's clothing but inwardly are ravenous wolves").

Despite some ostentatious redactional approaches to these texts, there is nothing intrinsically off base suggesting that Jesus himself used these images.

As to the question of authenticity of the saying, Hill considers 7:15 inauthentic based on his conclusion that "the verse envisages the situation in the early Church."[208] Yet, such warnings against false prophets commonly occurred in OT, and there is no difficulty finding reports of earlier false prophets (Jer 6:13-15; 8:8-12; Ezek 13:22-27; Zeph 3:4).[209] Further, Keener notes that Jesus' criticism of religious leaders in 23:25 (ἔσωθεν/ἁρπαγῆς) fits his depiction of false prophets in 7:15 (cf. also 'lawlessness' in 7:23; 23:28).[210]

The results of the investigation to identify the 'wolves' in the verse are confusing; it is better to leave the warning unspecified and to keep it open and susceptible to diverse applications.[211] Even if the picture envisioned in the verse implies false Christian prophets exclusively,[212] to rule out the probability that Jesus foresaw the continued existence of his newly formed community for a sustained period is simply an absurdity based on the criterion of the double dissimilarity.[213]

---

[207] Hays, *Echoes of Scripture*, 32, articulates the criteria of 'recurrence' and 'thematic coherence.'

[208] Hill, *Matthew*, 151; Luz, *Matthäus*, 4:522-523, detects intense Matthean redactions from Q.

[209] Cf. Carson, *Matthew*, 190, also underscores the Evangelist's care in preserving historical distinctions in 7:13-14.

[210] Keener, *Matthew*, 251, n.243.

[211] For the various attempts to identify the historical counterparts of the wolves, see Otto Böcher, "Wölfe in Schafspelzen," *TZ* 6 (1968): 405-426; Hill, "False Prophets and Charismatics: Structure and Interpretation in Matthew 7:15-23," *Bib* 57 (1976): 327-348.

[212] Later, Hill, "False Prophets," 348, identifies 'the false prophets' with none other than the Pharisees in Matthew's time.

[213] For the criterion of 'the double similarity-cum-double dissimilarity' which replaces the old criterion of the difference of 'the third quest,' refer to Gerd Theissen and Annette Merz, *The Historical Jesus: A Comprehensive Guide* (Minneapolis: Fortress Press, 1996), 11; Stanley E. Porter, *The Criterion for Authenticity in Historical-Jesus Research: Previous Discussion*

In this respect, we detect a more careful assessment in Davies and Allison's statement: "The motifs are traditional and the phrases prefabricated."[214] In addition, Luke's preservation of the parable of the shepherd seeking the one lost sheep (Luke 15:4-5//Matt 18:12-14) enhances the probability that the use of the image in that particular context in which the shepherd's mercy is frequently expressed toward downtrodden and lost sheep, originated from Jesus himself (cf. Matt 9:10-13, 36; 10, 6, 16; 15:24).

Our inquiry in this text concerns how this image of false shepherds in sheep's clothing echoes the Davidic Shepherd tradition, and it becomes clear that this concern is related to the issue of the identity of the false prophets in sheep's clothing. What is the referent of the false shepherds in this verse? To begin by stating the conclusion, the indefinite nature of the false prophets in the verse restricts speculations about the identity of Matthew's opponent group.[215]

The indefinite relative pronoun οἵτινες and the gnomic present ἔρχονται in the verse strongly suggest the persistent existence of false prophets into the future, continuous with the past.[216] In the biblical tradition, wolves are used to emphasize such distinct characteristics as deception and exploitation, a plain example being the sheep/wolves image in Matt 7:15, which reveals the machinations of deceitful false prophets and their exploitative effects.[217] Likewise, deception and exploitation characterize the false teachers in the early Church (Acts 20:29; *Did.* 16:3; *Ign. Phld.* 2:2; *2 Cle* 5:2).[218]

The same characteristics ascribed to wolves are ascribed to the false shepherds of Israel in Ezek 22:27-28: "Her officials within her [are] like wolves tearing their prey; they shed blood and kill people to make unjust gain; her prophets whitewash these deeds for them by false visions and lying divinations." On this point Gundry is right when he states that "the ravenousness of the wolves in sheep's clothing refers to the luxurious living of the false prophets at the expense of their followers and to other undue exercise of their authority."[219]

Likewise, the wicked shepherds YHWH will confront in Ezek 34:1-16 share the same characteristics: "The Lord YHWH says, alas! Israel's shepherds, they have been 'shepherding' only themselves!; you eat the curds, clothe yourselves with the wool and slaughter the choice animal" (vv. 2-3). Thus the surprising

---

and New Proposals (JSNTSup 191; Sheffield: Sheffield, 2000), 63-102, points out the methodological flaw of the criterion of difference.

[214] Davies and Allison, *Matthew 1*, 702.

[215] Luz, *Matthäus*, 4:524, opts for "die Falschpropheten Christen, nicht Juden" listing diverse views.

[216] Carter, *Matthew*,188, offers the verb 'come' as an indicator of outsiders, but this is too speculative.

[217] The identity of 'the false prophets' in the passage, Hagner, *Matthew*, 1:183 suggests, is best understood generally rather than specifically.

[218] Cf. Luz, *Matthäus*, 1:404 depicts them as 'money grabbers' in view of *Did.* 11:6.

[219] Gundry, *Matthew*, 129.

symbolic use of an unclean animal applied to Israel's shepherds as illustrated in Ezek 22:27-28 is also confirmed by the glaring faults of the wicked shepherds in Ezek 34:2-3. They might be counted among the sheep of God's flock, but by their actions, they reveal who they really are. It is noteworthy that in *Animal Apocalypse*, the wolves symbolically represent the nations – one of the unclean wild beasts, identified in particular with the Egyptians (89:9, 12-14, 16, 25).[220]

The saying of Jesus in Matt 7:15 hammers the wedge even deeper between Jesus' disciples and their contemporary leaders of Israel. Likewise Jesus' confrontation with the false shepherds of Israel may be confirmed by this text. This aspect matches with the fragmentary picture of Jesus as the eschatological Shepherd sent for the harassed flock who are lost and without a shepherd (cf. 2:6; 9:36).

3.3.2 Matt 12:11-12: Saving a Sheep on Sabbath

In Matt 12:11-12, Jesus defends his action of healing a man on Sabbath with an illustration of saving a sheep from a pit on Sabbath. The preceding incident (12:1-8) of Jesus defending his disciples who picked and ate grain from a field on Sabbath, presents a similar incident of conflict between Jesus and the Pharisees on the matter of doing good on Sabbath (vv. 2, 5, 8). There seems to be no critical evidence for disputing the historical roots of this story in Jesus' ministry.[221]

As mentioned earlier, the particular argument using the sheep image in Matt 12:11-12 occurs only in Matthew (cf. Mark 3:1-6; Luke 6:6-11). While the tradition of the saying itself is accepted, Gundry argues that Matthew needs a figure of speech for needy people; as a case in point, Matthew chooses the man with a withered hand, and notes how the Evangelist finds useful help from the parable of the lost sheep (18:12-14; Luke 15:1-7).[222]

Yet, we have no reason to think otherwise: Jesus needs a figure of speech for needy people, so he remembers the story he once told about the lost sheep; if Matthew could do it, why not Jesus? In addition, the presence of the sheep image in the original saying is more probable if seen in light of Luke 6:5 ('ass' or 'ox' instead of 'sheep'). Further, it is equally valid to argue that Jesus' use of the sheep image and his argument based on *qal wahômer* (Matt 12:11-12), absent in Mark 3:1-6, are simply preserved by Matthew for certain reasons.[223]

The question of the authenticity of this conflict story is in fact interlocked with the issue of the presence and characteristics of the Pharisees in Jesus' time, which we will closely

---

[220] Also, H. L. Strack and P. Billerbeck, *Kommentar zum Neuen Testament aus Talmud und Midrasch* (Munich: C. H. Beck, 1926), 1:574, cites *Tanhuma* 32b: "The Roman emperor Hadrian purportedly said to Rabbi Jehoshua: 'There is something great about the sheep (Israel) that can persist among the seventy wolves (the nations).' He replied: 'Great is the Shepherd who delivers it and watches over it and destroys the wolves before him.'"

[221] Luz, *Matthew*, 2:186-187, finds a Semitic background in the languages in vv. 11-12a; also, see Robert A. Guelich, *Mark 1-8:26* (WBC 34a; Nashville: Thomas Nelson, 1989), 132.

[222] Gundry, *Matthew*, 226.

[223] On the historicity of Matt 12:11, see Davies and Allison, *Matthew*, 2:316-317.

examine in the next chapter. Yet, many would readily acknowledge the existence of the conflicts between Jesus and the Pharisees, at least on the level of intra-Jewish strife within the Pharisaic circle.[224] Furthermore, multiple attestation supports the thesis that Jesus did have Sabbath conflicts (cf. Mark 2:23-28; Jn 5:1-9; Luke 13:10-17).[225]

As this story in 12:11-12 can be fittingly classified as a 'controversy/conflict story,'[226] Jesus' use of the sheep image in 12:11-12 should be understood in this context. Moreover, this incident presents the association of the sheep image with Jesus' healing. In the following discussion, we will examine these two aspects of Jesus' use of the sheep image in the pericope: the conflict and the healing.

(1) The Conflict Motif and Sheep Image: Jesus' use of the sheep image in his argumentation appears accidental, but not in Matthew's context. A close look at the setting of this incident increases the comparability of Jesus with the eschatological Shepherd of Ezekiel 34. For instance, one of the underlying issues of this incident is the escalating conflict between Jesus and the Pharisees (12:9, 14). It is challenging to pin down what kind of group the Pharisees were at the time of Jesus, and what they stood for on the matter at hand, yet it is likely that the Pharisees were populists who relied on "the popular constituency that traditionally followed them" (cf. Josephus, *J.W.* 17.165).[227] Some suggest that even the strictest of the Pharisees could not have found fault with Jesus' behavior (cf. *m. Yoma* 8:6; *Mek.* on Exod 22:2; *m. Šabb.* 22.6), but the question as to Jesus' healing in Matthew perhaps indicates that the dominant opinion in the Mishnah was that one should not heal on the Sabbath (cf. CD 11:13-14).[228]

A critical interpretive question confronts us: Why did Jesus not heal the man at a time when all would applaud, that is, any day except the Sabbath?[229] Jesus healing the sick on the Sabbath in Matt 12:9-14 appears intentional, moreover, confrontational. The message delivered by this incident was already made clear in the preceding verse of Matt 12:7: "God wants mercy not sacrifice " (Hos 6:6), which was probably one of the favorite agendas of the historical Jesus in his

---

[224] Cf. G. Stanton, *A Gospel of a New People: Studies in Matthew* (Louisville: Westminster Press, 1993), 98-102; Marcus J. Borg, *Conflict, Holiness and Politics in the Teaching of Jesus* (New York: E. Mellen Press, 1984), 140-143; Keener, *Matthew*, 352.

[225] Witherington, *The Christology of Jesus*, 66.

[226] Boris Repschinski, The *Controversy Stories in the Gospel of Matthew: Their Redaction, Form and Relevance for the Relationship Between the Matthean Community and Formative Judaism* (Göttingen: Vandenhoeck & Ruprecht, 2000), 107-116, notes, Matthew abbreviates the miracle and subordinates it to the controversy; see also, Keener, *Matthew*, 350-351.

[227] Cf. Keener, *Matthew*, 359.

[228] Davies, *Matthew*, 94; Carson, *Matthew*, 284; Keener, Matthew, 357-358, examines rabbinic rules for the case of using medicine on Sabbath finding it permissible (cf. *m. Shab.* 22:6; *m. Yoma* 8:6; *m. 'Abod. Zar.* 2:2).

[229] Davies and Allison, *Matthew* 2:318.

confrontations with the Jewish leaders (Matt 9:13; 22:40; 23:23).[230] YHWH's compassion for his flock and the subsequent clash with the false shepherds of Israel are the critical elements of the OT Davidic Shepherd tradition.

In Matt 12:9-14, which can also be classified as a 'controversy/conflict story,' Jesus' choice of the sheep metaphor is poignant both for the crowds and for the Pharisees, and is a particularly appropriate reply to the Pharisees' accusation, since they themselves are meant to be the shepherds of the helpless flock. To rescue a distressed sheep on the Sabbath was acceptable, Jesus assumes, in the eyes of the crowd; any one of them would do it.

This indicates that the Pharisees, as the shepherds of Israel, fail to measure up to the standard of mercy acceptable to and from the view point of the very people whom they claim to lead.[231] By means of Jesus' metaphor of the sheep in a pit, the Pharisees are revealed as the ones who would leave the crowd, – the sheep – vulnerable and exposed to danger for the sake of their concerns about ritual purity (cf. Matt 9:36). Hence, the Pharisees began to plot to get rid of him (v. 14).

Jesus' choice of the sheep metaphor – no other animals but sheep – particularly echoes the traditional theme of YHWH's confrontation with the false shepherds of Israel in Ezekiel 34. Whether the Pharisees can be regarded as the majority of the shepherds of Israel is another issue to be examined in the next chapter, but Jesus' use of the sheep image in 7:15 can easily trigger the expectation of the eschatological Shepherd for the flock, who is expected to show true concern for them. If this is true, the function of the sheep image in 7:15 somehow transmits the conflict between Jesus and the Pharisees into the realm of YHWH's own confrontation with them at the eschaton.

(2) Healing and the Sheep Image: Besides this confrontational setting and Jesus' emphasis on mercy, the occasion of Jesus' 'healing' the sick is suggestive of a link with the picture of the promised Shepherd in Ezek 34:1-16. Many scholars consider the genre of the pericope (12:9-14) to be merely a controversy story or pronouncement story, yet the pericope concludes not with the saying but with a miracle and its effects upon the Pharisees.[232]

The motif of healing lies at the climax of the story and presents itself as the main cause of Jesus' confrontation with the careless and self-centered shepherds of Israel. In short, it is out of God's mercy that Jesus confronts the shepherds of Israel as he heals the sick man/sheep. As the introduction to the incident in his healing of the sick in Matt 12:10-13, Jesus' emphasis on mercy is underscored as he cites Hos 6:6 in Matt 12:7: "I desire mercy, not sacrifice" (cf. Matt 9:13;

---

[230] Luz, *Matthew*, 2:188.

[231] For the significance of the motif of mercy/compassion in 12:9-14, see Don Verseput, *The Rejection of the Humble Messianic King: A Study of the Composition of Matthew 11-12* (Frankfurt am Main/Bern/New York: Peter Lang, 1986), 184-185.

[232] Heil, "Healing," 278.

9:36). Jesus' quote of Hos 6:6 placed in v. 7 introduces the following conflict story in vv. 9-13.

Furthermore, in the immediate context of Matt 12:11-12, Jesus mentions David's case in his apology for his disciples' picking grain on Sabbath in vv. 3-6, in which he introduces himself as 'one greater than the temple' (v. 6). The Davidic motif occurs again later in vv. 22-23, that is, another healing incident with Jesus involving a mute and a blind man: μήτι οὗτός ἐστιν ὁ υἱὸς Δαυίδ (v. 23). While examining this case is a task reserved for the next chapter, it is noteworthy to say here that Jesus' use of the sheep image in the healing context may be relevant to the identity of Jesus being more that David. Cousland notes that Matthew places Isa 42:1-14 (12:18-21) between two accounts of healing (12:9-14 and 12:22-24); the second of these compels the crowd to ask: "Is not this man the Son of David?" (v. 23).

Granted that the Davidic title in v. 24 expresses Matthew's own interpretation of Isa 42:1-4 as Lindars suggests,[233] the Son of David is identified with the Isaianic Servant of YHWH. Yet, as we shall argue in detail in next chapter, the more probable background is Ezekiel 34, which, as Cousland points out, also refers to 'my Servant David' (34:23)[234] The Davidic Shepherd is then designated as YHWH's Servant both in Matt 12:12-24 and Ezek 34:1-16. In fact, in Ezekiel 34 we see that this is precisely what was anticipated at the time of Israel's eschatological restoration: the coming of the compassionate Shepherd for the troubled sheep. Despite the metaphorical expression, the actual setting and elements of the incident do not miss the mark of the activities expected of the promised eschatological Shepherd.

One discrepancy should be noted, however. In Ezek 34:1-16, it is YHWH himself who confronts the wicked shepherds by healing the sick of his flock, and is said to raise up his Davidic Servant/Shepherd as the one shepherd over the rescued flock. Yet Matthew implicitly claims that *Jesus does both – as the Servant of YHWH and as the Son of David*; further, he also assumes the tasks of YHWH at the time of Israel's restoration.

3.3.3 Matthew 18:10-14: A Shepherd Seeking One Lost Sheep

Introduced by the command, ὁρᾶτε μὴ καταφρονήσητε ἑνὸς τῶν μικρῶν τούτων (18:10), Jesus tells a parable about a shepherd seeking one lost sheep in vv. 12-14.[235] Once again in this parable we see that Jesus preferred the sheep/shepherd image.

---

[233] Lindars, *Apologetic*, 262.

[234] Cousland, *Crowds*, 190.

[235] At the end of v. 10, a few Western texts such as D L^mg W and Majority text have an additional sentence: ηλθεν γαρ ο υιος του ανθρωπου σωσαι το απολωλος.

Luke 15:3-7 presents a similar parable, but the question of the source remains undecided (cf. *Gos. Thom.* 107). Common to Matthew and Luke is the core of the story line (v. 12 in Matt; v. 4 in Luke), but the emphasis of each version differs from the other. The terminology of repentance (μετανοέω; μετάνοια, v. 7) is missing in Matthew's version; Matthew's emphasis on the one among these little (ἐν τῶν μικρῶν τούτων, v. 14) is overshadowed by the theme of the shepherd's joy in Luke's version.[236] There is no reason to rule out the possibility, in principle, that Jesus told this parable more than once during his ministry, used with different emphases.

It is difficult to support that the shepherd/sheep images in the parable could be invented by the early church knowing that Jesus' descriptions of shepherding with its setting in the parables are realistic, reflecting the practices common to a Palestinian background.[237] In fact, the authenticity of the saying in Matt 18:10-14, including particularly the phrase ἐπὶ τὰ ὄρη in v. 12, is scarcely doubted.[238] Luz suggests that Matthew and Luke obtain this parabolic material from the oral tradition independent of each other.[239] Jesus' preference of the shepherd/sheep images in this parable is well suited with other sayings of Jesus that communicate his identity and mission through the image (7:15; 12:10-12; 25:31-46; cf. 9:36).

Does Jesus' use of the shepherd/sheep image echo the Davidic Shepherd tradition? As we examine the purpose and function of this parable in Matthew 18, we will argue that in at least three aspects the images in this pericope echo the tradition, particularly Ezekiel 34: (i) the mountain (ὄρη, v. 12) motif; (ii) the idea of seeking the sheep that wandered off (ζητεῖ τὸ πλανώμενον, v. 12); and (iii) the theme of restoring/maintaining justice within the community.

(1) The Mountain Motif: Compared to the instances of Matt 7:15 and 12:11-12, it is recognized with more confidence that the parable in 18:10-14 reflects Ezekiel 34.[240] While Bussby asserts the authenticity of ἐπὶ τὰ ὄρη philologically and archeologically,[241] Terence L. Donaldson hints at Ezekiel 34 (vv. 13, 14) for possibly influencing Matthew's ὄρη instead of Luke's ἔρημος ('wilderness,' 15:4); the imprint made by Ezekiel 34 in the parable is unmistakable.[242]

---

[236] Luz, *Matthew*, 2:438; Jeremias, *Parables*, 39-40, underscores that Matthew's version summons "those who share God's concerns" to pursuit the lost sheep.

[237] Keener, *Matthew*, 452.

[238] F. Bussby, "Did a Shepherd Leave Sheep upon the Mountains or in the Desert?: A Note on Matthew 18:12 and Luke 15:4," *AthR* 45 (1963): 93-94; cf. B. B. Scott, *Hear Then the Parable: A Commentary on the Parables of Jesus* (Minneapolis: Fortress Press, 1989), 405-410, introduces *Gos. Thom.* 107 as a variation.

[239] Luz, *Matthew*, 2:439.

[240] Keener, *Matthew*, 425.

[241] Bussby, "Leave a Sheep," 93-94.

[242] T. L. Donalson, *Jesus on the Mountain* (JSNTSup 8; Sheffield: JSOT Press, 1985), 219; also, J. D. Derrett, "Fresh Light on the Lost Sheep and the Lost Coin," *NTS* 26 (1979): 34; Carter, *Matthew*, 356-366. Cf. For the parable in *Gos. Thomas*, see Giles Quspel,"*Gospel of Thomas* Revisited," in *Colloque international sur les textes de Nag Hammadi* (ed. B. Barc; Quebeck: Presses de l'Université,1981), 233; also, W. L. Peterson, "The Parable of the Lost Sheep in the Gospel of Thomas and the Synoptics," *NovT* 23 (1981): 132-133.

In Ezekiel 34, the mountain motif is particularly associated with the hope of restoration; it is the place to which the eschatological Shepherd will lead the scattered flock, and on which the Shepherd would feed them in full: "I will tend them in a good pasture, and in the mountain [ἐν τῷ ὄρει, LXX] heights of Israel [which] will be their grazing land" (v. 14).

If this echo is likely, then the ὄρη on which the shepherd in Jesus' parable in Matt 18:12-14 tends his ninety nine sheep represents a place with safety and abundance. Thus, the idea of a shepherd abandoning the rest of the flock on the ὄρη is inappropriate.[243] Bussby's philological and archeological argument that the term for ὄρη in Matt 18:12 indicates a 'walled compound' has yet to be seriously challenged. The implication is that the care-provider for the remaining flock, as others suggest, may well be assumed in the story.[244]

Perhaps the phrase ἐπὶ τὰ ὄρη in Matt 18:12 better echoes the mountain (ἐπὶ τὸ ὄρος [sg.], Ezek 34:14) where the Shepherd would gather the scattered flock and feed them, though not 'all mountains' (ἐν παντὶ ὄρει, Ezek 34:6), which eventually refers to πάσης τῆς γῆς in which the whole flock is scattered. Even for the narrative of the First Gospel, Jesus' parable of the shepherd seeking the sheep that wandered off in Matt 18:12-18 matches well with the Gospel's presentation of Jesus as the eschatological Shepherd in view of the restoration of the flock (cf. 9:36; 10:1-6).

(2) Seeking the Sheep which Wandered off: Not only are the sheep/shepherd image and the specific elements from the metaphorical network detected in both texts, it also seems that the languages of *search/seek* (ζητέω), *lost/astray* (ἀπόλλυμι/πλανάω) echo the tradition.[245] While Luke's presentation focuses on repentance and the shepherd's joy,[246] in the case of Matthew's version, the emphasis lies on the shepherd's 'persistent searching' for those who are 'going astray' into sin and away from the church.[247]

France observes that in Luke the parable is aimed at Jesus' opponents who objected to his 'evangelical' concern with undesirables, but in Matthew the parable is specifically addressed to the disciples to remind them that God's pastoral care is extended to 'all his little ones.'[248] In this respect, it is indeed striking that *An. Apoc.* 89-90, despite its rich use of the sheep image, is entirely

---

[243] Scott, *Parables*, 415, thus mistakenly identifies the sheep left 'on the mountain' in Matt 18:12-14 with the scattered sheep 'on all mountains' in Ezek 34:6; thus, argues that the shepherd in Matt 18:12-14 'abandoned' the ninety nine sheep.

[244] Bussby, "Mountains," 94; concerning the alternatives, we learn that Jeremias, *Parables*, 133, says 'someone else' other than the shepherd is likely appointed to care for the flock left behind, whereas, Kenneth Bailey, *Poet and Peasant* (Grand Rapids: Eerdmans, 1976), suggests it would be the shepherd's family.

[245] Scott, *Parables*, 413 (italics mine).

[246] The term μετάνοια (Luke 15:7) does not occur in Matthew's version.

[247] Hill, *Matthew*, 274.

[248] France, *Matthew*, 286.

devoid of the shepherd searching out the lost! Thus, the emphasis on 'seeking' those going astray and the complementary concern for the members within the community can truly characterize Jesus' saying of the parable.

The emphasis on 'one sheep' in Jesus' parable could be "the result of the influence of the story recorded in 18:12-18, or merely a Semitism."[249] Less likely, the one sheep hardly points to a 'poor farmer.'[250] Yet, the 'one' sheep (πρόβατον ἕν) that strayed in Matthew is best compared to the 'scattered' sheep in Ezekiel. The 'one-ness' set against the many safe in the pen, underscores the status of being scattered, i.e., failing to stay in a good pasture.

Thus, the key to the parable's interaction with Ezekiel 34, as pointed out earlier, is the shepherd *seeking the strayed*. W. Zimmerli, interpreting Ezek 34:6, notes, 'seek' is a crucial key-word of the shepherd image in the NT" (Matt 18:12; cf. Luke 15:4).[251] Interestingly, while it is 'the lost' (τὸ ἀπολωλός) in Luke 15:4, the sheep in Matt 18:12 is τὸ πλανώμενον (cf. v. 13), 'the one astray' or 'the outcast,' i.e., the sheep that is in danger of being lost. Thus verse 14b says, "one of these little ones may not be lost (ἀπόληται)."

Here Gundry is mostly accurate concerning the characteristics of the shift: "This shift from Jesus seeking sinners to an ecclesiastical attempt to retain members means that the one sheep does not represent somebody lost (i.e., unbeliever), but somebody in danger of becoming lost through straying (i.e., a profession disciple in danger of apostasy)."[252]

In Ezekiel 34, YHWH the eschatological Shepherd promises to seek the lost and bring back the outcast or strayed sheep (vv. 4, 16). The Hebrew verb for 'bring back' (אָשִׁיב) is rendered by the LXX as ἐπιστρέφω. In Matt 18:12, the equivalent verb is ζητέω, but surprisingly, the whole idea of concern for the strayed in vv. 12-14 is represented well by the idea implied in the term ἐπιστρέφω, particularly in view of the immediate context of Matt 18:15-18 (cf. vv. 21-35).

The verb ζητέω fits best for those 'lost' sheep as we find in Luke 15:4 (cf. Ezek 34:4, 16), but in Matt 18:12-14, the verb ζητέω is applied to the potential πλανώμενον. To bring back those who stray, one must first seek them. There appears to be more to this particular combination, however. The 'shift' Gundry points out – "from Jesus seeking sinners to an ecclesiastical attempt to retain members" – is noteworthy. Similarly, McKnight maintains, "Matthew's use of the parable of the lost sheep at 18:10-14 has Jesus' ministry of mercy in view but is directed at contemporary wayward Christians."[253]

---

[249] Hagner, *Matthew*, 1:333.
[250] Luz, *Matthew*, 2: 187.
[251] Zimmerli, *Ezekiel*, 215. See also, בָּקַר ('to seek') in Ezek 34:10, 12 and בקשׁ in 34:16.
[252] Gundry, *Matthew*, 365.
[253] McKnight, *Vision*, 228.

Nevertheless, this shift does not have to be interpreted exclusively in terms of Matthew's redactional activity out of his concerns for his community. The process of establishing justice within the community can be seen as a part of the restoration itself as YHWH the eschatological Shepherd in Ezek 34:17-23 is depicted as concerned about the issue within the community, even before the time of the Davidic Shepherd-Appointee.

In fact, both tasks of the eschatological Shepherd seeking the lost and bringing back the outcast (Ezek 34:4, 16) are an integral part of the process of his restoration of justice within the community. Significantly, the Davidic Shepherd appointed by YHWH supposedly continues to maintain this restored justice among the rescued flock by leading them according to the laws and decrees of YHWH, i.e., by teaching them (Ezek 37:24-25).

(3) Justice within the Community: Both contexts of Matthew 18 and Ezekiel 34 (esp. vv. 17-22) testify to the significance of this theme: restoring and maintaining justice within the community. As justice is violated within a community, a critical sign of its disintegration is seen eventually in the victimization of the little ones of the community. The urge for seeking the lost and the strays underscores the significance of restoring justice for all the members in the community (cf. Ezek 34:1-17; 17-22). The shepherd's preference for 'one lost sheep' may be an effective way of demonstrating his concern for all in his flock, perhaps equally illustrated in the story of prodigal son (Luke 15:25-32).

Similarly, in Matthew 18 the urge to receive (vv. 1-5), protect (vv. 6-9), and respect the little ones of Jesus' disciples (vv. 10-11) precedes the parable of seeking those who have strayed (vv. 12-14). The parable is followed by the delineation of a just protocol and procedures for members to make accusations against other members (vv. 15-18), and the foundational need for forgiveness among the members of community (vv. 21-35). Jesus' concern for the whole community is apparent in the immediate narrative context of 18:10-14.

The task at hand in Matt 18:12-14 is better understood as the one prescribed for the Davidic Shepherd/King particularly in the whole context of 18:1-35. Nevertheless the language still reflects the task of YHWH the eschatological Shepherd, and his mission of compassion. In other words, the task of the Davidic Shepherd to maintain the rescued flock is depicted in terms of YHWH's action of seeking the lost, perhaps in view of the continuation and completion of the restoration process, the mission of compassion.

At any rate, the 'shift' does not necessarily indicate a shift from Jesus to Matthew's community. Ezekiel 34 provides the sufficient rationale for the shift primarily from the task of YHWH, the eschatological Shepherd to the task of the Davidic shepherd. Again, the picture of Jesus communicated through the shepherd/sheep images in Matt 18:12-14 proves to be more complex than the aptly distinguishable figures of YHWH and his Davidic agent in Ezekiel 34 and 37.

3.3.4 Summary

Besides those explicit citations and intended allusions, some less intentional shepherd/sheep images in the Gospel echo the tradition. The deceiving prophets in 7:14 can refer to false shepherds of old or to those contemporary with Jesus and Matthew. The background of Ezek 34:1-16, supported by other explicit textual interactions in the Gospel, serves as a caution against the tendency to confine false prophets to a particular group set in opposition to Matthew's community.

The episode in Matt 12:11-12 illustrates the identity of Jesus as the eschatological Shepherd who intentionally confronts the Pharisees as he heals the sick on the Sabbath. The picture of the shepherd in Jesus' parable of seeking the one lost sheep in Matt 18:10-14 does not seem to be coincidental with his identity as the Shepherd; Jesus is indeed the one who came to seek the lost, inaugurating the time of the promised restoration by YHWH the eschatological Shepherd.

3.4 Conclusion

The first and foremost conclusion we can draw from our examinations in this chapter is that the First Gospel displays remarkable textual evidence for its intense interaction with the Davidic Shepherd tradition. Matthew's explicit citations from Mic 5:2 in Matt 2:6 and Zech 13:7 in Matt 26:31 lay a firm foundation for our thesis that the First Gospel is deeply conversant with the Davidic Shepherd tradition.

From our analysis of both citations, Matthew's renditions focus on the shepherd figure – either of the Davidic (Mic 5:3-4) or of the smitten (Zech 13:7). The clarification of the characteristics of these shepherd figures can immensely be illuminated by various revisions of the tradition in the Second Temple period as well. Jesus is introduced in Matt 2:6 as the eschatological Leader whose leadership is to be distinctively manifested in terms of his mission as the eschatological Shepherd.

Yet, this Davidic Shepherd figure envisions the nations within the scope of his eschatological flock. The context of Matthew 2 strongly suggests that Jesus will be his Davidic Shepherd-Appointee who would represent YHWH's own leadership to be God's representative – surely not Herod the Great or even the leaders of the nation – for the sake of the harassed and downtrodden flock (9:36).

The citation from Zech 13:7 surprises the readers: Jesus takes up the role of the smitten shepherd as the representative of the wicked leadership of Israel at the eschaton (cf. CD-B 19:7-11). Moreover, the shepherd image that continues in Matt 26:32 (cf. "YHWH turning his hand upon the little flock," Zech 13:7) suggests rich connotations that can illuminate Jesus' role as the Davidic

3.4 Conclusion

Shepherd, originally ascribed to the role of YHWH in Zechariah's context (13:7). After the judgment of YHWH is rendered upon the shepherd (thus the whole flock) in 26:31, the restoration is depicted in terms of the risen Jesus who continues to lead his flock to the promised future.

It seems that in the First Gospel Jesus embodies the diverse figures, and thus assumes the varied roles within himself illustrated in the OT shepherd images, echoing also those in the tradition developed in the Second Temple period. Many allusions (Matt 9:36; 10:6, 16; 15:24; 25:31-46) and images (7:15; 12:9-14; 18:10-14) in the Gospel, mostly indebted to Ezekiel's rich shepherd/sheep images, suggest that Jesus assumes the role of YHWH the eschatological Shepherd whose mission is to seek, heal, gather, and judge his scattered yet rescued flock; what a marvel is this mission characteristic of unparalleled compassion and authority.

Compassion is the fundamental motivation for YHWH to bring the restoration. Likewise, Jesus assumes the role of the shepherd with compassion for the sheep without a shepherd (9:36), which signals the certainty of the renewal of the scatted flock despite their sin (2:21), and their eventual loathing for their own Shepherd (26:31). Along with this compassion, Jesus as the eschatological Shepherd for 'the sheep without a shepherd' executes the mission for 'the lost sheep of the house of Israel' with matchless authority (Matt 8-9). It is this authority of the eschatological Shepherd that Jesus conferred to his disciples in their mission to the lost flock (10:6, 16; 15:24).

The authority and compassion in Jesus' mission are characteristically pastoral, especially the healing of the sick and the seeking of the lost (cf. 7:15; 12:9-14; 18:10-14). Jesus' mission as the eschatological Shepherd is inherently confined within Israel, but aims at eventually fulfilling the vision of raising up the new Davidic Leader/Prince for Israel who will shepherd them as announced in Matt 2:6 (cf. Mic 5:1-4).

In Matt 25:31-46, the shepherd image applied to Jesus takes another surprising turn. As the most probable background for the shepherd and sheep/goats images, Ezekiel's scene of YHWH establishing justice within the rescued community in 34:17-22 paints the Gospel's final judgment scene (25:31-46) of the eschatological community under 'one Shepherd,' who rules eventually over both Israel and the nations. The vision for the future Davidic Shepherd/King/Judge in Ezek 34:23-24 (37:24-25), along with the judgment scene in Ezek 34:17-22, can hardly be doubted as the critical tradition with which the Evangelist is intensively conversant in the parable in Matt 25:31-46.

Those allusions and images, drawn mostly from Ezekiel, communicate Jesus as the eschatological Shepherd engaged in his mission to the house of Israel. The parable of the sheep and goats in 25:31-46 depicts the scene of the consummation of the reign of Davidic Shepherd/King/Judge on the throne at the momentum of *parousia*. On the other hand, Matthew's rendition of Mic 5:2

appears to look forward to the establishment of the Davidic Shepherd reign in the near future at the end of Jesus' ministry.

Yet, in the passion narrative of the Gospel, the Evangelist turns to the smitten shepherd (Zech 13:7) in 26:31, thus communicating the death of Jesus in the context of YHWH's judgment and renewal of his own flock as envisioned in Zechariah 9-14 and Micah 2-5 and Ezekiel 34-37 as well. Jesus embodies the diverse shepherd images in the tradition in the First Gospel, whether it be of YHWH, the Davidic Shepherd, or even the smitten one.

The complexity of Matthew's use of the shepherd image applied to Jesus can hardly be untangled without understanding the diverse aspects as well as the commonly shared pattern implied in the Davidic Shepherd tradition. Two critical citations at the beginning and the end of the Gospel, and various allusions and images echoing the tradition, taken together all reflect that Jesus and the Evangelist significantly dialogue with the tradition.

The intertextual echo evolves around the story about the future of the nation, and of all nations in the end; it is perhaps about the kingdom of God the Shepherd and his Shepherd-Appointee. Yet, for a comprehensive reading of the Gospel narrative in light of this tradition, in the following chapter we will pursue more thematic studies. In particular, we will pay closer attention to Jesus as the eschatological Shepherd, and its implications as to the christology of the First Gospel.

Chapter 4

Seeking the Lost and Healing the Sick: Jesus as the Eschatological Davidic Shepherd

In addition to the Gospel's explicit textual interactions with the Davidic Shepherd tradition, a number of implicit themes of the tradition shape the Gospel's presentation of Jesus. As previously suggested, the Son of David in Matt 1:1 is the Davidic Shepherd who will rule [ποιμανεῖ] over Israel (2:6). He is the compassionate eschatological Shepherd for the lost sheep of the house of Israel (9:36; 10:1, 6; 15:24) and the Shepherd-King/the Judge who is to come (25:31-45). But first, the shepherd is to be smitten by God (26:31). These are but a few visible peaks that emerge from the underlying theme of the restoration throughout the narrative.

In the chapters of this study in which the focus is on the OT Davidic Shepherd tradition and its development in the Second Temple period, we have identified the profile of the eschatological Shepherd and his Davidic Shepherd-Appointee.[1] First and foremost, the restoration of Israel to be wrought by these figures constitutes the fulfillment of the theocratic revision of the Davidic monarchy. Accordingly, it involves the return of God's presence to Zion and the inevitable confrontation between YHWH and the shepherds of his own flock. Further, the characteristics of this theocratic rule are distinctly 'pastoral': he comes to seek the lost and the outcast, heal the sick and bruised, strengthen the weak, and feed to the hungry.

Once YHWH's shepherd ministry among his harassed and downtrodden flock is completed, judgment will be executed upon the shepherd who is smitten, followed by the appointment of a new Shepherd over the renewed people. Thus YHWH appoints 'one Davidic Shepherd/King' over the renewed Israel (Ezek 34, 37), or over Israel together with the nations (cf. Mic 5:1-4; *An. Apoc.* 89-90; *Ps. Sol.* 17-18). Noticeably, this Davidic Shepherd is the Prince in the midst of the flock whose main role is to lead them into the righteousness of God's law, securing the perpetual presence of the Lord in their midst. That is the pattern envisioned in the tradition, the pattern which Matthew echoes in his presentation of Jesus and his mission. Thus, these various themes of restoration are, we propose, common both to the Davidic Shepherd tradition (mainly of the OT) and Matthew's Gospel. To do justice to a variety of these themes, in this chapter

---

[1] For details, refer to ch.1.6; ch. 2.2.7. "Conclusions: the Profile of the Shepherd Revisited."

we set our sights on two major activities by which Jesus is identified as the promised eschatological Shepherd, i.e., seeking the lost and healing the sick.

Concerning the theme of Matthew's adaptation and modification of the OT's revision of the Davidic monarchy – more precisely, the Evangelist's interpretation of the Davidic Shepherd tradition presented in the Gospel – will be discussed as we examine those two controversial themes. The themes of God's presence in the narrative (cf. 1:23; 28:16-20) and the appointment of the Davidic Shepherd/King over the renewed Israel and the nations (cf. 28:16-20) are concerns more compatible with the following chapter on Matthew's narrative strategy and the Davidic Shepherd tradition.

It seems appropriate to put the motif of feeding the flock under the rubric of those two distinguished tasks of the Shepherd. A common convention in Matthean scholarship is to analyze Jesus' confrontation with the Pharisees from a redactional viewpoint with its emphasis on the historical locus of 'Matthew's community' in its relation to contemporary Judaism. In this chapter, we propose that the conflict is better understood in terms of Jesus' identity and mission to Israel with its eschatological and christological significance.

On the other hand, the theme of Jesus healing the sick, tightly interlocked in the Gospel with his messianic mission, has become a stumbling block for many. The well-known puzzle of Matthew's christology – that is, the phenomenon of the Matthean 'therapeutic Son of David' – still remains an anomaly.[2] When considering the identity and background of Jesus as the 'therapeutic Son of David,' serious doubts are raised about the axiom that no evidence exists in Judaism that links the Davidic messiah with healing.

Yet, what might be described as Matthew's bizarre combination, we propose, is not necessarily the invention of the early churches. Instead, we suggest that in both cases – seeking the lost and healing the sick – Jesus takes up the roles of the promised Shepherd, as the First Gospel demonstrates by embracing the OT Davidic Shepherd tradition in its full measure among the Synoptics, bringing it to its climax. In the following section, we will take up these themes in order.

4.1 Conflict as Divine Reversal: Jesus' Seeking the Lost (Matt 9:10-13)

Jesus' proclamation that he came to call sinners – an action played out by sharing meals at table with them – brings him face to face with the harsh

---

[2] Cf. G. N. Stanton, "Origin and Purpose of Matthew's Gospel," in *Aufstieg und Niedergang der römischen Welt. II. 25.3* (ed. W. Haase and H. Temporini; Berlin: De Gruyter, 1985), 1889-1951.

opposition of the Pharisees in Matt 9:10-13. Also, Jesus' propaganda for seeking sinners is often related to his habitual usage of the shepherd/sheep images (12:11-12; 18:12-14; cf. 7:15). Both the confrontation motif and the shepherd/sheep images will be the clues with which we launch our investigation of Matthew's particular presentation of Jesus in its relation to the Davidic Shepherd tradition. In the following sections, we will discuss first the literary context of Matt 9:10-13 and its form. Then as we examine various exegetical issues, we will attempt to establish their links with the Davidic Shepherd tradition.

Yet, before we can deal with the specific exegetical issues of the text, the challenge of methodology must be tackled. In the case of Matt 9:10-13, the methodology that one employs to flesh out the meaning of Jesus' confrontation with the Pharisees in the context of Jesus' meal table with sinners often exposes the controlling role of the ideological baggage inherent in the methodology. It soon becomes clear that the methodology itself creates its own frame of reference into which the implications of the event are projected.

4.1.1 The Question of the Setting for Matt 9:10-13

By redefining first-century Judaism in terms of 'covenantal nomism,' E. P. Sanders has minimized the conflict between Jesus and the Pharisees.[3] The well-known Pharisaic pursuit of 'accuracy' (ἀκρίβεια) was in fact,[4] Sanders contests, a minor aspect if seen in the larger context of "Common Judaism" characterized by covenantal nomism. The traditional view of the considerable influence of the Pharisees in pre-70 Palestinian Judaism has been vigorously challenged.[5] Sanders argues that the Pharisees were a minor sect before 70 and relatively (compared to the priests) of little significance to the public. He maintains that even their religious tolerance toward other interpretations of the Torah, including that of Jesus, was indeed remarkable. Sanders considers covenantal nomism 'the lowest-common-denominator' for all religious groups of the Jewish people during that period. He thus argues that Palestinian Judaism before 70 was "a rich, diverse, and multifaceted society with a good deal of restless change; there are no set rules about which of these qualities counted most

---

[3] E. P. Sanders' *Paul and Palestine Judaism: A Comparison of Patterns of Religion* (Philadelphia: Fortress, 1977), proposes that 'Common Judaism' is characterized by 'covenantal nomism,' which is the 'lowest-common-denominator' for all religious groups of the Jewish people at that period including Jesus and the Pharisees; idem, *Jesus and Judaism* (Philadelphia: Fortress, 1985), 336-337; idem, *Jewish Law from Jesus to Mishnah: Five Studies* (London/Philadelphia: SCM/TPI, 1990), 90; idem, *Judaism: Practice and Belief 63 BCE-66CE* (London/Philadelphia: SCM/TPI, 1992), 457.

[4] On ἀκρίβεια as a slogan for the Pharisees' movement, see A. I. Baumgarten, *The Flourishing of Jewish Sects in the Maccabean Era: An Interpretation* (Leiden: Brill, 1997), 56, 133.

[5] Sanders, *Paul*, 419-428.

strongly; the different individuals and groups had different degrees of influence at various times and on various issues; who ran what?; it varied."⁶

Therefore, any conflict within the rubric of Common Judaism, including that between Jesus and the Pharisees, is now to be seen as a far less serious reason for breaking ties with the mother (Jewish) community. Likewise, Saldarini, reevaluating the nature of the conflict between Jesus and the Pharisees, defines the 'Jesus-movement' as basically an 'intra-Jewish dispute,' and Matthew's community/readers as *intra-muros* within a diverse Judaism.⁷ The Jewishness of Jesus' conflict with the Pharisee is certainly worthy of attention.

Nevertheless, it seems that the tendency of minimizing the conflict between Jesus and the Pharisees has gone too far reacting against the so-called 'Lutheran pre-holocaust' interpretation of pre-70 Judaism. Perhaps, in reality, Sanders' description of pre-70 Judaism better resembles our postmodern pluralistic culture which nourishes diversity and tolerance as its foundational virtues.⁸ The history of research of the historical Jesus demonstrates that various reconstructed pictures of the historical Jesus often turn out to be mirror images reflecting the spirit of the age in which the respective research is inspired and conducted. Sanders' proposal does not seem to be an exception.

In fact, Sanders, commenting on Matthew 23, states that Jesus' confrontation with the Pharisees suggests that there could have been only a few Pharisees who strictly attended to minutia, and neglected weightier matters. After all, "human nature being what it is, one supposes that there were some such."⁹ Accordingly, the conflict between Jesus and the Pharisees means little more than another

---

⁶ Sanders, *Practice*, 490.

⁷ Saldarini, *Christian-Jewish,* 44-67 ("Matthew's Opponents: Israel's Leaders"), 84-123 ("Matthew's Group of Jewish Believers-in-Jesus"). Also see his view on the Pharisees, idem, *Pharisees, Scribes, and Sadducees in Palestine Society: A Sociological Approach* (Wilmington: Glazier, 1988), 87-98, 90-103, which distinguishes the Pharisees' popularity from their political influence. Differing from Saldarini's view on Matthew's community/readers as *intra muros,* Stanton, *New People,* 113-145 ("Synagogue and Church"), sees Matthew's community as *extra muros* (cf. Matt 21:43).

⁸ For various critical responses to Sanders' view of the Pharisees, see D. R. de Lacy, "In Search of a Pharisee," *TynB* 43 (1992): 353-372; cf. J. Neusner, *Judaic Law from Jesus to the Mishnah: A Systematic Reply to Professor E. P. Sanders* (Atlanta: Scholars, 1993); R. Deines, *Jüdische Steingefäße und pharisäische Frömmigkeit* (WUNT 2/52; Tübingen: Mohr Siebeck, 1993) 19-21, 269-272, 280-282. Deines undertook a diachronic survey of the various views on the Pharisees. See the first volume of his work, *Die Pharisaer: Ihr Verständnis im Spiegel der christlichen und jüdischen Forschung seit Wellhausen und Graetz* (WUNT 101; Tübingen: Mohr Siebeck, 1997). Also, M. Hengel and R. Deines, "E. P. Sanders' 'Common Judaism', Jesus, and the Pharisees: A Review Article," *JTS* 46 (1995): 1-70; recently, Deines has reclaimed the Pharisees as the major religious party among the Jewish people before 70. See "The Pharisees Between 'Judaisms and 'Common Judaism'" in *Justification and Varigated Nomism, vol.1: The Complexities of Second Temple Judaism* (ed. D. A. Carson, Peter T. O'Brien, and Mark A. Seifrid; Tübingen: Mohr Siebeck, 2001), 443-504.

⁹ Sanders, *Practice*, 380-451.

opinion within a multi-faceted Judaism. However, as one follows these arguments and suggested solutions, one may end up questioning why the early churches had parted ways with the 'Common Judaism' in the first place. As we shall see later in our discussion of the topic, Jesus' confrontation with the Pharisees in the context of his table fellowship with sinners, this hypothetical framework of 'Common Judaism' easily rules out the likely eschatological and christological dimensions of the event.

Others attempt to read Jesus' table fellowship and his association with sinners primarily in light of Greco-Roman practices. Rhetorical devices such as the *chreia* form used to analyze texts like Matt 9:10-13 unintentionally lead one to situate Jesus' table fellowship within the ideological setting of Greek heroes; for instance, under the rubric of Cynic urban philosophers.[10] On the methodological level, this is a possibility; however, to determine what message was communicated through Jesus' table fellowship with sinners, Greco-Roman models are inadequate. Similarly, John D. Crossan's reconstruction of the historical Jesus as a peasant egalitarian who promotes 'open commensality,' does not seem any less theologically or ideologically driven than the traditional Lutheran view of Jesus' table fellowship as the manifestation of 'the religion of grace' set against that of the Pharisees' supposed 'merit-based legalism.'[11]

For the alternative view of Jesus seeking the lost, we approach the text as an 'intertextual echo' between Matthew and the Davidic Shepherd tradition.[12] Although the shepherd motif is not overt in Matt 9:10-13, its roots in and interaction with various motifs and themes of the Davidic Shepherd tradition are, we claim, verifiable. Thus, we will trace the textual and thematic interactions of Matt 9:10-13, in particular with Ezekiel 34. For the meaning of the passage in light of the tradition, at least four critical issues should be brought forward for scrutiny: (i) the identity of the sinners (Matt 9:10); (ii) the historical reliability of the influence of the Pharisees before 70 (v. 11); (iii) the nature of the conflict between Jesus and the Pharisees (vv. 11-13a); (iv) the mission and identity of Jesus (v. 13b). Before we look at each in the following sections, we will examine first the literary context of Matt 9:10-13.

---

[10] Kathleen E. Corley, "Jesus' Table Practice: Dining with 'Tax Collectors and Sinners,' including Women," *SBLSP* (1993): 445-459; Dennis E. Smith, "Table Fellowship and the Historical Jesus," NovTSup (1994): 135-162.

[11] J. D. Crossan, *The Life of a Mediterranean Jewish Peasant* (San Francisco: HarperCollins, 1991), 261-264, idealizes peasant society in general from which Jesus' 'egalitarianism' stems (263).

[12] As we discussed in 'Introduction,' Hays' criteria for hearing echoes, *Echoes of Scripture in the Letters of Paul* (New Haven: Yale University Press, 1989), 25-33, will prove to be helpful although they are not to be meticulously applied here; despite the rarity of the explicit textual interactions, we tune our ears to the echo; cf. Daniel Boyarin, "Inner Biblical Ambiguity, Intertextuality and the Dialectic of Midrash: The Waters of Marah," *Prooftexts* 10 (1990): 29-48.

4.1.2 A Controversy Story: The Literary Context of Matt 9:10-13

The texts that parallel Matt 9:9-13 are found in Mark 2:13-17 and Luke 5:27-32. Besides the multiple sources (Mark 2:15-17; Matt 8:11//Luke 13:29), the distinct forms attest the authenticity of Jesus' table fellowship: a conflict story (Mark 2:15-17), an interpretation of a similitude (Matt 11:19//Luke 7:34), the supper tradition in the parenesis in 1 Corinthians and in liturgical material embedded in the passion narrative in the Synoptic Gospels.[13]

Matthew 9:9-13 – along with Luke 5:27-32 – shares most of the components of the story described in Mark 2:13-17.[14] Yet, compared to Mark 2:13-17, Matthew reports the name 'Matthew' instead of Levi in v. 9, retains ὁ διδάσκαλος ὑμῶν, part of the Pharisees' question, and the quotation of Hos 6:6 along with the introductory clause πορευθέντες δὲ μάθετε τί ἐστιν. These unique Matthean features likely emphasize its form as a controversy story.[15]

As to the historicity of the event, and applying the criterion of dissimilarity, it is undeniable that Jesus on occasion ate with those whom others considered sinful.[16] R. Bultmann denounced the historical plausibility of the Pharisees' appearance at Jesus' meal partaken with sinners in Matt 9:10-13, saying, "it is unthinkable that they should be part of the company sitting at meal!"[17] Contrary to what Bultmann assumes, in Matt 9:10-13, nowhere is it stated that the Pharisees were actually sitting at the meal with sinners (v. 10). Ironically, this is exactly the point Jesus made: they *did not* associate themselves with sinners! As some suggest, the Pharisees might have waited and stood outside the door of the house, and then approached and asked the disciples who were busy handling the crowd.[18]

Matthew 9:10-13 is preserved as a controversy story. To be precise, the forms in Matt 9:9 and 10-13 display three different types: a call narrative (v. 9), a pronouncement story (vv. 10-13), and a brief 'I-saying' (v. 13b). R. Bultmann classifies Mark 2:14//Matt 9:9 as a biographical apophthegm [*Antwort*];[19] yet,

---

[13] R. L. Brawley, "Table Fellowship: Ban and Blessing for the Historical Jesus," *PRSt* (1995), 17-18; Sanders, *Jesus and Judaism*, 206-209, does not dispute that Jesus was notorious for his acceptance of 'sinners,' eating with 'tax collectors and sinners.'

[14] Luz, *Matthew*, 2:31; Hagner, *Matthew*, 1:236-237.

[15] Boris Repschinski, *The Controversy Stories in the Gospel of Matthew* (Göttingen: Vandenhoeck & Ruprecht, 2000), 74-75; Schlatter, *Matthäus*, 303; Davies and Allison, *Matthew*, 2:96; cf. Evert-Jan Vledder, *Conflict in the Miracle Stories: A Socio-Exegetical Study of Matthew 8 and 9* (JSNTSup 152; Sheffield: Sheffield Academie Press, 1997), 15-56, reviewing the history of the interpretation of Matt 8-9, emphasizes the theme of conflict in the whole section not only in 9:9-13.

[16] Sanders, *Jesus and Judaism*, 174; Keener, *Matthew*, 291; cf. Contra. Burton L. Mack, *A Myth of Innocence: Mark and Christian Origins* (Philadelphia: Fortress, 1988), 80, thinks that a later hellenistic meal is read into the meal narratives in the Gospels.

[17] Bultmann, *Tradition*, 18, n. 3.

[18] Likewise, R. H. Gundry, *Mark: A Commentary on His Apology for the Cross* (Grand Rapids: Eerdmans,1993), 129; Darrell L. Bock, *Luke* (BECNT; vol. 1: Luke 1:1-9:50; vol. 2: Luke 9:51-53; Grand Rapids: Baker, 1996), 1:491, views the Lukan parallel (5:30) as a result of literary compression.

[19] Bultmann, *Tradition*, 28.

4.1 Conflict as Divine Reversal

as J. Fitzmyer points out, it may not fit this classification, since no teaching is present.[20] On the other hand, Matt 9:10-13a definitely comes under the heading of controversy dialogue,[21] while the 'I-saying' (v. 13b) deserves separate attention.[22] In fact, the absence of the teaching section in the call narrative (v. 9) may be explained as we view verse 13b as its closing remark, which implies that to be called by Jesus is to join him in calling sinners. Jesus calls Matthew as a sinner (a lost sheep) to become his disciple (as an under-shepherd).[23]

Arland J. Hultgren notes that the closing remarks do not always appear in all of the controversy dialogues, and contends that what we have here in Matt 9:9-13 is "not simply controversy dialogue but narratives containing dialogue."[24] The question whether the emphasis of the pronouncement story is put on the closing remark or on the story itself, does not have to lead to an either/or choice. Here, the action of Jesus is the message he lived out, and conversely, his message is the action proclaimed. Likewise, the 'I-saying' at the end of the story serves as the punch-line that sums up the meaning of the action of Jesus' table fellowship as well as his calling the disciple. As Bultmann pointed out, its function is to underline that "the person of Jesus plays the substantial part."[25] The emphasis falls on Jesus as the subject of the verb ἦλθον; it is a christological claim in the context of conflict with the Pharisees.

Matthew's unique contributions in his retelling of the tradition (cf. Mark 2:13-17//Luke 5:27-32) reaffirms the theme of confrontation. The characteristic of Matthew's version is evident in that the Pharisees are singled out (without mentioning 'scribes,' cf. Mark 2:16; Luke 5:30), and that Matthew alone retains διδάσκαλος ὑμῶν ("your teacher," v. 11) uttered by his opponents, the Pharisees.[26] Moreover, only Matthew retains Jesus' reference to Hos 6:6: ἔλεος θέλω καὶ οὐ θυσίαν.

Tackling questions related to the quotation and the Pharisees is a task reserved for following sections. Here we will say that a comparison of Matthew's version with those Synoptic parallels highlights Jesus as the teacher *par excellence* set against the Pharisees. As we have argued above, the action

---

[20] Fitzmyer, *Gospel According to Luke (1-9)* (Anchor Bible 28; Garden City: Doubleday, 1981), 588.

[21] Davies and Allison, *Matthew 2:96*, calls it an 'objection story.'

[22] Bultmann, *Tradition*, 150-152, categorizes the 'I'-saying (Mark 2:17//Matt 9:13b) 'Dominical sayings,' and thus distinguishes it from apophthegms.

[23] As to the significance of discipleship in Matt 9:10-13, see Luz, *Matthew*, 2:2.

[24] A. J. Hultgren, *Jesus and His Adversaries: The Form and Function of the Conflict Stories in the Synoptic Tradition* (Minneapolis: Augsburg, 1979), 52-53.

[25] Bultmann, *Tradition*, 152.

[26] Other discrepancies include: (a) Matthew lacks both Mark's (2:13) and Luke's (5:27) introduction; (b) Matthew names 'Levi the son of Alphaeus' (Mark 2:14), 'tax collector Levi' (Luke 5:27) as 'Matthew'; (c) 'his house' (Mark 2:15; Luke 5:29) is described simply as 'the house'; and (d) only Luke includes the additional phrase, 'into repentance' (5:32).

and the message of Jesus in Matt 9:10-13 are the two sides of the same coin: Jesus acts out what he teaches. His calling the sinners, thereby implicitly healing the sick, matches with his teaching of mercy.[27]

To better understand this conflict necessitates an examination of a broader literary context of the pericope. As the literary framework of Matt 9:10-13, chapters 8-9 are characterized in a variety of of detailed terms: "the narrative section" paired with chapter 10 (mission discourse);[28] thus, "with the focus upon the significance of the action in the narrative section";[29] or subsumed under the heading "the proclamation of Jesus the messiah" (4:17-16:21).[30] Besides this emphasis on the deeds of the messiah, Davies and Allison note upon which there is a measure of scholarly agreement. For example, the people healed in Matt 8-9 are generally recognized either as the marginalized of Jewish society or as those lacking public status or power, as these examples attest:"a leper, an unnamed Roman, an unclean woman, an unnamed little girl, two blind men, and a dumb demoniac."[31]

Thus, a likely consensus is that Matthew 8-9 presents Jesus as the messiah in action, especially with concern for the marginalized. More pointedly, it is Jesus' ministry of healing the sick (thus implying Jesus calling sinners, cf. Matt 9:10-13) that combines these two characteristics, that is, the deeds of the messiah among the sinners, in these chapters.[32] A question arises: In these chapters, what christology does the picture of Jesus represent?

Matthew's Gospel introduces Jesus the messiah as υἱοῦ Δαυίδ (1:1). He is the one who will be appointed as the Davidic Shepherd (2:6; cf. Mic 5:1-2). The christology in Matthew 1-4 is that of the Son of God, and yet as the Son of David – that is, the one who represents his people and thus fulfills the promises and tasks given to them.[33] The narrative tells a story about the return of God's

---

[27] János Bolyki, *Jesu Tischgemeinschaften* (WUNT 2/96; Tübingen: Mohr Siebeck, 1998), 108, underscores "die Einheit der Sendung, der Worte (Predigt) und der Taten Jesu" in this table fellowship setting (cf. Mark 2:15-17).

[28] H. B. Green, "The Structure of St. Matthew's Gospel," *SE* 4 (1965): 50.

[29] D. L. Barr, "The Drama of Matthew's Gospel: A Reconsideration of Its Structure and Purpose," *TD* 24 (1976): 352; Davies and Allison, *Matthew* 2:1, says, "In (Matt) 5-7 Jesus speaks; in 8-9 he [most of the parts] acts."

[30] Kingsbury, *Structure*, 11-12.

[31] Davies and Allison, *Matthew*, 2:2.

[32] Carson, *Matthew* 1:197, aptly notes, "Matthew has shown Jesus peaching the gospel of the kingdom (4:17, 23) and teaching (chs. 5-7); now he records some examples of his healing ministry."

[33] Cf. Dale C. Allison, "The Son of God as Israel: A Note on Matthean Christology." *IBS* 9 (1987): 74-81; similarly, James D. G. Dunn, *Christology in the Making: A New Testament Inquiry into the Origins of the Doctrine of the Incarnation* (Philadelphia: Westminster Press, 1980), 49, notes, "one of the most striking features of Matthew's Son of God christology is his clear identification of Jesus with Israel (Matt 2:15; 4, 6) – Jesus is the one who fulfils the destiny of God's son Israel."

presence (1:23) among his own people in exile (cf. 2:15, 18). Jesus relives the failed life of Israel in the past. That is, Israel had failed in the wilderness, but this Son of God as God's Servant (3:17), embodying the task and destiny of Israel, succeeds in his obedience to the Father (4:1-11). The return of God's presence among his flock is demonstrated through Jesus' preaching, teaching (chs. 5-7), and healing (chs. 8-9). Concerning the christology of Matthew 8-9,[34] Luz suggests that for the First Evangelist chapters 8 and 9 tell who the Son of David is.[35]

Particularly, as we shall examine later in the section headed 'the therapeutic Son of David,' Matt 9:27-34 draws attention to this picture of Jesus as the healing Son of David (vv. 27-33), and also the motif of Jesus' conflict with the Pharisees (v. 34). The healing incidents in 9:27-33, and the telltale motif of conflict (v. 34), recalls the picture of Jesus as the one who came to seek/heal as well as to teach in Matt 9:9-13. In the middle of these two incidents (9:14-26), Matthew announces the groundbreaking newness that Israel experiences through Jesus' healing ministry, which eventually leads the crowd to shout in amazement, "Nothing like this has ever been seen in Israel" (9:34). A literary flow from 9:9-9:31 resembles the following:

A. mission of mercy, 9:9-13: *"I desire mercy, not sacrifice" (v.13)*
    B. new wine in new wineskins, 9:14-17
        B′. raising the dead to life, 9:18-26
A′. mission of mercy, 9:27-31: *"Have mercy on us, Son of David" (v.27)*

The immediate context of Matt 9:10-13 shows that Jesus' startling action of calling sinners [and implicitly healing the sick] is an integral part of the coming of the new age (9:14-17); even the dead are raised to life (9:18-26)! The literary structure suggests that Jesus' table fellowship with sinners (9:10-13) is the place where new wine (9:14-17) is offered to the sinners and the sick (9:10-13), and thus, after all, to the sheep without a shepherd (9:36). If Matt 9:35, along with 4:23, summarizes Jesus' ministry thus far in the narrative (viz., preaching, teaching, and healing), then the picture of the healing Son of David in chapters 8-9 is critical for the mission and identity of Jesus in Matt 9:10-13.

The transitional passage, Matt 9:35-38, links the picture of the healing Son of David in chapters 8-9 with the mission of the disciples among the lost sheep of the house of Israel (10:1-6). Thus, Jesus' confrontation with the Pharisees in Matt 9:10-13 is intensified as he, the Shepherd of 'the sheep without a shepherd' (9:36) sends his under-shepherds to the whole (symbolically) house of Israel in chapter 10. Eventually, the Pharisees will be replaced with the Davidic

---

[34] Kingsbury's Son of God Christology, "Observations on the 'Miracle Chapters' of Matthew 8-9," CBQ 40 (1978): 559-573, is one option, though exaggeration should be avoided. See, Davies and Allison, *Matthew* 2:4; Carson, *Matthew* 2:197.

[35] Luz, *Matthew*, 2:48.

Shepherd and his under-shepherds (cf. 2:6; 28:16-20). Matthew 9:10-13 announces that this fatal confrontation begins with Jesus calling sinners. Questions are thus raised: Who are the sinners?; why is it so critical to understand the mission and identity of Jesus?

4.1.3 The Sinners and the Lost Sheep of Israel (Matt 9:10)

Matthew 9:10 presents that Jesus is having table fellowship with 'tax collectors and sinners.' Affirming this as a historical reality has not proved to be a challenging task for many scholars, though they vary as to how it was done on the historical level, and as to what message was being communicated by this aspect of Jesus' ministry.[36]

As to the uniqueness of Jesus forgiving sinners, J. Jeremias underscores: "it is Jesus' message of God's love for sinners, and this was so offensive to the majority of his contemporaries that it cannot be derived from the thinking current in his environment."[37] Contrary to this, Sanders contends that Judaism at that time was indeed open, as much as Jesus was, to any Jew including ἁμαρτωλοί for their forgiveness and repentance.[38] The novelty of Jesus' message, according to Sanders, lies not in his message of forgiveness of sinners but in that he did not require his followers to repent, and "because Jesus may have well thought that they had no time to create new lives for themselves, but that if they followed him they would be saved."[39] Saying that Jesus abolished repentance itself is unacceptable, for the concept of restitution is never foreign to his teaching of forgiveness (Matt 5:23-24; 18:21-35). Moreover, if Judaism had offered the way of restoration to such sinners, as Sanders argues, then why did Jesus' conversion of the wicked not evoke a 'Hallelujah' from the Pharisees?[40]

---

[36] N. Perrin, *Rediscovering the Teaching of Jesus* (New York: Harper & Row, 1967), 102-108, accepts it as 'indubitably authentic'; W. Funk and W. R. Hoover, *The Five Gospels: The Search for the Authentic Words of Jesus: New Translation and Commentary* (New York: Polebridge, 1993), 164, acknowledge that it reflects Jesus' social habits; J. Dominic Crossan, *The Historical Jesus: The Life of a Mediterranean Jewish Peasant* (San Francisco: HarperCollins, 1991), 304, places Jesus' table fellowship with other features at the center of his program.

[37] Jeremias, *New Testament Theology* (New York: Charles Scriber's sons, 1971), 2.

[38] Sanders, "Jesus and the Sinners," *JSNT* 19 (1983): 21-23; cf. Young, "Some Queries," 74.

[39] Sanders, "Jesus and the Sinners," 26, thus says, Lukan εἰς μετάνοιαν cannot be authentic.

[40] Similarly, N. T. Wright, *Jesus and the Victory of God* (Minneapolis: Fortress, 1996), 149, says: "He (Jesus) ate with 'sinners' ... which of course, in his day and culture, meant not only merely social respectability but religious uprightness, proper covenant behavior, loyalty to the traditions and hence to the aspirations of Israel." cf. Contra., R. A. Horsely, *Jesus and the Spiral of Violence: Popular Jewish Resistance in Roman Palestine* (San Francisco: Harper & Row, 1987), 212-123.

Traditionally, Jeremias defends the majority view that the Pharisees stood in opposition to the majority of the people of the land (עַם הָאָרֶץ) whom the Pharisees considered 'sinners.' In contrast, Sanders protests that the Pharisees in general did not exclude the עַם הָאָרֶץ from their table fellowship, and suggests that the term ἁμαρτωλοί refers particularly to 'the wicked.'[41] Another way of defining ἁμαρτωλοί is to view them in association with 'Gentiles.' The tax collectors are coupled with sinners and prostitutes mainly because of their Gentile connection in such phrases as τελῶναι καὶ ἁμαρτωλοί (Matt 9:10), ὥσπερ ὁ ἐθνικὸς καὶ ὁ τελώνης (Matt 18:17), and οἱ τελῶναι καὶ αἱ πόρναι (Matt 21:32, 33).[42]

Some dismiss the possibility of identifying the group in its historical setting. Instead, they argue that the term is understood as a religious category, as the necessary counterpart of the 'righteous.' Thus it becomes "a term of rhetorical slander" for those seen from a sectarian group.[43] In Matt 9:10, however, the ἁμαρτωλοί refer to a group of people distinguishable from the tax collectors; otherwise, the tax collectors would not need to be singled out apart from the ἁμαρτωλοί. The term ἁμαρτωλοί might not simply be a linguistic device; rather, it seems to refer to a group still socially and religiously distinguishable in Jesus' time.

Despite the fluidity in defining the exact historical referent of ἁμαρτωλοί, it is possible to delineate certain boundaries that help identify the group. First, although frequently the term is used more or less as a synonym for Gentiles (Ps 9:17; Tobit 13:8; *Jub.* 23:23-4; *Ps. Sol.* 1:1; 2:1-2; Luke 6:33 ['sinners']// Matt 5:47 ['pagans']; Mark 14:41 pars; Gal 2:15), the group represented by the term may well refer to the Jews, and not to Gentiles. The Gentiles were literally lawless; not knowing the Torah, they naturally do not keep it.[44] Thus, the term

---

[41] Sanders, "Jesus and the Sinners," 29-30, 31 (n. 12); idem, *Jesus and Judaism*, 38; this view was initially proposed by I. Abraham, *Studies in Pharisaism and the Gospels: First Series* (Cambridge: University Press, 1917), 55. Cf. Against Sanders' narrow definition of *harubim*, N. H. Young, "Jesus and the Sinners: Some Queries," *JSNT* 24 (1985): 73, finds no distinction between *haburim* and the Pharisees in the NT.

[42] J. R. Donahue, "Tax Collectors and Sinners," *CBQ* 33 (1971): 59-60, suggests a possibility of the historical setting for this understanding of the tax collectors in Galilee certainly during A.D. 4-66; J. Gibson, "HOI TELŌNAI KAI HAI PORNAI," *JTS* 32 (1981): 429-433, agues that the phrase οἱ τελῶναι καὶ αἱ πόρναι (Matt 21:32, 33) particularly indicates their collaboration with Rome.

[43] For instance, D. A. Neals, *None But the Sinners: Religious Categories in the Gospel of Luke* (JSNTSup 58; Sheffield: JSOT, 1991), 68-97; D. E. Smith, "The Historical Jesus at Table," *SBLSP* (1989): 480-484; J. D. G. Dunn, "Pharisees, Sinners, and Jesus" in *The Social World of Formative Christianity and Judaism* (ed. Jacob Neusner et al.; Philadelphia: Fortress Press, 1988), 278.

[44] J. D. G. Dunn, *Jesus, Paul and the Law* (Louisville: Westminster Press, 1990), 74; Mikael Winninge, *Sinners and the Righteous: A Comparative Study of the Psalms of Solomon and Paul's Letter* (Stockholm: Almqvist & Wiksell International, 1995), 333, distinguishes 'the

'sinners' is likely to be understood within the boundary of the people of the covenant. In the Gospels, Jesus' table fellowship with Gentiles is not a feature of his ministry except in the future (cf. Matt 8:11-12). Thus the people referred to by the term ἁμαρτωλοί are Jews, not Gentiles.

Second, it does not seem to refer strictly to the עַם הָאָרֶץ, 'the people of the land.' The Pharisees probably looked down on those uneducated in the law. Granted the עַם הָאָרֶץ comprised the vast majority of Palestine Jewry in the first century. Yet the Pharisees could hardly have viewed עַם הָאָרֶץ as being excluded from the covenant. The nonobservant עַם הָאָרֶץ were judged by the Pharisees, as in the case of tithing, but clearly they were not excluded by them. It is noteworthy that Pharisees 'could,' although definitely not as a normality, attend banquets sponsored by one who belongs to עַם הָאָרֶץ if certain precautions were taken against their own impurity.[45] If the exclusive meal practices apart from the majority of the people in the land are to be considered, such a strict, sectarian style would be more true of Essenes (*J.W.* 2.128-133) or the Qumran community (*1QS*), but probably not true of the Pharisees.

The group ἁμαρτωλοί falls somewhere between the עַם הָאָרֶץ and the Gentiles – they are not the majority nor are they Gentiles. Further, if the ἁμαρτωλοί refers to a group conceptually distinguishable from the tax-collectors (Matt 9:10), one may consider the ἁμαρτωλοί as those guilty of flagrant immorality such as murderers, robbers, prostitutes, and idolaters (cf. Matt 21:32, 33) and/or as the wicked (i.e., those known by their occupations, which rob them of honesty, honor, and even the opportunity to repent and thereby restore their civil rights).[46]

Although it must be difficult to select a certain group of people as the referent of ἁμαρτωλοί from this wide range of people in this category, we can safely conclude that the historical 'sinners' are neither Gentiles nor do they represent the majority of the people in the land. The ἁμαρτωλοί probably refer to a group separated from the common people who respected the Pharisees' pursuit of accuracy in their handling of the Torah; thus the ἁμαρτωλοί were inclined to follow the Pharisees' leadership.

In Matt 9:9-13, Jesus confronts the Pharisees as to the issue of his mission to call the sinners. In this respect, the ἁμαρτωλοί as the object of Jesus' mission, as well as the figure of Jesus who seeks the sinners or the sick, echo Ezekiel's

---

sinfully righteous Jews' and 'the righteous sinners from the Gentiles' (cf. *Ps. Sol.* 17:25); the Gentiles by definition are the sinners, in this sense, they are rather lawless ones, those who can not commit sins since they do not have the law.

[45] Keener, *Matthew*, 294-296; Luz, *Matthew*, 4:33 (cf. Str-B 1. 498-499).

[46] Jeremias, *Jerusalem in the Time of Jesus* (Philadelphia: Fortress, 1969), 303-316. Based on rabbinic sources, he lists various kinds of despised occupations, in particular those involving frequent contact with women (barber), dishonesty (shepherd), exploiting the poor (physician).

4.1 Conflict as Divine Reversal   259

descriptions of the ministry of the eschatological Shepherd among his afflicted sheep.

In that vision of the restoration of Israel, YHWH proclaims that he will visit his flock on the day of Judgment, "on a day of clouds and darkness" (Ezek 34:12), to redeem and restore his afflicted flock. There, five groups are identifiable within the harassed sheep: (i) the weak (הַנַּחְלוֹת; LXX, τὸ ἠσθενηκός, Ezek 34: 4); (ii) the sick (הַחוֹלָה; LXX, τὸ κακῶς ἔχον, v. 4; τὸ ἐκλεῖπον, v. 16); (iii) the injured (נִשְׁבֶּרֶת; LXX, τὸ συντετριμμένον, vv. 4, 16); (iv) the outcast (הַנִּדַּחַת; LXX, τὸ πλανώμενον, vv. 4, 16); and (v) the lost (הָאֹבֶדֶת; LXX, τὸ ἀπολωλός, vv. 4, 16). In addition, apart from these groups of afflicted sheep, the fat (הַשְּׁמֵנָה) and the robust (הַחֲזָקָה) in verse 16 form a different group that YHWH judges for victimizing the rest of the flock. Yet these strong and robust sheep still belong to YHWH's flock. They are not of the nations but are part of YHWH's own (Ezek 36:20-22). YHWH promises to come to his own flock in order to sanctify his holy name that they defiled among the nations, even as he saves and sanctifies them, and eventually dwells among them (Ezek 34:23-24; 37:24-28). The eschatological visit of the Shepherd YHWH thus makes it clear that although the nations are in view, the restoration process will begin with his own flock, Israel.

In connection with Matthew's ἁμαρτωλοί (9:10), it is noticeable that these groups in Ezekiel 34 are divided and thus arranged according to the degree of their affliction, that is, in terms of the distance of each group from the center of the life of the community. If 'the weak' refers to the group that still survives around the center of the community, though treated "violently and harshly" by the robust and strong (Ezek 34:4), then 'the lost' must designate the group that can be located socially and religiously at the remotest fringe of the community. On the other hand, the outcasts (הַנִּדַּחַת; LXX, τὸ πλανώμενον) are to be differentiated from 'the lost.' Put another way, they are in peril of being lost, and as such, rank just above those who are lost. After all, we can see how the other groups such as 'the sick,' 'the injured' and 'the outcast' fall somewhere in the middle between 'the weak' and 'the lost.' The picture we come to see is that all of these groups are downtrodden (רָמַס, 34:18; LXX, καταπατέω) by the feet of rams and he-goats (34:17); thus they are constantly pushed by their hoofs and shoulders till eventually they are cast away (הָדַף, 34:21), outside of the pasture, i.e., YHWH's community of the covenant.

What constitutes 'the lost' in Ezekiel's time is hard to pinpoint. Nevertheless, Ezekiel's referent seems to resemble the group that Matthew's ἁμαρτωλοί represent (9:10). Interestingly, the eschatological Shepherd promises to *seek and find* the lost (בקשׁ, 34:16; LXX, ζητέω). Here, the verb ζητέω defines a task *only appropriate for 'the lost' sheep* (of the house of Israel). The reverse is also true: the verb helps identify any group as 'the lost.' According to Ezekiel's concept of Israel's restoration, this group – the lost – represents the most afflicted ones who perhaps live outside the community.

Matthew's ἁμαρτωλοί may be the equivalent of this group found in Ezekiel 34-37. While they are not the people of the nations (Gentiles), they are those who live on the periphery of the covenantal community. The verb ζητέω is clearly associated with 'the lost' in Jesus' saying in Luke 19:9-10. Zacchaeus was a chief 'tax collector' (v. 2) and thus a 'sinner' (v. 7), 'the lost' whom Jesus came to seek and save (ἦλθεν γὰρ ὁ υἱὸς τοῦ ἀνθρώπου ζητῆσαι καὶ σῶσαι τὸ ἀπολωλός, 10; cf. Matt 1:21). Likewise, Matt 9:13 says that Jesus came to 'call' (καλέω) sinners. The semantic function of the two verbs, ζητέω and καλέω in each context, is identical and indicates Jesus' ministry directed toward sinners.[47]

Matthew 18:10-14 supports this link. In telling the parable of the shepherd seeking the one lost sheep, Jesus implies that he came to do what his heavenly Father would do as well. Like the shepherd in Ezekiel 34-37, Jesus came 'to seek' (ζητέω) the lost sheep (v. 12). Interestingly, the one sheep that the shepherd would seek out is 'the strayed/outcast' (τὸ πλανώμενον, Matt 18:12; cf. τὸ πλανώμενον [הַנִּדַּחַת] in Ezek 34:4, 16) in peril, on the brink and about to be lost (ἀπόληται, Matt 18:14; cf. τὸ ἀπολωλὸς [הָאֹבֶדֶת] in Ezek 34:4, 16). The expressions in Matt 18:12, 14 indeed reflect sensitivity to Ezekiel's distinction made between the lost and the outcast. Thus the ἁμαρτωλοί – the outcast and the lost – as a group could be the equivalent of those guilty of flagrant immorality or even the equivalent of the wicked. To be precise, the concept of ἁμαρτωλοί is closest to the outcast – τὸ πλανώμενον – yet those who are at risk of being lost, just as the lost – τὰ ἀπολωλότα – can be considered the most extreme outcast (Matt 10:6; 15:24).

Since the ἀπολωλοί were the most victimized ones among the weak, the sick, the injured, and the outcast, the ἀπολωλοί, in a way, represent all of these afflicted groups of YHWH's flock, epitomizeing the misery of those other groups of the sheep. This could be true of the relationship of the ἁμαρτωλοί to the עַם הָאָרֶץ. If this is likely, the harassed and downtrodden crowd (τοὺς ὄχλους) in Matt 9:36 may not necessarily be sinners, but from Jesus' point of view, they are all afflicted as πρόβατα μὴ ἔχοντα ποιμένα.[48] In addition, this may inform us of Jesus' task within Israel, and of his self-understanding as their eschatological Shepherd who came to seek the lost sheep of the house of Israel while inevitably confronting their false shepherds.

---

[47] Cf. Polyki, *Tischgemeinschaften*, 108, renders καλέσαι as 'zurückzurufen' in this context (Mark 2:15-17//Matt 9:10-13) stressing the restoration of the sinners back to God's community.

[48] Likewise, the poor (οἱ πτωχοί, Matt 5:3) and 'those who labor and are heavy burdened' (Matt 11:28), do not necessarily and exclusively refer to 'the sinners' in its narrow sense. Cf. Davies and Allison, *Matthew* 1:442-443; 2:288; Luz, Matthew 2:172; Keener, *Matthew*, 168-169, 348-349; against Jeremias, *New Testament*, 109-113.

4.1.4 The Pharisees before 70 and the Shepherds of Israel (Matt 9:11)

Answers vary regarding the question of the Pharisee's identity and influence in and around the first century. The traditional view maintains that sinners constitute the majority of the people, yet at the same time asserts the popularity of the Pharisees among the masses. Sanders criticizes this view as contradictory, that is, "the Pharisees *excluded* most Jews from the 'true Israel' because the ordinary people [עַם הָאָרֶץ] would not follow them," and at the same time that "they [the Pharisees] controlled the people, so that everyone did what they said, especially with regard to religious observances." Sanders believes "both are in error." [49]

Our study of the identity of the sinners of Matt 9:10 in light of Ezekiel 34-37 enables us to side with Sanders that the sinners excluded by the Pharisees were *not* exactly the majority of the people; rather, 'sinners' refers particularly to the outcast and the lost living on the fringe of the covenantal community. Nevertheless, Sanders' efforts to minimize the influence of the Pharisees before A.D. 70 seems an overreaction. Also, it is an overstatement that "the Pharisees neither created zeal for the law nor decided when other people would display it." [50]

Were the Pharisees the leaders of the people in the time of Jesus? If they were, in what sense and to what degree? Again, the question comes back to the reliability of the sources. Despite many challenges and defenses from various sources, testimonies about the Pharisees from Josephus and the canonical Gospels emerge as the most dependable of witnesses; there are, in fact, hardly any other sources.[51] Since Sanders discredits the evidence of the Gospels, it is not surprising that he would claim that "there is *no* indication, after the time of Jannaeus, that the populace was guided by the Pharisees." Yet, this neither explains nor proves why such an influence – if what Sanders contends were true – suddenly died out after Jannaeus.

According to Josephus, the Pharisees' leadership was not as political as was the leadership of the priests (*A.J.* 13.408-410). To the official political powers, the Pharisees seem to have remained passive but not necessarily powerless. They could exercise political power when asked to join in the government (e.g., by Salome Alexandria, *J.W.* 1:110-111). On the other hand, the decrees they imposed on the people could easily be revoked by the Sadducees (e.g., by Hyrcanus, *A.J.* 13.293-298). Nevertheless, the Pharisees were indeed influential among those with political power because the Pharisees were supported by the masses: "The Pharisees are friendly to one another, and are for the exercise of

---

[49] Sanders, *Practice*, 449-450 (italics his).
[50] Sanders, *Practice*, 449 (also, 10-12).
[51] With regard to the clarification of the question of reliable sources, see J. P. Meier, "The 'Third Quest' for the Historical Jesus," *Bib* 90 (1999): 464-465; also, Martin Goodman, "A Note on Josephus, the Pharisees and Ancestral Tradition," *JJS* 50 (1999): 17-20.

concord and regard for the public" (*J.W.* 2.165). Likewise, while the Sadducees were able to persuade none but the rich, the Pharisees have the multitude on their side (*A.J.* 13.298). Thus, the Pharisees were "in a capacity of greatly opposing kings" (*A.J.* 17.41). More important than their political influence was their formative socio-religious force. The Pharisees determined the boundaries of what was and what was no longer Jewish through their pursuit of ἀκρίβεια in their reading, and doing the oral and written Torah (*A.J.* 17:41; *J.W* 1.110; 2.162).

Furthermore, the Pharisees' association with the Sadducees and the chief priests may not be discarded as anachronistic. Based on the appearance of the combined witnesses – Josephus, Matthew, and the Gospel of John (e.g., a Pharisee as a ἄρχων in 3:1; 7:45-52) – Urban C. von Whalde states, "Such geographically and ideologically distinct literary sources fulfill quite adequately one of the chief criteria of historicity, namely, 'multiple independent attestation.'" [52] In Matthew's Gospel, the overall portrayal of religious authorities is quite detailed, and includes the Pharisees along with Sadducees, scribes, elders, and chief priests. The Pharisees are mentioned in combination with the Sadducees five times (3:7; 16:1; 6:11, 12 [twice]), and with the chief priests twice (21:45; 27:62). Saldarini observes that Matthew portrays the Pharisees as a secondary (to the chief priests) political force in Jerusalem.[53]

Recently, R. Deines suggested that before A.D. 70 the Pharisees were not a holy remnant cut off from the majority of the people like *haberim*, but rather they stood between the *haberim* and the עַם הָאָרֶץ who supported them in political and official religious questions.[54] This picture fits both the descriptions of Josephus and the Gospels. Deines goes on to suggest that the various code words found in Qumran documents such as 4QMMT (e.g., the "seekers after smooth things" or "Ephraim") point to the Pharisees. In fact, the diversity within 'Common Judaism,' i.e., the so-called 'Judaisms,' does not go beyond Josephus' description of the three or four major religious parties before 70 (*A.J.* 18:2-6). Deines defines Pharisaism as 'normative' Judaism, considering it "the most fundamental and most influential religious movement within Palestine Judaism between 150 BCE and 70 CE."[55]

In this regard, Sanders seems to stand on a shaky ground when he asserts that there is *no* indication, after the time of Jannaeus and Salome Alexandria, that the populace was guided by the Pharisees. It seems much more difficult to

---

[52] Wahlde, "The Relationships between Pharisees and Chief Priests: Some Observations on the Texts in Matthew, John and Josephus," *NTS* 42 (1996): 521. Cf. Margaret Davies, "Stereotyping the Other: The 'Pharisees' in the Gospel According to Matthew" in *Biblical Studies/Cultural Studies: The Third Sheffield Colloquium* (ed. J. Cheryl Exum and Stephen D. Moore; JSOTSup 266; Sheffield: Sheffield, 1998), 415-432.

[53] Saldarini, *Pharisees*, 169-170.

[54] Deines, *Pharisaer*, 515-555; idem, "*Pharisaism*," 494-503.

[55] Deines, "*Pharisaism*," 460-491, 503.

silence the overwhelming indications that the Pharisees were in a leadership position before A.D. 70. Both the tradition preserved and the contributions edited in Matt 9:10-13 underscore the theme of the conflict between Jesus' and the Pharisees' leadership over the masses. In particular, the issue hinges upon the interpretation of the Torah. As we examined earlier, Matthew 9:10-13 is a controversy story. Already in the narrative of the calling of Matthew in verse 9, the emphasis is on Jesus' authority.[56] Jesus is distinctly a teacher. The Pharisees call him ὁ διδάσκαλος ὑμῶν, and pit Jesus and his disciples against their group, for they themselves were teachers of the law on behalf of the people. Most likely originating from the Levites and later from the temple scribes, the Pharisees took up the task of interpreting the laws and decrees for the people to apply in their daily lives.[57] Likewise, Jesus himself confronts the Pharisees with the authority of a new teacher of the Law, in which the Pharisees thought themselves experts: πορευθέντες δὲ μάθετε τί ἐστιν. Jesus challenges the Pharisees at both levels of interpretation and praxis of the Torah, thus claiming his leadership over the people.

The conflict between Jesus and the Pharisees is not to be read strictly in political terms. Rather, it seems more a matter of Jesus' increasing popularity that they enjoyed and upon which their power relied.[58] Neither are those shepherds YHWH confronts with harsh indictments in Ezek 34:1-14 strictly political leaders. W. Zimmerli suggests that the shepherds here are representative of their people; it could refer even to the whole 'house of Israel.'[59] Likewise, the alternative figure in Ezekiel's concept of restoration who is to replace the false shepherds – that is, the new David – is described in terms of נָשִׂיא ('prince,' e.g., 34:23; 37:24) and not only as מֶלֶךְ (e.g., 37:24).

The term נָשִׂיא in Ezekiel's vision particularly associates his intimacy with the people, though he is in touch politically as well (Ezek 44:1-3, 24; 45:8; 46:8). While מֶלֶךְ or shepherd may well be associated with the preposition 'over,' the נָשִׂיא is never said to be over the people but always 'in the midst of' them. As demonstrated by the Evangelist's rendition of Micah's מוֹשֵׁל (5:2) as ἡγούμενος and ὅστις ποιμανεῖ in Matt 2:6, it is a figure of a leader who influences people

---

[56] Boris Repschinski, *The Controversy Stories in the Gospel of Matthew* (Göttingen: Vandenhoeck & Ruprecht, 2000), 74-75, notes that Matthew achieves "a clearer christological focus of the controversy through the address 'teacher' of scripture (Hos 6:6)"; Schlatter, *Matthäus*, 303, thus states,"Ihr Interesse am Erzählten richtete sich nicht auf die Nebenfiguren, sondern einzig auf das, was Jesus tat."

[57] Joachim Schaper, "The Pharisees" in *The Cambridge History of Judaism: The Early Roman Period* (ed. W. Horbury, W. D. Davies and J. Sturdy; vol. 3; Cambridge: Cambridge University Press 1999), 402-427.

[58] Margaret Davies, *Matthew*, 95, underlines the Pharisees' lack of legal authority and power to destroy Jesus (Matt 26:57-68).

[59] Zimmerli, *Ezekiel 2*, 185, 214; also, R. M. Hals, *Ezekiel*, 250.

in his midst (34:1-16, 24; 37: 25) just as was the case with the Pharisees as leaders.⁶⁰

In this respect, granted that politics and religion can hardly be viewed separable in the first century Judaism, the Pharisees were not exactly political leaders *per se*. They were actually more influential at the popular level in a socio-religious context; this influence upon the masses remained in effect into the time of Jesus. Their influence did not wane or die out. It is at this level of leadership that Jesus confronted them.⁶¹ In Ezekiel's vision of the restoration of Israel, it is justified for YHWH to confront the shepherds of Israel on this same level. Thus the Pharisees were the shepherds to whom YHWH entrusted his flock. The historical descriptions of the Pharisees fit this role, which might explain why Jesus in Matt 9:10-13 confronts them as their influence on the masses becomes critical. Then, how does the theme of Jesus' conflict with the Pharisees (as the false shepherds of Israel) develop in Matthew's narrative, particularly in 9:10-13?

4.1.5 Jesus' Seeking the Lost and Conflict as Divine Reversal (Matt 9:11-13a)

Regarding the nature of the conflict, the traditional view contrasts Jesus' scandalous 'grace/forgiveness' with the Pharisees' 'work-righteousness/ legalism.' This position is best represented by Jeremias, and subsequently revised by Marcus J. Borg's antithesis of the Pharisaic holiness program and his view of Jesus' mercy as a redefinition of holiness.⁶² Dunn basically repeats the same traditional view as he contends that "Jesus' table-fellowship must be seen as both a protest against a religious zeal that is judgmental and exclusive and as a lived-out expression of *the openness of God's grace*." ⁶³

Sanders might bark at the notion that Pharisaism had nothing to do with the openness of God's grace. Neither is it likely that Jesus forsook the demand of

---

⁶⁰ Deines, "The Pharisees," 493, n. 172, likewise states, "The Φαρισαῖοι are clearly presented as the leading group of the Ἰουδαῖοι" (cf. John 7:11-13; 8:30; 10:19-20; 12:9, 17-18). Cf. R. W. Klein, *Ezekiel: the Prophet and the Message* (Columbia, S.C.: University of South Carolina Press, 1988), 123, defines the role of נָשִׂיא over against the past shepherds of Israel: "This prince ... will rule *among*, rather than over, the people. The oppressive shepherds, of course, had ruled *over* the people" (italics his).

⁶¹ Scot McKnight, "New Shepherds of Israel: An Historical and Critical Stdudy of Matthew 9:35-11:1" (Ph.D. diss., University of Nottingham, 1986), 30-59; McKnight, in the section on "The Pharisees in Matthew" (190-194), notes on Matthew's high view of the apostolate, the Twelve, 'the New Shepherds' of the downtrodden flock harassed by the Pharisees; Jesus' apostles are seen as the replacement of the Phrisees' leadership over Israel.

⁶² Marcus J. Borg, *Conflict, Holiness and Politics in the Teaching of Jesus* (New York: E. Mellen Press, 1984).

⁶³ Dunn, "Table-Fellowship," 268 (italics mine); idem, *Jesus Remembered* (Christianity in the Making, v. 1; Grand Rapids, Mich./Cambridge, U.K.: Eerdmans, 2003), 603-605, emphasizes more of the social-religious aspect of Jesus' table fellowship, i.e., his demolition of the 'boundaries' created by the Pharisees.

holiness when he redefined it as mercy as he offered table fellowship to sinners. Was the conflict indeed between the religion of judgment and the religion of grace? Wright thinks the meaning of the conflict is not the one traditionally assigned, but rather that it is attributed to 'eschatology and politics' rather than to 'religion or morality.' [64] While the second half, i.e., 'religion or morality,' requires an explanation since both parts seem inseparable, the first half of this statement appears illuminating. Particularly, the eschatological aspect of Jesus' meal practice, as well as its possible political implications within the nation Israel at that time, can hardly be captured if the practice is situated solely in Greco-Roman meal practices.

In fact, ancient meal practice demonstrates that table fellowship usually functioned as a socio-religious means of maintaining the boundaries of a community: inclusion and exclusion, ideologies of hierarchy, differentiation and conflict.[65] This general function can be applied without posing any serious difficulties to the Pharisaic practice of table fellowship. The earliest traditions concerning the Pharisees indicate a preoccupation with attention to table purity and food laws. Indeed, other divergent groups within Judaism pursued holiness in similar ways.[66] As it was with members of other holiness movements in the land of Israel, the Pharisees refused to eat with those who did not share their particular group's holiness concerns with food. The evidence of the Gospels (Matt 9:10-13; 15:1-2; cf. Mark 7:1-5) can be confirmed by rabbinic testimonies as to the Pharisaic meal practices: "If tax gatherers entered a house [all that is within it] becomes unclean; if a gentile or a woman was with them all becomes unclean" (*m. Toharot* 7:6); "Six things are a disgrace for the student ... (5) he should not recline at table with a common person" (*b. Berakot* 43b).[67] We might say that table fellowship served as a boundary marker between the clean and the unclean.

The message communicated by the sinners' presence at Jesus' meal table raises far more serious questions. Recent studies of Jesus' table fellowship tend to focus on the texts of Luke and Acts in the context of Greco-Roman

---

[64] Wright, *Jesus*, 372; McNight, "The God of Jesus" in *Vision*, 48-49, tries to combine both views of the traditional and the third quest of the historical Jesus.

[65] M. Douglas, "Deciphering a Meal," *Daedelus* 101 (1972): 61-81; Smith, "Table," 469-470; Blake Leyerle, "Meal Customs in the Greco-Roman World," in *Passover and Easter: Origin and History to Modern Times* (ed. P. F. Bradshaw and L. A. Hoffman; Notre Dame: University of Notre Dame Press, 1999), 29-45.

[66] J. Neusner, *From Politics to Piety: The Emergence of Pharisaic Judaism* (Englewood Cliffs, N.J.: Prentice-Hall, 1973), 86; Sanders, *Jewish Law*, 23-42; Dunn, *Jesus Remembered*, 602-603; B. D. Chilton, "The Purity of the Kingdom as Conveyed in Jesus' Meals," *SBLSP* (1992): 473-488; Crossan, *Historical Jesus*, 341-434; McKnight, *Vision*, 41-49. Cf. For the Greco-Roman Symposium, refer to F. Lissarrague, *The Aesthetics of the Greek Banquet: Images of Wine and Ritual (Un Flot d'Images)* (trans. A. Szegedy-Maszak; Princeton: Princeton University Press, 1990).

[67] McKnight, *Vision*, 42-44.

symposium literature, with the *Symposia* of Plato and Xenophon serving as archetypes. The message deciphered in this light often points toward the formation of a new community through table fellowship to offer reconciliation and solidarity among Jews and Gentiles, men and women, rich and poor.[68] In a similar way, Denis E. Smith particularly identifies Matt 9:10-13//Mark 2:15-17 as a Cynic type of *chreia*. This distinguishes itself from the Stoic *chreia*, in which the emphasis lies on moral teaching by its odd, extreme, and we might even say, a burlesque action of the central Sage-Hero that becomes the basis for a demonstration of Cynic ideals and values.[69] Smith thus contests that Jesus partaking in banquets, "not of hosting them, would be the activity of one who, in contrast to John the Baptist, existed in the urban world and affirmed that world." [70] Jesus' preference for the company of persons of dubious character such as "tax collectors and the sinners" can also be explained by this rural Cynic setting, not necessarily by any specific use by Jesus of table fellowship as a mode of proclamation.[71]

Smith may be correct that the table fellowship practices were "not capable of creating a community in themselves," [72] and that Jesus' eating with sinners would not imply that Jesus and his followers formed a separate table fellowship community right away.[73] Nevertheless, it would be a mistake to situate Jesus' table fellowship with the sinners in Matt 9:10-13 solely and even primarily within the framework of Greco-Roman meal practices, which would eliminate its possible mode of proclamation. The consequential characteristics of the historical Jesus involving 'open commensality,' an 'egalitarian Jesus,' and an 'urban Cynic hero,' can be misleading as we attempt to uncover the intent of the historical reality of Jesus' table with the ἁμαρτωλοί.

Therefore, as we concluded earlier, the ἁμαρτωλοί with whom Jesus shared his meal table in Matt 9:10-13 and elsewhere refers particularly *to Jews and not to Gentiles!* They are meant primarily to refer to *the outcast and the lost sheep*

---

[68] H. Moxnes, "Meals and the New Community in Luke," *SEA* 51 (1986): 158-167; D. E. Smith, "Table Fellowship as a Literary Motif in the Gospel of Luke," *JBL* 106 (1987): 613-638; David P. Moessner, *Lord of the Banquet: The Literary and Theological Significance of the Lukan Travel Narrative* (Minneapolis: Fortress Press, 1989), 2-3, 189, 316-317, presents Jesus as 'the journeying-guest Lord of the Banquet' in the central section of Luke, 9:51-19:44 in terms of New Exodus, and he underlines 'the meals as a focal point for the dynamic presence of God as salvation' and 'its significance as to the mission to the nations' (316-317).

Further, see J. H. Neyrey, "Ceremonies in Luke-Acts: The Case of Meals and Table-Fellowship," in the *Social World of Luke-Acts: Models for Interpretation* (ed. J. H. Neyrey; Peabody Mass: Hendrickson,1991), 361-387.

[69] Smith, "Table," 475.

[70] Against Crossan, *Historical Jesus*, 340, who thinks Jesus established rural missions.

[71] Smith, "Table Fellowship and the Historical Jesus," in *Religious Propaganda and Missionary Competition in the New Testament World* (Leiden: Brill, 1994), 162.

[72] Smith, "Table," 470.

[73] Smith, "Fellowship," 141.

of the house of Israel in Matt 9:10-13 (cf. Matt 10:6, 16; 15:24). Likewise, to be congruent with the tradition of Matt 9:10-13 preserved as a controversy story, the message communicated by Jesus' having table fellowship with sinners may be appropriately understood within the context of Jesus' *confrontation* with the Pharisees, the shepherds/leaders of Israel. What is at stake is nothing less than the future of Israel and of the nations. Also, Jesus' table fellowship with sinners can be characterized as *restorative* in view of YHWH as correcting Israel's past, if not characterized as formative.

The conflict between Jesus and the Pharisees, illustrated in the case of Jesus' table fellowship with the sinners (9:10-13), and sharpened particularly in Matthew's Gospel, does not need to be regarded as an anachronism created by the early churches. Parallel criticisms leveled at the Pharisees – similar to those in the Gospel (cf. Matt 23) – abound in contemporary Jewish sources. For instances, *b. Sotah* 22b criticizes the Pharisees' display of their religious devotion, particularly, "carr[ying] his piety on his shoulder"; *4QpNab* 2-4 II, 8-9 mentions "seekers after smooth things" as "lying interpreters," "seekers of deceit." [74]

Further, such conflict is common in the prophetic tradition in the Hebrew Bible, such as Elijah, Hosea, Jeremiah[75] – even more so than John the Baptist.[76] Undoubtedly, Ezek 34:1-6 illustrates a case in point. The tone of YHWH's indictments against the false shepherds of Israel in this text is no less bitter and harsh than was Jesus' warning against the Pharisees in Matthew 23. More parallels enhance the particular relevance of Ezekiel 34 to Matt 9:10-13 as we consider the comparability of the historical and theological frameworks of Ezekiel's time and those of the Pharisees before A.D. 70.

Ever since the nation experienced the disappointing failures of the Maccabean leadership, the hope for the revival of the monarchic ideal was revised (cf. *Ps. Sol.* 17-18). Pharisaism can be seen as one of the responses to the plight of the nation at the time, along with other divergent groups among pre-70 Judaism. For Judaism in this period, the Deuteronomic paradigm of sin-exile-return was most likely the biblical-historical framework undergirded by

---

[74] Moshe Weinfeld, "The Jewish Roots of Matthew's Vitriol," *BR* 13, no. 3 (1997): 31, lists texts such as *m. Hagigah* 2:1, *m. Yevamot* 8:7, *b. Shabbat* 31b that criticize the hypocrisy of the Pharisees for their lack of practice. See also Anthony J. Saldarini, "Understanding Matthew's Vitriol," *BR* 13, no. 3 (1997): 32-39, 45.

[75] D. A. Carson, "The Jewish Leaders in Matthew's Gospel: A Reappraisal," *JETS* 25, no. 2 (1982): 161-174; Carson, *Matthew*, 225.

[76] Wright, *Jesus*, 127, 382; likewise, R. Horsley, "Social Conflict in the Synoptic Sayings Source Q," in *Conflict and Invention* (ed. J. S. Kloppenborg; Valley Forge; Trinity Press International, 1995): 37-52.

these various groups in their reading of Israel's past, present, and future.[77] The Pharisees likely shared the conviction of the age that Israel had not yet fully recovered from the exile in the way which God had promised; the tendency to see the exile as an unfinished story was not an isolated phenomenon.[78]

Unlike the seer(s) of *An. Apoc.* 89-90, who hoped for a revelatory vision in place of the Mosaic vision for the restoration that they forsook, the Pharisees likely shared the thought of the age that the failure to obey the law of Moses was the principal cause of the nation's past plight and the suffering from the ongoing exile. Thus, the Pharisees influenced the masses – at least on the agenda-level – to believe that if all Jews adhere to and execute the Pharisees' program for the nation, then the expected restoration of Israel would come.[79]

Furthermore, the conflict between Jesus and the Pharisees' normative Judaism, as illustrated in the case of Jesus' table fellowship as to the treatment of sinners, can be seen in the context of their competing eschatological visions, with both of them laying claim to the future of the nation.[80] Yet, since the Pharisees' vision for the future of Israel (thus, eschatological and political) was executed in and by their own way of radicalizing the Mosaic ideal, this, in turn, would characterize their religion and morality as well.[81]

Likewise, Ezekiel's vision of the restoration of Israel (Ezek 34-37) is a thorough revision for the future of Israel under YHWH's leadership against the

---

[77] M. A. Knibb, "The Exile in the Literature of the Intertestamental Period," *HeyJ* 17 (1976): 253-279; D. E. Gowan, "The *Exile* in Jewish Apocalyptic," *Scripture in History and Theology: FS J. C. Rylaarsdam* (ed. A. L. Merrill and T. W. Overholt; Pittsburgh: Pickwick, 1977), 205-223; cf. O. H. Steck, *Israel und das gewaltsame Geschick der Propheten: Untersuchungen zur Überlieferung des deuteronomistischen Geschichtsbildes im Alten Testament, Spätjudentum und Urchristentum* (WMANT 23; Neukirchen-Vluyn: Neukirchner Verlag, 1967), 100-189; Deines, "Pharisees," 454.

[78] Refer to Martin G. Abegg, "Exile and the Dead Sea Scroll," 111-126; Louis H. Feldman, "The Concept of Exile in Josephus," 145-173 and J. Neusner, "Exile and Return as the History of Judaism," 221-238, and Craig A. Evans, "Aspects of Exile and Restoration in the Proclamation of Jesus and the Gospels," 299-328, in *Exile: Old Testament, Jewish, and Christian Conceptions* (ed. James M. Scott; Leiden: Brill, 1997). Also C. A. Evans, "Jesus & the Continuing Exile of Israel" in *Jesus & the Restoration of Israel* (ed. Carey C. Newman; Carlisle: Paternoster, 1999), 77-100. Related to Matthew's Gospel, Donald J. Verseput, "The Davidic Messiah and Matthew's Jewish Christianity," *SBLSP* (1995):104-106, lists texts that express the rise of a particular reform movement that marked the beginning of the recovery from the exile: *1 Enoch* 85-90, esp. 90:6-7, 93:1-10 + 91:11-17, esp. 93:9-10 [*4Q212* 4:12-14]; *Jub.* 1:7-18, 23:26; Tob 14:5-7; *As. Mos.* 4:8-9; *2 Apoc. Bar.* 68:5-7; *Sir* 36:16, "Gather all the tribes of Jacob and give them their inheritance, as at the beginning"; *4Q504* 1-2 v-vi.; Wright, *Jesus*, 127, 268-269, 476-478.

[79] Steven M. Bryan, *Jesus and Israel's Traditions of Judgment and Restoration* (Cambridge: Cambridge University Press, 2002), 87-88 argues in the similar vein.

[80] Cf. Dunn, *Jesus Remembered*, 604-605; Deines, "Pharisees," 493.

[81] Cf. Wright, *Jesus*, 372, tends to treat these two dimensions, religion and morality, separately.

backdrop of the recent failure of the Davidic monarchy and consequent exile. This is a context quite similar to the historical situation of pre-70 Judaism, due to the failure of the Hasmonean dynasty, triggered by the revived vision of the old monarchy. Moreover, obedience to the Law was a must in Ezekiel's blueprint for the future of Israel, just as it holds true for the Pharisees and other groups (such as Qumran sectarians) at the end of the Maccabean regime.

Nevertheless, unlike the Pharisees' practices, Ezekiel's concept of Israel's restoration *was not a simple return to the Mosaic ideal.* Rather, it involves a new heart, and a pouring out of the Spirit (e.g., 36:22-27) that would enable the renewed people to obey the law, and bring YHWH's blessings and peace back to the land. This coincides with the unification (37:15-23), revivification (37:1-14), and the restoration of Israel (37:24-28). Also, YHWH's establishment of the new Shepherd over the restored community is connected to their obedience of the laws and decrees, but in a new way (37:24-25; cf. 34:23-24). The settlement of the new community and YHWH's presence in their midst completes the restoration process (Ezek 40-48). After all, it is YHWH's zeal for his name that is to be hollowed before the nations (37:27-28; 48:35).

Jesus' mission to sinners is unique. Norman Perrin sees Jesus as inimitable, set apart "from both his contemporaries and even his prophetic predecessors."[82] Ben F. Meyer highlights the implication of Jesus' unique mission in the context of Israel's history: "Nothing, in fact, could have dramatized the gratuity and the present realization of God's saving acts more effectively than this unheard of initiative toward sinners; nor could anything have been more fundamental to the eschatological restoration of Israel."[83] With Jesus at table surrounded by sinners, confronting their shepherds, Israel is now about to be renewed (cf. Matt 8:28-9:36). Jesus' table fellowship exercises a restorative force. It rectifies the failed past while generating eschatological momentum in view of the history of the nation and the nations. Jesus seeking sinners is merely the inauguration of the process of restoration.

The literary structure of the immediate context of Matt 9:10-13 echoes Ezekiel's vision as well. At the heart of Ezekiel's restoration process is a vision of the revivification of the life of Israel in the image of resurrection from the grave (Ezek 37:1-14; cf. Matt 27:51b-53): the reversal of the ultimate curse. As he brings back his scattered sheep, the divine Shepherd cleanses them, gives them new hearts, and pours his Spirit on them so that they may finally be able to obey his laws and decrees (Ezek 36:22-32; 37:24-25). This means the complete reversal of the curse under the Mosaic covenant. Likewise, Matthew's narrative in 8:28-9:36 displays the radical newness of Jesus' ministry placed at the center of the section – "new wine in new wineskin" (9:4-17) – immediately

---

[82] Perrin, *Rediscovering*, 107; G. Vermes, *Jesus the Jew* (Philadelphia: Fortress Press, 1973), 224.

[83] Meyer, *Aims*, 161.

followed by Jesus raising the dead to life (9:18-26). This central section of newness is then surrounded by emphases on Jesus' mercy (9:9-13; 9:27-31) and his inimitable authority (9:1-8; 9:33).

Also, in Matt 8:28-9:36, the curses of the Mosaic covenant (Lev 26) are revoked by Jesus as he crosses over the Pharisees' paradigmatic categories of priestly purity: fellowship over food (9:10-13; cf. Lev 11); contact with an unclean woman with a sexually-related disease (9:21; Lev 12, 15); and contact with a corpse (9:25; Lev 19).[84] Jesus' conflict with the Pharisees – the theme that is explicit in the periphery of the section (8:28-34; 9:32, 34) – is inevitable. Furthermore, table fellowship epitomizes the initiation of the divine confrontation with Israel's leaders. Then the summary verses found in 9:35-36 hint at Jesus' 'compassionate' point of view – he sees the crowd as sheep without a shepherd.

Jesus does exactly what YHWH the eschatological Shepherd would do at the time of the restoration – seek the lost (Ezek 34:4, 16, אֶת־הָאֹבֶדֶת אֲבַקֵּשׁ; LXX, τὸ ἀπολωλὸς ζητήσω). As demonstrated previously, the lost constitutes the most afflicted of all the six groups presented in Ezek 34:4,16. They are those who have been pushed to the uttermost fringes of the covenantal community, and it is here that YHWH's restoration starts. Jesus proclaims that he came to seek sinners, i.e., 'the outcast and the lost' (Matt 9:10-13), heralding that the hoped-for restoration has finally begun!

The Evangelist is not alone in reflecting the eschatological shepherding mission assigned to Ezekiel's eschatological Shepherd in view of raising up the Davidic Shepherd figure over Israel and the nations. 4Q521 illustrates the case in point. Particularly, these eschatological pastoral activities are critical signs for identifying the messianic figure in view of the restoration of Israel. 'His anointed one' in 4Q521 is imbued with authority over the heavens and the earth (כי הש[מים והארץ ישמעו למשיחו) so they might listen to YHWH's appointee who will lead them by means of YHWH's precepts (l. 1; cf. Ezek 34:23; 37:25).

Yet while this anointed one can be compared to the Davidic Shepherd of Mic 5:1-4 (Matt 2:6), it is YHWH himself who is envisioned to bring about 'the eternal kingdom' to the pious (l. 7; cf. 2 Sam 7:14-15). This will be accomplished by the following activities of YHWH himself: "freeing prisoners, giving sight to the blind, straightening out the twisted (4Q521 l. 7; cf. Isa 35:4-6; 61:1-2). In his mercy (ובחסדו)...the Lord will perform marvelous, unprecedented acts (l. 9b-11); for he will heal (ירפא) the badly wounded and will make the dead live and proclaim good news to the poor (l.12; cf. Ezek 37:1-4; Isa 61:1-2); and he will satisfy the [poo]r (ודלים יישע, "et les [pauvre]s Il comblera"), lead the outcast (נתושים ינהל, "les expulsés Il conduira"), and feed the hungry/invite to the

---

[84] Cf. Chilton, "Jesus' Meals," 483-484.

banquet" (ורעבים יעשר), "et les affamés Il enrichira/invitera au banquet," l. 13; cf. Ezek 34:10, 14, 18)."[85]

Quite possibly, the anointed one will take upon himself the restorative roles of Ezekiel's eschatological Shepherd. Especially, with regard to 4Q521 l. 13, Émil Puech's emendation, נחושים ינהל as "les expulsés Il conduira," if compared to those by Zimmermann and Martínez/Tigchelaar, reflects explicitly the characteristics of the Ezekielian eschatological Shepherd, the Seeker.[86] The profile of the משיהו in 4Q521 carries out the distinctively pastoral tasks of healing the sick, feeding the hungry, even raising the dead, and particularly as Peuch makes clear, 'leading the outcast.' As 'his anointed' brings the outcast back to the community, all of these shepherding activities dovetail.

Significantly, this implies that YHWH will *reverse* all of the wrongdoings attributed to the shepherds of Israel. Just as the divine shepherd would do, Jesus *begins his mission of compassion by seeking the most victimized* of the flock – the outcast and the lost, i.e., ἁμαρτωλοί. This causes embarrassment and brings judgment upon the shepherds of Israel, just as YHWH's fierce indictments rained down on the leaders of Israel (Ezek 34:1-6; cf. Jer 23:1-6). Ezekiel's divine Shepherd exposes what Israel's leaders failed to do; they have *not* strengthened the weak, healed the sick, bound up the injured, brought back the outcast, nor sought the lost (Ezek 34:4). Instead, the wicked shepherds reversed the roles of sheep and shepherd and only fed and cared for themselves. Thus YHWH declares, "I will seek and find the lost, bring back the outcast, bind up the injured and the sick, and strengthen the weak" (Ezek 34:16). Jesus follows the same pattern, for he is told to *shepherd* Israel in Matt 2:6c: ὅστις ποιμανεῖ τὸν λαόν μου τὸν Ἰσραήλ.

Within the context of Matthew 9, we not only see Jesus seeking the lost and the outcast at the meal table – literally snatching some from the jaws of unclean spirits (vv. 10-13, 32-33; cf. 8:28-34) – we also see him *healing* the injured and the sick (vv. 1-7, 12, 20-22, 27-29; cf. v.35). In Matt 9:10-13, Jesus presents himself as the one whose priority is to heal the sick: οὐ χρείαν ἔχουσιν οἱ ἰσχύοντες ἰατροῦ ἀλλ' οἱ κακῶς ἔχοντες (v.12). Jesus, who came to call the sinners, the outcast and the lost, is the one who heals οἱ κακῶς ἔχοντες, just as Ezekiel's eschatological Shepherd would do: "I will bind up the injured and

---

[85] Émil Puech, *Qumrân Grotte 4* (DJD 25; Oxford: Clarendon Press, 1998), 10-11.
[86] Zimmermann, *Messianische Texte*, 344-345, reads l. 13, "[..., Vertrie]bene(?) führen und Hunger[nde(?)] reich machen(?)"; Martínez/Tigchelaar, *Scrolls*, 1044-1045, reads, "and [...]...[...] he will lead the [...]...and enrich the hungry."

strengthen the weak" (34:16).⁸⁷ Jesus as the eschatological shepherd heals and thus initiates the envisioned restoration of the whole community.⁸⁸

The echo of Ezekiel 34 and Matthew 8-9 continues. Matthew's οἱ ἰσχύοντες matches Ezekiel's "the robust and the strong" (וְאֶת־הַחֲזָקָה; LXX, τὸ ἰσχυρὸν, Ezek 34:17) in the flock; they are not the wicked shepherds/leaders *per se* whom YHWH eventually will remove from the flock (Ezek 34:10) nor are they Gentiles. Rather, in terms of Matt 9:13, they refer to 'the strong' (v. 12) and 'the righteous' (v.13). The eschatological shepherd will police (אַשְׁמִיד) them, meaning that he will shepherd the flock with justice (Ezek 34:17). Likewise, the 'strong' and 'the righteous' Jesus mentioned in Matt 9:12-13 are not to be excluded from his ministry.⁸⁹

The shepherds, however, fall into a different category in YHWH's eschatological handling of his people. The shepherds/leaders are to be replaced by YHWH's appointed new David: "I will remove them from tending My flock" (Ezek 34:10). In Ezekiel 34-37, the restoration begins by introducing the return of the theocracy through the Davidic-Appointee, the Davidic Shepherd. It is a theocracy, but with a new David. He is not merely a מֶלֶךְ of the Davidic monarchy – he is distinguished as a Shepherd-Appointee after YHWH the true Shepherd of Israel (34:23-24; 37:24-25).

In Matthew's Gospel, Jesus' conflict with the Pharisees is twofold; just as Matt 9:10-13 succinctly represent, through his seeking/healing and teaching. In Ezekiel 34, YHWH confronts the wicked shepherds of his own flock particularly as he reverses the shepherds' wrongdoings as to the well-being of the flock. Thus, his seeking the lost and healing the sick means his judgement upon them for their wrongdoings (Ezek 34:1-22). Also, YHWH eventually replaces the wicked shepherds of Israel with the righteous Davidic Shepherd/Teacher who can lead the flock into the way of righteousness (Ezek 34:23-24; Ezek 37:24-25). The First Gospel, as we shall examine later in the section for 'the therapeutic Son of David,' Jesus' healing ministry (thus involving seeking the lost and the outcast) propels hostility from the part of the Pharisees and the leaders of the nation (cf. Matt 7:15; 12:11-12; 18:10-14; 12:22-37; 21:12-17).

---

[87] Likewise, Christof Landmesser, *Jüngerberufung und Zuwendung zu Gott:Ein exegetischer Beitrag zum Konzept der mattäischen Soteriologie im Anschluß an Mt 9,9-13* (Tübingen: Mohr Siebeck, 2001), 104-108, finds Ezek 34 to be the background for the "Verbindung zwischen den Motivberufen Hirte und Arzt," and he also develops the conflict motif between Jesus and the Pharisees as the wicked shepherds in Ezek 34.

[88] J. J. Pilch, "Understanding Biblical Healing: Selecting the Appropriate Model," *BTB* 18 (1988): 60, notes that Jesus' 'cure' invariably involved establishing new self-understandings so that those formerly 'unclean' and excluded from the holy community now found themselves 'clean' and within the holy community.

[89] Meyer, *Aims*, 162, uses the parable of the prodigal son to argue his case for the inclusion of the righteous. Likewise, Wright, *Jesus*, 27-28; cf. Matt 10:6, 16; 15:24.

Further, as Matt 9:10-13 aptly demonstrates, Jesus confronts the Pharisees in terms of their teachings and practices (cf. Matt 5:1-7:29; esp. 7:28-29; 23:1-39). As the righteous Leader/Teacher, Jesus along with his disciples as his under-shepherds for the eschatological flock, eventually replace the Pharisees (Matt 10:1-6; 28:16-20).[90] Thus, these two aspects, Jesus' seeking/healing (the mission of YHWH the eschatological Shepherd) and teaching (the mission of the Davidic Shepherd), reveals the nature of Jesus' confrontation with the Pharisees as the shepherds of Israel.

The conflict in Matthew 9, also, demonstrates a different kind of leadership according to the fashion of YHWH as the true Shepherd of Israel. Put another way, this speaks of messianism (cf. 4Q521). Both in Ezekiel 34-37 and Matthew 9, we encounter the redefinition of Davidic expectation (cf. Matt 1:1, 17; 2:6 and 9:27) in terms of a shepherd image.[91] In Ezekiel 34-37 and 40-48, the נָשִׂיא also reflects a similar trend. For Ezekiel, the leadership of the future Davidic Shepherd is to be marked first by securing the return of the theocracy (34:1-16), second, by assigning a religious/political role for leading the renewed people to follow the decrees and laws of YHWH (37:24-25), and third, by representing his people in and through his intimate leadership among them (37:24-25; 45:16-17).[92]

In Matthew's Gospel, Jesus exactly fits all of these pictures: Jesus is *Immanuel* (1:23; 28:19-20), Jesus is the Teacher *par excellence* (9:11, 13; cf. 5:1-7:29); and Jesus saves his people from their sins (1:21) as he identifies himself with them (9:10-13).[93] Teaching, in particular, is a significant part of the Davidic Shepherd's role, manifestly characterizing Jesus' earthly ministry among all the other roles that recall YHWH as the eschatological Shepherd in Ezekiel 34-37. Yet it must be emphasized in the context of Matthew 8-9 that Jesus' seeking the lost while confronting the false shepherds of Israel is a critical sign of the arrival of YHWH the eschatological Shepherd.

Our examination of the nature of the conflict in the context of restoration brings us to another essential aspect of the conflict: Jesus' identity. When at the

---

[90] Scot McKnight, "New Shepherds of Israel," 30-59, 190-194.

[91] This tradition of redefining the Davidic leadership in terms of the shepherd images is apparent in *Ps. Sol.* 17-18, 4Q504, 4252, 4Q174, 4Q161, 4285, 1Q28b, 4Q521, CD-A 13 and CD-B 19. Refer to Ch.2.3.

[92] Duguid, *Leaders*, 50-55, points out the priest-like role and its representative responsibility of נָשִׂיא who is to gather the offerings required of the people so that eventually he can provide the cereal offering, sin offering, burnt offering and fellowship offering, "to make atonement (לְכַפֵּר) for the house of Israel" (45:16).

[93] Landmesser, *Jüngerberufung,* 107-108 (see also,131-132), likewise states, "Nicht mehr die Pharisäer sind in der Gottesgemeinschaft, sondern eben die Sünderin, denen sich Jesus in der Tischgemeinschaft zuwendet. Gerade damit erfüllt sich der Auftrag Jesu, der genau dazu in die Geschichte seines Volkes eingetreten ist, sein Volk von seinen Sünden zu erretten (Mt 1,21)."

end of the controversy story Jesus pronounces, "I have come to call, not the righteous but the sinners" (v.13), he draws them into the heart of the conflict: Who exactly does Jesus claim to be?

4.1.6 More Than a Prophet (Matt 9:13b)

The 'I-saying' in Matt 9:13b, οὐ γὰρ ἦλθον καλέσαι δικαίους, ἀλλὰ ἁμαρτωλούς is a christological statement intended to be a preventive/corrective to any misunderstandings. The same structure is found in Matt 5:17 regarding Jesus' fulfilling the Law (Μὴ νομίσητε...οὐκ ἦλθον καταλῦσαι ἀλλὰ πληρῶσαι); likewise, there is a parting of the ways in Matt 10:34 (Μὴ νομίσητε ... οὐκ ἦλθον βαλεῖν εἰρήνην ἀλλὰ μάχαιραν); and the higher priority of following Jesus over and above family ties and commitments in Matt 10:35 (ἦλθον γὰρ διχάσαι ἄνθρωπον). All three cases involve the decisive nature of Jesus' mission and its unequivocal consequences. Because of his coming and his mission, something happens not only to the Law and to the nation, but also to an individual's family solidarity. There is no allowance for a tendency to minimize the fatal impact of Jesus' mission. Thus his mission could/should never be thought of as a minor adjustment within the *status quo*.

Concerning the construction in Matt 9:13b, Gnilka states, "Der abschließende ἦλθον-Satz" substantiates Jesus' seeking the sinners as "eine Grundbestimmung der Sendung Jesu."[94] Similarly, Luz not only affirms the christological significance but links Hos 6:6 to Jesus' fulfillment of the law as announced in 5:17, thus, supporting the better righteousness in 5:20.[95] The controversy of Matt 9:10-13 gives prominence to Jesus as a new teacher with surpassing authority. The Pharisees claim Moses' seat but they fail to do what they themselves taught (Matt 23:1-3). In Matt 9:10-13, praxis is a prominent theme in Jesus' critique of the Pharisees: "go and learn" (v. 13a).

The critical difference between Jesus and the Pharisees does not stem from their attitude toward the holiness of God, but from their attitude toward the sins of the people, and more precisely, *toward sinners*. Motivated by mercy, Jesus was at table with the lost and the outcast, while the Pharisees shunned them for the sake of their pursuit of holiness. Jesus, as if he is mercy-personified, embodies divine compassion. Likewise, Jesus' reference to Hos 6:6 becomes the "explanatory word" of his healing activities (9:27; 15:22; 17:15; 20:30-31).[96]

From the Pharisees' point of view, having table fellowship with sinners, Jesus associate himself with the sinners, the lost, and the outcast and thereby

---

[94] Gnilka, *Matthäusevangelium* 1:333.
[95] Luz, *Matthew*, 2:33-34, not only affirms the chirstological significance but links Hos 6:6 as Jesus' fulfillment of the law as announced in 5:17, thus, supporting the better righteousness in 5:20.
[96] Luz, *Matthew*, 2:34.

threatens the Pharisees' concept of holiness, their interpretation of the Law.⁹⁷ A. G. Van Aarde notices the correlation of the names: 'sheep' (πρόβατον, Matt 9:36; 18:12), 'the lost sheep of Israel' (τὰ πρόβατα τὰ ἀπολωλότα οἴκου Ἰσραήλ, 10:6), 'the little children' (τὰ παιδία,18:3-5), 'the little ones' (18:6, 10, 14) – with all of whom Jesus identifies himself.

Moreover, Van Aarde points out that Matthew relates Jesus as a socially despised person according to the Pharisee's way of thinking (cf. Matt 12:24), but clarifies that Jesus did not experience alienation from God, but from God's presence.⁹⁸ Here it begins to crystallize: the nature of Jesus' leadership differs substantially from that of the Pharisees. Jesus identifies himself with sinners and reveals God's presence among them, while in contrast, the Pharisees separate themselves from sinners in order to preserve their holiness.

Jesus considers his table fellowship an act of mercy, the heart of God's will written in the Law: πορευθέντες δὲ μάθετε τί ἐστιν· ἔλεος θέλω καὶ οὐ θυσίαν (Matt 9:13ab). Hosea 6:6 complements this point of view: "I desire mercy not sacrifices" (v. 13b). It is not certain whether Jesus' table fellowship with sinners was a precedent for the Eucharist, especially since the latter could be seen in a sacrificial context;⁹⁹ nevertheless, it is interesting to note that his table fellowship as an act of mercy is set in contrast with offering sacrifices in the Temple. Wright notes that Hos 6:6 is quoted in *Aboth de R. Nathan* 4 (a third-century work), though it is attributed to Johanan ben Zakkai as an earlier comment on the destruction of the Temple, thus suggesting that here Jesus inaugurates a way of life no longer requiring the Temple.¹⁰⁰ In short, Jesus' identifying himself with sinners and the functional replacement of the offering of the sacrifices in the Temple characterize Jesus' table fellowship with sinners.

This echoes Ezekiel's concept of the shepherd; as Ian Duguid points out, it is the nature of the *occupant* not of the *office* of the leader for Israel that was being redefined in Ezekiel's revision of Israel's future.¹⁰¹ In Ezekiel 34-37, the Davidic Shepherd must be a righteous shepherd/prince in order to lead them into the way of the law of God. This requirement of the Davidic "Shepherd" – that is, his blameless nature – for the sake of the eschatological task to lead his restored people, whether united Israel (Ezek 37; cf. Zech 11) or Israel and the

---

⁹⁷ Cf. Moessner, *Banquet*, 325, notes the function of Jesus' drinking and eating with sinners, along with his baptism, to highlight "the fundamental solidarity of Jesus with the sinful *laos* of God."

⁹⁸ Aarde, "The Evangelium Infantium, the Abandonment of Children, and the Infancy Narrative in Matthew 1 and 2 from a Social Scientific Perspective," *SBLSP* (1992): 436, 447-448.

⁹⁹ Cf. B. Lang, "The Roots of the Eucharist in Jesus' Praxis," *SBLSP* 31 (1992): 467-472.

¹⁰⁰ Wright, *Jesus*, 335, 426-427; cf. J. Neusner, *A Life of Johanan Ben Zakkai* (Studia Post-Biblica 6, Leiden: Brill, 1970), 114, 130, sees the text in *Aboth de R. Nathan* 4 as a reflection of the debate in the time of Bar-Kochba whether the Temple should be rebuilt or not.

¹⁰¹ Duguid, *Leaders*, 47 (italics his).

nations as whole (*An. Apoc.* 90; *Ps. Sol.* 17-18), was certainly recognized in the texts contemporary with Jesus and Matthew. The Davidic Shepherd figure is characterized as 'free from sin' (*Ps. Sol.* 17:36, 41; cf. Ezek 36:25) and 'holy' (*An. Apoc.* 90:17-18, 38), thus assuming the role of sanctification or transformation of the rescued people at the eschaton.

In the light of this connection with sanctification, the 'I-saying' in Matt 9:13 deserves scrutiny. In the oracle of Ezek 34:1-16, the failures of the past leaders of Israel are confirmed repeatedly (vv. 2-6; v. 8; v. 10). Hence, YHWH's proclamation of His intervention (v.10) starkly contrasts the false shepherds' wrongdoings set against the deplorable backdrop of YHWH's scattered flock in utter hopelessness. The problem becomes obvious to YHWH as he cries out: וְאֵין דּוֹרֵשׁ וְאֵין מְבַקֵּשׁ ("there is *no one* searching; there is *no one* seeking," 34:6). This is what necessitates YHWH's divine intervention: "This being so" (לָכֵן; LXX, διὰ τοῦτο, 34:7) – since there is no one shepherd who seeks – "therefore, the Lord Yahweh says, Behold I am against the shepherds, I will seek my flock from their hands; *I Myself* will search" (34:10-11, 12).

In Ezekiel's vision, the motif of YHWH's presence opens and closes the whole restoration process. The absence of the one shepherd with a shepherd's heart for his flock (34:1-9) necessitates, for the sake of his name, the returning of the presence of the shepherd himself in the midst of his restored people (37:27-28; cf. 48:35). As in the case of Matt 5:17-48, Jesus' 'I-saying' in 9:13 is provocative; Jesus' ἦλθον is paralleled with YHHW's θέλω in the quote from Hosea 6:6. The emphatic 'I' implied in γὰρ ἦλθον καλέσαι in Matt 9:13b signals *the returning presence of the Shepherd* of Israel, who fiercely confronts the shepherds to whom he entrusted his flock (cf. Matt 1:23). Who then is Jesus as presented in Matt 9:10-13? With surprising authority, he teaches the experts of the Torah the will of God, and reverses what they have done for the nation by eating at the table with the lost and the outcast. J. P. Meier, in reviewing the third quest of the historical Jesus, presents a puzzling question:

If Jesus presented himself as an eschatological prophet who performed a whole series of miracles, what Old Testament figure or model would naturally be conjured up in the minds of first-century Palestinian Jews? In the Jewish Scriptures, only three prophetic figures perform a whole series of miracles: Moses, Elijah, and Elisha. Of these three, Moses never raises an individual dead person to life. And if we ask which of these three is expected to return to Israel in the end-time to prepare it for God's definitive reign, the answer from Malachi and Ben Sira through the intertestamental writings to the rabbinic literature is: Elijah.

I would therefore contend that it is not the early Marcan, Q, or Johannine traditions that first thought of Jesus in terms of *the miracle-working, eschatological prophet* wearing the mantle of Elijah, though they certainly may have developed this idea. The traditions coming from the historical Jesus strongly suggest that he consciously chose to present himself to his fellow Israelites in this light. *How this coheres – or whether it coheres – with the gospel traditions that portray Jesus as the awaited Davidic Messiah* or present him speaking himself as the Son of Man is a problem with which I must still grapple.[102]

Indeed, this association of the miracle-working Jesus with the Davidic messiah baffled many.[103] We offer the Davidic Shepherd tradition as the likely solution to this puzzle. Jesus performed miracles, fed thousands, healed the sick and raised the dead, and also was recognized as the Son of David in the Gospel. In Ezekiel, Davidic expectation merges with the shepherd image embodying YHWH's will to rescue and restore exiled Israel. The eschatological Shepherd will rescue, heal, feed, pour out his Spirit upon, and raise up his people from the grave. The new David, YHWH's Appointee, will teach and lead them to follow the will of their God, and to produce righteousness in the land. Peace returns as their true Shepherd returns to be in their midst.

Subtle distinctions exist here: YHWH rescues (נָצַל) his scattered flock, and appoints a new David as the one shepherd who saves (יָשַׁע) the rescued little flock by leading them with justice and mercy to full restoration.[104] In Matthew 9, however, *Jesus seems be doing both* (until the shepherd is smitten later in Matt 27:31, cf. Zech 13:7). Jesus rescues and restores as he seeks, forgives, and heals. If this view is correct, then it means that as Jesus sits at table with sinners, the eschatological theocracy has appeared, and *the conflict is final.*

This is the role Jesus acts out, and Matthew presents God's reign as having come in the midst of the people. Jesus assumes both of these Shepherd roles; through word and deed he claims in no uncertain terms that he is more than a prophet. The roles of YHWH the divine eschatological Shepherd and the Davidic Shepherd, distinguished in Ezekiel's vision of the future, are now indistinguishable in Jesus. Early in his narrative, Matthew intends to communicate who Jesus is, and enumerates the implications for Israel (1:1-17, 21, 23). Jesus, at table with the lost and outcast (9:10-13), counters the movement of the Pharisees, seen not necessarily as legalism but as a variation of the Mosaic system. Just as Ezekiel had done, Jesus now moves beyond the vision within the Mosaic covenant.

To conclude, the nature of the conflict is, therefore, both *eschatological and christological*. Jesus at table signals that the time for Israel's renewal has arrived, with her returning Shepherd in their midst (Matt 1:23; 18:19-20; 28:16-20): it is the return of the theocracy, YHWH's own revision of the former, failed

---

[102] J. P. Meier, "The Present State of the 'Third Quest' for the Historical Jesus: Loss and Gain," *Bib* 80 (1999): 482-483 (emphasis mine).

[103] Stanton, "Origin and Purpose," 1924; J. J. Collins, *The Apocalyptic Imagination* (Grand Rapids: Eerdmans, 1998), 263.

[104] Robert L. Hubbard Jr., *NIDOTT* 2 (ed. William VanGemeren; Grand Rapids: Zondervan, 1997, #3828), 556, says, "In general, the root *yš* 'implies bringing help to people in the midst of their trouble *rather than in rescuing them from it*" (Italics mine); also see Daniel I Block, *Ezekiel 25-48*, 357.

monarchy. The reversal of Israel's fortune is manifested in their midst through the Shepherd's seeking the lost and bringing back the outcast, healing the sick, feeding the hungry, thus gathering the scattered flock.

Consequently, the negligent shepherds, that is, the Pharisees in Jesus' time, are replaced by Jesus. He takes up the role of YHWH the eschatological Shepherd seeking primarily the lost and the outcast, i.e., the sinners. The time of restoration arrived, and the renewed Israel is with her Shepherd forever (Ezek 37:27-28; Matt 28:20). With the nations in view, YHWH does all of this for the sake of His name (Ezek 36:21-22), and thus Jesus teaches how to pray for His reign on earth: "Hallowed be Thy name; and Thy Kingdom come" (Matt 6:9).[105]

4.1.7 Summary

The theme of Jesus' seeking the lost is explicit in Matt 9:10-13. Two major elements of the profile for the eschatological and Davidic Shepherds are examined: (i) seeking the lost and (ii) healing the sick. Jesus' seeking the lost by frequently having table fellowship with sinners, we argued, must be understood primarily in its Jewish milieu. Jesus came to 'seek' the lost and bring back the outcast. These two groups – namely, the lost and the outcast within the covenantal community – identify ἁμαρτωλοί in the Gospel. Primarily they are Jews harassed and downtrodden by the fat and robust among the flock. Jesus' seeking the lost can thus be seen as the mission of YHWH to restore justice within his flock.

Thus the wild beasts that devour the sheep are none other than the Pharisees, the false shepherds of Israel. Jesus' confrontation with the Pharisees eventually results in God's judgment upon the shepherd of Israel (thus upon the whole flock, cf. Zech 11) as Jesus takes up this role as the 'smitten shepherd' in 26:31. God eventually replaces the false shepherds (represented by the smitten shepherd) with the one Davidic Shepherd in 28:16-20.

Both the shepherds and the flock are renewed through the representative suffering of the smitten shepherd. These events are initiated by divine reversal as Jesus shares meals with sinners (Matt 9:10-13). The conflict between Jesus and the Pharisees before A.D. 70 is eschatological in character, and final in its consequence. Jesus is more than a prophet; he takes upon himself the role of YHWH as the eschatological Shepherd as well as that of the smitten shepherd, thereby embodying the whole nation.

Likewise, the role of Jesus as the eschatological Shepherd illuminates his mission and identity. Also, as we shall examine in the following section, this model of Jesus as the eschatological Shepherd offers a critical clue to the picture of Jesus as the Son of David who does miracles, especially healing the

---

[105] Cf. For the 'name' of Jesus/God, see Matt 7:22; 12:21; 18:5, 20; 21:9; 23:39; 28:19; cf. 1:5; 24:5.

sick. In Matthew's Gospel, Jesus' healing activity, to a certain degree, characterizes his Davidic messiahship. This Matthean feature strongly supports our thesis that Jesus is the promised Shepherd(s). To this concern now we turn.

4.2 The Davidic Shepherd as the 'Therapeutic Son of David'

The Davidic Shepherd tradition can illuminate a critical aspect of Matthew's christology: Jesus as the therapeutic Son of David. In this section, (i) we will briefly present the significance of the Son of David in answer to scholarly controversies as to its function in the First Gospel; (ii) next, we will consider various proposals as to the identity of 'the therapeutic Son of David'; and (iii) finally, we will argue for the validity of the Davidic Shepherd figure as it relates to Jesus as the healing Son of David. This will be accomplished by a careful examination of each passage that involves the healing activities of the Son of David figure in the Gospel: 9:27-31; 12:22-24; 15:21-28; 20:29-34; 21:1-16; 22:41-46.

4.2.1 Jesus as the Son of David: Controversies

Despite the dissenting voices about the function and the purpose of Matthew's use of the title 'Son of David,' we maintain that Matthew is undeniably engrossed in this dimension of Jesus. As we tackle the controversies, we will first consider the title 'the Son of David.'

(1) The Son of David: In the opening verse of Matthew's Gospel, the Evangelist introduces Jesus Christ as the Son of David (1:1). The centrality of a text's introduction in ancient historiography is well recognized. M. D. Hooker suggests that each of the Evangelists consciously and deliberately spelled out at the beginning of his book both his purpose in writing and his interpretation of the story.[106]

For our purposes, this implies that Matthew's Gospel narrative is likely, though not exclusively, going to expound on how Jesus is the Son of David. In the first verse of the Gospel, the title υἱοῦ Δαυίδ is placed in the middle: Ιησοῦ Χριστοῦ υἱοῦ Δαυὶδ υἱοῦ Ἀβραάμ. The order is reversed but Δαυίδ comes also at the center between Abraham and Christ in the summary verse for Israel's past history in v. 17. The numerical value of the name Δαυίδ ($d + v + d = 14$) coincides with Matthew shaping his tripartite view of Israel's history divided into fourteen generations for each stage, from Abraham to David, David to the

---

[106] Morna D Hooker, "The Beginning of the Gospel," in *The Future of Christology* (ed. Abraham J. Malherbe and Wayne A. Meeks; Minneapolis: Fortress, 1993), 21, 28; cf. J. M. Gibbs, "Mark 1, 1-15, Matthew 1, 1-4, 16, Luke 1, 1-4, 30, John 1, 1-51: The Gospel Prologues and their Function" in *SE* 6 (ed. Elizabeth A. Livingstone; Berlin: Akademie,1973), 154-188.

exile, and from the exile to Christ. Matthew's implied perspective here may be theological, and probably 'deuteronomic' in its character: sin-exile-restoration.[107]

It is likely that the title, Χριστός gives off an aura of the hope for the end of exile, which is discernible in Matthew's patterning of Israel's past. Looked at this way, Matthew sees in the Son of David the Anointed, the one who inaugurates the restoration.[108] Matthew also likely demonstrates that Jesus is a legitimate offspring of David by virtue of being Joseph's son (1:18-20), and at the same time is the one who fulfills the Davidic hope by bringing the promised restoration.[109] By presenting Jesus as the Son of David (1:1) and *Immanuel* in the context of 1:18-25, Matthew also prepares the story of God's presence in and through the coming of the Son of David, which is a constant throughout the main part of his narrative.[110] The theme of the return of God's presence among his people, as we shall examine in chapter 5, should not be confined within the texts where the linguistic expressions appear (18:19-20; 28:19-20).

In fact, the theme of divine presence runs through the whole story of the Gospel as to Jesus' identity and mission. Particularly, the theme of Jesus' fulfilling the Davidic hope and ushering in the restoration courses through the series of Matthew's citations from the OT. In Matt 2:6, Jesus is the Davidic Shepherd who will rule over the new Israel in the future (Mic 5:3); in 2:15, Jesus relives Israel's past in their bondage in Egypt, but with the hope of coming out of Egypt as embodying Israel as the son of God (Hos 11:1).[111] The brutal massacre wrought by Herod in Bethlehem is interpreted by Matthew as the echo of Rachel weeping for her children on the way to the exile, yet even here as well there is an overwhelming hope of return (Jer 31:15 in Matt 2:18); Rachel may now stop weeping, for her deliverer has at last arrived.[112] Jesus embodies Israel's past history up to the exile, and moves beyond it.

Matthew's emphasis on Jesus as the Son of David and the fulfillment of Davidic expectation is not limited to the infancy narrative. The supposed 'redactional passages' involving the title 'Son of David' are but a few visible peaks that emerge from this picture of Jesus in the Gospel (1:1; 9:27; 12:23;

---

[107] Verseput, "Davidic," 104; similarly, Davies and Allison, *Matthew* 1:185, 187-188; Carson, *Matthew* 1: 69; Keener, *Matthew*, 77-78.

[108] Cf. Blaine Charette, *Restoring Presence: The Spirit in Matthew's Gospel* (Sheffield: Sheffield University Press, 2000), 22-23, notes the significance of the term 'the Anointed' in this respect.

[109] W. Barnes Tatum, "'The Origin of Jesus Messiah' (Matt 1:1, 18a): Matthew's Use of the Infancy Traditions," *JBL* 96 (1977): 534.

[110] Luz, "Eine thetische Skizze," 223-224.

[111] Similarly, Keener, *Matthew*, 107-109; cf. Gundry, *Matthew*, 34, underscores Jesus' divine sonship; Dunn, *Christology in the Making*, 49.

[112] Verseput, "Davidic Messiah," 109; see also, D. C. Allison, "The Son of God as Israel: A Note on Matthean Christology," *IBS* 9 (1987): 74; Davies and Allison, *Matthew*, 1:269; Keener, *Matthew*, 111.

15:22; 21:9, 15).¹¹³ To these texts, Matt 2:6 should be added as a clear allusion to the royal Davidic expectation; also Matt 27:37 ("This is Jesus the King of the Jews," cf. 2:2) should be seen as a continuation of the royal Davidic theme as we approach the end of the narrative.

Numerous places in the main body of the First Gospel manifest Matthew's keen interest in Jesus as the Son of David (9:27; 12:23; 15:22; 20:30, 31; 21:9,15; 22:42, 45) – a striking comparison as we notice the dearth of occurrences of the title in Mark and Luke (Mark 10:47, 48// Luke 18:28-39; Mark 12:35, 37//Luke 20:41, 44).¹¹⁴ Why does Matthew use the title Son of David more frequently than the other Synoptic Gospels? What are the functions and roles of Matthew's use of the title in his narrative?

(2) Controversies: To answer these questions, we turn our attention to the controversies surrounding the title. Traditionally, 'Son of David' in Matthew's Gospel is understood as the title for the earthly Jesus *set in contrast to* titles such as Κύριος and the son of Man for the exalted Jesus.¹¹⁵ To be sure, it is this contrast which is popular for many as they consider the function and purpose of the title in the Gospel. According to Jack D. Kingsbury, the title is limited in scope to Jesus' ministry to Israel in that it refers to Jesus, not as a political figure, but as one whose activity is strictly limited to healing.¹¹⁶ The antithetical scheme by which the Son of David is set against the universal resurrected Jesus, is apparent also in Kingsbury's argument that "the crowds and leaders of Israel neither see nor confess the truth of Jesus as the Son of David (i.e., a political and militant messiah), and the title 'Son of David' in Matthew's Gospel is apologetic, that is, it is used to underline Israel's offense and guilt for having repudiated its Messiah." ¹¹⁷

By teaming the crowds and the leaders together in their obduracy against Jesus as the Son of David, Kingsbury departs from the traditional position set by J. M. Gibbs who notably separates the crowds and their leaders in their reception and rejection of the Son of David.¹¹⁸ According to Kingsbury, Matthew highlights the 'healing' Son of David in order to demonstrate that Jesus

---

¹¹³ Stanton and Kingsbury limit the range of the presence of Matthew's emphasis on this title mainly within these passages. See Stanton's "Matthew's Christology and the Parting of the Ways," in *Jews and Christians: The Parting of the Ways A.D. 70 to 135* (ed. J. D. G. Dunn; Tübingen: Mohr Siebeck, 1992), 108; Kingsbury, "The Title 'Son of David' in Matthew's Gospel," *JBL* 95 (1976): 591.

¹¹⁴ Matt 22:45 is debatable; the Gospel of John makes no use of the title except in 7:42; no other NT writings express interest in this title to the extent Matthew does.

¹¹⁵ Kingsbury, "Son of David," 591, lists a number of German scholars who developed this contrast: beginning with Günther Bornkamm, Georg Strecker, Reinhart Hummel, Rolf Walker and Christoph Burger.

¹¹⁶ Kingsbury, "Son of David," 592, 601.

¹¹⁷ Kingsbury, "Son of David," 602.

¹¹⁸ J. M. Gibbs, "Purpose and Pattern in Matthew's use of the title 'Son of David'," *NTS* 10 (1963-1964): 463-464.

was not acknowledged by the Jews as the awaited Son of David as they understood it, and Kingsbury sees the purpose of Matthew' emphasis on the Son of David primarily in the negative, which is thus employed to accuse the Jews. In short, Jesus was not the Son of David the Jews understood and expected.

In a similar vein, others argue that Matthew uses the Son of David negatively in association with the blindness motif,[119] or that Jesus firmly rejected the title as constituting an inadequate description of his claims.[120] Affirming this line of argument, B. Gerhardsson states explicitly: "The Son of David is by no means the foremost title in the first Gospel for the earthly Jesus; and it gives only a very roughly correct picture of Jesus – it only says a superficial part of the truth about him."[121] For Gerhardsson, there are two reasons for the superficiality of the title. First, the Son of David in Matthew's Gospel is characteristically "a healer who humbles himself and serves his people," and accordingly this humble messiah does not fit, he argues, the political profile of traditional Jewish Davidic messianic expectation. Second, he contests that "the adversaries, the leading men of Israel, do not enter into dispute with him about this title."[122]

Yet, if Mic 5:2 in Matt 2:6 is understood as part of the Davidic Shepherd tradition as we have argued in the previous chapter, then Gerhardsson's second point withers, at least in Matthew 2, since Matthew's announcement of the coming Davidic Shepherd is set in a context of hostility, that is, the scene of Herod and the Jewish leaders in Jerusalem found in that chapter (cf. also, "Hosanna to the Son of David" in Matt 21:9, 15). As D. J. Verseput aptly points out, Matthew continues to underscore Davidic messiahship in key places throughout the story, "accenting it at crucial junctures and granting it the leading role in the conflict between Jesus and his adversaries," consistently fanning the flames of opposition from the Jews.[123]

Gerhardsson's first point that the healing Davidic son does not fit the political profile of traditional Jewish Davidic messianic expectation explains why many struggle to see Jesus as the Son of David. The healing Jesus and Jewish Davidic expectations are deemed incompatible. This presumption can easily be supported by the traditional contrast of the earthly, political Jewish Davidic messiah with the resurrected, universal, and spiritual Jesus as Lord.

Graham N. Stanton's view of Jesus as the Son of David is trapped in this traditional assumption. While maintaining that Jesus as the Son of God is rarely opposed by the Jewish leaders in Matthew's Gospel, Stanton acknowledges

---

[119] W. R. G. Loader, "Son of David, Blindness, Possession, and Duality in Matthew," *CBQ* 44 (1982): 570-585.

[120] Terence Y. Mullins, "Jesus, the 'Son of David'." *AUSS* 29 (1991): 125.

[121] Birger Gerhardsson, *The Mighty Acts of Jesus According to Matthew* (CWK: Gleerup,1979), 88.

[122] Gerhardsson, *Mighty Acts*, 87, 88.

[123] Verseput, "Davidic," 102, 112-114, also in "The Role and Meaning of the 'Son of God' Title in Matthew's Gospel," *NTS* 33 (1987): 536.

4.2 The Davidic Shepherd as the "Therapeutic Son of David"

nonetheless that no other major christological theme besides 'the Son of David' provokes as much sustained opposition from the Jewish leaders – despite Kingsbury's overemphasis upon the title 'the Son of God.'[124] In fact, Jesus' healing ministry as the Son of David is frequently linked to the motif of Jesus' conflict with the Jewish leaders (9:27/34; 12:23/24-32; 21:9,15/21:23-27).[125]

Why then do the Jewish leaders vigorously oppose Jesus as the Son of David? Stanton answers that Jesus as the Son of David in Matthew's Gospel should be understood as the early form of a "two parousias scheme" – first, the coming of the "harmless and humble" Son of David; and second, the coming of the "glorious son of Man as the judge."[126]

This scheme, Stanton argues, originates from the early church, evident in Justin Martyr's reply to his contemporary Jewish opponents.[127] Just as Justin Martyr encountered Jewish opponents, Stanton proposes, Matthew's Jewish contemporaries advocated that Jesus was not the Davidic messiah. Stanton points to Jewish criticism which Matthew encountered: (i) the disciples stole the body of Jesus from the tomb (28:1-15); (ii) Jesus was a magician and a deceiver (9:34; 10:25; 27:63-4); and (iii) Jesus was not the Davidic messiah (2:1-6; 9:34; 12:23; 21:9, 15). According to Stanton, Matthew replies by presenting Jesus as the Son of David, but characterized as harmless and humble, thus running contrary to "contemporary expectations of a triumphant" Davidic messiah – "the child Jesus" (2:2-6); "Jesus taking disease and infirmity upon himself" (8:17); "meek and lowly in heart" (11:29); "the servant of God" (12:17-21); and "the humble king" (21:5).[128] This might also explain, Stanton argues, why the Jews became indignant and reacted with hostility to the claim that Jesus is the Son of David. Thus the triumphant aspect of the Davidic messiah is, to say, postponed in the coming of the son of Man.[129] Eventually, Stanton comes to the

---

[124] Stanton, "Ways," 99-100.

[125] Cf. Hans Kvalbein, "Has Matthew Abandoned the Jews?: A Contribution to a Disputed Issue in Recent Scholarship," in *Mission of the Early Church to Jews and Gentiles* (ed. Jostein Ådna and Hans Kvalbein; Tübingen: Mohr Siebeck, 2000), 58-62, discusses the issue: "Jesus healing miracles-do they increase the guilt of Israel?"

[126] Stanton, "The Two Parousias of Christ: Justin Martyr and Matthew," in *From Jesus to John: Essays on Jesus and New Testament Christology* (ed. Martinus C. De Boer; Sheffield: JSOT Press, 1993), 183-195.

[127] O. Skarsaune, *The Proof from Prophecy* (Leiden: Brill,1987), 285, observes that in the NT, only Christ's glorious return is called a παρουσία (Matt 24:3, 27, 37, 39; 1 Cor 15:23; 1 Thess 2:19; 3:13; 4:15; 5:23; 2 Thess 2:1, 8; Jas 5:7-8; 2 Pet 3:4; 1 John 2:28) – never the first coming. Ignatius is the first to call Christ's coming in the flesh a παρουσία (*Philad*, 9:2).

[128] Stanton, "Ways," 111-112 and "Two Parousias," 192.

[129] Stanton, "Ways," 108, 114, mentions other NT types, which express similar christological tension comparable to Matthew's 'two parousias scheme': (i) 'the *incarnational* pattern' – "the one who was with God humbled himself among men and was exalted by God" (Phil 2:6-11; 2 Cor 8:9); (ii) 'the pattern of *reversal*' – "in raising Jesus God reversed the actions of those who put Jesus to death" (Acts 2:23-4; 3:13-14).

conclusion that this christological contrast evinces the social setting of Matthew's pockets of Christian communities that began to depart from the larger and hostile Jewish community.[130]

We agree with Stanton's observation of the correlation of the Son of David and the hostility of the Jewish leaders; however, the issue here is *not the humility but, rather, the authority of the Son of David*. In Matthew's Gospel, Jewish hostility escalates as the Davidic messiah's incomparable authority is exhibited particularly through his healing of the sick.[131] We have argued earlier that this ἐξουσία is that of YHWH himself as Jesus assumes the tasks of the eschatological Shepherd, seeking the lost and healing the sick to inaugurate the time of restoration for exiled Israel (cf. Matt 9:36; 10:1-6).[132] All of the four passages that associate the title 'Son of David' with Jewish hostility (1:1/2:3; 9:27/34; 12:23/24-32; 21:9, 15/23-27), point to the astounding authority of the 'therapeutic Son of David' rather than to his humility.

Further, Luz observes that the Evangelist does not always depict the Son of Man as the triumphant Judge, but frequently links the title with the suffering motif, characteristically his crucifixion (cf. 26:2).[133] The Final Judge in 25:31-46 may be an exception indeed, but even this imposing figure is also replete with shepherd/sheep images, especially compassion for the flock. The Son of Man as the Final Judge in Matt 25:31-46 is none other than the Davidic Shepherd who in the Gospel demands of his flock the very compassion he demonstrates for the lost sheep of the house of Israel. If we adopt our model for Jesus as the eschatological and Davidic Shepherd(s), then the contrast between the 'humble' Son of David and the 'triumphant' Son of Man (Stanton) can be harmonized in the figure of Jesus as both the compassionate eschatological Shepherd and Davidic Shepherd-Son of Man-Judge to come.

Second, Stanton does not present any other grounds for connecting 'healing' with the 'Son of David' except for a single reference to Isa 53:4 in Matt 8:17.[134] Those 'redactional passages,' however, fail to support his unwarranted assumption that the Jewish leaders became hostile because of the meekness of the Son of David. The Son of David in Matthew's Gospel indicate, as we shall examine, that the Jewish leaders are upset – and with unparalleled amazement from the crowds – by the astonishing authority Jesus manifested in his mission

---

[130] Cf. Stanton, *New People of God*, argues for the thesis that Matthew's Christian communities have recently parted company with Judaism.

[131] Joseph A. Comber, "Critical Notes: The Verb THERAPEUŌ in Matthew's Gospel," *JBL* 97 (1978): 432-434, aptly points out this link: "healings are the cause of the Jewish leaders' unbelief and rejection of Jesus" (433).

[132] Refer to Ch. 3.2.2. "Matt 10:6, 16; 15:24: The Lost Sheep of Israel."

[133] U. Luz, "The Son of Man in Matthew: Heavenly Judge or Human Christ?" *JSNT* 48 (1992): 3-21.

[134] Stanton, "Ways," 108-114.

of compassion by healing the sick and casting out demons from his distressed flock (cf. Matt 9:34; 12:22-24; 15:31).

Viewing the Gospel in light of early church experiences, Stanton's analysis may very well be a pinnacle of the redactional approach to the issue of the healing Son of David that is distinctive of Matthew's Gospel. Yet, to read Matthew's portrait of the Son of David from the vantage point of Justin's 'two parousias schema' may be anachronistic. In the final analysis, the traditional antithesis between the Son of David and the risen Lord needs to be reexamined. Further, Matthew's portrait of the Davidic messiah may well be read in light of the messianic hopes that were contemporary to Jesus' own time.[135]

The Son of David controversy would make much more sense, we propose, if the story is read from the vantage point of Israel's preceding history, just as the opening of the Gospel suggests, that is, "from Abraham, David, and the exile (Matt 1:1-18)." More important is that Jesus can be seen both as the eschatological Shepherd who heals and as the Davidic Shepherd/King/Judge-Appointee in the Gospel. The *two Shepherds schema* may explain the controversy involved in the title Son of David in the Gospel. With this hypothesis in mind, we will now survey the options that have been put forward that explain Jesus as the healing Son of David against the background closer to Jesus' contemporary historical and literary settings.

4.2.2 Various Proposals

It is a well recognized puzzle that Matthew's Gospel in particular links the title 'Son of David' with Jesus' healing ministry.[136] The Gospel of Luke does not show much interest in the 'title' Son of David, and Mark focuses on exorcism as the main activity of Jesus as the Son of David.[137] Dennis C. Duling notes, "Mark's more 'exorcistic' and wonder-working portrayal of Jesus has become in Matthew more 'therapeutic'; Matthew's Messiah of Deed is primarily [but not exclusively] a 'therapeutic' Messiah."[138] In Matthew, at least four (9:27; 12: 23; 15: 22; 21:15) of the six 'redactional passages' are associated with Jesus' healing activity. In addition, the Son of David saying in Matt 21:9 closely interrelates with the preceding healing account in 20:30-31, when the blind men twice recognize Jesus as the Son of David. Matthew's Gospel records that this occurred before Jesus entered Jerusalem, and this connection between healing and the Son of David is clearly affirmed once again in 21:14-15. Duling rightly

---

[135] Verseput, "Davidic," 102-116, notes that even the gentle Messiah "was not one which would have lacked for familiar motifs attractive to the Jewish ear" (116).

[136] For example, C. Burger, *Jesus als Davidssohn* (Göttingen: Vandenhoeck & Ruprech, 1970), 72-106; Gibbs, "Son of David," 446-464; Duling, "Therapeutic," 392-409; Kingsbury, "Son of David," 591-602; Luz, "Eine thetische Skizze," 223-226.

[137] Duling, "Therapeutic," 398-399.

[138] Duling, "Therapeutic," 398.

characterizes the Son of David in Matthew's Gospel "essentially as a 'therapeutic Son of David' (Matt 9:27-31; 12:22-24; 15:21-28; 20:29-34; 21:1-16; 22:41-46).[139]

This raises numerous questions. What is the source of this distinctively Matthean picture of Jesus? Why is the Son of David depicted with such characteristics? Stanton, in his monumental research on Matthean scholarship from 1945 to 1980, not quite convinced by Duling's proposal of Matthew's Son of David as Solomon the exorcist (1978), admits that he is "still puzzled by the evangelist [Matthew]'s association of the 'Son of David' with healing." [140] In fact, it is Duling who argued that the royal messiah in Jewish tradition does not fit the Son of David portrayal in the Gospels that associates "the title primarily with a figure who is so addressed by people in need of exorcism or healing." [141] An equally baffled Collins concurs: "How Jesus came to be viewed as the Davidic messiah remains a mystery." [142] The suggestions made for this puzzling association can be put into three categories: (i) Matthew's quotations from Isaiah's Servant passages in association with 'the Christian tradition' of Mark 10:46-50; (ii) the Solomon-as-exorcist tradition; and (iii) the Hellenistic "divine man."

4.2.2.1 The Servant: Early Christian Redaction

Duling argues that the sources of Matthew's therapeutic Son of David are primarily 'the early Christian tradition' in Mark10:46-52 and Matthew's expansion based on the picture of the Servant in Isaiah (Isa 53:4 in Matt 8:17; Isa 35: 5-6; 61:1 in Matt 11:5; cf. Isa 42:1-4 in Matt 12:17-21).[143] Why then does Matthew highlight the therapeutic Son of David? According to Duling, Matthew responds to the contemporary rabbinic belief in the 'futuristic-eschatological-Son of David' who is 'the (non-healing) Coming One' in Zech 9:9 (cf. *b. Sanh* 97b-99a), and at the same time responds to Mark's exorcistic Son of David, articulating the Servant theme from Isaiah 40-66.[144]

Matthew indeed cites Isaiah to explain Jesus' healing ministry (Matt 8:17/Isa 53:4; Matt 11:5/ Isa 35:4-6; Matt 12:17-21/Isa 42:1-4). As Lidija Novakovic argues, Matthew understands these texts from Isaiah as messianic and uses them for the purpose of demonstrating a direct link between Jesus' healings and his

---

[139] Duling, "Therapeutic," 392-410.
[140] Stanton, "Origin and Purpose," 1924.
[141] Duling, "Solomon, Exorcism, and the Son of David," *HTR* 68 (1975): 235.
[142] Collins, *Apocalyptic*, 263.
[143] Duling, "Therapeutic," 396; also, see A. J. Saldarini, *Matthew's Christian-Jewish Community*,180.
[144] Duling, "Therapeutic," 409-410.

4.2 The Davidic Shepherd as the "Therapeutic Son of David" 287

messiahship.¹⁴⁵ It is imperative, however, to emphasize that none of these citations involve the title 'Son of David,' and all of them are distanced from the Son of David passages in Matthew's narrative.

This is best illustrated by Matt 12:18-23. The Evangelist indeed links Jesus' healing ministry with the Isaianic Servant (v. 18). Yet as the narrative takes the readers back to the historical scene where Jesus heals the demon possessed man who is blind and mute (v. 22), the 'historical' crowd is puzzled as they ask, μήτι οὗτός ἐστιν ὁ υἱὸς Δαυίδ (v. 23). Isaiah may well have been a popular text in the early churches in terms of *Testimonia* known as the collection of the early church's proof texts from the Hebrew Bible.¹⁴⁶ As Cope notes, Matthew uses the text in such a way to make clear *to the reader* its application to Jesus.¹⁴⁷ But the citation from Isaiah is not known to the historical crowd in the narrative; the crowd in the historical scene remains puzzled. Considering that the often asserted antithesis between the Jewish royal Son of David and the early churches' Isaianic suffering Servant,¹⁴⁸ the rationale for the crowd's association of the healing Jesus with the Son of David should be found elsewhere than in the figure of the Isaianic Servant.

Last, as we examined earlier the correlation between the Son of David's healing and hostility in the Gospel, Matthew's emphasis on the Son of David's healing ministry does not defuse the political impact of this title; rather, the opposite would be the case.¹⁴⁹ Duling's contribution is his characterization of Matthew's depiction of Jesus as the therapeutic Son of David, but his proposal remains insufficient to clarify the matter as to the background of the very association of the two: healing and the Son of David.

As another candidate for the background of Matthew's therapeutic Son of David, besides the Markan exorcistic Son of David and the Isaianic Servant, it

---

¹⁴⁵ Lidija Novakovic, *Messiah, the Healer of the Sick: A Study of Jesus as the Son of David in the Gospel of Matthew* (WUNT 2.170; Tübingen: Mohr Siebeck, 2003), 79-80.

¹⁴⁶ Martin C. Albl, *"And Scripture Cannot Be Broken": The Form and Function of Early Christian Testimonia Collections* (Leiden: Brill, 1999), 185-189, defending Dodd's thesis (esp. 288-290), discusses Isa 42:1-4 in Matt 12:18-21 as part of *Testimonia*; cf. Dodd, *According to the Scriptures*, 126.

¹⁴⁷ O. Lamar Cope, *Matthew: A Scribe for the Kingdom of Heaven* (CBQMS 5; Washington D.C.: The Catholic Biblical Association of America, 1976), 32-52, argues likewise that Matthew in 12:18-21 "has used a citation text which he has freely shaped to suit his own purpose in applying the servant text to Jesus" (49). Similarly as to Matthew's editorial intention in this respect, see J. H. Neyrey, "The Thematic Use of Isaiah 42, 1-4 in Matthew 12," *Bib* 63 (1982): 457-473, who suggests that Matthew's citation of Isa 42:1-4, "points less to an ideal christological portrait of Jesus as the meek servant (Cope) and more toward the situation of Matthew's church in conflict with the synagogue" (459).

¹⁴⁸ Adrian M. Leske, "The Influence of Isaiah 40-66 on Christology in Matthew and Luke: A Comparison" *SBLSP* 33 (1994): 915-916.

¹⁴⁹ Cf. S. H. Smith, "The Function of the Son of David Tradition in Mark's Gospel," *NTS* 42 (1996): 538, argues for Mark that 'therapeutic' means 'ethical' not 'political.'

has been argued that the Son of David was a popular religious concept among first-century Palestine Jews, possibly associated with the Solomon-as-exorcist tradition. To this hypothesis we turn next.

4.2.2.2 Solomon as 'Son of David': A Jewish Legend

Many still prefer the Solomon-as-exorcist tradition as the way to understand the background of Jesus as the healing Son of David. Asserting that "we have no evidence that the Messiah was perceived as a healer in Palestine before 70 CE," James H. Charlesworth contests that "Bartimaeus (Mark 10:46-52) was thinking of Jesus as a healer after the order of Solomon."[150] Meier, while denouncing the option for a secondary Christian redaction, contests that the connection of the miracles of healing with the title 'Son of David' can be explained by the early picture of Jesus as 'a latter-day miracle-working Solomon' among the common people of the day.[151] Likewise, while affirming the significance of Matthew's connection of Jesus the Son of David with healing and exorcism, W. D. Davies contests that Matthew traces the descent of Jesus through Solomon, a Son of David, who later became famous as a mighty healer, exorcist, and magician (cf. Josephus, *Ant* 8:45-49).[152]

Further, S. H. Smith argues further that a Solomon-as-exorcist tradition had already been absorbed into Christian theology long before the time of Mark's Gospel, pointing to Mark 10:47-48 as representing this tradition in the early churches.[153] Anthony J. Saldarini, on the other hand, proposes a different kind of synthesis, namely, that this powerful figure comes from "the combination of Matthew's allusion to Isaiah and the Solomonic tradition"; he argues, "in some traditions, Solomon, who was the Son of David, is associated with powers of healing, exorcism, magic, and miracle; thus, *by allusion to Scripture (Isaiah) and popular tradition, Matthew gradually builds* a picture of Jesus as an authentic healer in Israel's tradition with access to the amplitude of divine power."[154]

---

[150] J. H. Charlesworth, "The Son of David: Solomon and Jesus (Mark 10.47)" in *The New Testament and Hellenistic Judaism* (ed. Peder Borgen and Giversen Søren; Peabody, Mass.: Hendrickson, 1995), 85-86; see also Davies and Allison, *Matthew 2*, 135; Loren L. Trotter, "Can This Be the Son of David?" in *Jesus and the Historian* (ed. F. T. Trotter; Philadelphia: Westminster, 1968), 82-97.

[151] John P. Meier, *A Marginal Jew: Rethinking the Historical Jesus* (vol.1: The Roots of the Problem and the Person; vol. 2: Mentor, Message, and Miracles; New York: Doubleday, 1991, 1994), 2:689.

[152] W. D. Davies, "The Jewish Source of Matthew's Messianism," in *The Messiah* (ed. James H. Charlesworth; Minneapolis: Fortress, 1992), 500; also see, Bruce Chilton, "Jesus *ben David*: Reflection on the *Davidssohnfrage*," *JSNT* 14 (1982): 88-112.

[153] Smith, "Son of David," 539.

[154] Saldarini, *Christian-Jewish*,180, 182 (italics mine).

4.2 The Davidic Shepherd as the "Therapeutic Son of David" 289

Yet, the presumed direct connection between the Isaianic Servant and Solomon the exorcist is unlikely. As an attempt to build a bridge between the "Christian therapeutic Son of David" and Judaism, Duling makes a hypothetical trajectory of the Solomon-as-exorcist tradition from 1 Kgs 5:9-14, Wis. 7:17-22, 11QPsAp^a, Josephus' *Ant* 8.2,5, and *TestSol* 20:1. A close look at these sources reveals, however, that only *TestSol* 20:1 explicitly involves the designation "Son of David" in connection with exorcism or perhaps healing: "Now it happened that one of the artisans, a dignified man, threw himself down before me saying, 'King Solomon, Son of David, have mercy on me, an elderly man.' I said to him, 'Tell me, old man, what you want.'" [155] We cannot help notice how the designation 'Son of David' is combined with the mercy motif. Yet besides plausible Christian influences (*TestSol*, dated A.D. 200-400), the designation 'Son of David' in this text may simply be vocative; furthermore, in the entire book of *TestSol*, this is the only occurrence when the vocative 'Son of David' is associated with Solomon (cf. *TestSol* 1:6-7).

The Testament's depiction of Solomon, however, is a far cry from a messianic figure. In Matthew 12:23 and 21:15 Jesus being called 'the Son of David,' with messianic connotations, evidenced in the context of healing, repudiates the Solomon connection. The *Testament of Solomon* depicts Solomon as an imbecile deceived and exposed by his own petitioners, who eventually became the most unwise idol-worshiper in the history of Israel (20:1-26:8).[156] Josephus' description of Solomon as an exorcist also concords with *TestSol*. Josephus highlights the wisdom of Solomon, especially his skills and the incantations to expel demons. The name Solomon is mentioned by those learners of his exorcism-techniques, e.g., by Eleazar (*Ant*. 8.2.5), but certainly with no messianic connotation.

The strongest objection may be that the name Solomon might be associated with wisdom but not with healing in Matthew's Gospel (12:42; cf. 1:6; 6:29), and the same holds true in Mark's Gospel which underscores the link between exorcism and Jesus as the Son of David.[157] In fact, Solomon is known as an exorcist and not as a healer *per se*; however, Matthew distinctively connects

---

[155] *OTP*, 1:982 (trans., D. C. Duling).

[156] *OPT*, 1:982-986: The end of the Testament says that Solomon sacrifices to Jebusite gods to obtain a beautiful Shummanite woman, and an old man tries to deceive Solomon in order to accuse his own son, but actually traps Solomon to put to death his own son. Here, the old man mocks Solomon by calling him 'Son of David.' However, as Solomon prays to God, begging for the power of exorcism, he asks that he might have 'authority' over the demons (Prologue;1:5); cf. the Qumran evidence for a connection between Solomon and the demons (11QPsAp^a), which is too scant to be taken even as conjecture.

[157] We may safely discount the injunction to be like the birds and flowers in 6:29.

Jesus as the Son of David with healing.[158] Meier bypasses this distinctive link between Jesus' *healing* and the Son of David as he argues for the Solomon-exorcist tradition in the case in Mark 10:46-52, though he mentions *Ant.* 8.2.5 as the [only] instance of Solomon's healing activity.[159]

Yet, such a thin reference bereft of any messianic note (as we examined in the case of Eleazar above) would not justify the Solomon-tradition as the likely candidate for the picture of the therapeutic Son of David in the Gospels. We safely agree with Luz's judgment: "The hypothesis [the Solomon-exorcist tradition] is difficult, however, since according to Jewish tradition Solomon does not heal." [160]

Further, H. B. Green, even though he suggests a Solomon-typology for Jesus in Matthew, ignores Jesus' healing activity as a significant reference to Solomon, and emphasizes instead Solomon's wisdom, temple, and the visitation of the nations (Matt 12:7, 42).[161] Unlike the case for Solomon, when Jesus is called [the] 'Son of David' in Matthew, it likely has messianic connotations. Further, Jesus heals the sick with no need for magical manipulations such as incantations. Matthew portrays Jesus as the healer without any Solomonic-style magical maneuvering (cf. Matt 15:29-31 [Mark 7:31-37]; Matt 8:22-26, omitted).[162] Rather Jesus frequently heals the sick merely by his 'word' (8:7; 9:29; 15:28; 17:18; cf. 12:13).[163]

Thus, the option for the Solomon-exorcist tradition falters. Perhaps the Jewish legend of Solomon-the-exorcist might echo in some of the healing incidents, though more by a stretch of the imagination, as in the example of Matt 15:22. Yet, Jesus healing in the temple and the children's praise of him as the Son of David in 21:14-15 assures us once again that the association is much more infused with messianic connotations.

As to the christological implications of the Solomon-exorcist option, Meier proposes that the Solomon-exorcist legend represents the 'extremely primitive

---

[158] Duling, "Therapeutic," 399-407, demonstrates this thesis through a redactional study; as does Kim Paffenroth, "Jesus as Anointed and Healing Son of David in the Gospel of Matthew," *Bib* 80 (1999): 550.

[159] Meier, *A Marginal Jew*, 686-690.

[160] Luz, *Matthew*, 2:47.

[161] H. B. Green, "Solomon the Son of David in Matthaean Typology" in *Studia Evangelica VII* (ed. Elizabeth A. Livingston; Berlin: Akademie-Verlag, 1982), 227-230.

[162] Duling,"Matthew's Plurisignificant 'Son of David' in Social Science Perspective: Kinship, Kingship, Magic and Miracle," *BTB* 22 (1992):109, 112; also, "Therapeutic," 397-398.

[163] With his touch in 9:29 as is also the case in 8:16 (where Jesus heals Peter's mother-in-law). In this respect, I cannot help but recalling the possible relevance of the staff motif of the shepherd image in the OT, e.g., 'the rod (שֵׁבֶט) of his mouth' in Isa 11:4, which the LXX renders as 'the word (λόγος) of his mouth' (cf. the principles of YHWH, the Shepherd's reign in terms of 'goodness'(טוֹב) and 'faithfulness'(חֶסֶד) in Ps 23; or the two staffs as 'favor' and 'union' in Zech 11:7).

Jewish christology' by which the association of Jesus' healing with the Son of David can be explained. He argues that this 'Jewish christology' is dismissed later by the Evangelists who understood the Davidic messiahship in "a completely different thrust" following Jesus' death and resurrection.[164] Thus according to Meier's assumptions, Jesus' healing activities have nothing to do with Davidic messiahship.

Not only is the validity of the Solomon-exorcist legend for a 'healing' Son of Daivd (as we have argued so far) faltering, but also a more fundamental problem is the understanding of Davidic messiahship, particularly in such "a completely different thrust" in our reading of the Gospels. It appears that to connect Jesus, healing, and the Son of David, we are forced to resort to the invalid legend of Solomon-exorcist.

This break between 'Jewish christology' and the Christian understanding of 'Davidic messiahship' calls for a remedy. This task requires several steps. First we need to demonstrate how healing is associated with Jesus as the Davidic Shepherd; and second, we need to show also how the Davidic Shepherd tradition in the Hebrew Bible and Judaism continue, for indeed they involve the 'intertextaul discourse' that the First Gospel presents. These will be our tasks in subsequent sections and the following chapter.

At this juncture, however, we will examine one last candidate which serves as the background of Jesus as the healing Son of David. We begin with Christopher Burger, who suggests that the figure of the healing Son of David in the Gospels is the result of correlating the royal Son of David with the Hellenistic 'divine man' in early Christianity.[165]

4.2.2.3 The Divine Man: A Greek Mediator

The θεῖος ἀνήρ ('holy man'; 'divine man') is a Hellenistic type of divine-human hero whose divinity expresses itself through the working of miracles.[166] Barry Blackburn, while noting that this concept was once introduced by Bultmann as an interpretive background for the miracle traditions used by Mark, lists a number of the miracle-working divine men in the pre-Christian era, e.g., Asclepius, Machaon, and Podalirius, the Sons of Machaon, Menecrates of Syracuse, and even Moses.[167] It has been noted, however, that the figure is not

---

[164] Meier, *A Marginal Jew*, 2:689-690.

[165] C. Burger, *Jesus als Davidssohn: Eine traditionsgeschichtliche Untersuchung* (Göttingen: Vandenhoeck & Ruprecht, 1970),169.

[166] Ralph J. Coffman, "The Historical Jesus the Healer: Cultural Interpretations of the Healing Cults of the Graco-Roman World as the Basis for Jesus Movements," *SBLSP* 32 (1993): 412-441, finds other Greco-Roman parallels, particularly Asklepios, Hygieia and Theos Hypsistos, who were known as savior-healers.

[167] B. Blackburn, *Theios Aner and the Markan Miracle Traditions* (Tübingen: Mohr Siebeck, 1991), 1-3, 24-72, notes that Moses was seen as a miracle worker, whom Artapanus elevated to semi-divine status.

a fixed expression, nor does it perform an institutionalized function in the pre-Christian era. The Hellenistic concept of the divine man is varied and can refer to any inspired man or to someone related to a god in some special sense, or it may simply refer to an extraordinary man.

Further, the early date given for the emergence of this concept is highly suspect. Saldarini indicates that the Greco-Roman literary figure of a wandering charismatic with miraculous powers was a product of the second and third centuries, not the first.[168]

Regardless of its instability, the Hellenistic origin of the therapeutic Son of David was pursued because of the unlikelihood of the Solomon-as-exorcist tradition. Burger hints at the direction the research takes: "the Davidic Messiah (in the Gospels) takes the function of the Hellenistic θεῖος ἀνήρ upon himself, which does not, it must be emphasized, fit him according to Jewish expectation."[169] The concept of the divine man is itself rather foreign even to the Hellenistic Jewish writers who did not stress miracle-working as a criterion for authenticity or divine approval, nor for the purpose of propagating their faith to Gentiles.[170] Although Josephus and Philo emphasize Moses' deeds and virtues, for instance, they did not elevate Moses as divine, probably because their monotheistic belief suppresses the tendency.[171] Evidence is simply too shaky and sketchy to accept this foreign and likely late dated concept could have afforded any facets of Jesus as the therapeutic Son of David in the First Gospel.

To summarize, Isaiah's Servant may indicate the messianic significance of Jesus' healing ministry for the readers of the Gospel, but not specifically the healing 'Son of David' for the crowd in the historical scene of Jesus healing in their midst. To some extent, the Solomon-as-exorcist tradition might echo some of the healing stories in the Gospel, but the tradition lacks messianic connotation both in *TestSol* and Matthew. Clearly it is more on the basis of his wisdom than on his healing that Jesus relates to the figure of Solomon. Last, the Hellenistic option of θεῖος ἀνήρ leaves us with very little. From the beginning, it was only introduced as an alternative figure based on the assumption that Jewish expectation of the Son of David does not match with Jesus' healing ministry. *A recourse is to challenge the axiom: As the healing Son of David, is Jesus thereby foreign to Jewish messianic expectations?*

4.2.3 The Therapeutic Davidic Shepherd: Ezekiel's Vision

To trace the source of the therapeutic Son of David in Judaism, we will discover that the Solomon-as-exorcist tradition is not the only option. In various sources

---

[168] Saldarini, *Christian-Jewish*, 181.

[169] Burger, *Davidssohn*, 169.

[170] Carl Holladay, *Theios Aner in Hellenistic Judaism: A Critique of the Use of This Category in New Testament Christology* (Missoula: Scholars, 1977), 238.

[171] Blackburn, *Theios Aner*, 72.

4.2 The Davidic Shepherd as the "Therapeutic Son of David"

from the Second Temple period, it is not rare to find the Davidic messianic figure associated with healing activity, and often with the shepherd image as well. It will be useful to sort those texts before we deal with the healing motif in its connection with the Son of David and shepherd image in Matthew's Gospel.

4.2.3.1 Echoes from Second Temple Judaism

The rich shepherd image associated with Davidic expectation in Ezekiel 34 exerts a powerful influence upon many texts in the shaping of their visions for the future. The instance of the extant *Apocryphon of Ezekiel* fragment 5 illustrates the case: "Therefore he says by Ezekiel; 'And the lame I will *bind up*, and that which is troubled I will *heal*, and that which is led astray I will return, and I will *feed* them on my holy mountain ... and I will be,' he says, 'their shepherd and I will be near to them as the garment to their skin.'"[172] Whether or not Matthew was acquainted with this text, the passage does indeed show that Matthew was not alone in his elaborations of the therapeutic shepherd.[173]

Turning to the Qumran writings, CD-A 13:7-12 presents the leader figure of the eschatological community, i.e., המבקר ('Inspector' [of Many/Congregation]; 'Seeker'; or 'Examiner'), and this המבקר takes up the eschatological pastoral tasks of Ezekiel's eschatological Shepherd.[174] Fragment 13, line 9a describes the compassion of this figure: "he [המבקר] shall have pity on them;" and l. 10 illustrates his pastoral tasks saying, "he will undo all the chains which bind them, so that they will be neither harassed nor oppressed in his congregation."

Characteristic to this המבקר is his mission to "seek/examine" (בקר) just as the eschatological Shepherd will seek the lost and wounded ones among the flock: "I [YHWH] myself will search for [אֲבַקֵּר] my sheep (Ezek 34:11, 12); I will search for the lost and bring back the strays" (v. 16a).[175] Knowing that the tasks of המבקר involve relieving the suffering of the oppressed ones among the flock, the inclusion of 'healing' as one of his main tasks would not be surprising. In this respect, line 9b reads, "and [he] will *heal* [וישקה] *all the afflicted among them like a shepherd his flock.*"[176]

The identity and relationship of המבקר to other messianic figures such as the Prince, the Branch, or the Messiah(s) of Aaron and Israel elsewhere in Qumran writings remain unclear. Yet the role of the Seeker to restore justice within the eschatological community – primarily restoring the oppressed ones – is identical

---

[172] *OPT* 1:488 (trans. J. R. Mueller and S. E. Robinson).
[173] Cf. Cousland, *Crowds*, 121.
[174] For the details, refer to Ch. 2, 2.3.8.1 "CD-A 13:7-12."
[175] Zimmerli, *Ezekiel* 2, 215-216.
[176] Martínez and Tigchelaar, *Scrolls*, 573, emend it to וישקה as in the translation in the text; for the idea of relief implied in וישקה while Baumgarten and Schwarz, *Scrolls* 2, 55, renders שקד as וישקוד ("to watch over"; "to show concern for").

with the task characteristic of the Prince and the Branch of David in 4Q521 2, 2:3-14 and 4Q252, 5:5-7.

Moreover, since this Seeker is to appear "in the appointed time when God visits the earth" (CD-A 13:22), he probably is a messianic figure.[177]

It is important to emphasize here that if the Instructor is indeed a messianic figure, then he is certainly a 'therapeutic' messiah. His roles to heal the sick and restore justice approximate those assigned to the eschatological and Davidic Shepherds in Ezekiel 34. Interestingly, CD-A 13:22 specifies these as regulations for the Instructor (משכיל, 'Master')[178] who will appear at the appointed time when God visits the earth. This Instructor can be, without much difficulty, identified with המבקר himself (cf. Matt 2:6). The echo can easily be captured in the figure of Jesus in Matt 9:36 (cf. 18:12). As the Davidic Shepherd (מוֹשֵׁל, Matt 2:6), Jesus heals the sick (9:35) and presents himself as the eschatological Shepherd, full of compassion for the harassed and oppressed flock (9:36). The Davidic connotations both in CD-A 13:7-12 and Matt 9:36 can be detected only indirectly through the mediation of Ezekiel's vision of the eschatological and Davidic Shepherds. All three texts, directly and indirectly, envision a Davidic messiah who heals the sick – a far cry from a militant warrior – and restores the community at the eschaton.

4Q521 fragment 2 is another critical example of a messianic figure linked to a healing ministry.[179] The משיחו ('his anointed') in 4Q521 frg.2, 2:1 likely refers to a singular messiah of God,[180] imbued with authority over the heavens and the earth (כי הש[מים והארץ ישמעו למשיחו, cf. Matt 28:16-20). Yet it is YHWH as the eschatological Shepherd who will bring about the eternal kingdom, and his healing activities are implied in his mission of "freeing prisoners, giving sight to the blind, straightening out the twisted" (l. 7; cf. Isa 35:4-6; 61:1-2). Strikingly, this eschatological restoration will be done "in his mercy (ובחסדו)"; "the Lord will perform marvelous, unprecedented acts" (l. 9b-11; cf. Matt 8:27; 9:8; 9:33).

Then, more explicitly, 4Q521 states with regard to the healing activity of the eschatological Shepherd: "*for he will heal* (ירפא, 'Il guérira') the badly wounded and will make the dead live and proclaim good news to the poor" (l.12; cf. Ezek 37:1-4; Isa 61:1-2).[181] This 'healing' mission will coincide with his other activities such as satisfying the poor, bringing back the outcast, and feeding the hungry (l. 13). All of these can be well attested by Jesus' mission of compassion

---

[177] Collins, "The Expectation of the End in the Dead Sea Scrolls" in *Eschatology, Messianism, and the Dead Sea Scrolls* (ed. C. A. Evans and P. W. Flint; Grand Rapids: Eerdmans, 1997), 90.

[178] Baumgarten and Schwartz, *Scrolls 2*, 55.

[179] Cf. Peter M. Head, *Christology and the Synoptic Problem* (Cambridge: Cambridge University Press, 1998), 185-186.

[180] Martínez, "Messianic Hopes," 168-169.

[181] Émil Puech, *Qumrân Grotte 4*, 10-11.

toward the 'harassed and downtrodden' flock in the Matthew's Gospel, especially chapters 8-9 (cf. 9:36).

Undoubtedly, the language and images in this text are Ezekielian. Mixed with the vision of Isa 61:1-2, the anointed one in 4Q521 frg 2 is depicted as the eschatological Shepherd in Ezekiel 34 and 37. Healing the sick, feeding the hungry (Matt 14, 15), and making the dead live (Matt 9:23-25; cf. 27:51-54) are all essential tasks for the eschatological Shepherd that are also clearly demonstrated by Jesus in Matthew's Gospel. How this anointed one can be Davidic is not explicitly stated, due perhaps in part to the fragmentary nature of the text of 4Q521. Nevertheless, the Davidic connotations can scarcely be overlooked with such a titular use of the term משיחו, whom YHWH the eschatological Shepherd would establish over his restored flock.

Further, a more obvious link between the healing ministry and the Davidic messianic figure can be found in 4Q504 fragments1-2 columns 4 and 5.[182] A cluster of the triple, critical successive designations – *David, shepherd, and prince* in column 5 lines 7-8 – occurs only in the texts of Ezek 34:23-24 and 37:24-25 in the Hebrew Bible. When David is remembered in the prayer of the community, he is recalled as the Davidic Shepherd. Legitimately refuting Pomykala's doubts, just as we argued in our examination of 4Q504, W. M. Schniedewind affirms that 4Q504 is "probably an allusion to 1 Sam 5:2 and Ezek 34:23." [183] Interestingly, the prayer that contains a recitation of the golden days of David in 4Q504 1-2, 4:5-14 is preceded by earnest supplications for the Lord's *healing* [תרפאנו] of their madness, blindness, and confusion [of heart] (1-2, 2:14), and illness, famine, thirst, and plague (1-2, 3:8). Thus, this procedure in 4Q504 – supplications for the the Lord's healing followed by the prayer for the ideal David – closely resembles the Ezekiel pattern in which the eschatological Shepherd comes to heal and later appoint his Davidic Shepherd.

It is the shepherd image drawn basically from Ezekiel 34, we argue, that provides Matthew's therapeutic Son of David with a coherent picture of a Davidic messiah who heals the sick. All of the texts illustrated above, in addition to *Apoc. Ezek.* fragment 5, involve Davidic expectation and the shepherd motif in the context of the restoration of Israel. This particular context, implanted in the Davidic Shepherd tradition, shapes Matthew's presentation of Jesus as the 'therapeutic' Son of David. The underlying web of concepts from this tradition is tightly interwoven in the narrative, and the verb θεραπεύω stands out clearly as one of the strands. Thus, before we examine the individual passage where the Son of David is associated with healing ministry, we will examine the use of the verb θεραπεύω in its association with various 'restoration' motifs that echo the Davidic Shepherd tradition, such as 'following,' 'feeding'

---

[182] For a detailed examination, refer to Ch. 2, 2.3.1 4QDibHam(a).
[183] Schniedewind, *Society and the Promise to David* (Oxford: Oxford University Press, 1999), 165.

and 'gathering,' Jesus as more than David/the Son of David, and last, the 'teaching' motif.

4.2.3.2 The Use of the Verb θεραπεύω and Restoration Motifs

Numerous passages in the Gospel employ the verb θεραπεύω, which characterizes Jesus' ministry among the crowds. They are frequently rich with messianic connotations (Matt 4:23, 24; 8:7, 16; 9:35; 10:8; 12:10, 15, 22; 14:14; 15:30; 17:18; 19:2; 21:14).[184] In five of these fourteen passages, Jesus' healing activity is explicitly associated with the Son of David title (9: 27-31; 12:22-24; 15:21-28; 20:29-34; 21:1-16; cf. 22:41-46).[185] This observation raises two questions: What are the characteristics of Jesus' healing ministry, and how are these two elements, namely, the title Son of David and healing activity, associated with one another?

In his short but illuminating article on the verb θεραπεύω in Matthew's Gospel, J. A. Comber observes that (i) after Matt 11:1, although the ministries of teaching and preaching cease, healing continues and remains a major function of Jesus; (ii) the healing ministry is the focus of bitter controversies between Jesus and the Jewish leaders; and (iii) the healings are frequently performed in the context of the crowds following Jesus.[186] Based on these observations, Comber proposes a redactional-historical hypothesis. He contends that the teaching and preaching to the Jews in Matthew's day ceased but the healing continued. Since healings are the cause of the Jewish leaders' unbelief and rejection of Jesus, he argues, the healing ministry thus becomes "a method used by Matthew to differentiate sharply between Jewish leaders and Jewish crowds." As Comber concludes, Matthew appeals therefore to 'the crowds' – a cipher for the Jewish people of Matthew's time – "not to follow the leadership of the Pharisees but to join the fellowship of the disciples of Jesus." [187]

It may be that Jesus' healing stories are transparent to the life situation in Matthew's day. As Comber does not force an anachronistic argument, it is assumed that the same observations should be, at least equally, valid for Jesus' day. Earlier we had rejected Stanton's proposal of the humble Son of David of the early church. Here we readily agree with Comber that healing propelled hostility from the Jewish leadership of Jesus' day. Unfortunately, Comber does not raise the question of why Jesus' healing ministry inflames such hostility.

In the narrative, Jesus' healing clearly functions to separate the Jewish leaders and the crowds. The underlying issue appears to be that of leadership or

---

[184] The Therapeutic Son of David is also found 9:12; 20:29-34 but minus the verb; the verb only occurs twice in Mark (3:2; 6:5).

[185] Duling, "Therapeutic," 405-410, includes 22:41-46 as a conclusive passage in terms of the motif.

[186] Comber, "THERAPEUŌ," 432-433.

[187] Comber, "THERAPEUŌ," 433-434.

4.2 The Davidic Shepherd as the "Therapeutic Son of David"

eventually messianism. For instance, the 'following' motif is apparent in those passages that involve the verb θεραπεύω. Jesus' therapeutic pastoral leadership results in the crowds following him (Matt 4:24; 12:10; 14:14; 19:2; 20:29-34; cf. 9:12-13). 'To lead' is the primary task of the shepherd in the biblical tradition (Gen 49:24-25/Ps 80:1; Exod 3:10, 11, 17; 6:6, 13, 26; 7:5; 12:4; 17:3; 23:23; Num 20:5; 27:16-17; Deut 4:38, 6:23; 2 Sam 7:5-7). The notion associated with the shepherd/sheep metaphor remains the same in the tradition during the Second Temple period; *An. Apoc.* is a case in point: "but the Lord of the sheep went with them as their *leader*, while all his sheep were *following* him" (89:22).

(1) 'Following' motif: To follow the shepherd is the natural consequence of the shepherd leading the flock. Hence, the crowds 'follow' (ἀκολουθέω) the healing Son of David, and the motif of 'the crowds' following' enhances the picture of Jesus as the eschatological Shepherd who gathers the scattered flock.[188] Importantly, he does it by seeking the lost and the outcast, and notably, by healing the sick. Thus those who are brought back and healed follow their eschatological Shepherd. Jesus' healing ministry and the crowds following him as their Shepherd are inseparable in Matthew's healing passages.

In the Hebrew Bible, the redemptive pastoral leadership in the Pentateuch is expressed in terms of restorative pastoral leadership, especially after the exile of the nation as shown in passages of the prophets (e.g., Mic 2:12-16; 5:1-4; Jer 23:3; 30:13-17; 31:7, 10; Isa 40:11; Ezek 37:15-23). It is Ezekiel's Shepherd described in 34:1-16, however, that explicitly details the healing ministry of the shepherd-leadership in the context of Israel's restoration. The eschatological Shepherd gathers the scattered flock by seeking and healing them. The tasks of the eschatological Shepherd in Ezekiel 34-37 thoroughly correspond to the situation of the broken covenantal community in exile.

The eschatological Shepherd's mission in Ezekiel 34 involves binding up (חָבַשׁ; LXX, καταδέω, in vv. 4, 16) the crushed or injured, healing (רָפָא; LXX, σωματοποιέω, v. 4) the sick, and strengthening (חָזַק; LXX, ἐνισχύω, vv. 4, 16) the weak. Since the most victimized ones are the lost and outcast, the Shepherd sets out to seek them first (cf. Matt 9:9-13); afterwards, he heals the sick to restore the community. Jesus' healing is a critical sign for Matthew and his readers to recognize the coming of Israel's promised Shepherd. This restorative Shepherd's leadership is what YHWH promised to demonstrate "on a day of cloud and darkness" (Ezek 34:12; vv. 11-19; cf. Mic 5:15; Zech 13:1-2), at the time of the restoration (cf. Matt 1-2). The return of the presence of YHWH the Shepherd primarily implies the return of healing among the flock; his absence is acutely exhibited in absence of healing (cf. Zech 10:2; Matt 9:36).

With this leadership accompanied by the following of the lost and the sick, YHWH confronts the wicked shepherds of Israel. Severe judgment is unavoid-

---

[188] Cousland, *Crowds*, 169-171.

able. Their leadership will be replaced by YHWH's appointment of his Davidic Shepherd. Healing as one of the critical tasks for Ezekiel's eschatological Shepherd is but one part of a bigger picture – the coming of YHWH himself as the eschatological Shepherd, and his appointment of the Davidic Shepherd over the restored community – that unfolds in detail in Ezekiel 34-37 and 40-48.

Healing, as a sign of the return of YHWH as Israel's eschatological Shepherd, is to lead YHWH to confront, judge, and eventually replace the wicked shepherds of Israel with all of their failed past and the exile, and finally to realize the promised future of Israel. The language of 'following,' 'compasion' and 'healing' mediated through the shepherd/sheep images hint at this underlying struggle between Jesus and the Pharisees to control the leadership in order to move Israel toward her future destiny.

In this regard, to speak about leadership in Israel's history is another way of speaking about messiahship, especially in view of the Davidic dynastic tradition. The leadership issue in Israel has been a profoundly theological topic for the Davidic kings and the people of Israel alike, particularly after the exile. If we were to 'log on' to this realm of Israel's leadership/messiahship, thus, the kingdom of God in that period, then the shepherd/sheep images would be the correct 'password.' Ezekiel's vision of the eschatological and Davidic Shepherds, in this way, finds its validity in our pursuit of the reason for the nexus between Jesus' healing and the Pharisees' hostility, to better grasp the conflict over the future of the nation and beyond. What is at stake is the authority of the leader, that is, the authority of the under-shepherd(s) of YHWH over Israel's future and over the nations.

(2) 'Feeding' and 'Gathering' motifs: The 'following' motif, in this way, focuses attention on Jesus as the shepherd of the flock, further related to other restoration motifs such as 'feeding' and 'gathering.' The feeding scenes in Matthew 14 and 15 display the identical procedure: (i) the crowds follow Jesus (14:13; 15:30); (ii) Jesus feels compassion for them (14:14; 15:32); (iii) Jesus heals the sick (14:14; 15:30-31); (iv) Jesus orders the crowds to sit down either on the grass (14:19) or the ground of the mountainsides (15:30); (v) then, Jesus feeds them abundantly, giving them more than enough, thus conjuring up an eschatological scene. Undoubtedly this is a pastoral image of Jesus; by feeding the flock, he fulfills the tasks as the shepherd for those without a shepherd.

All of these features of Jesus feeding the crowds strikingly match the tasks of YHWH as the eschatological Shepherd for Israel in Ezekiel 33-37 (cf. 34:2-3, 5, 14). Around the time of Jesus, this vision of Ezekiel's Shepherd may have been familiar, and popularly ascribed to many in Jesus' day. The *Apocryphon of Ezekiel* fragment 5 clearly links the 'feeding' motif with the coming of the eschatological Shepherd: "I will feed them on my holy mountain ... and I will be," he says, 'their shepherd and I will be near to them as the garment to their skin.'"

4.2 The Davidic Shepherd as the "Therapeutic Son of David"

Likewise, 4Q521 frg. 2 characterizes the leadership of the anointed by his feeding the hungry: "He (YHWH) will *lead* the [...] ... and *feed the hungry*" (ורעבים ישׂבר, 1. 13; cf. Ezek 34:10, 14, 18; cf. 4Q504 1-2, 4:14). More explicitly, the *Tg. Ezek* renders the 'shepherd' primarily as 'provider' (פרנס, i.e., 'to feed,' 34:3, 8, 10, 13, 14, 16), and not just in terms of 'to lead'; thus YHWH the eschatological Shepherd will 'reveal' himself (*Tg. Ezek* 34: 11, 20; cf. 36:9) first and foremost as the eschatological 'Provider.' YHWH confronts the false shepherds of Israel particularly for their failure to feed the flock: "And My people must eat the food left over by your servants, and must drink the drink left over by your servants (*Tg. Ezek* 34:20)."

Likewise the Davidic 'Leader' in *Tg. Ezek* 34:23 is *distinctly a 'Provider'* set in opposition against the wicked 'leaders' who exploit the people: "And I will set up over them one leader who shall provide for them, My servant David; he shall provide for them and he shall be their leader." Jesus feeding the crowds – against this background as well as in its narrative context – is not 'merely' about Jesus meeting human needs in general, but one should not bypass its significance in Matthew's presentation of Jesus the Shepherd; it is primarily 'eschatological and messianic' in the Gospel.[189]

No wonder the crowds, seeing all these shepherding activities of Jesus, were amazed and praised the '*God of Israel*' (15:31), which evokes YHWH's covenantal relationship with them (cf. Ps 41:13; 59:5; 68:35).[190] This may well indicate that Jesus' activities of healing and feeding the crowds with such compassion were recognized as signs of YHWH's faithfulness to his covenantal promises to his people. In Matthew, the crowds' response to Jesus' miraculous healing and exorcism refer not only to Israel (9:33), and the God of Israel (15:31), but also to the Son of David (12:13; cf. 21:15). Jesus' healing the sick is thus understood by the crowd in terms of *Heilsgeschichte*.[191] This heightens the comparability of the Ezekiel background. The revision of the Davidic expectation in Ezekiel 34-37, the return of the eschatological Shepherd and his Davidic-Appointee, matches the signs demonstrated by Jesus as the shepherd who heals and feeds.[192] This gives us a pause, however. How should Jesus be seen as he heals the sick and feeds the hungry? Is he indeed the eschatological or the Davidic Shepherd?

(3) '*More than David*' and the Servant Motif: As he examines Matt 22:41-46, Duling asks 'in what sense' Jesus is the Son of David, and then promptly

---

[189] The Matthean motif of Jesus feeding the multitudes 'on the mountains' or 'on the grass' enhances Ezekiel's pastoral image. See Donaldson, *Jesus on the Mountains*, 219.

[190] Against Carter, *Matthew*, 326, who thinks that the crowds misunderstand Jesus' identity (11:2-6).

[191] Similarly, Cousland, *Crowds*, 142.

[192] Cf. Marcus J. Borg, *The Meaning of Jesus: Two Visions* (co-author N. T. Wright, New York: HarperSanFrancisco, 1999), 96, suggests a Moses typology for the Jesus who feeds the multitude.

answers that Jesus is the 'healing' son of David. With his quick answer, however, Duling bypasses the crux of the question he himself raised. The puzzle Jesus poses when he converses with the Pharisees in 22:41-46 demands more in this respect. Jesus is, of course, the 'healing' Son of David. But, merely pointing out this characteristic does not explain *how* the Anointed One is more than the Son of David in 22:41-46. The key is not the fact that Jesus is the 'healing' Son of David, but the implied nexus between the Son of David and the healing messiah.

Adrian M. Leske argues, "the traditional royal Davidic role has been *superceded* by another, that of the *Servant* who is God's son (Matt 3:17)."[193] We have already argued that the title 'Son of David' in Matthew's Gospel is not necessarily antithetical to the resurrected Jesus, the Lord who was recognized as the Servant by the early church. Just as the chief priests and teachers of the law assume (2:6), and the genealogy in Matthew's Gospel (1:1-17) plainly demonstrates, the Christ came as the Son of David (1:1, 18-21); he is both Davidic and messianic, the Davidic messiah.

The point of Jesus' argument in Matt 22:41-46 is that the Christ is indeed the Son of David – and even more so, is the Lord of David himself. From this it can be understood that the Christ is then 'a particular kind' of Son of David, at least not like those kings of the Davidic dynasty in the past. Then, how, and in what sense, does Jesus' healing ministry qualify him as that particular kind of Davidic messiah?

The servant motif, not necessarily taken from Isaiah, may illuminate this subtle connection. The link between the Servant and the Davidic Shepherd is clear in Ezek 34:23-24 and in its recapitulation, 37:24-25. YHWH promises to raise up over the gathered flock "one shepherd" (34:23; 37:24; cf. Matt 2:6) who is "my Servant David" (34:23; 37:24), "My Servant David will be king over them" (37:24), and "My Servant David will be prince in their midst" (34:24; 37:25). As is already implied in Ezek 34:23-24 (MT), the *Midrash* on Ps 29:1 plainly demonstrates that the Davidic Shepherd, as YHWH's Servant, can take up the tasks of the eschatological Shepherd: "'What more am I to do for them?' asked God. 'Let my Servant David feed them,' as is said, 'And I will set up one shepherd over them, and he shall feed them, even my Servant David (Ezek 34:23)'; David answered, 'Thou are the deliverer; be Thou also the shepherd; Save thy people and bless Thine inheritance; feed them also, and carry them for ever (Ps 29:1)'." [194]

It is explicit that the task is transferrable to the Davidic Shepherd. At the same time, however, 'David' requests God to feed/shepherd them forever. While the distinction between God's role and that of the 'Servant David' is still

---

[193] Leske, "Isaiah 40-66," 909-910 (italics mine).
[194] William G. Braude, *The Midrash on Psalms* 1 (New Haven: Yale University Press, 1976), 380.

retained, 'David' remains 'my Servant.' As YHWH's Servant, likely, the Davidic Shepherd feeds YHWH's flock.

If this is correct, Matthew's use of the Isaianic Servant passages may provide a link, though indirectly, of Jesus as the Servant with Jesus as the Davidic Shepherd.[195] This is one of the ways *how* Matthew, as well as Jesus, modifies the Davidic Shepherd tradition. Matthew applies the tasks of YHWH the eschatological Shepherd to Jesus himself who is called the Son of David, the Davidic messiah, but as as the Servant (Ezek 34:23; 37:24; Isa 35; 61).[196] Jesus' identity is therefore a complex issue, but he is certainly *more* than the 'son' of David, as hinted at in Matt 22:41-46.

(4) The 'Teaching' Motif: The absence of the term related to Jesus' teaching task after Matt 11:1 presents another clue to the Ezekiel background. Comber proposes that its absence indicates a cessation of preaching and teaching of the Gospel in Matthew's time.[197] Comber explains this in light of Matthew's day while not ruling out the same for Jesus' own day. Indeed, terms such as 'to teach' and 'to preach' do not occur, whereas 'to heal' continues to appear after Matt 11:1.

Even before chapter 11, however, we see this tendency already revealed in Matt 9:36 and 10:1. The terms 'teaching' and 'preaching' in 9:36 are dropped in 10:1. Concerning a healing ministry, the identical phrase is repeated: θεραπεύων [θεραπεύειν, 10:1] πᾶσαν νόσον καὶ πᾶσαν μαλακίαν (9:35). In other words, as Jesus sends out his twelve disciples to the lost sheep of the house of Israel (10:6, 16), he asks them to heal the sick as well as to cast out demons, but 'to teach' is omitted. A question arises which does not appear to be insignificant: Why does Jesus only command healing but not teaching as he sends out his under-shepherds to minister as they seek out the scattered sheep of Israel?

At the end of Matthew's Gospel, the risen Jesus finally commands his apostles *to teach* the peoples from the nations, from whom they are to make disciples (28:19-20). Only after the smitten shepherd is raised again, does the appointed King, that is, the ever-present Davidic Shepherd in the midst of the restored flock assign his disciples to the teaching ministry (cf. Matt 26:31; 28:16-20). The Davidic Shepherd theme with Matt 28:16-20 is, as we shall examine later in the following chapter, nothing but vibrant and viable.

The Ezekiel pattern may be able to tell us why the teaching task of the new community is postponed and why healing must precede teaching in the

---

[195] Recently, Richard Beaton, *Isaiah's Christ in Matthew's Gospel* (Cambridge: Cambridge University Press, 2002), 141-173, argues, Matthew's use of Isa 42:1-4 implies more than 'a proof-text,' but with the preceding and subsequent context in Matt 12, by which 'the themes of Jesus' identity' are traceable.

[196] Jesus transfers the healing task to his apostles in ch. 10, but they themselves do not bear messianic identity. This can be also considered a serious modification of Ezekiel's vision.

[197] Comber, "THERAPEUŌ," 432-433; cf. J. D. Kingsbury, *The Parables of Jesus in Matthew 13* (Richmond: John Knox, 1969), 28-29.

Gospel.[198] In Ezekiel's vision (ch. 34), the eschatological Shepherd's own ministries are to rescue and restore the oppressed flock (34:1-16; 17-22) by "seeking, healing, and judging the flock" (34:1-22). This process of seeking and healing precede the teaching task by the Davidic Shepherd Appointee. It is only 'after' YHWH appoints him over the rescued community (34:23-24) that the Davidic Shepherd is supposed to lead them in the laws and decrees, and on to the path of righteousness (37:24-25). In short, the rescue of the endangered flock by YHWH precedes the sanctification of the flock by the Davidic Shepherd-Appointee.

To conclude, the varied features associated with the healing motif in the Gospel, i.e. following, feeding, gathering, and Jesus as the one more than David/the Son of David along with the teaching motifs, can be successfully explained as we conceive of Jesus – the therapeutic Davidic Shepherd – in terms of *the eschatological and Davidic Shepherds of Ezek 34-37*. It is Jesus' healing ministry as the Shepherd for 'the sheep without a shepherd' that has as its result the crowds' following him. This coincides with YHWH's eschatological confrontation with the wicked shepherds of Israel. The seriousness of the hostility on the part of the Jewish leaders reflects an overtone of anxiety over the arrival of the promised theocracy. The promised kingdom of David has dawned – and it has come modified through the Davidic Shepherd tradition. Even the technical eclipse of 'teaching' after Matt 11:1 (cf. 10:1) affirms the validity of the Ezekiel pattern of the coming of the eschatological and Davidic Shepherd(s) to Matthew's presentation of Jesus.

4.2.4. *The Healing Son of David in Healing Contexts*

There are five critical 'therapeutic Son of David' passages in the Gospel in which the 'healing' Jesus is called 'son' of David (υἱὸς Δαυίδ, 9:27),[199] 'the son' of David (ὁ υἱὸς Δαυίδ, 12:23; 21:15), or 'Lord, son' of David (κύριε υἱὸς Δαυίδ, 15:22; 20:30-31).[200]

For the titular approach as to the christology of the Gospel, the absence or presence of the article can be significant for detecting any messianic connotation in those passages. Is there any real distinction between 'Son of David' used with and used without the definite article? Alfred Suhl maintains that there is "a far-reaching distinction between the two and the occurrence of the title with the article betrays a false messianic understanding."[201] In his critique of Shul's position, Cousland aptly refutes Suhl's argument by pointing to the children

---

[198] One may raise a question of the figure of Jesus as the teacher with authority already in Matt 5-7; this can be seen, I suggest, as Jesus' proleptic role of the Davidic Shepherd-Appointee; the motif of ἐξουσία is heightened at the end of the Gospel in 28:16-20.

[199] A few manuscripts such as N, f^{13}, 892^c retain κυριε υιε instead.

[200] The case of 20:30-31 is fluid in its retention of κυριε; the MSS ℵ, D, θ, f^{13} omit κυριε.

[201] Suhl, "Der Davidssohn im Matthäus-Evangelium," *ZNW* 59 (1968): 73.

4.2 The Davidic Shepherd as the "Therapeutic Son of David"

praising the healing Jesus as 'the' Son of David; Jesus' quotation of Psalm 8 expresses a definite opinion about the children's utterance – it is "perfect praise." [202]

Chilton argues for another option, stating that Matthew maintains both – an articular and non-articular use of the designation. Although the titular usage of David's son in the context of healing is a Matthean invention (12:23; 21:15), for Matthew, the rationale for the usage of the title is not therapeutic.[203] For Chilton, the non-articular 'son of David' is a healer following the pattern of Solomon-as-exorcist, and the articular 'the' Son of David is the messiah following the pattern of the anointed Son of David in *Ps. Sol*.[204] This observations as to "Matthew's incomplete synthesis" could strengthen the Solomon-as-exorcist option in those passages where the article is not found, while affirming that Solomon-as-exorcist is not associated with any messianic connotation.

Nevertheless, it is doubtful that the absence of the article in those cases of the vocative (Matt 9:27, 20:30, 31) can be such a crucial element.[205] More importantly, Chilton's assumption – even if accepted as is – is still subject to the general critique of the relevance of the Solomon tradition to Matthew's therapeutic Son of David.[206] In particular, Chilton's assumption of Solomon as "a skilled healer" is an obvious exaggeration; as many have underscored, Solomon is known as a skilled "exorcist" who uses manipulative incantations. To view the healing Son of David as the anointed Son of David in *Ps. Sol* 17-18 remains a possibility, which demands further explanation Chilton fails to provide. Further, Chilton's argument for the unlikely distinction between the articular and non-articular use of 'Son of David' in the Gospel creates a more hypothetical problem as to why Matthew leaves "the synthesis" incomplete.

The Gospel of Matthew, however, opens with the introduction of Jesus as the Anointed, the Son of David (1:1), and it is likely that through the narrative the Evangelist retains and unfolds the implied association of the Son of David with its messianic connotations. Likewise, Luz aptly states that the Evangelist introduces the figure in the "charter document" of 1:2-16 as the descendant from "the royal line of David and thus not as an antitype of the 'wise' Son of David, Solomon."[207] Only the narrative that follows identifies the healing ministry as part of those messianic connotations. These preliminary observa-

---

[202] Cousland, *Crowds*, 176.
[203] Chilton, *ben David*, 94-95.
[204] Chilton, *ben David*, 98, 105.
[205] F. Blass and A. Debrunner, *BDF*, 81-82, explain the omission of the article in those cases (also, οἶκος Ἰσραήλ in Act 7:42) saying, "because in such cases (these designations in the vocative) the article does not appear in Hebrew either." Cf. Luz, *Matthew*, 2:47-49, identifies 'son of David' in 9:27 with Israel's Messiah (1:18-25; 4:23).
[206] Refer to the previous section, 4.2.2.2 "Solomon as the Son of David: A Jewish Legend."
[207] Luz, *Matthew*, 2:47-48 (the excursus on "Son of David in the Gospel of Matthew").

tions, however, call for a more detailed examination of each of the passages in the Gospel that involve both healing and [the] Son of David.

4.2.4.1 The Two Blind Men (Matt 9:27)

The episode in Matt 9:27-31 describes how two blind men follow Jesus, crying out and saying, ἐλέησον ἡμᾶς, υἱὸς Δαυίδ. Following Chilton's hypothesis that 'David's son' – without the article – at one time functioned not as a messianic title but to clarify Jesus' ability to heal, Davies and Allison postulate that 'David's son' here is the address applied to Jesus at the level of tradition when he was to heal or exorcize in a manner reminiscent of Solomon.[208]

The story of Jesus healing the blind is attested by multiple sources: Mark 8:22-26; 10:46-52; cf. Matt 15:20; Jn 9; Matt 11:5//Luke 7:22.[209] Although many assume that the tradition behind Matt 9:27-31 is to be traced in Mark 10:46-52, a much closer link with Mark 10:46-52 is found in Matt 20:46-52.[210] As we will duly argue for the case of Matt 20:46-52, those who view Matt 9:27-31 as 'a redactional creation' appeal too quickly to the possible literary dependence ruling out (in principle) the plausibility of the historicity.

For instance, Davies and Allison argue for Matthew's preference to telling stories 'in a triad format.' But this hardly proves that the Evangelist 'created' the story. Would the Evangelist have risked his integrity by telling such a significant story about such a significant figure (Jesus) for him and his community only for the sake of a pet literary device, known as the 'triad'? Why should readers (both ancient and modern) give up on the plain meaning of the sentence with its intended truthfulness by the Evangelist? Matthew 9:28a says, for example, ἐλθόντι δὲ εἰς τὴν οἰκίαν προσῆλθον αὐτῷ οἱ τυφλοι, which indicates that the incident occurred in a house and not on the roadside [cf. Matt 20:46-52]!

On the other hand, as to the skepticism of modern scholars concerning most or all of the miracles, one needs to discuss their Deistic presupposition that from the outset eliminates any transcendental phenomenon. In fact, we need to be reminded, as J. P. Meier pointed out, of the "all-too-ready acceptance of miracles by ordinary people" evident in the first century A.D.[211] Meier, however, contests that the most crucial evidence for the historicity of the story in Matt 9:27 is the "archaic conception of Jesus as a 'Son of David' like Solomon the miracle-worker."

---

[208] Davies and Allison, *Matthew*, 2:136.

[209] Ben Witherington, *The Christology of Jesus* (Minneapolis: Ausgburg Fortress Press, 1990), 170.

[210] For instance, Gundry, *Matthew*, 176; Keener, *Matthew*, 305, n. 112; Davies and Allison, *Matthew*, 2:133, view Matt 9:27-31 as 'a redactional creation'; likewise, Luz, *Matthew*, 2:46 argues that Matthew makes story into two; Heinz J. Held, "Matthew as Interpreter of the Miracle stories" in *Tradition and Interpretation in Matthew* (ed. G. Bornkamm, G. Barth, and H. J. Held; trans. Percy Scott; Philadelphia: Westminster Press, 1963), claims that Matt 9:27-31 is a redactional reworking of Matt 20:29-34; cf. Kurt Aland, *Synopsis Quattuor Evangeliorum* (Stuttgart: Württembergische Bibelanstalt, 1964), 557, 568, yet indicates the different degree of the resemblance among those stories; also, Hagner, *Matthew*, 2:585.

[211] Meier, *A Marginal Jew*, 2:509-521, 535-552, deals with the miracles employing the distinction between 'modern minds' (ch. 17) and 'ancient minds' (ch. 18); cf. Keener, *Matthew*, 305-306.

²¹² As to the thin evidence for this Solomon-exorcist legend, we have already argued its spuriousness.²¹³ This touches upon the identity of Jesus as the healing Son of David; even if the case is proved that he is the Davidic Shepherd instead of the Solomon-exorcist, the authenticity of the story can hardly be refuted since the Davidic Shepherd tradition is traceable both in the OT and Judaism, which is far more 'archaic.'

The option for the Solomon-as-exorcist strips bare any messianic connotation from the healing Jesus when he is called υἱὸς Δαυίδ, without an article. Besides the general critiques leveled against the validity of the Solomon-as-exorcist option for this passage as we have argued earlier, it is noteworthy that Jesus, commending *the faith* of the blind (9:29), prevents any casual understanding on the part of those who called Jesus υἱὸς Δαυίδ. Jesus takes it as their faith in him.²¹⁴

Characteristic of his mercy and healing ministry, Jesus is understood as υἱὸς Δαυίδ by the blind, and the understanding implied in his faith is affirmed as such by Jesus himself. It is hardly likely that Jesus affirmed that he is one like Solomon the exorcist.²¹⁵ Certainly, the Evangelist in 9:13 takes the title υἱὸς Δαυίδ as messianic (cf. 9:33b).²¹⁶

Jesus' identity as 'the healing υἱὸς Δαυίδ,' which this element of faith communicates, does not seem confined to this episode alone. As Luz points out, it is significant that the title 'Son of David' first occurs at the end of miracle cycle of Matthew 8-9; thus, "for him (the Evangelist) chapters 8 and 9 *tell* who the Son of David is." ²¹⁷

In fact, the whole narrative context of chapters 8-9 communicates the same message. The significance of the location of this episode (9:27-31) at the end of chapters 8-9 is widely recognized.²¹⁸ Verseput notes, "as we approach the end of this carefully composed section [chs. 8-9], the messianic character of the whole is placed into a distinctively Davidic garb by the final two pericopae (9:27-31; 31-34)."²¹⁹

Particularly, the echo of υἱὸς Δαυίδ by the two blind men in 9:27 is amplified by the exclamations of the amazed crowds who shout: "Nothing like

---

²¹² Meier, *A Marginal Jew*, 2:690.

²¹³ Refer to 4.2.2.2 "Solomon as 'Son of David': A Jewish Legend"; e.g., Luz, *Matthew*, 2:47.

²¹⁴ Carson, *Matthew*, 2:233; cf. Meier, *A Marginal Jew*, 686-690, owes an explanation for Matthew's and Jesus' understanding of the faith in Matt 9:27-32.

²¹⁵ Keener, *Matthew*, 306; Cf. Charlesworth, "Solomon," 85-86.

²¹⁶ Gnilka, *Matthäusevangelium* 1:345, also links the Davidic messiah with healing yet not without discussing the issue of the background for that association.

²¹⁷ Luz, *Matthew*, 2:48 (italic mine).

²¹⁸ W. G. Thompson, "Reflections on the Composition of Matt 8,1-9, 34," *CBQ* 33 (1971): 365-388; Luz, "Die Wundergeschichten von Matt 8-9" in *Tradition and Interpretation in the New Testament,* 149-165; Verseput, "Davidic," 109-110; Novakovic, *the Healer*, 79-80.

²¹⁹ Verseput, "Davidic," 111.

this was ever seen in *Israel*" (9:34; cf. 4Q521 2, 2:11). The particular emphasis on 'Israel' in v. 34 is verified as a critical indicator of Jesus' mission in the midst of the 'sheep without a shepherd' in v. 36, certainly corresponding to the announcement of Jesus' future appointment as the Davidic Shepherd of Israel in 2:6. Jesus' mission with the given priority to the 'lost sheep of the house of *Israel*' in 10:1, 6, crystalizes the implication of υἱὸς Δαυίδ, which is designated by the blind men (9:27) and is implicitly admitted by Jesus himself (9:29). In short, the narrative framework is obviously 'Davidic,'[220] but more accurately, it is the framework of the Davidic 'Shepherd.'

The composite picture emerging from 9:35 and 10:6 is likely reminiscent of the prophetic image used in Ezekiel 34 (and Jer 23:1-3) to describe the destruction of the exile. Verseput notes, Jesus of chapters 8-9 is thus "the divinely ordained Davidic agent" for the leaderless people "whose pre-exilic leadership has been cut off."[221] Yet, in addition to this observation, the shepherd image is necessary to be singled out in identifying this healing Jesus called υἱὸς Δαυίδ. The image is abundant and explicit (9:26; 10:1, 6), and is inseparable from Jesus' healing activity driven by his compassion for the harassed flock (9:35-36). The mercy that the blind men ask for as they call Jesus υἱὸς Δαυίδ (9:27) is rooted in Jesus' compassion for the harassed flock (v. 35).

In fact, four of the five instances of the verb, 'to have compassion' (σπλαγχνίζομαι; 9:36; 14:14; 15:32; 20:34; cf. 18:27), and five of the eight instances of the verb 'to have mercy' (ἐλεέω; 9:27; 15:22; 17:15; 20:30; 20:31; cf. 5:7 [x2]; 18:33) occur in conjunction with Jesus' healing ministry. In nearly every case, the titles Son of David and/or Lord appear as part of the semantic field.[222] The compassion of YHWH for his scattered flock is one of the profound motivations for Israel' s hope of restoration. We are to see Jesus' healing the blind men as part of the whole restoration process.[223]

The immediate context of Matt 9:27 attests to the dawning of a new era. Jesus' raising a dead girl in the preceding pericope (9:18-26) creates a pair with the new wine section (9:14-17). This theme of a new era of restoration is wrapped in the motif of compassion (9:9-13, 27), authority (9:1-8, 33), and conflict with the leaders of the people (8:28-34; 9:32, 34). In 9:35 we find in the summary that Jesus is identified as the eschatological Shepherd who came to

---

[220] Novakovic, *Healer*, 81, following Verseput's analysis, accepts this.

[221] Verseput, "Davidic," 112; Carter, *Matthew*, 228, fails to note an Ezekiel connection, though he does elsewhere in his commentary; Hill, *Matthew*, 180 and France, *Matthew*, 172, remain ambiguous as to the source of the Davidic messiah in these chapters.

[222] Duling, "Plurisignation," 112.

[223] Jesus healing human bodies may also imply his reworking the social-religious boundaries in his culture. Refer to John J. Pilch, *Healing in the New Testament: Insights from Medical and Mediterranean Anthropology* (Minneapolis: Fortress Press, 2000), 81-82; also, Douglas, *Purity and Danger*, 113.

4.2 The Davidic Shepherd as the "Therapeutic Son of David"

seek and heal the harassed sheep in 9:36. A detailed literary structure of 8:28-9:36 looks like the following:

A. Jesus' casting out the demons/rejection, 8:28-34
   B. authority of Jesus, 9:1-8: "Praise God, who gave such authority to men" (v. 8)
      C. mission of mercy, 9:9-13: *"I desire mercy, not sacrifice" (v. 13)*
         D. new wine in new wineskins, 9:14-17
         D'. raising the dead to life, 9:18-26
      C'. mission of mercy, 9:27-31: *"Have mercy on us, Son of David" (v. 27)*
   B'. authority of Jesus, 9:33: "Nothing like this has been seen in Israel" (v. 33)
A. Jesus' casting out the demons/ rejection, 9:32, 34
     9:35 - a summary of Jesus' mission: preaching, teaching, and healing (cf. 4:23)
     9:36 - Jesus identified as the eschatological Shepherd

The tight web of the episodes in Matt 8:28-9:36, woven together with various themes reminiscent of Ezekiel's vision of restoration (chs. 34-37), is concluded in the stark depiction of Jesus' compassion for the sheep without a shepherd (9:36). In the next verse (9:27), Jesus acts out what he proclaims to do in 9:13 – out of mercy, he heals the sick. The eschatological Shepherd to come (9:13) is the Son of David who heals. Thus at last the promised Shepherd has arrived: "I have come" (v. 13). Yet, Jesus' identity as the eschatological Shepherd, taking the role of YHWH upon himself, may be blurred by the blind men's designation of Jesus as υἱὸς Δαυίδ.

Here we cannot resist bringing to mind 4Q521. Virtually all of the motifs present in Matt 8-9 can be identified with the shepherding tasks of YHWH in 4Q521 2, 2:9-13: "in his [YHWH] *mercy* (ובחסדו)...the Lord will perform *marvelous, unprecedented acts* (l. 9b-11; cf. Matt 9:33); for he will *heal* (ירפא) the badly wounded (cf. Matt 9:27, 32, 34-35) and will *make the dead live* (cf. Matt 9:18-26) and proclaim good news to the poor (l.12); and he will satisfy the [poo]r (ודלים ישיע), lead the outcast (נתושים ינהל), and feed the hungry/invite to the banquet (l. 13)."[224] Furthermore, the motif of authority in Matt 9:8, "Praise God, who gave such authority to men" (v. 8), corresponds to the figure of משיחו ('his anointed') in 4Q521 2, 2:1, who is invested with authority over the heavens and the earth (כי הש[מ]ים והארץ ישמעו למשיחו). The parallel is unmistakable.

More importantly, the ambiguity between משיחו (l. 1) and the [divine] subject of those eschatological shepherding activities (l. 10-13) corresponds to the question of Jesus' identity as υἱὸς Δαυίδ, the one vested with unparalleled authority, and who amazes the crowds as he acts out his eschatological shepherding mission of compassion in Matthew 8-9. In the case of 4Q521, it is likely that the subject of the shepherding mission described in l. 10-13 is the Lord himself as 'his anointed' would follow him for the succeeding stage of the restoration (l. 5-14).[225] Yet, in view of other Qumran texts such as 1QSb 5:20-

---

[224] Puech, *Qumrân Grotte* 4, 10-11.
[225] Zimmermann, *Messianische Texte*, 364; Puech, DJD 25, 11.

29, CD-A 13:7-12 where the Lord's mission of compassion is taken up by figures like 'Prince' or 'Seeker/Inspector,' the plausibility of the functional identification of 'his anointed' with the tasks of YHWH the eschatological Shepherd may not be entirely ruled out.[226]

In Ezekiel 34, it is not the Davidic Shepherd but clearly YHWH himself who is to seek the lost, heal the sick, and confront the false shepherds. It is now Jesus who takes up the roles of the eschatological Shepherd in Matthew 8-9, though he is still called υἱὸς Δαυίδ by the crowds. As suggested in 4Q521, the distinction is critical, but who else would take up these tasks for YHWH except the David who is the Servant (Ezek 34:23; 37:24; cf. *Midrash* on Ps 29:1)?[227] Similarly as Jesus' disciples are Jesus' under-shepherds, his agents in their healing of the sick among the house of Israel (Matt 10), Jesus as the Servant and the Davidic Shepherd-Appointee by YHWH can perform YHWH's eschatological tasks for Israel. Only Jesus, and not the disciples, was able to impart his eschatological shepherding authority (10:1; 9:25) to his apostles: ἔδωκεν αὐτοῖς ἐξουσίαν. In this sense, Jesus as YHWH's Shepherd-Appointee *uniquely represents YHWH's eschatological shepherding authority* over the eschatological flock (cf. Matt 2, 1-5, 6; 15:22; 28:16-20).

In short, the absence of the article for the title υἱὸς Δαυίδ in Matt 9:27 hardly affects the clear implication of the messianic connotations applied to Jesus in the narrative context of the eschatological mission of compassion depicted in Matthew 8-9. To the blind men (and the crowds), in need of God's mercy, the expected healing must occur. It could only be performed through YHWH's Servant, the future Davidic Shepherd-Appointee who is distinctly the healing Son of David in the Gospel.

4.2.4.2 The Blind and Dumb Demoniac (Matt 12:23)

Matthew 12:22-24 is an episode of Jesus' healing a blind and dumb demoniac, followed by a question from the amazed (ἐξίσταντο) crowd that witness this miracle: μήτι οὗτός ἐστιν ὁ υἱὸς Δαυίδ. The flow of the story echoes that of the stories in 9:27-34: Jesus' healing the sick (9:27-33a) is followed by the crowd's amazement (v. 33b), which eventually provokes opposition from the Pharisees (v. 34).[228]

The opening verses (12:22-24) are quite close in content but less similar in wording to 9:32-34. The potential parallels to Matt 12:22-30 are found rather in Mark 3:22-30 and Luke 11:14-

---

[226] Collins, DSD 1,100; Martínez, "Messianic Hopes," 168-170; Cf. Wise and Tabor, "The Messiah at Qumran," 61-62.

[227] Daniel L. Block, "Bringing Back David: Ezekiel's Messianic Hope," in *The Lord's Anointed: Interpretation of Old Testament Messianic Texts* (ed. P. E. Satterthwaite et al.; Carlisle: Paternoster, 1995), 172, notes the under-shepherd role of 'David' for YHWH; cf. Cousland, *Crowds*, 188-190.

[228] Novakovic, *Healer*, 79.

15.²²⁹ Gnilka assumes some influence of each passage on the other, perhaps brought about in the process of oral transmission.²³⁰ Darrell Bock argues for one event represented by these passages that Mark and Luke have located topically and thinks Matthew and Luke used a similar source.²³¹

The historicity of Jesus' exorcism is firmly attested as authentic.²³² In this case, the criterion of embarrassment can be applied as well; the early Christians would not create such as story, which puts Jesus in an ambiguous light.²³³ With the report of an exorcism followed by three typical responses – the crowd's response to Jesus; the Pharisees' response to the crowd; and Jesus' response to the Pharisees – this is a controversy story, though with Jesus' response relatively extended.²³⁴

The preliminary exegetical issue is the implication of μήτι in the question raised by the crowd. Eventually, many notice the positive connotation of the term for the favor of the whole context of the story, especially acknowledging the crowd's approving response contrasted with the Pharisees' hazardous slander.²³⁵ The main clause of the term μήτι usually anticipates a negative answer, though the meaning of μή is slightly modified in Matt 12:23b.²³⁶ Regarding the use of μή in Matt 12:23b (cf. the similar use of μή in John 4:29),²³⁷ we observe at least three elements from the context: (i) the affirmation of the fact observed; (ii) the surprise the fact caused; (iii) and the further response or witness the fact demanded.

The crowd in Matthew 12 detects some qualities in Jesus that make them recall a messianic figure, the Son of David. The question is triggered by some sort of evidence strong enough to give them second thoughts. Further, the element of surprise reveals a positive link between healing and the title. All (πάντες) of the crowd are besides themselves in their astonishment, to turn to their leaders as if to seek assurance that Jesus' identity is just as they had concluded. Lastly, the crowd turning to the Pharisees, and calling for their leaders' response of what thexey had come to believe about the Son of David,

---

²²⁹ Hagner, *Matthew*, 1:340; Davies and Allison, *Matthew*, 2:233; Bock, *Luke*, 2:1067.

²³⁰ Gnilka, *Matthäusevangelium* 1:456.

²³¹ Bock, *Luke*, 2:1070.

²³² James D. G. Dunn, "Matthew 12:18/ Luke 11:20-A Word of Jesus?" in *Eschatology and the New Testament: Essays in Honor of George Raymond Beasley-Murray* (ed. W. H. Gloer; Peabody, Mass.: Hendrickson, 1988), 29-49.

²³³ Witherington, *The Christology of Jesus*, 164; Meier, *A Marginal Jew*, 2:625; Keener, *Matthew*, 361.

²³⁴ Davies and Allison, *Matthew*, 2:233-234.

²³⁵ Duling, "Therapeutic," 400; Davies and Allison, *Matthew*, 2:334-335; Gundry, *Matthew*, 231. Carter, *Matthew*, 272.

²³⁶ *BDF*, 427.1, lists John 4:29 as a syntactically similar case: μήτι οὗτός ἐστιν ὁ χριστός.

²³⁷ These three elements can easily be identified in the use of μήτι in the story of Samaritan woman in John 4:29; the woman finds Jesus knows her secrets, and she is surprised, turning to the people in the village.

places the leaders in the position of witnesses who must decide whether to approve or refute what the crowd witnessed. They are in the position of witnesses as well as judges of what was witnessed. As Luz underscores, the positive nuance from the crowd's guess is readily affirmed in that the Pharisee formulate their accusation 'in response to it' (ἀκούσαντες).[238]

The crowd witnesses the work that, to their knowledge, could possibly be done only by ὁ υἱὸς Δαυίδ, but it is in vain that they turn to their leaders for confirmation. The healing Jesus appears to be the expected Son of David to 'all' of the crowd, and the whole crowd loudly sounds its consensus. How was it that the entire crowd could relate the healing Jesus to ὁ υἱὸς Δαυίδ?

Most commentators respond to this question by relying on the traditional axiom that the Jews did not expect the political-warrior type of the Davidic messiah to heal or to exorcize demons. Thus Gundry turns to the redactional view claiming that "Matthew constructs the question from the standpoint of a Christian who believes in Jesus as both healer and Davidic Messiah."[239] Hill turns to the Isaianic Servant (Isa 53:4 at Matt 8:17).[240] Carter appeals to the tradition of 'Solomon the miracle worker,'[241] which would be, borrowing Gundry's expression, 'otiose' if its alleged Christian origin is accepted.[242] France simply calls it a 'puzzlement.'[243] Luz suggests a 'narrative Christology' while affirming the Markan origin of Matthew's therapeutic Son of David.[244]

Yet would it have been a puzzle to the crowd in Matt 12:23 as well? As we examined the function of μή in Matt 12:33 in light of the 'syntactically' similar case of John 4:26, it would have not been an entirely puzzling question. 'All' the crowd in Matt 12:23 were "amazed" (ἐξίσταντο), but were not necessarily clueless as many modern commentators prove to be. At least, the crowd was able to associate what they had witnessed – Jesus healing the sick with incomparable authority – with their concept or expectation of the Davidic messiah, ὁ υἱὸς Δαυίδ, whatsoever it may be. Then, they turned to their shepherds, the Pharisees, to confirm [or, implicitly witness] this Davidic healer who performs wonders in the manner of YHWH the eschatological Shepherd who would do these things in their midst. Pushed into a corner, the Pharisees are desperate, and twist the truth about Jesus' identity. By contrasting their exaggerated distortion with the crowd's witness, it is the narrator, and not the Pharisees, who renders a favorable verdict that confirms the crowd's best guess.

---

[238] Luz, *Matthew*, 2:202.
[239] Gundry, *Matthew*, 231.
[240] Hill, *Matthew*, 215.
[241] Carter, *Matthew*, 272.
[242] Gundry, *Matthew*, 231.
[243] France, *Matthew*, 208.
[244] Luz, *Matthew*, 2:47-49, 202; also, idem, "Thetische," 221-235.

4.2 The Davidic Shepherd as the "Therapeutic Son of David" 311

The Solomon tradition as an option fails to explain why the Pharisees turned desperate and why they resorted to wickedly slandering Jesus. If it were the Solomon connection that the crowd had in mind, then of course their leaders must have known what they were thinking. Moreover, if the Pharisees knew of the Solomon connection, it is baffling why they were driven to such a display of hostility against a mere Solomon-pretender lacking any messianic pretense. In addition, Jesus' own defense by means of the Spirit-terminology in 12:25-32 would hardly make sense if the Solomon tradition were in view, especially when Solomon only cast out demons by manipulative incantations, and never by the power of the Spirit.

Furthermore, the narrative context of Matt 12:23 presents a rich foil for the crowd's glimpse at Jesus' identity as the Davidic messiah. From 12:1 to verse 14, the three motifs – David (vv. 1-3), mercy (v. 7) and the sheep/shepherd (vv. 11-12) – tightly interlock. One of the points of the comparison implied in 12:3-4 is that Jesus as a descendant of King David behaves much like his ancestor.[245] Here, Jesus' reference to the historical David appears casual, but at the same time, it is an implicit claim that Jesus' status is comparable to that of David. As the narrative moves on, the historical David is left behind, yet with such Davidic connotation Jesus emphasizes 'mercy' and shows his extraordinary concern by 'healing' the sick (vv. 6-15).[246] It is the mercy of God that motivates Jesus, who, acting from his Davidic royal status, heals the sick as his most urgent mission to fulfill, even on the Sabbath. Further, Jesus' healing on the Sabbath evokes the parable of the sheep trapped in a pit, and Jesus speaks as if he is the expected Shepherd of Israel (vv. 10-12) facing hostility from her shepherds (v. 14).

Jesus' choice of the sheep metaphor in this context is no accident, as we have examined earlier, if we take seriously the numerous examples of the association of Jesus' mission of compassion with the sheep/shepherd metaphor in the Gospel (9:35-36; 10:1-6; 12:9-12; 14:13-21; 15:24, 29-39; 18:12-14). Here we see the healing shepherd: "many followed him, and he healed all their sick" (12:15). Next, Matthew takes a quote from Isa 42:1-4, which most likely was familiar to early Christians (vv. 18-21). The last part of this OT text already envisions the extension of the reign of this incomparable healer who is full of extraordinary compassion for the sick: "In his name the nations will put their hope" (cf. Mic 5:3-4; Matt 2:6 [Mic 5:2]). Then who is this healer?

Matthew's quote from Isaiah 42 spells out that he is 'the Servant' on whom God put his 'Spirit' (v. 18), and his is a mission of compassion in his ministry carried out among his own people (v. 19-20) before the nations flock to him (v. 21). Yet again we must ask: Who is this healer, not to Matthew's implied

---

[245] Davies and Allison, *Matthew*, 2:308.
[246] Kim Paffenroth's article, "Anointed," 553, emphasizes the comparison of David and Jesus as the one greater than David: "In Jesus' healings, he is shown to be greater than his father David."

readers but to the historical crowd in the time of Jesus? The quote from Isa 42 is provided by the narrator who indicates, "this was to fulfill ..." (v. 18), but the narrator takes the readers back to the scene of the historical Jesus in v. 22: "then [τότε], they brought him a demon possessed man who was blind and mute, and Jesus healed him."[247] Matthew's implied readers may have affirmed the healing Jesus as the Isaianic Servant; however, the narrative suggests 'that particular knowledge,' known to Matthew's readers, is not yet known to the 'historical crowd' in Jesus' time. The crowd *still wonders* who this compassionate healer could be! Hoping to gain approval from their leaders, in their stunned response, the crowd takes what proves to be the best guess: μήτι οὗτός ἐστιν ὁ υἱὸς Δαυίδ (v. 23).

Up to this point, the narrative sets a foil so that the question evoked from the crowd does not seem baseless. Earlier the name David was brought up (12:1-3), but Jesus is far more than the historical David, especially with his extraordinary compassion for healing the sick (12:4-8). The following sheep metaphor associated with Jesus' healing mission (12:9-14), thereby, subtly invites readers to presume that Jesus may be the true Shepherd who really cares about the sheep in their many troubles, namely, the crowd itself.[248] When at last we reach verse 23, the crowd is nearly persuaded to conclude that this healing Jesus could be the eschatological Shepherd for the harassed flock (9:36; 2:6), and thus perhaps the promised Davidic Shepherd. The text of Matthew yields some clues.

Refuting the Pharisees' distorted slander that stands in stark contrast to the crowd's positive reaction to this extraordinary healer, Jesus states by what authority he heals and why he does so (12:25-33). Interestingly, Jesus neither refutes the slander nor openly acknowledges the crowd's positive view of himself as ὁ υἱὸς Δαυίδ. According to Jesus' own testimony, his authority for healing resides in the Spirit, God's eschatological agent for the restoration of Israel (v. 28: εἰ δὲ ἐν πνεύματι θεοῦ ἐγὼ ἐκβάλλω τὰ δαιμόνια, ἄρα ἔφθασεν ἐφ᾽ ὑμᾶς ἡ βασιλεία τοῦ θεοῦ), and his mission's main goal – healing the sick on Sabbath – is *to gather* (συνάγω, v. 30). Healing is but part of the gathering task. Gathering the scattered is a prerequisite for YHWH's rebuilding of the nation, the twelve tribes, which is the main task of the new David (Ezek 34-37). The echo is loud and clear. Jesus in Matt 12:25-31 is most likely reminiscent of the eschatological Shepherd in Ezekiel 34-37: "For I will take you out of the nations; I will *gather* you from all the countries and bring you back into your own land ... and I will put *my Spirit* in you and move you to follow my decrees and be careful to keep my laws" (Ezek 36:22, 27; Italics mine).

---

[247] Cf. Cope, *Scribe*, 49; Neyrey, "Thematic Use of Isa 42,1-4," 459.

[248] Beaton, *Isaiah's Christ*, 163-172, explores the possible link of Matt 12 with Ezek 34 (cf. 4Q521) to reinforce his argument that Matthew uses Isa 42 to underline the theme of justice through healing as well as that of 'meekness and humility' known as the typical motif common to Isa 42:1-4.

Again, Jesus' healing is an essential part of his mission to *gather* the flock, and not to *scatter* them (Matt 12:30; cf. "[the Son of David] συνάξει λαὸν ἅγιον," *Ps. Sol.* 17:26). Gathering, seeking, and healing the scattered flock are the necessary tasks for the eschatological Shepherd. The verb ἄγω ('to lead'), here in the eschatological context of Matt 12:25-32, may well be taken as a technical term for YHWH's shepherding activities in Ezekiel 34-37: YHWH's 'seek-and-deliver' mission (Ezek 34:10, 12). This is illustrated in a series of redemptive actions set out in 34:13: "I will cause them to come out (LXX, ἐξάγω)"; "I will gather them (LXX, συνάγω) from the peoples and the countries"; "I will lead them into (LXX, εἰσάγω) their own land." [249]

In other words, Jesus' answer to the Pharisees in Matt 12:25-30 constitutes his own reply to the crowd's question- μήτι οὗτός ἐστιν ὁ υἱὸς Δαυίδ (v. 23). The crowd dimly traces the link between Jesus healing and the Son of David, but Jesus clarifies that he is fulfilling the task of the eschatological Shepherd yet as ὁ υἱὸς Δαυίδ. He heals now 'to gather' the scattered in the power of the Spirit, making it his priority to restore the twelve tribes (10:1-6), though eventually he would draw the nations to himself (Matt 12:21; Isa 42:1-4; Mic 5:1-4; cf. Matt 28:16-20). The next episode centers on a witness that we might consider a sneak preview of this future.

4.2.4.3 The Canaanite Woman (Matt 15:22)

By nature, a rumor tends to circulate despite best efforts to prevent it. Jesus often warns those healed not to spread the message about him, though it seems to be in vain (Matt 8:4; 9:30). The future to come is felt much nearer in the story of the Canaanite woman; this Gentile woman comes to Jesus asking him to heal her daughter. Unlike Mark (7:24-31), Matthew focuses on Jesus' mission of healing and not on exorcizing demons, and Matthew retains the designation, 'Son of David.' [250] The Canaanite woman cries out to Jesus for help: ἐλέησόν με, κύριε υἱὸς Δαυίδ ("Have mercy on me, O Lord, Son of David," 15:22).

The focus of this story of the Syrophoenician woman (Matt 15:21-28//Mark 7:24-30), lacking some of the typical motifs of a regular exorcism story, rests upon, as Meier points out, the "thrust-and-parry dialogue" between Jesus and Syrophoenician woman.[251] Further, Matthew's report differs from Mark's version significantly in Matt 15:22-25, 28: e.g., ἐλέησόν με, κύριε υἱὸς Δαυίδ (v. 22); οὐκ ἀπεστάλην εἰ μὴ εἰς τὰ πρόβατα τὰ ἀπολωλότα οἴκου Ἰσραήλ (v.

---

[249] The idea of feeding the flock is also part of the mission: "I will shepherd them (וּרְעִיתִים; LXX, βόσκω) on the mountains of Israel" (v. 13d).

[250] Duling,"Therapeutic," 402.

[251] Meier, *A Marginal Jew*, 2:659, notes that the motifs missing are such as: confrontation between Jesus and the demoniac, struggle of the demon to ward off Jesus, Jesus' rebut and command to come out, the amazement of the crowd); rather, the focus of the story.

24); κύριε, βοήθει μοι (v. 25).²⁵² Yet, Matthew's characteristics, which reflect more Jewish tenets,²⁵³ do not cause any serious conflict with Mark's report, nor do they necessarily prove literary dependency.

The historicity of the incident is favored by a number of unusual, concrete details to which we can point: e.g., a pagan Syrophoenician woman in the region of Tyre; a woman interceding for a demoniac daughter; also, the rejection motif (i.e., Jesus' addressing a sincere petitioner with such harsh, insulting language) is found nowhere else in the Gospel tradition.²⁵⁴ Yet Meier chooses to remains suspicious about the historicity of the story, and thinks the story "is so shot through with Christian missionary theology" (cf. ἄφες πρῶτον χορτασθῆναι τὰ τέκνα, Mark 7:27).²⁵⁵ The claim that Jesus had no idea or intention as to the strategy to deal with Israel first and then reach out to the nations, however, is to be taken as one's methodological/ theological assumption and must not be passed off as evidence.

Regarding the vocative κύριε as a clue to the christology in the story, the opinions vary. The prefixing of κύριε might indicate, as Gundry argues, that the woman hopes to win from Jesus an exceptional benefit "in view of his universal dominion."²⁵⁶ Some say this is even the informed Christian confession, but others think it quite ambiguous,²⁵⁷ or at best, no more than a polite address like 'sir.'²⁵⁸ The term κύριε appears to play a important role in this context, although it may not be an informed Christian confession.

In the Gospel, Matthew is consistent in his use of the title as the disciples address Jesus this way (8:25; 14:28, 30; 16:22; 17:4; 18:21). The sick do likewise when they approach the Lord for help (8:2, 6, 8; 9:28; 15:22, 25, 27; 17:15; 20:30-31, 33), and it is plausible that they do so with an emphasis on "the sovereign will of Jesus."²⁵⁹ Likewise, the episode in 15:21-28 centers around the pagan woman's request for Jesus to extend his sovereign will of healing beyond the limits of his own mission to Israel.

The key issue addressed in the episode itself is the possibility and meaning of Jesus healing the sick outside of the lost sheep of the house of Israel (cf. 10:1-6). Carter notes that the woman's petition 'challenges' Jesus' very identity and mission stating, "it confronts Israel's imperialist ideology."²⁶⁰ On the surface, it appears so. It is not likely, however, that overcoming "Israel's imperialist ideology" is totally novel to Jesus or to the Evangelist. It is well

---

²⁵² Gundry, *Matthew*, 310-311; Luz, *Matthew*, 2:336, ascribing these discrepancies to Matthew's redaction.

²⁵³ Davies and Allison, *Matthew*, 2:542, characterize the differences "more Jewish"; "potentially offensive to non-Jews," finding the parallels in Matt 8:5-13 and 9:27-31.

²⁵⁴ Meier, *A Marginal Jew*, 2:660; also, see Keener, *Matthew*, 415.

²⁵⁵ Meier, *A Marginal Jew*, 2:660-661.

²⁵⁶ Gundry, *Matthew*, 272.

²⁵⁷ Carson, *Matthew*, 2:354.

²⁵⁸ France, *Matthew*, 246; E. A. Russell, "The Canaanite Woman and the Gospels (Mt 15:2-28; cf. Mark 7:24-30)" 126 *StudBib* (1978): 265.

²⁵⁹ Luz, *Matthew*, 2:6, 339.

²⁶⁰ Carter, *Matthew*, 272.

4.2 The Davidic Shepherd as the "Therapeutic Son of David" 315

planned out in the vision of Israel's restoration, and the woman's request could imply how widely the rumor circulated of the divine fulfillment of the extension of the Shepherd's rule over the nations (Matt 12: 21[cf. Isa 41:3-4]; Mic 5:3-4 [cf. Matt 2:6]).

By addressing Jesus as υἱὸς Δαυίδ, the woman recognizes Jesus as Israel's messiah.[261] This is affirmed by Jesus' reply concerning his mission to 'the lost sheep' of Israel (15:24). It is noteworthy for our purposes, that the woman designates Jesus as υἱὸς Δαυίδ (v. 22) in terms of his mission as the eschatological Shepherd for the lost sheep of the house of Israel. Further, this very mission is clearly to be extended to the nations by the Davidic Shepherd. In this respect, the presence of κύριε becomes more understandable. There is no need to see it as Christian terminology, especially due to the unwarranted assumption that the Jewish tradition does not include a healing Davidic messiah. In this respect, the implied expectation in the pagan woman's request that Jesus would heal her daughter was not an impossibility.

Viewing the theme of 'Glaube' as the most distinctive message in Matthew's presentation of the story in this passage, Hubert Frankemölle argues that Matthew omitted Mark's πρῶτον (7:27), since for Matthew Jesus as 'YHWH's messiah' promised for his own 'Bundesvolk' is sent 'only' (as in exclusively) to the Jews, and not necessarily 'first' to them. The reason why, he contests, is that "Jesus nur zu Israel gesandt war, dieses ihn abgelehnt hat, die Heiden aber aufgrund ihres Glaubens Zugang zum Heil fanden."[262] Nevertheless, this 'Pauline' Matthew as a 'redactor' betrays the bare facts presented in the story. The Gentile woman calls Jesus as 'Son of David,' Israel's messiah.

It is important to note that in the narrative, the motif of Israel's rejection of her Shepherd, though it might be implied (cf. Matt 15:1-20), is neither explicit nor required for her faith. The woman places her 'Glaube,' the Evangelist reports, in the υἱὸς Δαυίδ who heals among Israel, and not necessarily the 'rejected' messiah; that is, asking for the bread 'crumbs' is not the bread itself. Jesus' allowance of her request is certainly an anomaly, but it can readily be seen as a foretaste of what is to come as the Son of David will be established as one Shepherd over both Israel and the nations (cf. 2:6; 28:16-20). The woman's request simply hastens the coming of the universal reign of the Davidic Shepherd.

---

[261] Hill, *Matthew*, 254.
[262] Hubert Frankemölle, *Jahwebund und Kirche Christi: Studien zur Form und Traditiongeschichte des Evangeliums nach Matthäus* (Münster: Verlag Aschendorff, 1973), 114-115; similarly, Heinz J. Held, "Matthew as Interpreter of the Miracle Stories," in *Tradition and Interpretation in Matthew* (ed. G. Bornkamm, G. Barth and H. J. Held; Philadelphia: Westminster Press, 1963), 197-200.

Another exegetical difficulty in this episode has been identified as Jesus' initial refusal of her request for mercy.²⁶³ Novakovic comments on the complexity of the issue: "The behavior of Jesus does not coincide with *the general trait of his character – compassion*; on the other hand, Matthew takes pains to show that there is a specific reason for Jesus' initial attitude, which does not effect his basic disposition toward those who come to him in expectancy of mercy."²⁶⁴ The surprising effect of Jesus' initial refusal casts doubts on Jesus' position regarding issues such as gender and ethnicity.²⁶⁵ More important, it can raise a question about the consistency of Jesus' compassion toward the sick regardless of gender or ethnicity; how could such a compassionate Jesus utter such harsh things to the Canaanite woman?

Yet it is to be noted that the terminology of 'compassion or mercy' is rather strictly bound to the theme of Jesus' mission to Israel in the Gospel. Whenever the compassion motif occurs, what is specifically in view is Israel's restoration. Healing the sick among the lost sheep of Israel is also part of this vision initiated and motivated by divine compassion. We might even suggest that it is not necessarily the 'general' trait of his character.

Rather, *the compassion is a characteristic of the eschatological Shepherd for his own flock*. The compassion/mercy motif functions thus as the key to Jesus' mission of restoration for the lost sheep of the house of Israel. Still, Jesus as the Davidic Shepherd heals the Canaanite's daughter since he is also to be the Shepherd over the nations (Mic 5:3-4; Isa 41:3-4). This is only *proleptic* for what is to become true of the 'one' Shepherd over both Israel and the nations (cf. Matt 25:31-46), which is not without parallel in Jesus' contemporary thoughts (*Ps. Sol.* 17-18; *An. Apoc.* 90:17-18; 4 Ezra 2:34; cf. Ezek 34:23; 37:24).²⁶⁶

4.2.4.4 Two Other Blind Men (Matt 20:30-31)

Preceding Jesus' entry into Jerusalem, the story of Matt 20:29-34 resembles that of Mark 10:46-52, though the 'compassion of the Son of David' is distinctly emphasized in Matthew's version. Instead of Mark's υἱὲ Δαυὶδ ['Ἰησοῦ], ἐλέησόν με (10: 47, 48),²⁶⁷ Matthew retains ἐλέησον ἡμᾶς, κύριε, υἱὸς Δαυίδ (20: 30, 31), bringing the compassion of the Son of David into the spotlight.²⁶⁸

---

²⁶³ Cf. Elaine Wainwright, "The Matthean Jesus and the Healing of Women" in *The Gospel of Matthew in Current Study* (Grand Rapids: Eerdmans, 2001), 74-95.

²⁶⁴ Novakovic, *Healer*, 84 (italics mine).

²⁶⁵ Keener, *Matthew*, 416-417, notes the change of Matthew's 'Canaanite' (the ethnic issue) instead of Mark's 'Greek' (the class issue), and underscores the language of 'dogs' as certainly unpleasant one (Eurip. *Orestes* 260; Hom. *Il.* 8.527; 9.373; *Od.* 17.248; 22.35).

²⁶⁶ Gnilka, *Matthäusevangelium*, 2:31, notes on vv. 23-24 stating, "Als Hirt Israels erweist sich Jesus als der Messias," yet quoting Isa 53:6, Mic 2:12, and Zech 9:16 (LXX).

²⁶⁷ The name Ἰησοῦ is dropped in the second cry.

²⁶⁸ Matthew's vocative nominative υἱὸς Δαυίδ, instead of υἱὲ Δαυίδ in Mark may indicate Semitism; refer to *BDF*, 147.4.

4.2 The Davidic Shepherd as the "Therapeutic Son of David"

In accordance with this tendency, Matthew does not record Jesus' reference to the faith of the one healed (cf Mark 10:52); rather his compassion (σπλαγχνισθείς) is put forward as the real motivation for and as the cause of the healing (20: 34).[269]

The closest parallel is found in Mark 10:46-52. It is argued by many that one story of Jesus healing the blind (Mark 10:46-52) becomes two in Matt 9:27-31 and Matt 20:29-34 ('redactional doublet').[270] Should the possibility of literary dependency disregard the historical probability of the frequency of Jesus healing the blind (numerously attested in the Gospels)? Luz remarks, "the duplication of the story causes greater problems"– but to whom (to Luz or to Matthew?). Assuming these are separate occasions would present 'less' problems. Matthew clearly indicates different settings for the two occasions: ἐλθόντι δὲ εἰς τὴν οἰκίαν (9:28); παρὰ τὴν ὁδόν near Jericho (20:29-30).[271]

As to the historicity of the occasion, the story of Bartimaeus in Mark 10:46-52, as Meier argues, would be one of the strongest candidates for the report of a specific miracle going back to the historical Jesus. Here are his reasons. This is the only case in the Synoptics in which we are given names for the direct recipients [*bar Tim'ai*/son of Timaeus] of a miracle performed by Jesus (an argument from discontinuity).[272] Second, this named individual is tied to a precise place (the road outside Jericho leading up to Jerusalem), and third, the precise time of year is indicated (shortly before Passover).[273]

Observing the Matthean emphasis, Duling argues that Matthew uses Mark's healing Son of David to 'oppose' the Jewish concept of the Son of David as the militant warrior, who is yet to come and who will fulfill the promise of eternal reign on the throne of Israel.[274] Similarly J. M. Jones argues that the crowd, by rebuking the blind men who shout and beg the Son of David to heal them, reveals its "ignorance of this messiah's purpose (to replace Rome with the empire of Israel)."[275]

Yet as in the case of Matt 15:22, the crowd by itself was liable to associate the Son of David with his healing activity, possibly guessing it as part of the Davidic Shepherd's mission to the lost house of Israel. Hence, to contrast the crowd and the blind men may not be necessary. After all, the crowd frequently

---

[269] M. Müller,"The Theological Interpretation of the Figure of Jesus in the Gospel of Matthew: Some Principal Features in Matthean Christology," *NTS* 45 (1999): 172, underscores the theme of compassion on the Christological themes in Matthew's Gospel, yet without explicating its association with healing.

[270] Luz, *Matthew*, 2:46; Davies and Allison, *Matthew*, 3:104; Hagner, *Matthew*, 2:585.

[271] Keener, *Matthew*, 488, underscores the local and temporal settings for the case of Matt 20:29-31.

[272] Meier, *A Marginal Jew*, 2:688, suggests that the name Bertimaeus (Son [*bar*] of Timaeus) in Mark 10:46 likely reflects Markan Gentile audience, the name which is absent in Matthew's and Luke's versions.

[273] Meier, *A Marginal Jew*, 2:690.

[274] Duling, "Therapeutic," 407, 410.

[275] John M. Jones, "Subverting the Textuality of Davidic Messianism: Matthew's Presentation of the Genealogy and the Davidic Title," *CBQ* 56 (1994): 268.

brings the sick to Jesus (4:24; 8:16; 9:2, 32; 12:22-23). The two blind men can be seen as representative of the whole crowd already 'following' Jesus, their new Shepherd: ἠκολούθησεν αὐτῷ ὄχλος πολύς (20:29; cf. 9:27-31).

As has been argued thus far, in many instances Matthew associates healing and following (12:15; 14:13-14; 19:2), and healing is one of the dominant motifs in the crowds following Jesus.[276] In the case of 20:29-34, however, the two blind men join the crowd, following the Son of David only 'after' they have been healed (v. 34). In other words, besides recipients, so to speak, they might also be considered eyewitnesses of Jesus' compassion.[277]

This compassionate Shepherd overlaps with the picture of the humble king of Zech 9:9 announcing the return of YHWH to Zion in 20:29-21:17. The shepherd/sheep motif in Matt 20:29-34 proves to be significant. It prepares the readers to decipher the implications of Jesus' entry into Jerusalem as the humble Davidic Shepherd who will later be the wounded healer as he assumes the role of the smitten shepherd of Zech 13:7 in Matt 26:31. Thus the 'merciful' healing-Davidic Shepherd, witnessed by the two blind men who repeatedly call out to him in 20:30, 31, eventually places himself under YHWH's judgment that is to be inflicted on the wicked shepherd of Israel.

On the surface, Matthew explains this sacrificial role of Jesus and its connection with his healing ministry in terms of Isaiah's suffering Servant (e.g., Isa 53:4 in Matt 8:17). Yet the rich shepherd/sheep image from the OT Davidic Shepherd tradition, especially Ezekiel, provide the background for Jesus in his historical context.

Hence, a statement such as, "Matthew challenges a common [Jewish] messianic conception,"[278] merely repeats the untested axiom that Matthew's healing Son of David is antithetical to Jewish Davidic messianic expectation, thus asserting that the healing Son of David is a Christian alternative that subverts the Jewish expectation of the Son of David. Not only is it a challenge to define what "a common Jewish messianic conception" might mean during the time of Jesus,[279] but it is also inappropriate to base the compassionate healing Son of David completely on a Christian understanding in that it opposes any kind of Jewish messianic figure. The Davidic Son has already been subjected to profound revisions in the OT and in the Second Temple period as well, especially with the employment of shepherd/sheep images.

---

[276] Luz, *Matthew*, 2:550, notes on the historicity of the story stating, "it is emphasized that *back then* two blind men who called out to Israel's Messiah were given the light of sight; it is only as a report about an actual healing that this story becomes transparent for the experiences of the readers" (italics his).

[277] Cousland, *Crowds*, 164-166; cf. Gundry, *Matthew*, 405, underlines 'two witnesses.'

[278] Jones, "Subverting," 268.

[279] Cf. Collins, *Scepter and Star*, 40-51.

To miss this connection leaves one without any other compelling options but to accept the Solomon-as-exorcist tradition in Matt 20:29-34.[280] Needless to say, Solomon was not known as a healer, nor was he known for compassion, but he was known as an exorcist.[281] Matthew's description of Jesus is definitely alien to the legendary Solomon. His description bears close resemblance to the humble Davidic 'messianic' Shepherd/King in Zech 9:9 as set up in the following episode in 21:1-17. Besides its weak link with Jesus' healing activity, the figure of Solomon completely lacks the compassionate and messianic features of Jesus as the Son of David in Matt 20:30-31.

4.2.4.5 Healing in the Temple and the Children's Praise (Matt 21:14)

In Matthew 21:1-19, the crowd continues to follow Jesus as he enters the city of Jerusalem. Now the 'very large' crowd (ὁ δὲ πλεῖστος ὄχλος) going ahead of as well as following behind him, shout, ὡσαννὰ τῷ υἱῷ Δαυίδ (vv. 8-9; cf. v. 14).[282] Obviously, the two blind men Jesus healed in 20:29-34 who began to follow him (v. 34) join this larger crowd. Not only the blind men but also the whole crowd proclaim Jesus as the lowly 'Coming One,' the Davidic Shepherd (Zech 9-14). This humble messianic king is still the Son of David who heals, the therapeutic Son of David.[283] As Jesus enters the city, a key question is raised about his identity (v. 10), which is later readdressed more precisely in terms of the authority that would define who he really is (v. 23). The episodes inserted between these two questions – "Who is he?" (v. 10) and "By what authority are you doing these things?" (v. 23) – potentially offer clues.

What does Jesus do? As Jesus enters the temple, he rectifies the problem of the temple having become a 'den of robbers' (vv. 12-13). Moreover, immediately following the incident in the temple, Jesus heals the blind and the lame there: Καὶ προσῆλθον αὐτῷ τυφλοὶ καὶ χωλοὶ ἐν τῷ ἱερῷ, καὶ ἐθεράπευσεν αὐτούς (v. 14).

Matthew 21:14-17 has no parallel in Mark, while Luke 19:39-40 contains something similar: the Pharisees ask Jesus to rebuke his disciples for praising him. Gnilka thinks of it as a Matthean redaction.[284] Davies and Allison further speculate that the Matthean version shares the same tradition of an objection story by which Luke's version is explicable as well.[285] Yet, Matthew's report of Jesus healing the blind and the lame in the temple has no parallel even in Luke's version. Hagner presents a typical 'redactional' view: "the pericope [Matt 12:12-17]

---

[280] Charlesworth, "Solomon," 72-87; also, Novakovic' *Healer* inspired by Charlesworth's article.

[281] Charlesworth, "Solomon," 83, acknowledges that Solomon is not called a healer in the pre-70 Jewish tradition nor in the NT.

[282] Note the comparative πλεῖστος; it is merely ὄχλος πολύς in 20:29.

[283] Duling, "Therapeutic," 404-405.

[284] Gnilka, *Matthäusevangelium*, 2:207.

[285] Davies and Allison, *Matthew*, 3:133.

blends several types of material together under the common theme of things that take place 'in the temple'.[286]

Considering the purity laws prohibiting the entrance of those with physical abnormalities (1QSa 2:8-9; 1QM 7:4-6; 12:7-9; cf. Lev 21:17-18; 2 Sam 5:8 LXX),[287] Jesus' deeds performed in the temple must have been regarded as serious challenges against the authorities. Matthew seems to highlight that Jesus' healing the sick in the temple was witnessed by the chief priests and scribes (21:15). Such a controversial incident could hardly have been created by the Evangelist himself. Jesus healing in the temple yet plays a critical role in Matthew's presentation of Jesus; this is likely why he preserved the story.

The close connection between the cleansing of the temple and the coming forward (προσῆλθον) of the blind and the lame to Jesus in the cleansed temple seems to suggest that the sick, i.e., the outcast and the lost, are *finally found, brought back to, and healed at the center* of YHWH's covenantal community.[288] Undoubtedly multiple layers of implications emerge from Jesus' act of cleansing the temple in vv. 12-13. Luz lists various possibilities:[289] (i) political action;[290] (ii) prophetic sign-action [Zeichenhandlung] such as (a) restoration of true cult, (b) destruction of the present temple, (c) resistance against the economic power of the temple aristocracy.[291] Additionally, Keener lists more: a protest against unholiness[292] and a protest against ethnic segregation based on the fact that the selling occurred in the outer court, beyond which Gentiles could not travel (cf. Jos. *Ant.* 12. 145; 15.417).[293]

None of these options falls short of bearing some significance upon Jesus' action in the temple, and as Luz points out, "Jesu Einzug betrifft die ganze Stadt Jerusalem und den ganzen Tempel." [294] Yet of significant messianic import in Matthew is that the Evangelist associates Jesus cleansing the temple (vv. 12-13) with his healing the blind and the lame who came to him 'at the temple' in the

---

[286] Hagner, *Matthew*, 2:599.

[287] Gnilka, *Matthäusevangelium*, 2:208; Keener, *Matthew*, 502.

[288] Carter, *Matthew*, 420, highlighting God's 'inclusive [universal] and merciful empire,' misses out on the covenantal and national significance of the event both for Israel's and the nations' destiny.

[289] Luz, *Matthäus* 18-25, 185-187.

[290] Luz, *Matthäus*, 185. Jesus at the temple is seen as 'a messianic pretense' such as 'the Egyptian' who led, according to Josephus, thirty thousand supporters on the Mount Olive ready for a sudden attack to seize Jerusalem. Keener, *Matthew*, 498, turns down this option as he sees Jesus' action as "more symbolic than efficacious; also, for the probably minimal impact of the action, see E. P. Sanders, *Jesus and Judaism*, 11.

[291] Sanders, *Jesus and Judaism*, 61-71.

[292] For instance, Borg, *Holiness*, 171-172; Ethelbert Stauffer, *Jesus and His Story* (trans. Richard Winston and Clara Winston; New York: Alfred A. Knopf, 1960), 67.

[293] Keener, *Matthew*, 495-501.

[294] Luz, *Matthäus* 18-25, 187.

4.2 The Davidic Shepherd as the "Therapeutic Son of David"

following verse 14, the link which is absent in the synoptic parallels (Mark 11:15-18; Luke 19:45-47).[295]

Likely, the return of the outcast and the lost into the cleansed temple symbolizes nothing less than part of the fulfillment of Israel's restoration. Jesus cleansing of the temple thus may well indicate that the royal Son of David has finally arrived at the center of the community to restore justice/righteousness.

In this respect, Matthew's omission of πᾶσιν τοῖς ἔθνεσιν (Mark 11:17) can be seen as an indication of the Evangelist's vantage point of Israel's restoration.[296]

Again, the return of the lowly king to the city of Jerusalem coincides with the return of the outcast and the lost. Surely, *the theme of reversal* could not be more evident than at this moment. Dramatically those outcasts are finally brought back to the center of the community, and returned to them are the blessings, e.g., healing. Now, the exilic curses are reversed even at the center of the community – in their temple – through the restoring presence of Jesus, the healing Son of David, in their midst.

According to Ezekiel's vision of the restoration in chapter 34, YHWH's return to his own flock is initiated by his seeking the lost and healing the sick. The negligent and wicked leadership of Israel's shepherds resulted in a destructive strata of injustice within the community: the strong and fat sheep at its center, the weak, the sick, and the down-trodden in the middle, and the outcast and the lost banished to the periphery (Ezek 34:4).

YHWH's return reverses this strata of exile: He seeks the lost, brings back the outcast, heals the sick, binds up the down-trodden, strengthens the weak, but polices the strong and the fat (v. 26). *His return implies the return of the lost and the outcast back to the center of the community.* The promised divine reversal is wrought not by sword but by the compassion and the authority of the Davidic Servant and Shepherd. YHWH's reversal is now about to be completed. Is this not what is happening in Matt 29:29-34 and 21:1-17?[297]

In returning to the question of Jesus' identity, Matthew's narrative communicates that his identity is to be defined in terms of the titular use of 'Son of David,' but particularly as the healing eschatological Shepherd. More precisely,

---

[295] Gnilka, *Matthäusevangelium* 2, 208, yet refers to the story of David in the temple in 2 Sam 5:8.

[296] Keener, *Matthew*, 500, thus thinks the omission weakens the option of 'protest against ethnic segregation'; Luz, *Matthäus*, 3:187, explains it from the viewpoint of Matthew's community after A.D. 70.

[297] YHWH as the eschatological Shepherd promises, "I will gather" (וְאָסַפְתִּי) the nations (Zech 14:2). The restoration of the lost and the outcast among the sheep of the house of Israel coincides with the return of God's presence to the temple. This is to be manifested, according to Zech 14:21, by the casting out (this time) of the money changers and the merchants from the temple: "And on that day there will no longer be a Canaanite (כְּנַעֲנִי, 'traders' or 'merchants') in the house of the Lord Almighty."

he does what YHWH promised to do as the eschatological Shepherd of Israel, and at the same time, he is also hailed as the Davidic appointee.[298] This can be supported first by 'the inclusion' of the titular use of the Son of David (vv. 9, 15) and second, this Son of David is, in particular, 'the healing' Son of David.

Unlike in Mark 11 where Jesus' entry is separate from his cleansing of the temple as he goes out to Bethany (v. 11), Matthew does not indicate any break in the events. We see a succession of episodes from the time Jesus enters the city (v. 10): his entrance is greeted with and accompanied by the crowd's cry, ὡσαννὰ τῷ υἱῷ Δαυίδ (v. 9); next, Jesus cleanses the temple (vv. 12-13); the outcasts come back into the temple and Jesus heals them (v. 14); and finally, we reach the climax: Praise emits from the mouths of children, who shout and once again we hear: ὡσαννὰ τῷ υἱῷ Δαυίδ (v. 15).[299] The inclusion of the identical proclamation from both the crowd and the children (vv. 9, 15) – their exaltation of Jesus as the Son of David – is affirmed by Jesus himself as he quotes Psalm 8 in v. 16.[300] Both the crowd and Jesus are in accord, and with one voice reply to the question of the identity of Jesus (v. 10). He is the Son of David who seeks the lost and heals the sick; and this 'therapeutic Davidic Shepherd,' finally in the temple in Matt 21:16, accepts the praise worthy of YHWH as he himself fulfills the task of YHWH as the promised eschatological Shepherd among the lowly ones.

Truly, this Son of David draws the outcast and the lost back to the center of the community and heals them as he cleanses the temple in view of renewing the whole covenantal community (vv.11-14). Gundry asserts that in this way Matthew "combines royal messianism (Jesus as son of David) with prophetic messianism (Jesus as miracle worker) [under the Christian influence]."[301] A problem persists. David is associated with the prophet's predictive abilities, but not with the miracle-working abilities of prophets like Elijah or Elisha, either in the OT or pre-70 Judaism. A much more relevant figure associated with prophetic messianism is Moses rather than David.[302] We are left to wonder if the

---

[298] France, *Matthew*, 302, notes that the praise of Ps 8:2 quoted by Jesus is offered to God and not to the Messiah.

[299] The second cry of 'Hosanna, the Son of David' is absent in Mark 11 and the same is true of Jesus' healing the blind and lame in the temple; rather the whole crowd is amazed *by his teaching* in Mark (v.18).

[300] Carson, *Matthew*, 2:443, aptly notes that Jesus is not only acknowledging his messiahship but also justifying the praise of the children by applying to himself a passage of Scripture applicable only to God. Cf. Davies and Allision, *Matthew*, 3:141-142, who link the quote with Moses typology (cf. Exod 15:2) while neglecting the significance the explicit context of Matt 21:15-16, that is, Jesus' quote of Ps 8:2 is his answer to the children's praise of 'Hosanna, the Son of David.'

[301] Gundry, *Matthew*, 412-413.

[302] W. A. Meeks, *The Prophet-King: Moses Traditions and the Johannine Christology* (NovTSup 14; Leiden: Brill, 1967), 100-285; Novakovic, *Healer*, 110-112 lists such relevant texts as Deut 18:15; Num 24:17; 1 Macc 4:46; 1QS 9:11; CD-A 7:18; 4Q175 l.5-8; 4Q174 1-2,

association is a novelty invented by Matthew for his apologetic concerns (Duling), or if it is strictly Markan early church influence (redaction criticism)?

The theme of David as the father of Jesus occurs in 22:41-46 again (cf. 12:1-4). While paying attention to Jesus healing as the Son of David, Kim Paffenroth asserts that Matthew shows Jesus to be more than 'David,' but not necessarily more than 'the Son of David,' thereby affirming Jesus unambiguously as the Son of David.[303] While employing a typological concept, Paffenroth suggests that Jesus as the Son of David still differs from David and is thus 'more than David.' As a powerful warrior, David killed the figuratively blind and lame and excluded them from his house (2 Sam 5:6-8), but his son Jesus is a powerful healer who cures the literally blind and lame within his 'house,' the temple.[304] Paffenroth rightly affirms that Matthew does not deny that Jesus is the Son of David (22:41-46), invalidating the contrast between 'Jesus the Christian healing Servant-Messiah' and 'the militant Son of David.' Thus Jesus is the Son of David, he argues, but in a particular way.

Nevertheless, Peffenroth's typological interpretation of Jesus as the antitype of the historical David is so fragmentary and historically remote from Jesus and Matthew that it barely explains Jesus' identity as depicted in Matthew's narrative. In particular, Matthew's quotation from Zech 9:9 (Matt 21:5) indicates that Matthew is already interacting with the Davdic tradition developed long after the life of the historical David.

Matthew's consistent and coherent employment of the sheep/shepherd images – 'healing' and 'following' in 20:29-34; 21:9, 14-15 (cf. 2:6; 9:35-36) to name two – point toward the mission of the eschatological Shepherd and his Davidic Shepherd-Appointee. This model not only solves the problem of the puzzling association of the Son of David with healing, but provides readers of Matthew's narrative with a meaningful framework of the reference, particularly the Hebrew texts of Ezekiel, Zechariah, and Micah – that is, the OT Davidic Shepherd tradition.

4.2.5 Summary

In the First Gospel, the title Son of David is emphasized, but disputes arise over its implications. The title is understood as being restricted to Jesus' ministry to Israel, and related to Israel's rejection of her messiah. Moreover, the characteristically Matthean picture of Jesus as the healing Son of David remains puzzling with regard to its background. Various proposals, such as the Isaianic Servant, the Greek divine man, and the Solomon-exorcist legend, we have argued, fail to meet the criteria set in the Gospel and likewise in the traditional sources. In particular, the Solomon-exorcist view falters. Solomon was not known as a

---

21:11.
[303] Paffenroth, "Anointed," 553-554.
[304] Paffenroth, "Anointed," 553.

healer, but was renowned for his exorcism and wisdom. Importantly, no messianic connotations are attached to the figure while the Son of David is messianic in the Gospel.

For the background of Jesus as the 'healing' Son of David in the Gospel, we turn to the Davidic Shepherd tradition (CD-A 13:9; 4Q521 2, 2:12; 4Q504 1-2, 2:14; *Apoc. Ezek.* 5; cf. Zech [LXX] 12:10), noting in particular Ezekiel 34. Our investigation makes it necessary to challenge the axiom that the Son of David in Judaism is militant and therefore cannot be associated with miracles such as healing the sick. The shepherd image functions to transform the traditional view of the Son of David as a non-militant warrior, that is, the righteous Prince and Leader/Teacher. This revision of the Davidic messiah, however, already takes place in the OT (Ezek 34). As we tackled the texts of Matthew where the Son of David heals the sick (Matt 9:27; 12:33; 15:22; 20:30-31; 21:14), we concluded that Matthew's descriptions of the healing Son of David fit the mission tasks reserved for the eschatological Shepherd in Ezekiel 34.

Furthermore, it is not difficult to understand why it is the Son of David who assumes the role of the eschatological Shepherd in the Gospel. The Son of David, who will be the Davidic Shepherd-Appointee, is referred to by YHWH as 'My Servant David' (Ezek 34:23-24; 37:24-25; cf. *Midrash* on Ps 29:1). Who would fulfill the mission of the eschatological Shepherd for YHWH if not it is his Servant? YHWH's role of rescuing the flock, i.e., seeking, healing, gathering, etc. (cf. Ezek 34:1-17), and the role of the Davidic Shepherd-Appointee frequently overlap in a host of Qumran texts, while the distinction between the two is still maintained (Two Shepherds schema).

4.3 Conclusion

Our study of Jesus' mission to seek the lost (Matt 9:10-13) and healing the sick in chapter 4 ends with the conclusion that Matthew's Gospel presents the earthly mission of Jesus as the mission of YHWH as the eschatological Shepherd for his lost sheep of the house of Israel. Jesus' mission of seeking the lost is the critical sign of the inauguration of the promised Shepherd's theocracy over his flock. Ezekiel 34 and 37 help us to understand the significance of Jesus' mission to seek the sinners as well as their identity primarily as the lost and the outcast. Jesus' conflict with the Pharisees signifies YHWH's judgement upon Israel's shepherds that eventually results in his raising up the new Davidic Shepherd over the restored flock. This process of Jesus' confronting Israel's leaders is in nature eschatological and final. Further, Jesus' healing the sick demonstrates that he takes up the role of YHWH as the eschatological Shepherd for the sheep without a shepherd.

4.3 Conclusion

The well-known axiom that the Davidic messiah was not expected to be a healer baffles many as they realize that Matthew portrays Jesus distinctly as the healing Son of David. These two features of Jesus in the Gospel – namely, Jesus being the Son of David and Jesus being one who heals – are now fixed, but remained unsolved elements of the christological claims of the Gospel. In the past, the titular approach has dominated the studies of Matthew's christology.

As to the issue of Jesus' identity as 'the therapeutic Son of David,' a problem with the titular approach is its alliance with the unwarranted, antithetical premise of 'the healing [Christian] Jesus as the Servant' in Matthew's Gospel pitted against 'the [Jewish militant] Son of David' in Judaism. Thus it seems that the only possible background left for Matthew's Jesus as the healing Son of David is the Solomon-as-exorcist tradition, which as we have clearly demonstrated, has little relevance to the Davidic messianic figure in the Gospel.

The axiom that the Davidic messiah was not expected to be a healer raises serious doubts. Is there really no evidence linking the Son of David with healing in the OT and/or pre-70 Judaism? As we argued earlier, texts such as 4Q521 2, 2:7-13, CD-A 13:9-10, 4Q504 1-2, and *Apoc. Ezek.* 5, and most importantly Ezekiel 34, are firm indications for Matthew's Gospel that this particular association was no anomaly but rather was an integral part of the mission of the eschatological Davidic Shepherd.

We conclude that Davidic expectation, particularly after the exile, cannot be fairly described in terms of a one-sided militant enthusiasm. Expectation of the new David was modified repeatedly and consistently in texts such as Micah 2-5, Ezekiel 34-37, and Zechariah 9-14. Yet we maintain that the shepherd image is the key to unlock the modified visions of the coming Davidic rule. Moreover, our study shows that it is also true to Matthew's Gospel in that it embraces all the critical portions of the Davidic Shepherd tradition at the narrative's critical junctures (2:6; 9:36; 26:31).

Ezekiel 34-37 in particular provides Matthew's narrative with rich images of the eschatological Shepherd who seeks the lost, brings back the outcast, and above all heals the sick. Most important, Jesus who accomplishes the pastoral roles of the eschatological Shepherd is also called the 'Son of David.' Once again we see how the shepherd image binds together two unlikely but associated concepts – the Son of David and healing.

Matthew's own narration that introduces the healing Jesus as the Isaianic Servant is meant to be heard by the implied readers and not by the historical crowd. The crowd in the narrative is equally baffled, yet probably would not be frustrated by our axiom; indeed, they were able to make the connection between the Son of David and healing. Further, they as well as their leaders, the Pharisees, read the signs of the restoration of the nation to be wrought by this healing Son of David. Leading the blind and the lame, Jesus enters the city of Jerusalem and the temple, and the crowd – 'the children' included – shout, 'Hosanna, to the Son of David!'

The story does not end there, however. From the beginning of the narrative, it is not meant to be a militant victory of the Davidic messiah; rather, it is meant to be the arrival of the compassionate eschatological Shepherd who has come for the harassed and down-trodden sheep without a shepherd. After Jesus' entry into Jerusalem, the shepherd image is pushed forward to the climax of the story of the Son of David as the king of the Jews. Jesus as the representative of the shepherds of Israel in the past is to be smitten (26:31), and his healing ministry reaches its climax when the smitten shepherd himself is resurrected with the saints (27:51-53). Finally, the Davidic-Shepherd is appointed with all authority in view of the extended reign over the nations (28:16-20).

Of course, these readings remain open to serious inquiry, as we shall see in the task that awaits us in the next chapter. One of the strengths of the Ezekiel pattern for Jesus as the healing Son of David in the Gospel is its matchless comparability and coherence as compared to other sources and traditions. This is evidenced not only in the micro aspects of Jesus' mission of seeking and healing, but is also evidenced in the macro framework, which the Evangelist of the First Gospel uses to communicate both the identity and the task of Jesus as the therapeutic Davidic Shepherd. It is through his reign that eventually YHWH's holy name will be known and exalted among the nation.

Chapter 5

The Rise of the One Davidic Shepherd: Echoes in Matthew 27:51b-53 and 28:16-20

Matthew faithfully embraces the Davidic Shepherd tradition, and it is our suggestion that the First Gospel's adoption of the shepherd and sheep images is best understood in the framework of the comprehensive vision of the tradition. In previous chapters, we examined Matthew's textual interaction with the Davidic Shepherd tradition as well as Jesus' ministry of 'seeking the lost and healing the sick' as the two major features of the mission of the promised Shepherd.

Now we come a step closer to the complete picture of Jesus as the eschatological Davidic Shepherd as presented in Matthew's Gospel. Eventually, we hope to present that particular picture of Jesus in Matthew's narrative as a whole in the next chapter. There, we will assess Heil's thesis that Ezekiel 34 provides the First Gospel with narrative strategy in terms of the shepherd and sheep images.[1] Yet, in order to reach this point, we set our aim in this chapter to examine two more significant texts, Matt 27:51-53 and 28:16-20.

These two passages do not involve the picture of the 'therapeutic' Davidic Shepherd any more. The healing ministry of Jesus as the eschatological Davidic Shepherd is characteristic to his mission of compassion among the lost flock of Israel. After the incident of the smitten shepherd in Matt 26:31, new aspects of the Davidic Shepherd emerge as the narrative reaches its final climax at 28:16-20. We suggest that these texts, Matt 27:51-54 (cf. Ezek 37:1-14; Zech 14:4-5) and 28: 16-20 (cf. Ezek 34:23-24; 37:24-25; Mic 5:3-4), echo the Davidic Shepherd tradition, in a such way that the echoes in these texts complete Matthew's intertextual dialogue with the tradition in his presentation of Jesus as the Davidic Shepherd.

Admittedly, we find neither quotations nor undisputable textual allusions in those two passages in the Gospel. But, we should note that Matthew's more explicit textual interaction with the traditi on in terms of quotations (2:6; 26:31) and allusions (9:36, 10:6, 16; 15:24, 25:31-36) in other places in the narrative, would easily satisfy such criteria as 'recurrence' and 'thematic coherence' as to the credibility of the echoes that we claim in those texts.[2]

In fact, we proposed that echo refers to the intertextual discourse invoked by the codes and conventions intended by the author, which enable both author and

---

[1] J. P. Heil, "Ezekiel 34," 698-708.
[2] Hays, *Echoes of Scripture*, 29-32.

readers to engage in dialogue as they participate in reinterpreting the tradition.³ Along with the shepherd/sheep 'images' for the cases of echoes in Matt 7:15; 12:11-12; 18:12-14 and the 'theme' of the therapeutic Son of David (9:27; 12:23; 15:22; 20:30-31; 21:14), we classify Matt 27:51b-53 and Matt 28:16-20 under the category of echoes to the Davidic Shepherd tradition. The rationale for this classification is that we find particularly Ezekielian pattern in these texts. To articulate the echoes in terms of this Ezekiel's pattern thus will be our task in this chapter.

5.1 The Rise of the Shepherd and the Resurrection of the Saints

Matthew 27:51b-53, another Matthean *Sondergut* text, narrates a series of breathtaking and earthshaking events. As Jesus' death is marked by the rending of the veil in the Temple (v. 51a), the earth shakes, the rocks and tombs split open, and the saints rise from the tombs, who then appear in 'the holy city' [τὴν ἁγίαν πόλιν] after Jesus' resurrection (vv. 51b-53).

This is a neglected passage; Raymond E. Brown notes that the passage appears to embarrass many.⁴ Some argue that it is a displaced resurrection account, originally connected to 28:2.⁵ Not only its placement but also its descriptions are puzzling. Who are these 'saints?' Why were they raised in this momentum of Jesus' resurrection (v. 52)? What does 'the holy city' refer to, and what does it mean for the saints to enter it (v. 53)? How, to whom, and why did they show themselves? And, if they were risen from the dead immediately after Jesus' death (v. 52), and entered the city after Jesus' resurrection (v. 53), what were they doing between these events?⁶ Where did they go afterwards? None of these questions yield simple answers.

5.1.1 Historicity and Sources

As regards the historicity of the events, Hill's warning against 'a host of pseudo-problems' sounds prudent if one hopes not to lose sight of any meaning in the text.⁷ If one questions the possibility of the miracles in the Gospel, then the

---

³ Refer to Introduction, "The Methodology."

⁴ R. E. Brown, "Eschatological Events Accompanying the Death of Jesus: Especially the Raising of the Holy Ones from Their Tombs (Matt 27:51-53)" in *Faith and Future: Studies in Christian Eschatology* (ed. J. P. Galvin; New York: Paulist Press,1994), 43.

⁵ D. Hutton, "The Resurrection of the Holy Ones (Matt 27:51-53): A Study of the Theology of the Matthean Passion Narrative" (Th.D. diss., Harvard Divinity School, 1970).

⁶ Carson, *Matthew* 2:581-582, differentiates the two foci, i.e., the earthquake (at Jesus' death) and the breaking open of the tombs with the rising of "the holy people" (at Jesus' resurrection). Cf. J. W. Wenham, "When Were the Saints Raised?" *JTS* 32 (1981): 150-152.

⁷ D. Hill, "Matthew 27:51-53 in the Theology of the Evangelist," *IBS* 7 (1985): 76.

issue is one of world-view, exemplified by Deism. Yet, as Matthew reports various miracle stories in the narrative, it is no strain to agree with D. Wenham that the majority of Matt 27 has all the appearance of being in intention 'a straightforward description of historical events' with little or not hint given of any changed intention in verse 51 or elsewhere in the chapter.[8]

While defining the language of the text in such terms as 'the semi-poetic format of the material' or 'a dramatization in quasi-poetic format,' Brown conjectures that the text reflects 'the unnuanced, prejudiced, theological judgments found among the ordinary people' late in the first Christian century.[9] While this makes for an attractive theory, the poetic description does not necessarily exclude its possible historical ground; these two are not mutually exclusive. For instance, the poetic description sung by Moses with the Israelites and Miriam in Exod 15:1-21 does not blur, but rather matches with the preceding historical narrative in Exod 13-14.

Similarly, Donald Senior points out that this type of poetic format is found also in Matt 7:25, 24:29 where the materials do not appear to be drawn from popular folk stories.[10] Further, the earthquake is witnessed by the centurion, and the resurrected saints are said to have appeared to many. Unless the Evangelists were all mistaken in supposing that the events of the passion and resurrection were of more than human significance, Wenham argues, it would be most unwise to ascribe symbolic significance to any and every unusual and supernatural occurrence in the narratives (e.g., to the rending of the veil).[11]

Besides Brown's proposal for Matt 27:51b-53 as an 'unnuanced popular folk story,' the speculations about the origin of the text present various alternative sources; from the early church, the *Gospel of Peter,* the OT, and intertestamental Jewish apocalyptic materials.[12] The association of the text with the 'descent into hell' motif in 1 Pet 3:19 and in the early creeds are no longer taken

---

[8] D. Wenham, "The Resurrection Narratives in Matthew's Gospel," *TynBul* 24 (1973): 43.

[9] Brown, "Holy Ones," 44, 55; in addition, see *The Death of the Messiah: From Gethsemane to the Grave* (New York: Doubleday, 1994), 1:61.

[10] Don Senior, "Revisiting Matthew's Special Material in the Passion Narrative," *ETL* 70 (1994): 421.

[11] Wenham, "Resurrection," 43-44; furthermore, on p. 46 he notes contemporary observations of the earthquake (*Josephus, Jewish War* 4.299), a cleavage in the masonry of the temple porch (a letter of Jerome, 120.8), and the Holy Place left open for forty years before the fall of Jerusalem (*B. Yoma* 39b): also, Keener, *Matthew*, 687, presenting similar evidence, warns against "the modern presuppositions about the possibilities of the event's historical reliability of the Gospel traditions as a whole."

[12] As to the intertestamental materials, R. L. Troxel, "Matt 27.51-4 Reconsidered: Its Role in the Passion Narrative, Meaning and Origin," *NTS* 48 (2002): 30-47, suggests *1 Enoch's* 'Apocalypse of Weeks.'

seriously.[13] Whether *GP* is argued to represent a tradition independent of the synoptic Gospels (Hutton) or even influential to the Gospels (Crossan, especially on Matt 27:51b-53), the arguments for the priority of *GP* over the synoptic Gospels present many difficulties. Based on a glaringly subjective criterion – 'the argument from a better-flowing narrative' – Crossan complains that in Matthew the eschatological effects of the great fear are recorded without apparently having any effect on anyone, and thus regards Matt 27:51b-53 as an awkward adaptation from *GP* 6. 21, which reads: "And then they [the Jews] drew the nails from the hands of the Lord and laid him on the earth; and the whole earth shook and there came a great fear."[14]

Refuting this proposal, Brown corrects Crossan's statement by simply pointing out that Matt 27:54 indeed records the response of the centurion and those who were with him: "they feared exceedingly, saying, 'Truly this was God's Son.'"[15] Brown drives home his point that Matthew narrates "more effectively than does *GP!*" while repudiating Crossan's interpretation based on Crossan's own susceptible criterion.[16] Interestingly, pleading for the priority of Matthew's scene, Brown suggests that it is much closer than *GP* to a group of eschatological Scripture passages such as Joel 4:16 (the earth shaken), Ezek 37:12 (graves opening, raising from the graves, and bringing people home to the land of Israel), and 1 Cor 15:20 (the Christ being the first fruit).[17] Certainly, Matt 27:51b-53, reflecting these patterns in the OT eschatological visions, depicts what occurred at the momentum of Jesus' death and resurrection.

5.1.2 Traditions behind the Text

Among the OT traditions possibly standing behind this seemingly enigmatic scene, there is a broad consensus that the text of Matt 27:51b-53 is largely indebted to the portrayal envisioned in Ezek 37:1-14.[18] Along with this Ezekiel passage, such elements as the earthquake, the splitting of the rocks, and the Lord's appearance with all the saints (πάντες οἱ ἅγιοι, LXX) are suggested to

---

[13] Senior, "Matthew's Special Material in the Passion Narrative," *ETL* 63 (1987): 278; cf. Luz, *Matthäus* 26-28, 359, is still supportive of the Christology of "Christi Höllenfahrt," while viewing "the saints" as the Gentile; thus, Jesus as "Erlöser der Welt."

[14] John Crossan, *Gospels*, 141-142.

[15] Brown, "The Gospel of Peter and Canonical Gospel Priority," *NTS* 33 (1987): 330.

[16] Brown, "Gospel Priority," 331.

[17] Peter J. Kearney, "He Appeared to 500 Brothers," *NovT* 22 (1980): 274, lists similar NT passages such as Rom 1:3-5, Acts 6:15, 7:55-56; Eph 4:8-13; 1 Tim 3:10-14.

[18] Troxel, "Origin," 42; Gnilka, *Matthäusevangelium* 2:477; Senior, "Passion," 280, 282, lists the names of Riebl, Maisch, Hill, Gundry; cf. Luz, *Matthäus* 4:357, 364, recognizes linguistic links such as ἀνοίγω and μνημεῖα.

derive their roots from Zech 14:4-5.[19] The Dura-Europos panels (A.D. 245-256), discovered in the synagogue at Dura on the upper Euphrates near ancient Mari, are known as a Jewish interpretation of the vision of Ezek 37:1-14.[20] This archeological find which expresses the lingering hope of Ezekiel's vision of revivifying those exiled indicates that Matthew is not alone in his reading of Ezek 37:1-14 in terms of the eschatological resurrection of the dead, i.e., of the body rising from an opened grave, which, at the same time, could be read as a parable of the restoration of Israel.[21]

Whether Ezek 37:1-14 depicts the eschatological resurrection of the dead, or merely expresses metaphorically the restoration of Israel from the exile, to choose between two aspects of the symbolic expression in the text seems more an interpretive question. As many church fathers found the option for the final resurrection of the dead was in this Ezekielian text,[22] the language taken from Ezek 37 can be seen as indicating a literal, individual resurrection.[23]

Yet, on the other hand, many argue that Ezekiel 37 is merely a parable for Israel's future restoration from exile, seeing this view as 'typically Jewish' in contrast and opposed to the prevailing Christian resurrectionist view.[24] Moshe Greenberg states that the dry bones are of the whole house of Israel and the metaphoric status of the dry bones is assured by the exiles' use of the expression with reference to themselves; they cannot mean literal death, and "Ezekiel's vision concretizes the metaphor, but it remains a metaphor for all that."[25]

---

[19] Dale C. Allison Jr., *The End of the Ages Has Come: An Early Interpretation of the Passion and Resurrection of Jesus* (Philadelphia: Fortress Press, 1985), 40-46, underlines the significance of Zech 14:4-5, and he also lists the texts associated with the earthquake in the context of eschatological events: Joel 2:10; Hag 2:6; I Enoch 1:3-9; 4 Ezra 6:13-16; *2 Apc. Bar.* 70:8. Roger David Aus, *Samuel, Saul and Jesus: Three Early Palestinian Jewish Christian Gospel Haggadoth* (Atlanta: Scholars Press, 1994), 129, denies the historicity of Matthew's text and suggests Dan 12 but believes it basically derives from the Judaic tradition of 1 Samuel 28.

[20] For dating the panels and other archeological concerns, see Clark Hopkins, *The Discovery of Dura-Europos* (ed. B. Goldman; New Haven/London: Yale University Press, 1979), 141-145, 168-174.

[21] Harald Riesenfeld, *The Resurrection in Ezekiel XXXVII in the Dura-Europos Paintings* (Uppsala: Lundequistska, 1948), 37.

[22] Zimmerli, *Ezekiel 2*, 272-274, for this traditional view refers to W. Neuss, *Das Buch Ezechiel in Theologie und Kunst bis zum Ende des XII. Jahrhunderts* (Münster: Aschendorf, 1912).

[23] J. Grassi, "Ezekiel xxxvii 1-14 and the New Testament," *NTS* 11 (1964-1965), 162-164: in *b. Sanhedrin* 92b, R. Eliezer says, "the dead that Ezekiel revived got up on their feet, sang a hymn, and died": Luz, *Matthäus* 26-28, 365, views this as the resurrection of the "Gerechte," the event implicitly reminiscent of "der endzeitlichen Auferweckung der Gerechten"

[24] C. H. Kraeling, *The Synagogue, The Excavations at Dura-Europs: Final Report*, VIII/1 (New Haven: Yale University Press, 1956), 179.

[25] Greenberg, *Ezekiel*, 750.

Nevertheless, even those who view Ezek 37:1-14 as 'really a parable' often leave open the possibility of both metaphorical and literal meanings of the opening of the graves in verse 12.[26] Likewise, when considering the paintings of the synagogue at Dura-Europos, the question of whether the artist meant to convey a metaphoric vision or a reality also remains ambivalent.[27]

In this instance, any exclusive interpretation runs the risk of being one-sided. All three cases – Ezek 37:12-14, the Dura-Europos panels, and Matt 27:51b-53 – demonstrate that both motifs of resurrection and restoration can coexist. It may be more a matter of emphasis than difference in each text or panel. Likewise, Ezekiel's vision primarily depicts the revivification of the life of the nation as a whole, but hints at physical resurrection vividly pictured in the metaphor of dead bones being raised to life. The Dura panels focus on the vision of Israel's restoration as depicted in Ezekiel 37, yet still express lingering expectations of a future resurrection of the dead. In contrast, Matt 27:51b-53 associates 'the physical death and the resurrection of Jesus' with the rising of the dead saints in accordance with the pattern of the hope of Israel's restoration in Ezek 37:12-14.

Our point is this: The Christian resurrectionist view, which reads the resurrection motif from Ezekiel's vision while forsaking any element of restoration motif, misses a critical implication of the event of Jesus' death and resurrection in light of Israel's restoration in Matt 27:51b-53. In the case of Ezekiel's vision, the restoration motif is primary; the resurrection is peripheral, and only implicitly present.[28] Yet Matthew's reading of Ezekiel's vision is invoked and dominated by the event of Jesus' death and resurrection.[29] Here we argue, the restoration motif is to be seen as consequential.

The Dura paintings put the resurrection motif in the spotlight, thus siding with Matthew' emphasis. The critical difference between Matt 27:51b-53 and the Dura panels, however, is uncovered in the key figure in the respective eschatological scenes. It is the prophet Ezekiel himself in the Dura paintings, upon whose head God's hand is laid.[30] In Matthew's text, it is Jesus, 'the King of Israel' (27:42) in the preceding pericope and 'the Son of God' (27:43) in the following verse, whose death coincides with God raising up the saints.

---

[26] Greenberg, *Ezekiel*, 750-751, illustrates Rashi's and Kimhi's cases; cf. Zimmerli, *Ezekiel* 2: 219; Groningen, *Messianic Revelation*, 773.

[27] Kraeling, *Dura*, 185-194.

[28] Riesenfeld, *Dura-Europos*, 3-28, presents the idea of the resurrection from the dead before Israel's exile, and also in Ezek 37:1-14 traces various OT passages and traditions, especially the New Year festivals.

[29] Gnilka, *Matthäusevangelium*, 2:477, states, "Mt verknüpft die Überlieferung von der Totenerweckung nach Ez 37 mit dem Tod und der Auferstehung Jesu"

[30] Hopkins, *Discovery*, 169; cf. Grassi, "Ezekiel," 163, believes that the figure in pink on the side of the mountain is the Davidic Messiah.

If both the resurrection and the restoration motifs in Ezek 37:12-14 are germane to Matthew's scene as well as to the Dura panels, then the same can be said of Zech 14:4-5. Dale C. Allison, repudiating Senior's view of Matt 27:51b-53 as resulting from Matthean creative and free redactional activity, contends that Zech 14:4-5 is the main pre-Matthean resource from which Matthew draws the paradigm.[31] According to Allison, the language of Matt 27:51b-53 neither excludes nor verifies the hypothesis of a redactional origin. Terms such as ἐσείσθη and πέτραι could be regarded as Matthean, but these might come from the tradition as well (Joel 2:10; Ezek 37:7 LXX; Hag 2:6; Zech 14:5; 1 Enoch 1:3-9; *As. Mos.* 10:4; *4 Ezra* 6:13-16; *2 Apoc. Bar.* 70:8; *T. Levi* 4:1).[32] There is no persuasive reason, he argues, to dampen the probability that the Evangelist relied on Zech 14:4-5 instead of creating these ideas on his own.

The Dura paintings also affirm the Zechariah connection, in that they display a mountain at the center of the resurrection scene. Reisenfeld notes that the mountain can indicate 'the valley of Dura" known as the place where Ezekiel revived the dry bones, according to a few Jewish texts, which might explain a reproduction of the event in the synagogue located in Dura-Europos.[33] Yet the panels reveal not so much a valley as a mountain dramatically split in two with an olive tree atop each of the cloven halves. He suggests that the mountain in the panels is the Mount of Olives symbolically located at the center of the holy city.

This image is associated with the belief in resurrection shared by Jews in the Dispersion, such as those residing in Dura-Europos. A clear example is found in the Targum of the Song of Songs 8:5: "When the dead rise, the Mount of Olives will be cleft, and all Israel's dead will come up out of it, also the righteous who have died in captivity; they will come by way of a subterranean passage and will emerge from beneath the Mount of Olives."[34] This text likely represents an interpretation of Zech 14:4: "On that day his feet will stand on the Mount of Olives, east of Jerusalem, and the Mount of Olives will be split in two from east to west."[35] In addition, Zech 14:5 retains the language of the '[all the] holy ones [with the Lord]' (πάντες οἱ ἅγιοι, LXX) with whom those resurrected

---

[31] Allison, *The End of the Ages*, 40; cf. Senior, *The Passion According to Matthew: A Redactional Study* (BETL 29; Louvain: Louvain University Press, 1975), 207-223.

[32] Allison, *The End of the Ages*, 41.

[33] For the Jewish texts, see Reisenfeld, *Dura-Europos*, 30, lists *Tg. Ps.-Jon.* to Ex. 13.17; *Tg. Cant.* 7. 10; b. *Sanh.* 92b; *Pirqe R. Eliezer* 33.

[34] Riesenfeld, *Dura-Europos*, 31, refers to *Tg. Zech* 14:4, but the text scarcely reveals the idea of the resurrection of the dead; the only variation in *Tg. Zech* 14:4 from MT's text is "he shall reveal himself in his might" instead MT's "his feet shall stand"; see also, K. J. Cathacart and R. P. Gordon, *The Targum of the Minor Prophets* (Wilmington: Michael Glazier, 1989), 223.

[35] Allison, *The End of the Ages*, 43.

in Matt 27:51 can be identified. The reference to 'the earthquake' amplifies the echo.

Yet the influence of Zechariah's vision is more apparent and even comprehensive in Matthew's scene. In addition to the quotation from Zech 13:7 in Matt 26:3 ('the smitten Shepherd,' cf. Mark 14:27), Matthew cites Zech 9:9 ('the gentle king,' Matt 21:4-5), Zech 11:12-13 ('throwing the coins to the potter,' Matt 27:3-10), and alludes to Zech 11:12 ('thirty silver coins,' Matt 26:15), and most likely to Zech 14:4-5 ('Mount Olives, earthquake, the holy ones') in Matt 27:51b-53.[36] Considering this series of citations and allusions taken from the texts of Zech 9-14 in the Gospel, it is tempting to ask what Matthew's intention might be as he draws from Zechariah 9-14 to interpret Jesus' death and resurrection, particularly in 27:51b-53.

Most likely, Zechariah 9-14 was well known in the early church as part of the *Testimonia*,[37] and Matthew's consistent interaction with Zechariah's vision indicates perhaps more than just proof-texts. To advocate the Zechariah connection with Matt 27:51b-53, Allison risks renouncing the widely accepted association of Ezek 37:12-14 with Matthew's scene. Nevertheless, observing that the Dura panels draw upon both Ezekiel 37 and Zech 14:4-5, Allison left a short yet meaningful note saying, "there appears to have been *a pre-Christian exegetical tradition that brought Zech 14:4-5 into association with Ezekiel 37.*"[38] We surmise that this association might be explained by the Davidic Shepherd tradition, especially since Micah 2-5 is likewise included. This would certainly illuminate the riddle involved in Matt 27:51b-53, which we will incorporate in our discussion when we examine Matt 28:16-20. As in the Dura panels, the traditions appear to have already been merged and formed into a unified vision for the future of the nation.

5.1.3 The Davidic Shepherd Tradition

The language and descriptions found in Matt 27:51b-53 indeed resist being reduced to one particular OT source, either Zech 14:4-5 or Ezek 37:12-14. Nor does it appear that Matthew's scene is an isolated, abrupt, and later insertion. We argue rather that it is a critical part of the passion narrative which represents a dramatic *divine reversal* followed by the tragic judgment upon the smitten shepherd in Matt 26:31 (Zech 13:7), who is none other than the humble king who enters into the city of Jerusalem in 21:4-5 (Zech 9:9) accompanied by those he healed (Matt 20: 29-34; cf. 21:6-9, 14-15).

The scene in Matt 27:51b-53 expresses the climactic moment of the divine reversal as envisioned in the Davidic Shepherd tradition. Now, the reversal

---

[36] Troxel, "Origin," 42-43, discredits Allison's claim, accepting "the saints" in Matt 27:52 as the only element that can possibly be dependent upon Zech 14:15.

[37] Dodd, *According to the Scriptures*, 64-67; Albl, *"And Scripture Cannot Be Broken,"* 30.

[38] Allison, *The End of the Ages*, 44, n. 5 (italics mine).

wrought by the eschatological Shepherd reaches its climax: from death to life. To explain this, it is noteworthy that in Matthew's passion narrative, it is primarily Zechariah 9-14 and not Ezekiel 34-37 or Micah 2-5 that dominates and guides Matthew's presentation of Jesus' passion and resurrection. Nevertheless, as the story of the eschatological Shepherd in the Gospel nears its conclusion, the symphony of the various trends within the Davidic Shepherd tradition begins to be heard in a harmonious chorus. What we hear in Matt 27:51b-53 is a prelude to the finale. It is distinctly staccato, one of the highest pitches in the drama of the eschatological and Davidic Shepherd in Matthew's Gospel. Our thesis is that the text of Matt 27:51b-53 echoes Zechariah's vision, and evidently also Ezekiel's and even Micah's vision of the eschatological break-through (Mic 2-4).

A close look at Matt 27:51b-53 reveals Matthew's synthesis of Ezek 37:12-14 and Zech 14:4-5. The earthquake, the cloven rocks (v. 51b), and probably the term οἱ ἅγιοι may be ascribed to Zechariah. No less vivid are the rest of the depictions and these cannot be traced to Zech 14:4-5, but rather to Ezek 37:12-14: 'the tombs were opened' (v. 52); 'many bodies of the holy ones were raised' (v. 52); 'coming out from the graves' (v. 53); and perhaps, '[they] entered into the holy city' (v. 53). More so than Zechariah, therefore, Ezekiel echoes the later part in verses 52-53, and depicts the quick pace of this eschatological event. Zechariah's influence, in contrast, is mostly evident as the setting of the drama in the first half of verse 51.

Widely recognized as a consensus, the depictions in verses 52-53 are indebted to the portrayal of resurrection found in Ezekiel 37.[39] Many observe that the initial phrase of Matt 27:52, καὶ τὰ μνημεῖα ἀνεῴχθησαν echoes the phrase, ἐγὼ ἀνοίγω ὑμῶν τὰ μνήματα in Ezek 37:12 (LXX), while the LXX's continuation of that verse, ἀνάξω ὑμᾶς ἐκ τῶν μνημάτων ὑμῶν is echoed in the phrase, καὶ ἐξελθόντες ἐκ τῶν μνημείων in verse 53.[40] Yet these observations seem partial. A more significant resemblance is the basic pattern of the events common for both Matt 27:52-53 and Ezek 37:12-14. Both texts present 'the identical sequence of the events,' i.e., (i) beginning with the opening of the tombs, (ii) raising the dead and bringing them out of the tombs, and (iii)

---

[39] Allison, *The End of the Ages*, 40-46. Cf. Luz, *Matthäus* 4:357, doubts the relevance of Ezek 37, which he does not see as an apocalyptic or eschatological text; but Luz does acknowledge the exilic connotations of Ezek 37 (357), while viewing Jesus as "Christi Höllenfahrt" in Matt 27:51b-53. Luz neglects to explain how these features fit together. On the other hand, David C. Sim, *Apocalyptic Eschatology in the Gospel of Matthew* (SNTSMS 88; Cambridge: Cambridge University Press, 1996), 111, describing the scene as "the terrifying apocalyptic events" (200), affirms the text's "clear dependence upon Ezek 37:12-13, also Zech 14:4-5," and in addition, Dan 12:2.

[40] Troxel, "Origin," 36; cf. Brown, *The Death of the Messiah* (New York: Doubleday, 1994), 1123, 1140; Wolfgang Schenk, *Die Sprache des Matthäus* (Götthingen: Vandenhoeck & Ruprecht, 1987), 232.

bringing them to their final destination, either the land of Israel (Ezekiel) or the holy city (Matthew). While all three of these stages are absent in Zech 14:4-5, they are present in Ezek 37:12-14, even with apparent linguistic similarities.

The divine promises, "I [will] open (ἐγὼ ἀνοίγω, LXX) ... will lead [you] out (ἀνάξω) ... and lead [you] in (εἰσάξω)," in Ezek 37:12-14 are now responded to in Matt 27:52 with the divine passives, "the graves] were opened (ἀνεῴχθησαν), [the saints] were raised (ἠγέρθησαν), and a participle," their coming out [of the tombs] (ἐξελθόντες)" and an indicative, "[they (with the resurrected Jesus)] entered (εἰσῆλθον) [the holy city]" in verse 53:

	Opening the tombs	Raising the Dead	Bringing into the Land/City
Ezek 37:12-14 :	ἀνοίγω	→ ἀνάξω	→ εἰσάξω
Matt 27:52-53 :	ἀνεῴχθησαν	→ ἠγέρθησαν	→ ἐξελθόντες ... εἰσῆλθον

As regards the first two of the tripartite divine promises of Ezek 37:12, namely, ἀνοίγω and ἀνάξω, Matthew retains the divine passive, ἀνεῴχθησαν ('they were opened') and ἠγέρθησαν ('they were raised"). Concerning the last promise, namely, Ezekiel's εἰσάξω ("I will lead [them] into [the land]"), Matthew's text responds differently: "as they come out [after he (Jesus) was raised], they [thus, the risen Jesus and the resurrected together] entered the holy city." The difference is delicate but appears to be of importance. While the tasks of opening the tombs and raising the dead are still ascribed to YHWH in the Ezekiel text, this time in Matthew's Gospel, as to entering into the land (or the holy city), *the subject changed from YHWH to 'they,'* i.e., *the risen Jesus and those resurrected* who enter into the promised destination. Who then led them into the city?

If we can rely on the relevance of Ezek 37:12, then the shepherd-leadership of those saints into the land in Ezek 37:12 can be said to *have been transferred to* the risen Jesus in Matt 27:53. The use of the verb ἄγω in *Ps. Sol.* 18:5 illustrates the case in point: καθαρίσαι ὁ θεὸς Ἰσραὴλ εἰς ἡμέραν ἐλέους ἐν εὐλογίᾳ εἰς ἡμέραν ἐκλογῆς ἐν *ἀνάξει* χριστοῦ αὐτοῦ ("May God cleanse Israel for the day of mercy and blessing, for the day of election as he *brings forth/back* his anointed one"; italics mine). Charlesworth notes, "the author was referring ..., as seems more probable, to the return of the Messiah, who is like the wonderful King David"; the function of this kind of eschatological 'leading' is 'transferrable from God to Messiah and then back again to God.'[41]

Is this not perhaps involved in Jesus' promise, μετὰ δὲ τὸ ἐγερθῆναί με προάξω ὑμᾶς [εἰς τὴν Γαλιλαίαν] in Matt 26:32? Here the composition and the context of μετὰ δὲ τὸ ἐγερθῆναί με strikingly resemble those in the phrase μετὰ τὴν ἔγερσιν αὐτοῦ in 27:52. This may well indicate the critical role of Jesus' leadership as he leads those raised from the dead into the holy city. The promise

---

[41] Charlesworth, "From Messianology," 30.

5.1 The Rise of the Shepherd (Matt 27:51b-53)

of Jesus' continual shepherd-leadership after his resurrection implied in προάγω in 26:32 matches the picture of Jesus the Leader in 27:52, regardless, for now, whether it be the holy city where they appear to many, or Galilee where they set out on a mission beyond to the nations. It is likely that Matt 27:52 projects a triumphant entry into Jerusalem (cf. 21:1-9) – whether 'the holy city' in verse 53 is represented by or identified with Jerusalem. At the center of the entrance scene stands 'the risen' Jesus, who was introduced as the humble king (21:5; Zech 9:9), then as the smitten shepherd (26:31; Zech 13:7). Now he is followed by those risen from the dead, and not merely by those who were healed (20:29-34; 21:14-16).

Matthew 27:51b-53 shows that the passion narrative replete with Zechariah's tragic images of the suffering shepherd(s) in Zechariah 9-13 finds its conclusion of hope but mainly in accordance with Ezekiel's vision (Ezek 37) while fulfilling Zechariah's hope (Zech 14) as well. The combination of Ezekiel's vision with that of Zechariah in the scene in Matt 27:51b-53 is anything but a coincidence. Interacting with these traditions, Matthew replies to the riddle of the smitten shepherd of Zechariah, i.e., to the question how the restoration of Israel is realized. Jesus is both Zechariah's smitten shepherd and the Davidic Shepherd. He assumes the role of YHWH the eschatological Shepherd who is to set free his flock from exile. This constitutes the entire vision of Ezek 34:23-24, 37:24-25, and Zech 14:1-19. The hopeful vision in Zechariah 14 follows the tragic striking of the shepherd and the scattering of the flock in Zech 13:7-9. Likewise, Jesus the smitten shepherd in Matt 26:31, after his resurrection, leads the risen saints, the scenes replete with descriptions of Zech 14:4-5.

Matthew continues to interact with the Davidic Shepherd tradition. Within the tradition, Zech 14:1-9 shares the common theme of YHWH's eschatological battle and its consequential breakthrough with texts such as Mic 2:12-13 and 4:11-13 (cf. Zech 12:8-9; Ps 2:7-9). In Micah 2:12-13, the king (מֶלֶךְ) is a shepherd figure who passes through the gate with his gathered people going 'before' (לִפְנֵיהֶם, cf. Exod 13:21; Num 27:16-17; Zech 12:8) them at their head (LXX, καὶ ἐξῆλθεν ὁ βασιλεὺς αὐτῶν πρὸ προσώπου αὐτῶν ὁ δὲ κύριος ἡγήσεται αὐτῶν). Yet while this shepherd-king is the Breaker (הַפֹּרֵץ), the picture is of a gathered flock with the shepherd at its head breaking through the gate together as a whole in order to go out (עָלָה הַפֹּרֵץ לִפְנֵיהֶם פָּרְצוּ, v.13a).

Thus, the role of 'the Breaker,' that is, 'the one who makes a breach' (הַפֹּרֵץ) is closely tied to the movement of the flock. The death of Jesus (as the smitten shepherd) in Matt 27:51b has the immediate consequence of the outbreak of the saints raising from their graves. Yet the collective appearance of the raised saints in Matt 27:52-53 necessitates that the risen Jesus be at their head as they enter the holy city.⁴² Just as it is uncertain who breaks through and leads the flock through the gate in Mic 2:12-13 – be it YHWH or more likely the Shepherd/King/Breaker – Matthew's usage of the divine passive in 27:52-53 reflects this subtle distinction.

The context of Mic 4:11-13 is YHWH's return to Zion (4:6-8). The kingship is to be restored to the people of Zion. Here the picture envisioned is more detailed than what is envisioned in Mic 2:12-13: Following the break through the gate, the Breaker/Shepherd returns to Zion leading his flock. Micah 4:6-5:5 presents the Lord's plan in a sequentially reversed order: the kingship of YHWH returns to Zion (4:6-9), the eschatological breakthrough in a style similar to Ps 2:9 (4:11-13), and the announcement of the Davidic Shepherd (5:1-4). It is interesting to note that the theme of returning to Zion (cf. Matt 27:53) is followed by the eschatological breakthrough (cf. Matt 27:51b-52), which assumes the rulership of the Davidic Shepherd (cf. Matt 2:6; 28:16-20). The eschatological breakthrough and the entrance into the holy city in Matt 27:51b-53, in light of both Zechariah 14 and Ezekiel 37 as its main source, appear to follow this pattern laid in the Davidic Shepherd tradition.

This proposal may be reinforced as we find another critical contact point between Ezek 37:12-14 and Matt 27:54, namely, the Roman centurion's recognition of Jesus as the Son of God: 'Surely he was the Son of God.' Scholars such as Witherup and Troxel lay an emphasis on this verse as the key to unlock the narrative reading of the scene found in 27:51b-53. Witherup insists that the primary purpose of the scene is "to vindicate Jesus' role in salvation history as the obedient and faithful Son of God who by suffering, death, and resurrection fulfills his Father's will to perfection."⁴³ Similarly, while suggesting the influence of the visions of the holy ones in 1 Enoch 93:6, Troxel argues that the saints' resurrection immediately following Jesus' death refutes

---

⁴² The referent of "the holy city" remains bewildering. In Ezekiel's context, when Ezekiel describes the eschatological regathering of the house of Israel, the prophet mentions both inner, religious renewal and outer, environmental renewal, i.e., the land (34:1-22; 36:16-38; 37:15-23). The spiritual regeneration of Israel coincides with the renewal of the land (= return to the state of Eden, Ezek 36:35) when the covenant is renewed; the Spirit is given; and the people follow the laws and decrees of the Lord (Ezek 36:24-28). Further, in Ezek 40-48, "the city" named "the Lord is there" is not even referred to as "Jerusalem," let alone as the "city of David." See ch.1.4.1 in our study. Cf. Craig A. Evans, "Aspects of Exile and Restoration in the Proclamation of Jesus and the Gospels," in *Exile: Old Testament, Jewish, and Christian Conceptions* (ed. James M. Scott; Leiden: Brill, 1997), 302, recalls Josephus' reports of two Jewish men in the first century: Theudas (*Ant.* 20.5.1, 97-98) and the Egyptian Jew (*War* 2.13.4-5, 258-263; *Ant.* 29.8.6, 167-172). Both promised signs and wonders in view of the Exodus, and more importantly, both guaranteed their followers "an entrance into Jerusalem."

⁴³ R. D. Witherup, "The Death of Jesus and the Raising of the Saints: Matthew 27:51-54 in Context" in *Faith and the Future: Studies in Christian Eschatology* (ed. J. P. Galvin; New York: Paulist Press, 1994), 585.

5.1 The Rise of the Shepherd (Matt 27:51b-53)

the earlier denials that Jesus was the Son of God.[44] Placing the emphasis on Jesus' identity as the Son of God as confessed by the Roman centurion, both scholars tend to play down the traditional view of Matt 27:51b-53 as the announcement of the eschatological moment, the dawn of the new age.

We are faced with a few subsidiary issues as we attempt to reply to this proposal, e.g., the issues such as the salvation-history in the Gospel and Matthew's understanding of the title Son of God. Within the context of Matt 27:32-54, however, the Son of God (vv. 40, 43) is used interchangeably with 'King of Jews' (vv. 37, 42 [of Israel]). The emphasis on Jesus as the Son of God does not separate him from the story line of Jesus' involvement in Israel's history and destiny in Matthew's Gospel. In the unfolding drama, Jesus as their eschatological and Davidic Shepherd, and as the Son of God recognized by the Gentiles (v. 54), is indeed the King of the Jews. The Roman centurion and those with him acknowledged Jesus as God's Son. This testimony from the Gentiles is generated by their having witnessed the entire breathtaking events that coincided with Jesus' death, and implicitly his resurrection (vv. 51-53): ἰδόντες τὸν σεισμὸν καὶ τὰ γενόμενα. The customary Matthean phrase for the crowd's surprise evoked when witnessing the wonder of Israel's God is well expressed in the similar expression such as ἐφοβήθησαν σφόδρα which is often accompanied by the testimony from the lips of the crowds, λέγοντες ἀληθῶς θεοῦ υἱὸς ἦν οὗτος (cf. Matt 9:33; 12:13; 15:31).

Those scholars who attempt to overturn the traditional view of Jesus' death and resurrection as the eschatological turning-point, often fail to recognize that Ezek 37:12-14 fits the entire pattern laid out in Matt 27:52-54. This raises the question why some scholars stop short at verse 53. The witness of the Roman centurion and of the on-lookers echoes Ezek 37:13: "Then, [you my people - missing in LXX], will know (γνώσεσθε) that I am the Lord, when I open your graves and bring you up from them." Not only the procedure of 'opening the tombs,' 'bringing them out,' and 'leading them into,' but *the motif of the recognition of God's Holy Name by the nations (or Israel herself or both) is identical both in Ezekiel 37 and Matthew 27.*[45] In short, the point of the Gentiles' witness and confession in 27:54 is their recognition of the presence of Israel's God, Ἐμμανουήλ ('God with us,' 1:23), not Jesus' filial obedience *per se*.

To pinpoint the eschatological significance on one particular moment in the narrative – Jesus' birth (1:23), the beginning of his public ministry (4:17, 23), or his death and resurrection (27:51-53) – may be a matter of one's theological perspective. Yet, what is implied in Matt 27:51-54 is, at least, the culmination of the ministry of Jesus as Israel's eschatological Shepherd among the sheep

---

[44] Troxel, "Origin," 54-47.

[45] For a detailed study of the theme of YHWH's zeal for his name, refer to the Ezekiel section in ch 1; whether the witnesses are Israel or the nations matters little. In Matt 27:51b-53, it is the nations who witness God's presence and mighty work.

without a shepherd. While it may be a high point, it cannot compare with the climax that is to come in the narrative. The history of Israel is fundamentally transformed by this eschatological breakthrough – the resurrection of the smitten shepherd, his breakthrough from the grave, and his entrance into the holy city followed by a procession of risen saints. This is the triumphant vision that sets a foil for the continued mission of the renewed eschatological flock in Matt 28:16-20.

5.1.4 Summary

The scene of the resurrection of the saints in Matt 27:51b-53 echoes Ezek 37:12-14 as well as Zech 14:5. We found an identical procedure of the sequential events in Matt 27:52-54 and Ezek 37:12-14: (i) 'opening the tombs,' (ii) 'raising the dead,' (iii) 'bringing them into the land/city,' (iv) and even 'acknowledging God's name by the onlookers/nations.' The difference we observe in Matthew's text is yet acute in the change of the subject: from YHWH in Ezek 37:12 to 'they,' i.e., the risen Jesus and those resurrected in Matt 27:52-53. This means, we argued, that the shepherd leadership in Ezek 37:12 is transferred to the risen Jesus in Matt 27:53. Once Jesus is raised by God, it is the risen Jesus who leads the saints into the city (cf. 'the Shepherd as Breaker,' in Mic 2:12-13).

The identical pattern in Ezekiel and Matthew appears enhanced by the fourth element of the same pattern: peoples' acknowledging God's name. We have observed that as Ezek 37:14 presents the motif of the recognition of God's holy name by the onlookers/nations, Matt 27:54 mentions the Roman centurion's recognition of Jesus as the Son *of God*. Jesus as the Shepherd in the Gospel continues after 2:6 through numerous allusions and echoes to the tradition in the main body of the Gospel, and is evident even in the narrative following the scene of the smitten shepherd in 26:31. The risen Jesus leads the resurrected saints into the holy city in 27:51b-54, and continues to lead his disciples who, though now scattered, were instructed to go to Galilee (26:31-32).

Through the ministry of the compassionate eschatological Shepherd, God's presence was witnessed by the crowds of Israel. Yet through God's reversal of the tragedy of the smitten shepherd of Israel and his flock as a whole, this return of God's presence is confirmed and testified by the Gentiles as well.

5.2 The Divine Presence and the One Davidic Shepherd-Appointee (Matt 28:16-20)

The next stage is the perpetuation of the Davidic Shepherd's reign through which the divine presence is to be secured in view of the nations. This is the next step according to the Ezekielian pattern, and we will examine its relevance

5.2 *The Divine Presence and One Davidic Shepherd-Appointee (Matt 28:16-20)* 341

to the closing of the First Gospel: Matt 28:16-20. Though many scholars scrutinized this well-known pericope in search of the tradition and redaction behind the text, they have yet to reach a consensus. John P. Meier asked, "Did the evangelist use previous traditions or is it his own redactional creation?"[46]

The way this question is framed, however, leaves us a false dilemma that rules out the possibility of the authenticity of the text.[47] The question of the authenticity of Jesus' own commission *for the Jewish mission for the Gentile* is significant especially in regard to the question of the 'Ursache der Heidenmission' of the early church. Did not the earthly Jesus have the Gentile-mission in view at all? Does it seem likely that the vision of the Gentile-mission was totally new, apart from Jesus tradition, for the early churches?[48] These questions underscore the importance of christology in the Gospel, especially here in Matt 28:16-20.

Moreover, the missiology implied in the text is inseparable or, better put, is controlled by christology which one chooses for the figure of Jesus in the text. As we shall argue in detail, we propose that Jesus is presented as the One Davidic Shepherd over both Israel and the nations, i.e., πάντα τὰ ἔθνη. We will first discuss various proposals for the background of Matt 28:16-20, examine the significance of the setting described in verses 16-17, and present an analysis of verses 18-20 in the light of Ezek 34-37.

5.2.1 Tradition and Setting

As to the tradition behind the text, suggestions vary: (i) Dan 7:13-14 with the emphasis on 'authority' (Michel);[49] (ii) Ps 109:5-6 with the emphasis on 'the submission of the nations' (Hahn; cf. Stuhlmacher: Ps 80; 87; 96:2-3);[50] (iii) the

---

[46] J. P. Meier, "Two Disputed Questions in Matt 28:16-20," *JBL* 96 (1977): 407, 424.

[47] Likewise, following Martin Dibelius (*Die Formgeschichte des Evangeliums*, 1961), Rudolf Bultmann *(The History of the Synoptic Tradition*, 1963), J. D. Kingsbury, "The Composition and Christology of Matt 28:16-20" *JBL* 93 (1974): 573-584, regards the passage as purely redactional, that is, unhistorical; Luz, *Matthäus* 4:436, says, Matt 28:18b-20 is not "ein Logion des Herrn" but "von Matthäus komponiertes Logion im Herrn."

[48] E. Schnabel, *Urchristliche Mission*, 345-347, presents an extensive argument for the case of the authenticity of Matt 28:16-20, which offers less problematic and simpler answers to the early churches' assumption of the legitimacy of the Gentile mission. Likewise, Peter Stuhlmacher, "Matt 28:16-20 and the Course of Mission in the Apostolic and Postapostolic Age," in *Mission of the Early Church to Jews and Gentiles* (ed. Jostein Ådna and Hans Kvalbein; Tübingen: Mohr Siebeck, 2000), 22, 34, notes, the phrase πάντα τὰ ἔθνη both in Matt 28:19 and Luke 24:46 points to the old tradition; also, see Gundry, *Matthew*, 596.

[49] O. Michel, "Der Abschluss des Matthäusevangeliums," *EvT* 10 (1950/1951): 16-26, 22-23.

[50] F. Hahn, *Mission in the New Testament* (Naperville, Ill.: Alec R. Allenson, 1965), 55-56; Stuhlmacher, "Matt 28:16-20," 24-30, combines the theme of "the eschatological restoration of Greater Israel" with the motif of the conversion of the nations against the background of the Feast of Tabernacles in Matt 28:16-20, also along with the influence of Dan 7:13-14.

OT commissioning passages such as Gen 12:1-4, Exod 3:1-10, Josh 1:1-11, Isa 49:1-6 with the emphasis on 'authority' and 'teaching' (Hubbard);[51] and (iv) 2 Chron 36:23 with highlights on the parallels in the literary form of Matt 28:18-20 (Malina).[52] Otto Michel asserts that like Dan 7:13-14, Matt 28:16-20 represents a coronation text (cf. Phil 2:9-11), suggesting that these texts exhibit an underlying unity, and share common motifs like the authority of Christ, his Lordship, and the eventual recognition of all the people of the earth. The type of christology informing the text, according to Michel, is that of Kyrios or the Son of Man.[53] Despite the language of ἐξουσία present in both texts, this option is widely refuted on the basis the fact that many important features of the texts do not correspond with one another. There are four examples: the coming of the clouds of heaven, δόξα, the themes of teaching and the divine presence, and the critical title, Son of Man.

More pointedly, the picture depicted in Dan 7:13-14 is too final compared to that in Matt 28:18-20. In Matthew's closing scene, Jesus is still at work with his disciples; even with all authority of heaven and earth, he is not yet in his final glory, and we find him engaging the ongoing mission on earth through his perpetual presence among his disciples. While the Danielic Son of Man is about to receive the kingdom, Jesus on the Mount in Galilee launches the movement to the nations, yet has no intention of leaving them alone in their mission. Jesus at the last scene of Matthew's Gospel is not on the clouds yet, but on the way toward it.

Similarly, Luz notes that Jesus' promise of his presence with his disciples to the end of the age in verse 20b points back to 'die Geschichte des irdischen Jesus'; thus, he argues that one finds neither 'Son of Man' nor κύριος, nor 'the Son of God' as the triad baptismal formula is mentioned in passing.[54] It is more likely that the risen Jesus with his restored flock in Matt 28:16-20 moves toward fulfilling the vision of the coming of the Danielic Son of Man.[55] Daniel 7:13-14,

---

[51] B. J. Hubbard, *The Matthean Redaction of a Primitive Apostolic Commissioning: An Exegesis of Matthew 28:16-20* (SBLDS 19; Missoula: Scholars Press, 1974).

[52] B. J. Malina, "The Literary Structure and Form of Mt 28:16-20," *NTS* 17 (1970): 87-103; cf. Chris Manus, "King-Christology': The Result of a Critical Study of Matt 28:16-20 as an Example of Contextual Exegesis in Africa," *Scriptura* 39 (1991): 26-29, presents a succinct survey of some previous studies on this passage. Cf. For some other unsubstantial options: Frankenmölle, *Jahwehbund*, 53-59, argues that 2 Chr 36:23 is also a covenant-renewal (*Bundeserneuerung*); W. Trilling, *Das wahre Israel: Studien zur Theologie des Mattheäusevangeliums* (Leipzig: St. Benno-Verlag, 1959), 33-34, sees Matt 28:18-20 as "divine utterance" (*Gottesrede*).

[53] Michel, "Abschluss," 22-23.

[54] Luz, *Matthäus*, 4:456. Against Keener, *Matthew*, 716-717, who picks up Matthew's christology in 28:18-20 from the baptismal formula, i.e., Jesus as the divine Son.

[55] Gnilka, *Matthäusevangelium* 2:508, expresses the tension saying that it is not "die Parusie" that is mentioned in v. 20b, yet "die Parusieerwartung" is not out of sight.

5.2 The Divine Presence and One Davidic Shepherd-Appointee (Matt 28:16-20)

however attractive the linguistic affinity, fails on the level of thematic context and is no real help in discovering the theological framework of the passage.[56]

Likewise, Kingsbury rightly notes that the languages related to the coming Son of Man in the Gospel do not contain the motif of exaltation as a christological element specific to them. Instead, he claims, it is the Son-of-God christology that fits the description of Jesus in Matt 28:16-20 as well as in the entire narrative.[57] According to Kingsbury, Matthew stresses the continuity between 'Jesus of Nazareth' and 'the resurrected, exalted Jesus' of the post-Easter church (the so-called horizontal dimension), and also between Jesus' unique relationship with the Father 'then' and 'now' (the so-called vertical dimension).[58]

As neat and tidy as this structural (horizontal and vertical) analysis appears to be in terms of the Son-of-God christology, few will believe that the Son of God is the overarching christological title in the Gospel, and further that the Son of God is the prominent figure in Matt 28:16-20.[59] In general, Kingsbury's titular approach provides few compelling arguments for other parts of the text elsewhere than the designation 'Son' in the trinitarian baptismal formula in verse 19.[60]

Others fit different parts of the text into various traditions. Hahn highlights the motif of the submission of the nations, particularly referring to Ps 109:1 LXX (110:1) inclusive of the entire Psalm. Although acknowledging the presence of the enthronement theme in Matt 28:16-20, Hahn stresses the motif of the final lordship of the heavenly Christ over the world powers. Thus we see, he argues, the OT theme of the subjugation of the nations is now replaced by the emerging strategy to bring the Gospel to Gentiles in the last days.[61] Psalm 110 is a well-known OT reference to the early church;[62] thus Matthew's allusion can neither be improbable nor surprising. Yet as Schuyler Brown aptly points out, the extension of the missionary scope to include all nations "is not made to depend on Jesus' exaltation."[63] The scene in Matt 28:16-20 indicates that the full exaltation of the resurrected Jesus is not yet a reality. The risen Jesus would not leave his disciples, and even guarantees his enduring presence among them as they should continue the mission he commands them to do.

---

[56] Cf. Donaldson, *Mountain*, 177.
[57] Kingsbury, "Composition," 582.
[58] Ibid., 583.
[59] Verseput, "The Role and Meaning of the 'Son of God' Title in Matthew's Gospel," *NTS* 33 (1987): 532-556; idem, "The Davidic Messiah and Matthew's Jewish Christianity," *SBLSP* 36 (1995): 102-116.
[60] Refer to Luz's approach to the narrative christology, "Eine Thetische Skizze," 221-235.
[61] Hahn, *Mission*, 55-56.
[62] Albl, *"And Scripture Cannot Be Broken,"* 216-236.
[63] S. Brown, "The Two-fold Representation of the Mission in Matthew's Gospel," *ST* 31 (1977): 32.

Malina says that Matt 28:18-20 closes the Gospel with the decree of Jesus looking to the expansive growth of the circle of disciples, in the same manner that 2 Chr 36:23 closes the Jewish Scriptures with the decree of Cyrus looking to the reconstruction of Jerusalem.[64] Two particular themes – authority and God's presence – are said to be common to these texts: "the Lord, the God of heaven, has given me (Cyrus) all the kingdoms of the earth"; "the Lord his God be with him" (anyone of his people going up Jerusalem to rebuild the temple, 2 Chr 36:23). Jesus' possessing the fulness of authority, according to Malina, is the key to understanding the nature of his commission since "no reason is given for this commission except that Jesus now has the fulness of authority."[65] Jesus' presence promised in verse 20b is seen not simply as a promise of constant help and assistance, but most significantly as "a sign of Jesus authority and victory."[66]

Regardless of what these implications suggest, there is one major dissimilarity, as Meier insists: the complexity of the decree of Cyrus in that God gives a command to the king, and in turn, the king gives a command to the people. In Matthew, this structure simply is not found.[67] The case for 2 Chron 36:23, nevertheless, is not quite rendered invalid. The complexity of the commissioning-process comes about by God's appointment-process of Cyrus as his agent for a special assignment. Isaiah 44:28 informs YHWH's decree of appointing Cyrus to be 'my shepherd' (רֹעִי), the one who will accomplish God's will. The exact phrase, 'my shepherd' (רֹעִי), occurs only in Zech 13:7 to denote the smitten shepherd (cf. Matt 26:31). The application of the phrase to Cyrus is an anomaly since the OT avoids using the phrase for the kings of Israel except the Davidic Servant in Ezek 34:23 ('my servant David [עַבְדִּי דָוִיד] ... one Shepherd'[רֹעֶה אֶחָד]).

Malina's proposal withers as we realize that his points of comparison between Matt 28:16-20 and 2 Chr 36:23 are insufficient; only the motifs of 'authority' and 'divine presence' apply. Other major motifs, in particular 'teaching' in Matthew's commissioning scene, do not quite match Cyrus' task of rebuilding the Jerusalem temple.[68] Also, the theme of the extension of Jesus' reign over the nations appears overtly universal compared to the vision of Cyrus' task presented in 2 Chr 36:23. Cyrus as a Gentile called to rebuild the Jerusalem temple may insinuate YHWH's universal intention for the Jerusalem temple in his sovereign economy. Nevertheless, while it is the Gentile Cyrus who is sent to restore the Jewish temple in 2 Chr 36:23, it is Jesus, who is called

---

[64] Malina, "Structure," esp., 101-103.
[65] Ibid., 102.
[66] Ibid., 103.
[67] Meier, "Disputed," 419.
[68] For this emphasis, refer to Oscar S. Brooks, "Matthew xxviii 16-20 and the Design of the First Gospel," *JSNT* 10 (1981): 2-18.

to be the king of the Jews in Matthew's commissioning scene, sends his disciples to all nations. The direction of the movement in each commission is not identical, but is rather successive. The movement of the mission toward Jerusalem implied in 2 Chr 36:23 does not match with Matt 28:16-20 where the eschatological community moves toward all nations.

Lastly, although it is not a question of tradition but that of genre, Hubbard's proposal for understanding the setting of the last scene of the Gospel attracted many. He proposed the OT commissioning *Gattung* for the structure of Matt 28:16-20 with the formal elements of circumstantial 'introduction' (v. 16), 'confrontation' between the commissioner and the commissioned (17a and 18), 'commission' proper (19-20a), and 'reassurance' from the deity (20b).[69] This view – detailed, yet vaguely applicable to the last scene of the Gospel – is largely refuted because of the fundamental uncertainty of its distinctiveness as a special *Gattung* based on the relevant OT texts.

Likewise, the applicability of the 'commissioning *Gattung*,' if there is any, does not seem confined to any particular commissioning passage; it is loose to fit "any other narrative where an authoritative X meets Y for a significant exchange."[70] A skeptic might also ask how Matthew's numerous interactions with the prophetic sections in the OT in the main narrative would fit the alleged adaptation of the OT commissioning *Gattung* to the closing scene of the Gospel.[71] It seems that Hubbard's mainly literary – noticeably incomplete – parallels do little to inform us of the christology involved.[72]

5.2.2 The Significance of Matt 28:16-17

Returning to the text of Matt 28:16-20, there are at least four integral elements in Matthew's closing scene (28:16-20) in view of various OT traditions: (i) Jesus' authority (v. 18); (ii) teaching (v. 20a); (iii) outreach to the nations (v. 19); and (iv) the divine presence (v. 20b). None of the OT traditions suggested above, however, satisfy all four of these main themes, and these suggestions

---

[69] Hubbard, *Commissioning*, 69-72. Here, other formal items such as "protest" and "conclusion" are missing; H. K. McArther, *CBQ* 38 (1976): 107-108, reviewing Hubbard's study, points out that the item, "reaction," is frequently lacking, and the item, "protest," is occasionally absent in the relevant OT passages.

[70] McArther, "Book Review," 108.

[71] Luz, *Matthäus*, 4:432, suggests the *Gattung* of vv. 16-20 as "eine Erscheinungsgeschichte"; similarly, K. Smith, "Matthew 28: Resurrection as Theophany," *ITQ* 42 (1975): 259-271, views Matthew's scene following the pattern of Sinai theophany (Exod 19); Carson, *Matthew* 2:592, points out that the commission in Matt 28:18-20 is addressed to a group of disciples not to individuals as in the OT, and particularly in the context of a new covenant (Matt 26:26-29); thus, he concludes, "this pericope does not easily fit any known literary form and must not be squeezed into a poorly fitting mold."

[72] Meier, "Disputed," 422-424.

largely neglect verses 16-17, which are tightly interwoven to the thrust of verses 18-20.

Verse 16 provides the geographical setting of the risen Jesus' encounter with his disciples: "then, the eleven disciples went to *Galilee*, to the *mountain* where Jesus had told them to go." The closest reference to Galilee is found in 28:10: Jesus says to the women to tell his disciples, "go to Galilee; there they will see me." This is, in fact, the risen Jesus' affirmation of the message of the angel given to the women to tell the disciples in verse 7: "He has risen from the dead and is going ahead (προάγει) of you into Galilee." Again, it is not difficult to recognize from the wording of verse 7 that this angelic message simply repeats Jesus' instructions when he promised in 26:32, "But after I have risen, I will go ahead (προάξω) of you into Galilee."[73] Our point is that *this shepherd imagery in 26:31-32 continues in 28:16.*[74]

The identity of the risen Jesus as the very smitten shepherd in 26:31 (cf. Zech 13:7) is carried through the language of ἐγείρω, προάγει, and Γαλιλαία (cf. 26:32) in 28:7, 10. In Matt 28:16, among these three items, only Γαλιλαία appears. This is understandable since Jesus is already 'risen' (ἐγείρω) and faces the challenge 'to lead' (προάγει) them into a new phase of their journey. The doubt expressed in verse 17 serves as a carefully intended catalyst for giving Jesus an opportunity to reclaim his leadership over his disciples and more. This is likely a continuation of the shepherd imagery reenacted in verse 16 to catch up with what remains in the promise given in 26:32.

Another critical clue to the theological import of the geographical setting in 28:16 is found in the term ὄρος. For most commentators, the significance of this mountain motif in this verse, if any, would be that it is typically the locus of revelation (cf. 17:1, 5; cf. 5:1).[75] In responding to this superficial observation, Donaldson scrutinizes ὄρος in the Gospel in terms of Zion or its overtones. He argues convincingly that in the Second Temple period ὄρος frequently functions as the site for eschatological events and that Mount Zion in particular is at the center of expressed hopes for the eschatological restoration of the people of

---

[73] Gnilka, *Matthäusevangelium*, 2:506.

[74] Recently, Schtuhlmacher, "Matt 28:16-20," 25-26, observed this feature, pointing out that the "verb προάγειν can be applied to the notion of the Messianic Shepherd from John 10:4, 27...like the one God leads, Israel as a shepherd (cf. Ps 80:2) and as a King, goes ahead of His people through the gate, in order to reestablish Israel (cf. Mic 2:12-13)."

[75] For instance, G. Strecker, *Der Weg der Gerechtigkeit: Untersuchungen zur Theologie des Matthäus* (Göttingen: Vandenhoeck und Ruprecht, 1966), 208; Kingsbury, "Composition," 582; Hubbard, *Commissioning*, 73. Cf. Kevin Smyth, "Theophany," 267-268, proposes the Sinai connection arguing that the phrase in Matt 28:16 is "practically a replica of Ex 34:4 (cf. Ex 19:7)"; both Exodus' Sinai and Matthew's mountain are not merely places of revelation but of "theophany and covenant."

God. He suggests that this is true of Matthew's description of the scene on the ὄρος in 28:16.[76]

Can we find a Sinai connection or a Moses typology when considering Matthews' mountain motif and christology in Matt 28:16-20? It is possible, but it is more likely that it should be captured through the prism of Zion-theology and Davidic connotations which we believe to be historically and theologically more contemporary to the event. If this is correct, then the rich association of post-exile Zion theology with its pastoral imagery corroborates the intense atmosphere of the eschatological momentum in verses 16-20.[77]

5.2.3 An Analysis of Matt 28:18-20

If Jesus' promise to go ahead of his disciples to Galilee in 26:32 is fulfilled in 28:16-17, and if the mountain in verse 16 conveys overtones of Zion in Galilee, then verses 18-20 is to be read in terms of the *reversal* of the judgment that had fallen upon the smitten shepherd of 26:31. Thus the *restoration* of Israel is represented by the risen Jesus – the once smitten shepherd of Israel.[78] What then happened to the scattered flock in 26:31? What would be like the reversal of the judgment?

When analyzing Matt 28:16-20, two main issues emerge: (i) determining what is the dominant christology in this final scene, and (ii) understanding how the closing of the Gospel functions in the whole narrative. We believe that Jesus as the Davidic Shepherd, rather than the Danielic Son of Man or the Son of God, explains better the closing scene, i.e., the definitive momentum to inaugurate the Davidic Shepherd's reign as announced in 2:6 (Mic 5:1-2). It is set up through the mission of the eschatological Shepherd in the main body of the narrative (Ezek 34-36), and culminates in the Evangelist's interaction with Zechariah 9-14 in the Gospel's passion narrative.

The four main themes of Matt 28:18-20 – 'authority,' 'teaching,' 'outreach to the nations,' and 'the divine presence' – are prominent in Ezek 34:23-24, 37:24-25. Matthew 28:16-20 functions as the conclusion of the passion narrative, replete with shepherd images from Zechariah 9-14, and remarks about the climax of the promises of Ezekiel's Davidic Shepherd who is YHWH's Davidic appointee-Servant. Finally, this section takes the readers back to the beginning

---

[76] Donaldson, *Mountain*, 41-48, 171-180.

[77] Donaldson, *Mountain*, 47-48, illustrates OT passages like Ezek 34:26-31 and Jer 31:10, 12-14. In this regard, Jesus feeding the crowds on the mountain (15:30) can be seen as bearing *proleptic* significance, which awaits the new messianic being finally inaugurated on the mountain in Galilee in 28:16.

[78] It is hard to resist recalling Isa 53:5-6 and 7-8, where the suffering of the smitten Servant (cf. "my servant David" in Ezek 34:23; 37:24) is explained in terms of sheep/flock imagery, and the explicit statement that "YHWH has laid on him the iniquity of us all"; yet this line of investigation would require another chapter.

of the narrative, to Micah's promise of the Davidic Shepherd who will shepherd (the restored) Israel in 2:6 (Mic 5:1-2).

Hence we propose that Matt 26:18-20 reflects the fulfillment of the composite vision of the inauguration of the Davidic Shepherd's reign not only over restored Israel but also over the nations as pictured in Ezek 34:23-24, 37:24-25, and Mic 4:2; 5:3-4 (cf. Mic 5:1-2 in Matt 2:6). The risen Jesus is now One Shepherd over all, YHWH's Appointed Davidic Son/ Servant, the Leader over both Israel and the nations, and the perpetual Prince among them. In the following, we will examine these four contact points – authority, reign over the nations, teaching, and presence – that display Matthew's final interaction with the Davidic Shepherd tradition.

In Matt 28:18-20, the motif of authority is tied to Jesus' extended reign over the nations. However, we have already indicated that Malina's proposal of 2 Chr 36:23 with the note of the enthronement theme in Matt 28:18-20 is insufficient despite the similarities. Others suggest a close resemblance to enthronement hymns (Dan 7:14; Phil 2:9-11; 1 Tim 3:16; Heb 1:5-14).[79] Thus, Hubbard is correct when he stresses that Matthew's is more of a down-to-earth scene that focuses on the commission of Jesus rather than the person of the Christ, and that focuses on the making of disciples for the exalted Jesus rather than the exaltation itself. Here the picture is not of the Christian community proclaiming the greatness of the Christ, but of Jesus charging his disciples with the missionary task for the nations.[80]

On the other hand, Donaldson turns to Second Temple Judaism and finds that the mountain – especially Mount Zion – was intimately tied to the eschatological enthronement of YHWH or his Anointed One. Besides the mountain motif, he mentions four integral elements for 'Zion theology in eschatological perspective': (i) the gathering of scattered Israel (Jer 31:1-25; Isa 35; Ezek 34-37; 40-48); (ii) the pilgrimage of the nations (Isa 29:8; Mic 4:11-13); (iii) eschatological blessings (Isa 25:6-10; Jer 31:12; Ezek 34:26-31; Mic 4:6-7); (iv) a new law/a new giving of the law (Isa 2:2-3; Jer 31; cf. Ezek 37:24-25); (v) enthronement (Ezek 17:22-24; 34:23-31; Mic 5:3-4); and (vi) pastoral imagery (Ezek 34:26-31).[81] As he points out, the motifs of enthronement and pastoral imagery are closely interrelated, and with Ezekiel 34-37 as the outstanding common source based on his findings. As Donaldson exegetes Matt 28:18-20 for the tradition behind the enthronement motif in the passage, however, he turns to Ps 2:8-9 (cf. Rev 2:26) instead of to Ezekiel 34, while not impossible is neither appropriate nor sufficient.[82] In this respect, Donaldson's approach is

---

[79] Michel, "Abschluss," 22-23; J. Jeremias, *Jesus' Promise to the Nations* (London: SCM Press, 1958), 38-39.
[80] Hubbard, *Commissioning*, 9.
[81] Donaldson, *Mountain*, 41-48.
[82] Donaldson, *Mountain,* 181.

5.2 The Divine Presence and One Davidic Shepherd-Appointee (Matt 28:16-20)

worthwhile but his exegesis of the passage falls short and is not coherent with his research findings on Zion theology in general.

As we have demonstrated earlier when explicating the significance of the setting laid out in Matt 28:16-17, it is Ezekiel 34 and not necessarily Ps 2:8-9 that best fits the following scene in verses 18-20: the rise of the Davidic Shepherd, the One Shepherd over Israel and the nations. Ezekiel 34:23-24 is recapitulated by 37:24-25 from a bit of a different angle and with some clarifications. Yet both texts retain the term 'one shepherd' (רֹעֶה אֶחָד): "I will raise up over them [the restored flock] one shepherd, my servant David" (34:23); "My servant David will be king over them and they will all have one shepherd" (37:24). Since it is likely that the shepherd image of Matt 26:31 continues in 28:16-17, the scene in verses 18-20 shares a close affinity with Ezekiel's picture of YHWH appointing the Davidic Shepherd at the end of Ezekiel 34, the chapter in which YHWH as the eschatological shepherd confronts the false shepherds of Israel, seeks the lost, heals the sick (vv. 1-16), and judges the flock with justice (vv. 17-22). Ezekiel's programmatic vision – the returning presence of the eschatological Shepherd is followed by YHWH's appointment of the Davidic Shepherd for the restored flock in view of YHWH's universal recognition among the nations – finds, we argue, its fulfillment in Jesus on the mountain in Galilee in the First Gospel.

According to Ezekiel's vision in Ezekiel 34-37, the distinction is clarified between YHWH raising his people from the grave – primarily meaning exile symbolized by death – and YHWH appointing 'my servant David,' 'the one Shepherd' over the restored flock. These two represent successive stages (37:1-15 and vv. 16-25; 34:23-24) preceded by YHWH's eschatological shepherding among his people (34:1-16). As we find in Matthew's narrative structure, the climax of Ezekiel's vision of the restoration is not YHWH revivifying the dead bones (37:1-15; cf. Matt 27:15b-53), nor is it YHWH seeking the lost and healing the sick, thereby restoring the community (34:1-6). All of these restorative tasks culminate in the establishment of YHWH's theocratic rule when YHWH replaces the wicked leadership of the community with his own Davidic Shepherd.

This final stage of YHWH appointing his servant David and placing him as the One Shepherd and the king (מֶלֶךְ) 'over' the restored flock (34:23; 37:24) opens up the climactic part of the vision in 34:25-30 and 37:24-28, which entails elements comparable to those found in Matt 28:18-20. These are: the 'covenant of peace' (34:25), which is the 'everlasting covenant' (37:26); YHWH increasing the number of his eschatological flock (37:26); the joining of nations most likely into 'the sheep of my pasture, my people' (34: 31); the perpetual presence of the sanctuary, and YHWH's tabernacle among them forever (37:26b-27a, 28b); and the nations' eventual recognition of YHWH's holy name (34:30; 37:28a). The components of the closing of Ezekiel's vision in Ezek

34:23-31 and Ezek 37:24-28 are not only identical with Matthew's scene, but also satisfactorily comprehensive, though some modifications are evident in Matthew's ending. In both visions, the reign of the Davidic Shepherd – authorized by God himself – is to be extended through teaching the people, both Israel and the nations, the laws and decrees (as fulfilled and interpreted through Jesus in Matthew), and eventually, to bring the nations to the recognition of YHWH's holy name. The following chart summarizes the elements common to both Ezekiel 34-37 and Matthew 28, and identifies certain procedures in both closing visions:

Ezek 34:23-31/37:24-28 *Matt 28:18-20*

(i) Authority: *appointing the Shepherd/King* *Jesus given the authority*
ἀναστήσω...ποιμένα ἕνα ἐδόθη μοι πᾶσα ἐξουσία
("I will place...one shepherd," ("all authority is given me," v. 18)
34:23; 37:24, LXX)

(ii) Nations: *nations' acknowledging God's name* *baptizing into the divine name*
וְהִרְבֵּיתִי אוֹתָם (37:26c, MT) μαθητεύσατε πάντα τὰ ἔθνη,
("I will increase them") ("making disciples all nations")
וְאַתֵּן צֹאנִי צֹאן מַרְעִיתִי אָדָם (34:30) βαπτίζοντες αὐτοὺς
("the flock of my pasture, ("baptizing them")
people, you")
αἱ γνώσονται τὰ ἔθνη εἰς τὸ ὄνομα (v. 19ab)
("the nations will know," 37:28) ("into the name [of the Father...]")

(iii) Teaching: *leading by teaching & doing the law* *teaching & doing what Jesus taught*
ἐν τοῖς προστάγμασίν μου διδάσκοντες αὐτοὺς
πορεύσονται ("teaching them")
("they shall walk in my laws")
καὶ τὰ κρίματά μου φυλάξονται τηρεῖν
("they will keep my decrees") ("to keep/to do")
καὶ ποιήσουσιν αὐτά (37:24b) πάντα ὅσα ἐνετειλάμην ὑμῖν (v.20a)
("And they will do them") ("all that I have instructed you")

(iii) Presence: *presence of Prince/Sanctuary* *Jesus' presence among disciples*
  /Tabernacle
Δαυιδ ἐν μέσῳ αὐτῶν ἄρχων καὶ ἰδοὺ ἐγὼ μεθ' ὑμῶν εἰμι
("My Servant David Prince ("Behold, I am with you")
[נָשִׂיא] will be in their midst," πάσας τὰς ἡμέρας ("always")
34:24b; 37:25c)
θήσω τὰ ἅγιά μου ἐν μέσῳ ἕως τῆς συντελείας τοῦ αἰῶνος
αὐτῶν εἰς τὸν αἰῶνα ("till the end of the days," 20b)
("My Sanctuary among them
forever," 37:26, 28)
ἔσται ἡ κατασκήνωσίς μου ἐν αὐτοῖς
("My tabernacle will be with [upon, MT] them," 37:27)
καὶ γνώσονται ὅτι ἐγώ εἰμι κύριος
ὁ θεὸς αὐτῶν καὶ αὐτοὶ λαός μου (34:30)

("[the nations] will know that I am the Lord
their God and am with my people [them, MT]")
Cf. וְשֵׁם־הָעִיר מִיּוֹם יְהוָה שָׁמָּה׃
("The name of the city, from that day, will be
THE LORD IS THERE," 40:35)

5.2.4 The Authority of the Davidic Shepherd

The language and the themes of both texts strongly suggest that Ezekiel 34-37 is the most comprehensive text in terms of all four integral components of Matthew's ending in 28:18-20 – authority, teaching, the nations, and the divine presence. Likewise, these four elements are essential to Ezekiel's vision. The divine ἀναστήσω in Ezek 34:23 (37:24) resonates with Jesus' proclamation of the divine ἐδόθη in Matt 28:18. It is difficult to resist suggesting that the authority granted by God to Jesus is none other than that of ποιμένα ἕνα (the one shepherd) envisioned in Ezekiel 34 and 37. In the narrative of the Gospel, the prophecy of Mic 5:1 may well find its true fulfillment in this momentum: ποιμανεῖ τὸν λαόν μου τὸν Ἰσραήλ (Matt 2:6). The long-awaited Davidic Shepherd, whose coming is heralded in Matt 2:6, proclaims to his "eleven" disciples – the re-gathered and restored flock – that he is now the Shepherd with sovereign authority. We suggest that he is the eschatological Shepherd over Israel and the nations.

The passive ἐδόθη indicates that Jesus' authority is granted by none other than God himself. Thus verse 18 is meant to be Jesus' reply to those who doubt as well as to those eleven who worshiped him in v. 17.[83] In other words, in v. 18 the risen Shepherd claims his authority and leadership "over" his re-gathered flock, the eleven. Further, the narrative context helps explicate the nature and function of ἐξουσία in vv. 16-20. This ἐξουσία is granted by God to the risen Shepherd who once was smitten and later raised by God. The events of Jesus' death and resurrection are accompanied by the eschatological and *proleptic* signs of Israel's restoration in terms of entering the holy city (27:51b-53),[84] but this is not the focal point of Matthew's ending. Matthew moves beyond Jesus' death and resurrection. The risen Jesus is now, we argue, the Davidic Shepherd (Ezek 34, 37; Mic 5:3-4), appointed by God and placed "over" the re-gathered flock.

Placed over the restored flock and assigned a new role and task, Jesus illuminates what the resurrection of the smitten shepherd would mean to the Evangelist. Through his death and resurrection, the risen Jesus, as Matthew depicts, led the saints into the holy city (27:51b-53). Yet, in order for the final

---

[83] As to the referent of οἱ δέ, either some or all of the eleven, refer to Hagner, *Matthew*, 2:884.

[84] Keener, *Matthew*, 686; cf. Oscar Cullmann, *The Early Church* (ed. A. J. B. Higgins; London: SCM, 1956), 168; Hagner, *Matthew*, 2:686.

consummation to take place in its full scale of what definitely happened (aorist) in 27:23, the risen Jesus must continue to lead his re-gathered little flock into the promised future. Not exactly Jesus' death or resurrection, but his ἐξουσία that is granted subsequent to God raising him up from the dead, thus explains how Jesus would fulfill his task of shepherd-leadership over the new community comprised of Israel and the nations.

The aorist of the ἐδόθη does little to pinpoint when it was given to Jesus. Yet according to the procedure of Ezekiel's vision of the restoration, YHWH is to raise up his servant David (one shepherd/ the king, 34:23-24; 37:24-25; cf. *Ps. Sol.* 17:21) only after he fulfills the tasks of the eschatological Shepherd – to seek the lost, heal the sick, and judge the fat and robust. This occurs only after he accomplishes the restoration of his flock (34:1-16; 17-23; cf. 4Q521 2, 2:10-14). Then he puts in place the new eschatological leadership in accordance with his own theocratic shepherd-leadership. In Ezekiel's vision, YHWH's choice was clearly in view of the panoramic history of the monarchic experiences: no other would suffice than one like David.

Likewise, it is the Son of David (Matt 1:1), particularly the therapeutic Son of David, who accomplishes the tasks of Ezekiel's eschatological mission of compassion among his flock (esp. 9:10-13). In this respect, it is the ἐξουσία granted to the risen Jesus (cf. 4Q521 2, 2:1), by which Jesus represents God the Shepherd over both Israel and the nations (cf. הָאָדָם, Ezek 34:31). Thus, the authority is comprehensive and matchless to fulfill God's will in heaven and on earth (cf. Matt 6:10).[85] Further, M. Christudhas argues that the ἐξουσία of Jesus in Matthew, "in its origin, it is neither derived from nor delegated by any secondary sources." Thus, it is "different from the judicial or rabbinic or prophetic model;" yet, "it is the same as that of God's," and the authority of Jesus is also "*connected with his consciousness of the mission or his messianic consciousness.*"[86] This authority – in keeping with our proposal – is indeed YHWH's himself as "the eschatological Shepherd" in the midst of the harassed and downtrodden flock, which is clearly demonstrated in chapters 8-9 (cf. 7:29; 9:6, 8; 10:1; 21:23-27; cf. 11: 27).

The passion narrative, flush with the image of the suffering Shepherd of Zechariah 9-13, indicates that Matthew also interacts with Zechariah's revision of Ezekiel's hope for the Davidic Shepherd.[87] Thus, it is necessary for the Evangelist to complete the story of the eschatological Shepherd in the Davidic

---

[85] Luz, *Matthäus*, 4:442.

[86] M. Christudhas, *The ἐξουσία of Jesus in the Gospel of Matthew* (New Deli: Regency Publications, 2000), esp. 103-131, 237, 239 (italics mine).

[87] Cf. S. Talmon, "The Concepts of MĀŠÎAH and Messianism in Early Judaism," in *The Messiah: Developments in Earliest Judaism and Christianity* (ed. J. H. Charlesworth et al.; Minneapolis: Fortress Press, 1992), separately notes "the failure" of Ezekiel's vision (103), and later Zechariah's revision (114).

Shepherd tradition. Zechariah's theme is the sin of the flock; Zechariah's solution is the smitten shepherd, that is, the suffering of the mysterious shepherd. Accordingly, in Matthew's Gospel, Jesus' fulfillment of the role of Zechariah's smitten shepherd (26:31) opens the way to the fulfillment of Ezekiel's vision: YHWH establishes the Davidic Shepherd-kingship over Israel, and now even over the nations. Moreover, for the Evangelist of the First Gospel, this fulfills Micah's vision for the Davidic Shepherd's extended reign over the nations (Mic 5:3-4) not only over Israel (5:1): καὶ στήσεται καὶ ὄψεται καὶ ποιμανεῖ τὸ ποίμνιον αὐτοῦ ἐν ἰσχύι κυρίου καὶ ἐν τῇ δόξῃ τοῦ ὀνόματος κυρίου τοῦ θεοῦ αὐτῶν ὑπάρξουσιν διότι νῦν μεγαλυνθήσεται ἕως ἄκρων τῆς γῆς (Mic 5:3, LXX, italics mine). Micah's extended vision in this verse – YHWH's establishment of the Davidic Shepherd for the sake of God's name to the ends of the earth – resounds with echoes in Matt 28:18-20.[88]

The authority granted to Jesus is the authority of the eschatological Shepherd, which is splendidly demonstrated during Jesus' mission of compassion among the harassed and down-trodden flock of Israel (9:36; esp. chs. 8-9). Yet, the transfer of the eschatological Shepherd's authority upon the Davidic Shepherd was officially completed only after Jesus assumed the role of Zechariah's smitten shepherd (26:31-32). Thus, the fundamental enmity of the flock against their Shepherd and its consequential exilic status was resolved, and the new covenant for the new future founded. From the onset (2:6), the Evangelist had in mind the picture of YHWH's appointment of his Servant David, the one Shepherd ruling over both Israel and the nations. The Evangelist's announcement of the coming of the Davidic Shepherd in Matt 2:6 (Mic 5:1) is fully activated in the risen Jesus in the Gospel's final scene (28:16-20).

5.2.5 One Shepherd over Israel and the Nations

Determining whether πάντα τὰ ἔθνη refers to "all peoples" (both Jews and Gentiles) or to "all Gentiles," that is, whether the phrase includes or excludes Israel, remains one of the critical issues of Matt 28:16-20. Luz presents alternative explanations for each respective option: "Is the destruction of Jerusalem as God's punishment for Israel's rejection of Jesus and his messengers a final judgement, so that the judgement by Jesus in his parousia takes place only as to the Gentiles? Or, does the Son of Man set right with all peoples, Jews and Gentiles, only at his return?"[89] The key to the meaning of ἔθνη in v. 19, Luz thus suggests, can be found in the implication of the judgment upon Israel. If the Son of Man figure is pictured in Matt 28:16-20, and if the judgment

---

[88] Even the language of Matt 28:18-20 is reminiscent of those languages underlined in Mic 5:3 (LXX).

[89] Luz, *Matthäus*, 4:448.

that has fallen upon Israel is the final one, then it is likely that the risen Jesus' commission in v. 19 can suggest the exclusion of the mission to Israel.

While observing the "*grundsätzlich* universalistisch" mission to "allen Völkern," Luz notes that Matthew no longer has great hopes for that mission (cf. 22:8-10; 23:39-24:2; 28:15), and the Great Commission "schließt eine weitere Israelmission zwar nicht explizit aus."[90] Working on the assumption that Matthew's community as a minority living in "heidnischen" Syria faces a majority hostile to Jesus, Luz argues that they embrace the task to reach out to the Gentiles.[91] Thus πάντα τὰ ἔθνη in v. 19 should be rendered, "all Gentiles."[92]

As we have previously argued, however, the Son of Man figure does not fit Jesus in Matt 28:16-20. Despite the eschatological judgment upon Israel (cf. 26:31), it is likely that the Evangelist continues to develop the motif of the restoration of Israel in terms of the renewed leadership of Jesus as the once-smitten shepherd in 26:31, though now as the Shepherd-Appointee in 28:16-20. As the involvement of the renewed Israel within the great commission by the Davidic Shepherd need not be seen as subsidiary but as necessary, the judgment upon Israel cannot be considered final.

Of significant import is the christology of Matt 28:16-20. As to the judgment upon Israel, the sheep, along with goats (25:31-46), within one flock of πάντα τὰ ἔθνη (both Jews and Gentiles), the Evangelist clearly laid out the vision of the Son of Man who is also the Shepherd as the King and the Judge, yet only at his second coming. In Matt 25:31-46, the risen Jesus, newly appointed over the flock of the eleven in Matt 28:16-20 is still, so to speak, on the way to that far-end eschatological momentum. Compared to the vision of the second coming of the Son of Man, what is envisioned in Matt 28:16-20 is but the beginning of that final eschatological momentum.

Luz emphasizes that the Great Commission does not explicitly prohibit the disciples' mission to the Jews; therefore, the continuous mission to Israel may have been assumed. In speaking to this issue, E. Schnabel underscores three salient points: (i) if the unambiguous directions given by Jesus when sending his disciples "to Israel" in Matt 10 are to be considered, it is unlikely that the Gentiles replace Israel in Matt 28:19; (ii) in light of the narrative structure of the

---

[90] Luz, *Matthäus*, 4:451 (italics his), notes also (447-449) that ἔθνη in the LXX is ambivalent, meaning either "alle Völker" or "alle Heiden"; the same is true of the First Gospel, e.g., "Gentiles" (10:5; 20:19; cf. 4:15; 6:32; 10:18; 20:35; cf. 2:1-12; 24:7); the case of 21:43 appears questionable, "all peoples" (24:9; 25:32).

[91] Luz, *Matthäus*, 4:451.

[92] Luz, *The Theology of the Gospel of Matthew* (Cambridge: Cambridge University Press, 1995), 139-140; however, Luz, "Has Matthew Abandoned the Jews?: A Response to Hans Kvalbein and Peter Stuhlmacher concerning Matt 28:16-20" in *Mission of the Early Church to Jews and Gentiles* (ed. Jostein Ådna and Hans Kvalbein; Tübingen: Mohr Siebeck, 2000), 65, says, "In 28:16-20 Matthew does not categorically [in a salvation-historical sense] exclude a continuation of the mission of Israel."

5.2 The Divine Presence and One Davidic Shepherd-Appointee (Matt 28:16-20) 355

Gospel, the disciples sent to Israel are not reported to have returned, implying that the mission to the Jews is an unfinished task (cf. "Paul in the synagogues" in Acts; Rom 9-11);[93] (iii) nevertheless, the commission to go to all peoples "hebt die Einschrängkung auf Israel" (10:5) while not excluding Israel, means "Absorbierung Israels in die Völkerwelt, die eine Relativierung der Vorzugsstellung Israels," that is, "Mit dem Abschluss des Heilswerkes des Messias Jesus in Kreuzestod, Auferweckung und Erhörung hat Israel seine besondere heilsgeschichtliche Rolle erfüllt."[94] Likewise, there is no lack of support for the suggestion that the πάντα τὰ ἔθνη of Matt 28:19 includes Israel.[95]

Thus, between Matt 10:5-6 and Matt 28:16-20, we find a continuity regarding the mission to Israel. It is also true that a fundamental change has occurred in the latter passage. How can we harmonize the universal scope of the mission without excluding the particular, Israel, within that vision? Are the commissions of mission in Matt 10:5-6 and 28:16-20 contradictory? If not, in what sense can those seemingly contradictory directions be comprehended? Thus far we have refuted any proposal that has at the center of its argument the figure of the Danielic Son of Man or the highly exalted Son of God as Kyrios. To answer the question of missiology in Matt 28:16-20 is to answer the question of christology in that pericope: What christology would be able to embrace these seemingly opposing aspects of the commission?

Recently Stuhlmacher has attempted to read Matt 28:16-20 in light of the theme of "the eschatological restoration of Greater Israel" (cf. Jer 30:1-31:40; Mic 2:12-13; Ps. 80, 87, 96), yet, he continues to appeal to the figure of the Son of Man in Dan 7:13-14.[96] It is noteworthy that Stuhlmacher argues for the continuation of the shepherd motif of Matt 26:31 (cf. Mark 14:27) in Matt 28:16-20.[97] Further he refutes Luz's interpretation of Jesus' "going ahead" to Galilee (προάγω, 26:32) as the sign of the Evangelist's concerns about Gentile mission (4:15-16). For Stuhlmacher, Jesus going on to Galilee (26:32; 28:16-17) is "a symbolic act" involving "the symbolic restoration of [Greater] Israel

---

[93] The assumption of the cessation of evangelizing the Jews even in Matthew's time is implausible. See Hans Kvalbein, "Has Matthew Abandoned the Jews?: A Contribution to a Disputed Issue in Recent Scholarship," in *Mission of the Early Church to Jews and Gentiles* (ed. Jostein Ådna and Hans Kvalbein; Tübingen: Mohr Siebeck, 2000), 56, notes, "Justin still attests to the continuation of an active mission to the Jews circa A.D. 160 (cf. Dial. 39:1-2); see also, Davies and Allison, *Matthew* 3, 684.

[94] Schnabel, *Urchristliche Mission*, 358; cf. Hagner, *Matthew*, 2:887.

[95] Gnilka, *Matthäusevangelium*, 2:507-508; Stuhlmacher," Matt 28:16-20," 17-43; Hans Kvalbein, "Has Matthew Abandoned the Jews?" 54-57; von Axel von Dobbeler, "Die Restitution Israels und die Bekehrung der Heiden: Das Verhältnis von Mt 10,5b.6 und Mt 28,18-20 unter dem Aspekt der Komplementarität, *ZNW* 91 (2000): 18-44; Hagner, *Matthew*, 2:885-886; Davies and Allison, *Matthew*, 3:684.

[96] Stuhlmacher, "Matt 28:16-20," 24-30.

[97] Ibid., 25-26.

after the catastrophe of judgment upon the shepherd and his flock."[98] On the other hand, Sthulmacher associates the πάντα τὰ ἔθνη with the sovereignty of the Danielic Son of David (Dan 7:13-14); thus, the rendition of ἔθνη as "Gentiles" would "improperly restrict the dominion of the exalted Son of Man."[99]

Stuhlmacher's combination of "the divine Shepherd" (Ps. 80) with the Danielic "Son of Man," however, appears incomplete, and does not fit the context of Matt 28:16-20. This raises the question: By which tradition in the OT or anywhere else can we harmonize these two figures?[100] The tradition that links "the divine Shepherd" (Ps 80) with "the Son of Man" (Dan 12) can rarely be attested in the OT. Likewise, it is a challenge using those traditions to explain the link between Jesus' particular mission to the lost sheep of Israel (Matt 10:5-6; 26:31) and his universal mission to all nations (Matt 28:16-20) in Matthew's narrative and theological structure. This is why Stuhlmacher stumbles over his explanation of the motif of Jesus' promise of "divine presence" as that of the Son of Man. The vision of the promise of the risen Jesus' presence *on the way* to the glorified future of the coming of the Son of Man in the clouds, is obviously *not* attested either in Psalm 80 or in Daniel 12.

In short, the christology of Matt 28:16-20 is located *somewhere in between* Israel's eschatological Shepherd (for the lost sheep of the house of Israel; Matt 10:1-6) and the futuristic Son of Man (the triumphant futuristic reign over all; Dan 7, 12). We propose that the essential link between these two figures, conspicuously missing in Stuhlmacher's proposal, is none other than the Davidic Shepherd-Appointee, the Prince, and the Teacher (cf. Matt 2:6). The tradition behind Matt 28:16-20 is likely Ezekiel's vision along with Micah 2-5, *via* Zechariah 9-14 in Matthew's passion narrative (cf. Matt 26:31-32), namely, the Davidic Shepherd tradition, rather than Psalm 80 and Daniel 12. We shall examine later not only the motif of the "divine presence" (cf. Ezek 37:24-25) but also the distinctive occurrence of the motif of "teaching" or "obedience to the Law" (cf. Ezek 34:24; 37:25) in Matt 28:19-20, challenging the probability of Stuhlmacher's view.

While Stuhlmacher's reading of Matt 26:31 is valuable and certainly supported, in part, by our thesis of the shepherd image in the Gospel, a baffling

---

[98] Ibid., 26, also, Stuhlmacher sees the eleven as "the core of new Israel" (29).

[99] Ibid., 27.

[100] Linking the figure of the Son of Man (in his second coming) with shepherd imagery is not impossible in the Gospel (25:31-46), yet doing so betrays the context of Matthew's narrative (2:6; 9:36, etc); Luz, "A Response to Hans Kvalbein and Peter Stuhlmacher," 67, rightly critiques Stuhlmacher's proposal, and thus sounds critical: "On what ground does Stuhlmacher bring exactly these texts into play?"; likewise, Axel von Dobbeler's proposal for the "Komplementarität" between Matt 10:5-6 and 28:18-20, in "Die Restitution Israels und die Bekehrung der Heiden," would fall under Luz's critique. Dobbleler's reliance mainly upon Isa 42:6 and 49:6 bears only partial significance, and fails to present a consistent christology.

question remains. How do we explain the joining of the nations in the restored Israel? Oskar Skarsaune phrases it simply, "to this restored Israel other sheep, from the Gentiles, *are to be added.*"[101] Yet, which christological and theological framework can explain this essential continuation in light of the radical newness we find in Matt 28:16-20?

Hans Kvalbein, siding with Stuhlmacher against Luz's position as to πάντα τὰ ἔθνη in Matt 28:16-20, argues that "the mission to the Gentiles is a continuation and a fulfillment of the ministry to Israel, not a replacement of it."[102] Then, Kvalbein turns to the Abraham motif, i.e., Jesus as the son of Abraham (Matt 1:1), and links this designation of the OT covenant with Abraham in Gen 12:1-2 while Kvalbein completes the picture with Jesus as the healing Son of David for the afflicted among Israel.[103]

Kvalbein's thesis seems appropriate in some sense but his argument would frustrate many interpreters of the First Gospel, since the Abraham motif is rarely developed in the narrative. In fact, the evidence he presents, i.e., Rom 1:3; 9-11; 1 Thess 2:13-16, appears shaky, taken far afield from Matthew's context.[104] Further, Kvalbein's arguments fall short of the mark when he associates the Abraham motif with the healing Son of David who merely "favored" Israel.[105] In the final analysis, Kvalbein denies his own thesis a legitimate argument. Once again, we face the issue of the consistent christology of Matt 28:16-20, which would work "both" in the narrative of the First Gospel as a whole, and in the light of the tradition-historical background.

The christology of Matt 28:16-20 has roots in the Davidic Shepherd tradition, especially in the figure of the "One" Davidic Shepherd over both Israel and the nations according to Ezekiel 34-37 (cf. esp. Mic 5:3-4). Ezekiel's vision implies the concept of πάντα τὰ ἔθνη as both Israel and the nations in the context of YHWH desiring to continue to gather his flock now under this newly established Davidic Shepherd leadership. Thereby Ezekiel implies that "the people from the nations" join the expanding flock when he says in 34:31, "My flock," this time repeating the same צאן, yet omitting "my," and simply using "flock, [of] my pasture, people [אדם]." The movement from the restored little flock toward the nations' recognition of YHWH's holy name is unmistakable both in Ezekiel 34 and 37 (from "My flock" to "people, you") and Matt 28:16-20 (from "the eleven" to "all nations"). It is noteworthy that both texts show concern for

---

[101] O. Skarsaune, "The Mission to the Jews: A Closed Chapter?: Some Patristic Reflections Concerning 'the Great Commission'," in *Mission of the Early Church to Jews and Gentiles* (ed. Jostein Ådna and Hans Kvalbein; Tübingen: Mohr Siebeck, 2000), 70, rephrases Stuhlmacher's reading of Matt 26:31 (italics mine).

[102] Kvalbein, "Has Matthew Abandoned the Jews?" 55.

[103] Ibid., 57-61.

[104] Ibid., 57-58.

[105] Ibid., 58-60, under the subsection titled: "Jesus' healing miracles: Do they increase the guilt of Israel?"

the divine name as well. For Ezekiel 34 and 37, the nations must recognize YHWH's holy name; in Matthew 28 the disciples are called to baptize the people from the nations in the divine name (cf. Matt 6:9).

Admittedly, the One Shepherd in Ezekiel 34 and 37 refers to the one who will rule primarily over the united Israel, *all twelve tribes* (37:15-24). The universal dimension is only implicit in the ambiguous ending, "You are my flock, the flock of my pasture, people (הָאָדָם) you!" Although *Tg. Ezek* 34:31 renders הָאָדָם strictly as 'the house of Israel,' other prominent Davidic Shepherd texts in the Second Temple period such as *An. Apoc.* 90: 17-18 and *Ps. Sol* 17-18 explicate what is only implicit in Ezek 34:31: the universal reign of the One Shepherd both over Israel and the nations.

The similar development is reflected in Matt 28:18-20. Up to this point, Jesus is the risen Shepherd of the eleven (οἱ ἕνδεκα) in vv. 16-17, while Matthew's detailed expression, οἱ ἕνδεκα certainly recalls the betrayal of Judah and the incident of the smitten shepherd in 27:3-10, 31;[106] the eleven can symbolically represent the re-gathered flock, the renewed Israel.[107] They come and worship (προσεκύνησαν) the risen Jesus, their Shepherd-king in v. 17.[108] Those who doubt only function as a catalyst for Jesus proclaiming the legitimacy of his leadership over them.

The idea of increasing the number of the members of the flock is apparent in both texts as well; now these 'eleven' as the restored flock of Israel are charged to make disciples from the nations. In Matthew 28:18, Jesus as the risen Shepherd clarifies, so to speak, how the Davidic Shepherd in Ezekiel 34 and 37 would fulfil the vision of extending his reign over the nations (וְהִרְבֵּיתִי אוֹתָם ["I will increase them," 37:26c]; Καὶ γνώσονται τὰ ἔθνη εἰς τὸ ὄνομα ("the nations will know," 37:28, LXX]). Matthew's (Jesus') solution is the disciples, namely, the under-shepherds as proleptically realized already in Matthew 10 for the mission to Israel.

The result is that *Jesus' identity as the eschatological and Davidic Shepherd in the First Gospel – this unified christological picture of Jesus – best explains the seemingly contradictory mission of Jesus*. Jesus as the eschatological Shepherd of Israel ministered once to the lost sheep of the house of Israel, and

---

[106] Davies and Allison, *Matthew*, 3:680.

[107] Zimmermann, *Messianische Texte,* 127, notes that the eschatologically reconstituted twelve represents "ganzen Volkgemeinschaft,"; in Matthew at 28:16-17, to be precise, they are the eleven.

[108] Matthew's usage of the verb προσεκύνεω most likely shows the exaltation of Jesus as the king of the Jews, the Shepherd of Israel. The verb occurs 14 times in the Gospel (Mark, 2x; Luke 2x; John 8x): Matt: 2.2 (the magi); 2:8 (Herod); 2:11 (the magi); 2:11 (the magi); 4:9 (the Tempter); 4:10 (Jesus); 8:2 (a leper); 9:18 (a ruler); 14:33 (disciples); 15:25 (Canaanite woman); 18:26 (king and servant); 20:20 (mother of James and John); 28:9 (two Marys); and 28:17 (the Eleven). Using γονυπετέω (27:29) instead of Mark's προσκυνέω (15:19), Matthew avoids employing the verb in the mocking scene. Cf. Davies and Allison, *Matthew*, 1:237.

5.2 The Divine Presence and One Davidic Shepherd-Appointee (Matt 28:16-20)

now Jesus is raised up by God – truly as the risen and appointed One – over one flock consisting of Israel and the nations. Thus Jesus' authority of God's kingdom is still Davidic, yet it is now at the same time universal. Finally he stands as the Shepherd-Appointee (cf. Mic 1-4; Matt 2:6) who uniquely represents YHWH the Sovereign over all nations, for the sake of His holy name.

Last but not least, this Davidic Shepherd-Appointee will be none other than the Shepherd/King/Judge as the Son of Man in his second coming, who will gather πάντα τὰ ἔθνη (25:32; cf. 12:30) to judge them, and to claim his own flock among them. Until then, the risen Jesus will be the prominent Prince-Teacher ever present in their midst, leading them to follow his teachings of the Law of God.

5.2.6 The Davidic Shepherd as the Prince-Teacher

The motif of teaching in the visions of both Ezekiel and Matthew plays a crucial role. Donaldson suggests Isa 2:2-3 for the Matthean motif found in 28:16-20. Though Isa 2:2-3 could provide instructive background, a major challenge presents itself. The nature of the mission envisioned here is characteristically centripetal – i.e., the nations come to Zion, God's holy mountain – whereas in 28:16-20 the direction of mission is characteristically centrifugal – i.e., the disciples are sent out from the mountain to the nations.[109] Simply put, in Ezekiel 34 and 37 the notion of the nations coming to Zion is absent, whereas the extension of the Davidic Shepherd's reign is explicit.

The Shepherd of the restored flock is to reign even over the nations. The centrifugal movement in both visions of Ezekiel and Matthew, we suggest, is neither accidental nor trivial. As One Shepherd over both Israel and the nations, Jesus in Matt 28:18 is finally entitled to command his disciples 'to go' to the nations. It may point to the continual shepherd-leadership of the Davidic Shepherd who leads or characteristically often 'goes ahead' of his flock (Mic 2:12-13, 4:11-13; Matt 26:32). The image of the shepherd-leader is comparable with the centrifugal movement regarding the nations in both Ezekiel and Matthew.

The question of the difference between Ezekiel's 'laws and decrees of YHWH' and Matthew's 'all that I [Jesus] have instructed,' can be related to the significant issue of Jesus and the Mosaic Law in the Gospel (cf. Matt 5:17-48).[110] In Ezekiel's vision, it is clearly YHWH's laws and decrees that Israel

---

[109] Donaldson, *Mountain*, 183-186, supporting Isa 2, toils to rectify the discrepancy; yet his appeal to Matthew's redactional emphases not on "going" but on any other elements is insubstantial.

[110] This issue touches upon an entirely different realm – Jesus and the Law – that involves highly debated texts such as Matt 5:17-48; 11:12-13. Here we will name a few works: R. Banks, *Jesus and the Law in the Synoptic Tradition* (London: Cambridge University Press, 1975); W. D. Davies, "Torah in the Messianic Age and/or the Age to Come," (JBL Manuscript

failed to keep in the midst of the nations, and the authority that the One Shepherd/King would exercise in the matter of interpreting the Law is not directly mentioned. On the other hand, what is in view in the phrase πάντα ὅσα ἐνετειλάμην ὑμῖν (Matt 28:20a) is believed to be the Law, significantly fulfilled through and interpreted by Jesus himself (cf. Matt 5:17-20; 11:13; 22:34-40). This may suggest that Jesus is the Davidic Shepherd as the Teacher of the law *par excellence*; implying that he is the unique (the one who takes up the role of the eschatological Shepherd), authorized (the smitten shepherd who is raised up and given with the authority) Teacher of the Law.

In this regard, it is noteworthy that in Ezek 37:24ab, the task of teaching (24b) explicates the nature of the leadership of the Davidic Shepherd/Prince (24a). The image of the Shepherd who is expected to lead the eschatological people is squarely shaped by his role of the eschatological Teacher. With little difficulty we can attest to the particular association of the image of the Davidic Shepherd and the Teacher in the Davidic Shepherd tradition of the Second Temple period (cf. Sir 18:13). The Son of David, or better yet, the Davidic Shepherd depicted in *Ps. Sol.* 17:41 is the Leader-as-Teacher: "he will *lead* [ἄξει] *all in holiness*" (17:41).

Similarly, the eschatological messianic white bull in *An. Apoc* appears after the Lord of the sheep judges the evil shepherds, and transforms both Israel and the nations into holiness (90:17-18). The Qumran texts illustrate an identical trend – a substantial part of the eschatological paradigm assigned to the Davidic messiah revolves around his role as Teacher of the Law. The anointed of righteousness (משיח הצדק) is identified with the Branch of David (צמח דויד) in 4Q252 fragment 5 l. 3-4, who will obey the Law with his people of the community (l. 5-7).[111] The role of the Prince (נשיא) of the congregation in 1Q28b 5:21-22 amounts to the figure of the Prince/Teacher of Ezekiel 34 and 37, who judges the poor with justice (l. 21c; cf. Isa 11:4) and leads the people in perfection before him on all the paths (l. 22).[112]

7; Philadelphia: SBL, 1952); J. P. Meier, *Law and History in Matthew's Gospel: A Redactional Study of Mt. 5:17-48* (Analecta Biblica; Rome: Biblical Institute, 1976); G. Barth, "Matthew's Understanding of the Law" in *Tradition and Interpretation in Matthew* (Philadelphia: Westminster, 1963); D. J. Moo, "Jesus and the Authority of the Mosaic Law," *JSNT* 20 (1984): 16-28; S. Westerholm, "The Law in the Sermon on the Mount: Matt 5:17-48," *CTR* 6 (1992): 43-56; Recently, Paul Foster, *Community, Law and Mission in Matthew's Gospel* (WUNT 2.177; Tübingen: Mohr Siebeck, 2004); Roland Deines, *Die Gerechtigkeit der Tora im Reich des Messias: Mt 5,13-20 als Schlüsseltext der matthäischen Theologie* (WUNT; Tübingen: Mohr Siebeck, 2004).

[111] It is interesting to note that 4Q252 is dated around the time of the Herodian regime (30 B.C.- A.D. 70). For a detailed discussion, refer to ch. 2.3.2 4QpGen (4Q252).

[112] Furthermore, 1Q28b 5:26b-29 describes the expansion of this Prince/Branch's powerful rule over the nations alluding to Mic 4:13 (l. 26b) and Mic 5:1-5 (l. 27-28).

5.2 The Divine Presence and One Davidic Shepherd-Appointee (Matt 28:16-20) 361

In texts such as 4Q521 2, 2:3-14 and CD-A 13:7-12, the role of establishing justice/righteousness within the eschatological community is assigned to the Instructor/Inspector of the community, i.e., the Seeker of the flock (מבקר, cf. Ezek 34:11, 12), whose role is apparently influenced by YHWH the eschatological Shepherd in Ezekiel 34. Despite the variety of the Teacher figures in the Qumran texts, none of them approximates the Teacher in Matt 28:18-20 in that the risen Jesus promises to be present in the midst of his disciples, and yet commands them as his under-shepherds to teach the people from the nations. Jesus' delegating his role of teaching to his disciples is not observable in those texts in the Second Temple period, nor is it found in Ezekiel's Davidic Shepherd/Prince in the scenes of Ezekiel 34 and 37. The 'princes' (נשיאים) in Ezekiel's vision, however, appear later in 45:8, 9 where Ezekiel 40-48 unfurls a blueprint for the settlement of the eschatological cultic community.

The more distinctive feature of Jesus in view of the contemporary visions of the Teacher would be his unparalleled stance toward *what* should be taught in the community. For the Son of David of *Pss. Sol* 17-18, it is doubtlessly the Torah, the Law of Moses. For the messianic white bull of *An Apoc* 90:17-18, it is probably the revelatory truths that transform the people to be holy, but is scarcely involving the Law of Moses. For Jesus, while it is still the Law, he claims it should be taught as filtered through him, that is, as it was fulfilled, interpreted, and further lived out by him. This is an astounding, unique claim compared with the typical Teacher figure in light of those texts in the Second Temple period.[113]

The characteristic to πάντα ὅσα ἐνετειλάμην ὑμῖν (Matt 28:20a) seems to point to the totality of what is involved in relationship of the Law with Jesus – that is, his fulfilment, interpretation, and his living out of the Law. If Jesus can be seen as the Davidic Shepherd in 28:16-20, then it is likely that his ministry of compassion as he takes up the mission of the eschatological Shepherd for the lost house of Israel may be at the core of what should be taught in 20a. If any continuity between the mission of the disciples and the mission of Jesus is intended in the phrase πάντα ὅσα ἐνετειλάμην ὑμῖν, then it readily recalls the mission of the eschatological Shepherd by which the nations will be judged before the Shepherd-Judge in Matt 25:31-48 (cf. Matt 9:13; 12:7).[114]

This suggestion is strengthened since both christology and mission in 28:18-20 maintain the coherency with the narrative as a whole.[115] Furthermore, this is

---

[113] Perhaps this constitutes the nature of Jesus' ἐξουσία in v. 18.

[114] Cf. C. Lendmesser, *Jüngerberufung*, 132-133, 149, likewise argues that Jesus' mission for the sinners' "Zuwendung zu Gott" as declared in 9:9-13 continues throughout Jesus' mission to Israel in Matt 10 and to the nations in 28:16-20, i.e., as the essence of Matthew's soteriology.

[115] Pheme Perkins, "Christology and Mission: Matthew 28:16-20," *Listening* 24 (1989): 302-309.

where Donaldson's proposal of Psalm 2 or Isaiah 2 for the tradition behind Matt 28:18-20 withers. Those traditions do not yield a coherent christological picture either in 28:18-20 or in view of the relationship of the closing (vv. 18-20) with the main narrative of the Gospel. Jesus as the Davidic Shepherd, Leader and Prince (Ezek 34:23-30, 37:25-28; Matt 28:16-20), who takes up the mission of compassion the eschatological Shepherd presumed for himself (Ezek 34:1-16; cf. Matt 9-25), would readily connect both the main body and the closing of the Gospel.

A discontinuation is evident as well. As attested in Ezekiel's pattern for YHWH, the eschatological Shepherd precedes the establishment of the Davidic Shepherd/Prince and Teacher (37:24-25). A similar pattern may be found in Matt 10:1-5 and 28:16-20; e.g., with such catchwords as πορεύεσθαι and ἔθνη. Yet, the striking differences, besides the target groups of the Shepherd's mission – whether the lost sheep of the house of Israel (Matt 10) or both Israel and the nations (Matt 28) – involve the absence (Matt 10) /presence (Matt 28) of 'the teaching' motif, i.e., of 'obedience to the Law.'[116]

Why does the risen Jesus command his disciples 'to teach' those who join his eschatological community, but did not command them 'to teach' when he sent them to Israel, to the lost sheep of the house of Israel? Noticing that Jesus never says explicitly that he gives the disciples the ἐξουσία to teach in 10:1-6, S. Byrskog comments on the characteristics of Jesus' teaching ἐξουσία especially in view of 28:19-20:

Jesus' didactic exousia defines the identity and activity of the adherents themselves. The final and all-embracing exousia is the prerogative (οὖν) of the mission to the nations, involving the disciple's teaching ministry (28:19f). Not until Jesus has been given all exousia is it possible to extend the mission to all the nations. ... Jesus is in 28:19f the risen one, always to be with the disciples. And the teaching of the disciples represents a direct manifestation of Jesus' own exousia.[117]

It is likely that the teaching ἐξουσία remains exclusively bound to Jesus' person especially in 28:19-20 in which Jesus assures his disciples of his presence among when he commands them to teach. The mode as well as the expression of Jesus exercising his authority in 28:19-20 differs from those depicted when giving the ἐξουσία to his disciples on their mission to Israel in 10:1-6. In both instances (10:1-6 and 28:19-20), however, Jesus does not impart his ἐξουσία of teaching to his disciples. In other sections of his study of Jesus' didactic authority in the Gospel, Byrskog points out that Jesus being the only Teacher (23:7-10) indicates that "Matthew enhances Jesus' unique status as teacher with

---

[116] Cf. Luz, *Theology*, 16-17, 140, notes these similarities and differences.

[117] Samuel Byrskog, *Jesus the Only Teacher: Didactic Authority and Transmission in Ancient Israel, Ancient Judaism and the Matthean Community* (Stockholm: Almqvist & Wilsell, 1994), 283-284.

5.2 The Divine Presence and One Davidic Shepherd-Appointee (Matt 28:16-20)

God himself, the heavenly Father, as the point of reference." Furthermore, he indicates, "The adherence to Jesus as teacher relates to the confession of the one and only God."[118] Jesus' ἐξουσία is descriptive not only of his mission but also of his identity and status as the Shepherd.

It is also noteworthy that among the activities of healing, preaching, and teaching (cf, 4:23; 9:35), only teaching is prominent and tied to Jesus' ἐξουσία in 28:19-20 in view of all the nations. On the other hand, in Matt 10:1-6 Jesus' ἐξουσία is associated with healing and preaching (v. 7),[119] which are clearly activities in which the disciples engaged when among the lost sheep of the house of Israel. How can we explain this subtle transition in terms of Jesus' missionary activities and his authority?

Ezekiel's pattern suggests that Jesus' role in Matthew 10 is that of YHWH the eschatological Shepherd for the lost sheep of Israel, but the risen Jesus in Matthew 28 must be the Davidic Shepherd, the One Shepherd over both Israel and the nations. The task of teaching is characteristic of the Davidic Shepherd while the seeking and healing (thus, preaching, Matt 10; cf, chs. 8-9) belong to the mission of compassion of YHWH the eschatological Shepherd. This affirms the relevance of the Ezekielian pattern to Jesus' mission toward 'all nations' in Matthew 28 as well as to 'the lost sheep of the house of Israel' in Matthew 10.

Equally important is the eschatological Teacher whose role is primarily to 'sanctify' the eschatological community described in such terms as 'holy' (*An. Apoc.* 90:17-18) and 'free from sin' (*Ps. Sol* 17:36). This qualification is required of the Teacher since his role is not to rescue – which if executed, could be militant – but to lead an already rescued people on the path of righteousness (cf. Ezek 34, 37; Isa 11:1-4). At this point, Matthew's narrative assumes that Jesus as the smitten shepherd has already inflicted the Lord's judgment on the sins of Israel (26:31; 27:51b). Might we then view Matthew's birth narrative of Jesus' sinless conception through the Holy Spirit in 1:18-20 as closely tied to his task of saving 'his people from their sins' in verse 21?[120]

If it is correct to read this connection in light of the Gospel's contemporary understanding of the holy Teacher who is to lead the eschatological people into holiness, then Jesus' role of *saving* his people from their sins (1:21) can be seen as commencing with the rise of the Davidic Shepherd at 28:18-20 (cf. 4Q174 1-2, l. 11-12). After his identification with the smitten shepherd, Jesus is raised

---

[118] Byrskog, *Teacher*, 299-300.

[119] As to the distinction between "preaching" (κηρύσσω Matt 10:7) and "teaching" in this context, I presume that the key difference lies between the role of YHWH the eschatological Shepherd (rescue/proclamation of the eschatological theocratic rule among the lost flock) and that of the Davidic Shepherd (restoration/sanctification).

[120] Cf. Dunn, *Christology in the Making*, 50, notes, while dismissing the significance of the Son of David theme in the First Gospel, that Matthew has extended the understanding of the divine sonship by dating it from his conception and attributing that to the power of the Spirit and by depicting Jesus' sonship of his mission which fulfilled the destiny of God's son Israel.

by God from the dead along with the saints whom the risen Shepherd led into the holy city (27:51b-53). These themes – viz, Jesus as the sinless Teacher, conceived by the Holy Spirit in the dawn of salvation for restored Israel and the nations – are followed by the final motif of the restoration of the divine presence both in 1:21-23 (*Immanuel*, 'God with us') and 28:18-20 ('I [Jesus] will be with you').

5.2.7 Jesus as the Eschatological Locus of the Divine Presence

As we consider the theme of the divine presence, the comparability of Ezekiel's shepherd/Prince and Matthew's Jesus is substantially reinforced. The theme of the divine presence opens and closes both respective narratives. Ezekiel's vision of Israel's restoration in chapter 34 opens with the painful recognition of the absence of the true shepherd among the flock: 'there is no one searching; there is no one seeking' (v. 6), 'because there is no shepherd' (v. 8; cf. Matt 9:36). The divine presence motif imprints also the macro structure of the book of Ezekiel: YHWH's glory departs from the temple (chs. 1-11) and returns to the rebuilt temple (chs. 40-48).[121] The vision of restoration in Ezekiel 34 and 37, however, presents the return of the eschatological Shepherd for the scattered flock and his Davidic Shepherd in their midst. Their return entails the restoration of the flock, which leads eventually to the rebuilding of the eschatological community with the temple in their midst (chs. 40-48).

For Matthew's Gospel, the significance of the theme of the divine presence has often been hailed in terms of 'a framework,' 'ein Grundthema' or 'the kernel of the Matthean narrator's ideological perspective,' or the 'theologische Leitidee.'[122] David D. Kupp, having surveyed two major research projects devoted to this topic by H. Frankemölle and Van Aarde, states that these redactional studies assume that the Evangelist is 'historicizing' post-Easter experiences or being 'transparent' to his community, in order to guarantee God's covenantal faithfulness, or the continuous mission of the Son of God (Jesus) through that of the sons of God (disciples) for 'Matthew's churches.'[123]

Kupp's main critique of these previous works, despite Frankemölle's helpful investigation of Matthew's indebtedness to the OT presence theme,[124] is that they have not fully examined the divine presence theme within Matthew's

---

[121] For a detailed analysis, refer to chapter 1, 1.4.3 "The Structure of Ezek 34-37"

[122] France, *Matthew*, 48; Luz, *Matthäus* 1:105; A. G. Van Aarde, "God Met Ons: Dié Teologiese Perspektief van die Matteusevangelie," (Th.D. diss., University of Pretoria, 1983), cited from David D. Kupp, *Matthew's Emmanuel: Divine Presence and God's People in the First Gospel* (SNTSMS 90; Cambridge: Cambridge University Press, 1996), 24-25; Frankemölle, *Jahwebund*, 122.

[123] Kupp, *Emmanuel*, 24-25.

[124] Frankemölle, *Jahwebund*, 331-342.

5.2 The Divine Presence and One Davidic Shepherd-Appointee (Matt 28:16-20)

narrative itself.[125] Restricting his research to the parameter of the 'I am with you' formula, Kupp's research of the theme in the OT and Second Temple Judaism does not yield much help, especially when he takes the formula mostly from the Deuteronomist and Chronicler, including 2 Chr 36:23 (YHWH's appointment of Cyrus).[126] Kupp's findings from his study of the theme within Matthew's narrative itself, as well as in 28:20, deserve mention:

> After all, what is presence at the end of Matthew? – not the authoritative magnificent activities of the anointed Messiah king, the divine Son of God, but the promises of empowerment and of Jesus' divine 'witness' to a group of doubting disciples who will face persecution and troubles; not convincing theophanic phenomena, but identification with the μικροί. Jesus' resurrection is powerful, but the irony is strong: God manifests his divine authority and presence not in Jerusalem, but in Galilee, not on a throne, but in an upside-down kingdom; not in the capital city with political and military strength, but on a wilderness mountain with teaching and healing; not in divine appearances, but through future 'gatherings' of followers; not in power and wealth, but in poverty and humility; in alliance not with the power-brokers, but with 'little ones'; not with the thousands, but with 'two or three'; not in the Temple, but in his risen Son.[127]

In a passionate and even poetic voice, most of Kupp's conclusions seems to complement the characteristics of Jesus primarily as 'the eschatological compassionate Shepherd of Israel' in the main body of the narrative where Jesus acts particularly as the healing Son of David. Yet, if there is any progression among Jesus' mission as the eschatological Shepherd (cf. Ezek 34:1-16), the smitten shepherd (Matt 26:31; cf. Zech 13:7), and the Davidic Shepherd (cf. Ezek 34:23-24; cf. Mic 5:3-4) enthroned as the One Shepherd over both and the nations in Matt 28:20 (2:6; cf. Mic 5:1-2), then Kupp's characterization remains only partially valid to the picture of Jesus as the Davidic Shepherd in 28:20.

The presence theme in the narrative closing in the Gospel envisions the reign of the Shepherd extended beyond the broad horizon of the nations, and not merely likened to that of Jesus' presence among 'the [harassed and oppressed] sheep without a shepherd' (9:36) confined within the territory of Israel. God's unexpected presence among little ones, as Kupp underlines, certainly remains characteristic, but only as a part of the theme, especially in the closing of the Gospel. Jesus as the Davidic Shepherd now openly declares that he is invested with 'all authority' (cf. Matt 21:41-46). For the tradition behind the presence motif in 28:20, Kupp turns to the 'I am with you' formula taken mostly from the Deuteronomist and the Chronicler. More than likely, however, the text echoes the development of the theme within the *Heilsgeschichte* in the OT, particularly within the post-exilic Davidic Shepherd tradition, historically and theologically

---

[125] Kupp, *Emmanuel*, 22, 26.
[126] Ibid., 109-156; for similar research results, see Frankemölle, *Jawehund*, 73-77.
[127] Kupp, *Emmanuel*, 242.

much more akin to the framework of the Gospel, and not merely on the level of the presence-formula (cf. Ezek 34:30).[128]

The description of 'his anointed one' in 4Q521 is a case in point that draws a distinction between the characteristics of God as the eschatological Shepherd and the Davidic messiah as his Shepherd-Appointee. The משיחו ('his anointed') is imbued with authority over the heavens and the earth (והארץ ישמעו למשיחו כ[י הש]מים) so they might listen to YHWH's appointee who will lead them by means of YHWH's precepts (4Q521 l. 1; cf. Ezek 34:23; 37:25). The eschatological pastoral tasks of mercy listed in 4Q521 l. 5-12 – freeing prisoners, giving sight to the blind, straightening out the twisted, healing the wounded, raising the dead, proclaiming the good news, feeding and inviting the poor, bringing back the outcast – are assigned to YHWH who will eventually establish his anointed over his restored flock. He is the one who will guarantee the Lord's perpetual presence through his role as Teacher of the law.[129]

If this is correct, then Jesus' presence promised to his disciples and his ever-increasing eschatological community in Matt 28:18-20 may well correspond to the rich language and concepts of the theme in the closing of the vision of the Davidic Shepherd tradition. Among the tradition, we find the archetype of the pattern in Ezek 34:23-24 and 37:24-25. In these texts, it appears that the presence theme takes its shape from at least three distinguishing and contributing aspects: (i) Prince (נשיא, 34:24b; 37:25c), (ii) 'my sanctuary' (τὰ ἅγιά μου, 37:26, 28), and (iii) 'my tabernacle' (ἡ κατασκήνωσίς μου, 37:27).

(1) The Prince/Teacher in Their Midst: Matthew 28:18-20 can be neatly divided into three subsections: v. 18 (receiving all authority from God); v. 19-20a (the commission), and v. 20b (Jesus' presence with the disciples). Jesus' authority is highlighted as he claims his leadership over the disciples; yet when he characterizes his relationship with his disciples, it is Jesus' presence that is emphasized. The commission of the risen Jesus is supported by these critical relationships, first with God, and secondly with his people. From God, he receives all authority; to his disciples, he promises his presence.

To be precise, for Ezekiel it is the Shepherd or the King whom YHWH appoints 'over' the rescued flock (34:23; 37:24), though characteristically it is the title of the Prince by which the same Shepherd maintains his intimate presence 'in the midst' of the flock (34:24; 37:25). It should be noted in both cases, however, that it is the same 'My servant David.' This subtle distinction between the Shepherd/King and the Prince is still maintained during the Second Temple period in texts such as 11QPs[a] 28:11-12a, which entail the double titles

---

[128] Even in Ezek 34:30 where the formulaic expression is detected, it is clear that the presence motif closely interrelates with the nations' recognition of the Lord as well as Israel's recognition. Refer to Ch.1.4.5.

[129] For a detailed discussion of the interchangeability of the roles between these two figures, refer to Ch. 2.2.7.

5.2 The Divine Presence and One Davidic Shepherd-Appointee (Matt 28:16-20)

of YHWH's eschatological agent, 'prince [of the people] and the ruler [over them]' (cf. *An. Apoc.* 89:46, 48b, 52). Accordingly, it is tempting to suggest that 'the Prince of the Congregation' in the Qumran writings reflects this basic idea of the presence of the eschatological Teacher especially 'with' or 'in the midst of' the rescued eschatological community (4Q161, 1Q28b; cf. the Instructor in CD 13), while the Davidic Branch in the same texts tends to involve more of the task of extending his reign over the community (cf. 4Q521, 4Q252).

Our point is that the presence of Jesus with the disciples in Matt 28:20 may correspond particularly to the role of the Prince in the Davidic Shepherd tradition while maintaining the vision of the Davidic Shepherd to extend his reign over the nations. Yet the immediate implication of Jesus' presence with the disciples in 28:20 would highlight his role of Teacher within the rescued eschatological community. We suggest that this is why Jesus' promise of his presence follows the commission of teaching all that he taught. The task of 'teaching' stands out in Jesus' commission to the disciples, supported by his presence with them forever. The community over which Jesus will reign in the future is thus characteristic of the words Jesus taught. That is, what is envisioned for the future community of the Davidic Shepherd among the nations in 28:20 is thus characteristic of the centrality of the teaching of the risen Jesus as *the Teacher in their midst*.

In the main narrative of the Gospel, the divine presence awestruck the crowds before the rescue mission of the compassionate, eschatological Shepherd. Jesus' presence as the Prince in 28:20, however, relates more to salvation, so to speak, in terms of the sanctification of the community (cf. 1:21, 23; Ezek 37:23; 4Q174 1-2, 11-12).[130] Now teaching is the most prominent task, strikingly absent when Jesus sent out his twelve disciples to the lost house of Israel (to rescue them) in 10:1. Sanctification or salvation in this sense for the eschatological community is assumed to be accomplished by διδάσκοντες αὐτοὺς τηρεῖν πάντα ὅσα ἐνετειλάμην ὑμῖν ('teaching them to do all that I have instructed you'), which in effect is similar to the transformation into holiness of the eschatological people (cf. *An. Apoc.* 90:17-18; *Ps. Sol* 17-18).

(2) The Sanctuary in Their Midst: This analysis leads us to the next clue in the expression, θήσω τὰ ἅγιά μου ἐν μέσῳ αὐτῶν εἰς τὸν αἰῶνα ('my Sanctuary [מִקְדָּשׁ] among them forever,' Ezek 37:26, 28, LXX). In the vision of Israel's restoration in Ezekiel 34-37 and 40-48, the vision of a new sanctuary requires a new residence in the land preceded by the coming of the new Davidic Shepherd. In this respect, 4Q174 1-2. 1:6b is illuminating. The Branch's

---

[130] The role of the prince in Ezek 40-48, unlike the Shepherd/King, is less political and more cultic; he is the one who takes upon himself the representative responsibility of gathering the gifts and offerings required of the people in order to provide the cereal, sin, burnt, and fellowship offerings "to make atonement (לְכַפֵּר) for the house of Israel" (Ezek 45:16-17). This could be relevant to the picture of Jesus as the Prince/Teacher of righteousness.

extended reign toward the nations is already hinted at in the phrase, 'the sanctuary of man' (מקדש אדם, 1.6b). This mention of מקדש אדם after the destruction of the temple of Israel because of her sin may well echo YHWH's claim for his own flock, eventually, 'people (הָאָדָם) you!' in Ezekiel's vision of restoration (Ezek 34:31; cf. *An. Apoc.* 90:17-18). Jesus' promise of his presence in the midst of his eschatological community of the disciples, likewise, is situated against the backdrop of his extended reign over the nations. Now, this divine presence as the eschatological sanctuary dwells among אדם through Jesus.

This particular emphasis of the presence of the sanctuary is closely related to its critical function of sanctifying the community. If Jesus' presence among his community in Matt 28:20 involves this role of the sanctuary among the eschatological people according to Ezekiel's vision, this amounts to the claim that Jesus himself is *the eschatological locus* of the divine sanctifying presence for his community among the nations. Similarly, Matt 18:19-20 implies that Jesus will be the center of the community as the sanctuary of divine presence in much the same way the Temple functions as the locus of divine presence for those who pray for God's favor upon its sanctuary, the holy of the holies (cf. 1Kings 8:27-53).

(3) The Tabernacle on the Move: Furthermore, if the risen Jesus' presence functions as the eschatological sanctuary, then it is equally possible that it functions as the מִשְׁכָּן ('tabernacle,' κατασκήνωσις [LXX], Ezek 37:27). In the context of Ezekiel 34 and 37, the sanctuary and the tabernacle – both of YHWH – are interchangeable in terms of YHWH's presence among the rescued flock to be led under the leadership of the Davidic Shepherd/ Prince. The divine presence through the medium of the Prince, the sanctuary, and the tabernacle at the closing of Ezekiel's vision is promised to be perpetual as well. Noticeably, the phrase 'forever' (לְעוֹלָם) occurs five times in chapter 37 (vv. 25 [2x], 26 [2x], 28), thereby recalling the perpetuity of the Davidic promise in 2 Sam 7:12-14 – especially in the text of the Gospel's ending. It is fully charged with the messianic enthronement motifs of the authority, the mission, all nations, and the perpetual presence (cf. 4Q504 1-2, 4:6-8; *Ps. Sol.* 17:35).

Perhaps these two expressions of the divine presence with his people, the sanctuary and the tabernacle, represent its complementary dimensions, "one for His transcendent nature and the other for His condescending immanent presence."[131] In Matthew's ending, however, we link Jesus' presence with the concept of the sanctuary, aided by the contextual clue of 'the teaching' and the presence as the sanctifying/transforming center among the nations. In this way, Jesus' presence as the tabernacle is illuminated by the other contextual clue, i.e., Jesus' commission given to his community 'to go' or '[to keep] going'

---

[131] Block, *Ezekiel: 25-48*, 421.

5.2 The Divine Presence and One Davidic Shepherd-Appointee (Matt 28:16-20)

(πορευθέντες, v. 19a) to the nations. In other words, Jesus' disciples is *to be on the move* to the nations until the age is consummated, and has run its full course (ἕως τῆς συντελείας τοῦ αἰῶνος, v. 20b).

Similarly, at the stage in which the Davidic Shepherd is enthroned with authority and mission in view of the nations in Ezekiel 34 and 37, the presence theme is depicted in terms of the Prince, the sanctuary, and the tabernacle. It is not yet expressed in the language of 'the city' as in Ezek 40:35, the place which indicates the completion of the settlement of the eschatological community with the divine presence at its center. Likewise, Jesus envisions the movement of his community ever going to the nations until the age comes to its completion. His promise of the presence thus matches in particular the idea of the tabernacle, God's dwelling presence upon his people on the way, not in the city (cf. Rev 21).

Interestingly, *Tg. Ezek* 37:27 renders MT's מִשְׁכָּנִי as *Shekinah* '(cf. 36:5, 20), reflecting the Lord's glorious presence apart from the First Temple (cf. Ezek 10). In this respect, the presence of the risen Jesus is promised to his followers on the Mount in Galilee (28:16) rather than on Mt. Zion where in the narrative the Second Temple still stands. More pointedly, in 28:20, Jesus' presence is guaranteed to *the community 'on the move'* and not merely to the little ones.[132] Finally, the Shepherd returns to his flock who once was 'without a shepherd' (9:36), and all that is entailed in the expectations of the divine return. The throne is set down not on Zion but in the midst of the eschatological community on the move toward the nations.

5.2.8 Summary

In Matt 28:16-17, Jesus appears to his eleven disciples, now as the Shepherd/Leader who restores his leadership over his once-scattered flock. As to the tradition behind the text of 28:16-20, we argued that Ezekiel 34 and 37 present the comprehensive elements involved in the picture we find at the end of the Gospel. First, all authority is given to Jesus in Matt 28:18 while the Davidic Shepherd is placed as one Shepherd by YHWH in Ezek 34:23; 37:24. Second, Jesus commands the disciples to make disciples from all nations in Matt 28:19 while YHWH promises to increase the flock of the Davidic Shepherd as the nations come to know him (34:30; 37:28). Third, Jesus commands his disciples to teach the nations to keep all that he has commanded in Matt 28:20 while the Davidic Shepherd is expected to lead the one flock to walk in the laws and decrees of YHWH so they will keep them in Ezek 37:24. Last, Jesus assures them of his presence among his new community in 28:20 while YHWH proclaims that the Davidic Prince will be in the midst of the one flock, and his tabernacle will be with them (34:24, 30; 37:25-28).

---

[132] Cf. Kupp, *Emmanuel*, 242.

5.3 Conclusion

Considering the Gospel's other explicit textual interactions with the tradition – especially the picture of Jesus as the eschatological Shepherd in the main body of the Gospel – Jesus as the eschatological Davidic Shepherd at Matt 27:51b-42 and 28:16-20 becomes all the more plausible. In fact, Matthew's narrative follows exactly the pattern laid out in Ezekiel's vision, i.e., the Ezekiel pattern: the coming of the eschatological Shepherd and the consequential appointment of the Davidic Shepherd over the rescued flock.

According to our analysis, Matthew's enigmatic depiction of Jesus' death and resurrection in Matt 27:51b-54 turns out to be an integral piece of the whole picture of Jesus as the eschatological Davidic Shepherd. The four critical linguistic and thematic resemblances found both in Matt 27:51b-54 and Ezek 37:1-14, i.e., "opening the tomb, raising the dead, bringing them into the city, and the recognition of God's name by the peoples," strongly suggest that the Evangelist, picking up the shepherd motif in Matt 26:31-32, continues to tell the story of the Shepherd in Matt 27:51b-54. This is a climactic scene in Matthew's Gospel as to the shepherd image applied to Jesus. At this point of his resurrection depicted in terms of the restoration, the risen Shepherd completes his role as YHWH the eschatological Shepherd for his own flock. The restoration of Israel reaches, proleptically, its ultimate end: entering the city.

Yet, this passage functions as a great exhortation in the Gospel for the one flock of Israel and the nations under the one Davidic Shepherd in Matt 28:16-20. Despite the climax depicted in Matt 27:51b-54, the restoration process is not finished completely. Rather, the disciples as the under-shepherds of the one Davidic Shepherd-Appointee are called to gather the flock, now including the peoples from all nations. The Shepherd in their midst now is none other than the Shepherd/King/Judge who will judge his own flock at the end (Matt 25:31-46).

The comparability of Ezekiel 34 and 37 with Matt 28:16-20 is highly credible. All the common themes – the authority, the commission of teaching, the nations, and the promise of the divine presence – provide solid foundations for our thesis that the Gospel of Matthew embraces the Davidic Shepherd tradition. Likewise, it interprets the identity and mission of Jesus from the vantage point of Ezekiel's vision of the appointment of the Davidic Shepherd. In light of the tradition both in the OT and the Second Temple Period, a profound difference is Jesus' commission of teaching his words, instructions, and the promise of 'his' presence with the eschatological community; it was the Prince, the sanctuary, and the tabernacle of the Lord in Ezekiel's vision.

Yet, Jesus' presence as the Davidic Shepherd in 28:20 may be characterized as 'salvific' or 'sanctifying' through the disciples' teaching and obeying (v. 19) all that Jesus 'taught' rather than seen as 'rescuing,' which involves 'healing and casting out demons' as demonstrated in Jesus' mission as the eschatological

5.3 Conclusion

Shepherd of Israel (cf. 10:6-8). This distinction may shed light on the close connection between the salvific significance of the name Ἰησοῦς (1:21) and the divine presence announced in the name Ἐμμανουήλ (v. 23) at the beginning of the Gospel as well. Likewise, the ending and the opening of the Gospel correspond to each other.

Now it is time to review and put together all the findings from our investigation of the theme in the narrative in order to be able to assess Matthew's interaction with the Davidic Shepherd tradition in view of the macro-structure of the Gospel.

Chapter 6

Matthew's Narrative Strategy and the Davidic Shepherd Tradition

Thus far, we have examined all individual textual interactions, allusions, images, and echoes of the First Gospel in its discourse with the Davidic Shepherd tradition. Sporadically, we observed that those interactions with the tradition form a thread of narrative strategy; for instances, the King/Shepherd/Judge in 25:31-46; the smitten shepherd in 26:31-32; and the risen one Davidic Shepherd-Appointee in 28:16-20 (cf. 27:51b-53), as well as the inclusion of theme of the Davidic Shepherd both in 2:6 and 28:16-20. This type of observation raises a further question: How is the Davidic Shepherd theme reflected in the narrative structure in the Gospel as a whole?

Our attempt to investigate the shepherd/sheep images in the Gospel in light of the entire Davidic Shepherd tradition up to Matthew, in fact touches upon the question of the theological *scheme* for understanding and communicating the nature of Jesus' messianic ministry. Davies and Allison, in their monumental commentary on Matthew's Gospel, overlook a coherent explanation of Jesus' messianic mission in terms of the narrative as a whole. Instead, they reluctantly suggest that Isaiah 61 and 2 Samuel 7 would be the best possible tradition behind the blueprint according to which Jesus unfolds his messianic ministry, while at the same time voicing doubts.[1]

We propose that the Davidic Shepherd tradition with all its quotations, allusions, images, themes, and patterns provides the Gospel with a critical and coherent literary and theological framework through which Jesus' identity and mission are meant to be understood and communicated. This does not mean that the picture of Jesus as the Davidic Shepherd is the only or even the most prominent aspect of Matthew's christology. But, missing the christology of the Davidic Shepherd in the First Gospel would result in that Matthew's comprehensive and coherent usage of the shepherd image, together with Jesus' identity as the healing Son of David, remain puzzling. In the following, we will briefly survey investigations into Matthew's structure, and will assess the implications of the Davidic Shepherd tradition as to Matthew's narrative strategy with its significant consequences upon the structure of the Gospel.

---

[1] Davies and Allison, *Matthew*, 2:601; cf. Allison, *The New Moses: A Matthean Typology* (Minneapolis: Fortress Press, 1993), presents Jesus as the New Moses in the First Gospel; likewise, Davies, "The Jewish Sources of Matthew's Messianism," in *The Messiah*, 494-511, presents Jesus as 'the Greater Moses' in the First Gospel.

6.1 Matthew's Grand Scheme

Many would denounce the idea that there is any 'grand scheme' intended in terms of the narrative structure of the First Gospel.[2] Despite little consensus on this matter, the issue of the macro-structure of Matthew's Gospel has been attractive to many for its possible correlation with the Evangelist's theological/christological intentions.[3]

W. D. Bacon's proposal of 'the five-fold discourses' (i.e., chs. 5-7, 10, 13, 18, 24-25) for the Gospel illustrates a foundational case in point.[4] Bacon argues that the Gospel, with its five central discourse sections, is designed to introduce a new Pentateuch for the early churches and present Jesus as the 'new Moses': "The Torah consists of five books of the commandments of Moses; Matthew is a 'converted rabbi,' and as such is a Christian legalist; he has followed the plan of aggregating his teaching material from all sources into five great discourses corresponding to the oration codes of the Pentateuch."[5] We note that Bacon made a correlation between the Gospel's grand scheme and the christology of the Gospel. Bacon's case for Jesus as the new Moses is based on the five-fold discourses. Thus the portrait of Jesus as a new Moses stands and falls, in Bacon's theory, depending on the legitimacy of his analysis of the five discourses.

In fact, it is repeatedly pointed out that there are more than five discourses

---

[2] Cf. Gundry, *Matthew*, 10; Davies and Allison, *Matthew*, 1:61; David R. Bauer, *The Structure of Matthew's Gospel* (Sheffield: Sheffield Academic Press, 1988), 11, 135.

[3] Jack D. Kingsbury, *Matthew: Structure, Christology, Kingdom* (Minneapolis: Fortress, 1975), 1-37, analyzes the previous studies of Matthew's structure and explores those relations among those subjects, i.e., the structure, the salvation-history and the christology.

[4] Bacon, *Studies in Matthew* (New York: Holt, 1930); idem, "Jesus and the Law: A Study of the First 'Book' of Matthew," *JBL* 47 (1928); also, in his earlier article, "The 'Five Books' of Matthew Against the Jews," *The Expositor* 15 (1918): 56-66. J. A. Findlay, "The Book of Testimonies and the Structure of the First Gospel," *The Expositor* 20 (1920): 388-400, claims that the five-fold division of Matthew is drawn from a hypothetical "Book of Testimonies," a theory greatly influenced by Rendal Harris' proposal; M. S. Eslin, "The Five Books of Matthew: Bacon on the Gospel of Matthew," *HTR* 24 (1931): 67-97, explains that Harris, based on the phrase, Ματθαιος ειργει των Ἰουδαιων θρασος. Ὡσπερ χαλινοις πεντε φιμωσας λογοις ("Matthew squashed the boldness of the Jews as if he muzzled [it] with five bridles of books [Testimonies]") in a sixteenth century manuscript discovered around 1925 titled Ματθαίου μονάχου, identifies these 'Testimonies' (the chains of proof-texts) with 'the Oracles' or 'Logia' ascribed to Matthew by Papias of Hierapolis. While discarding 'new Moses' typology for Matthew's Jesus, Bacon's theory of 'five discourses' still finds its supporters in recent analyses; for instance, A. G. van Aarde, "Matthew's Portrayal of the Disciples and the Structure of Matthew 13:53-17:27," *Neot* 16 (1982): 21-74; also, H. J. Bernard Combrink, "The Structure of the Gospel of Matthew as Narrative," *TynBul* 34 (1983): 61-90.

[5] Bacon, *Studies in Matthew*, 81.

in the Gospel (cf. chs. 11 and 23).[6] More importantly, the critical weakness of Bacon's proposal is also found in his treatment of the infancy narrative (chs. 1- 4) and the passion narrative (chs. 26-28) as appendixes, i.e., as negligible Prologue and Epilogue respectively.

The significance of the infancy narrative would best be illustrated by D. L. Barr's modification of Bacon's theory. While discarding a "new Moses' typology for Matthew's Jesus but retaining some of Bacon's analysis of the five discourses,"[7] Barr further stresses the interconnection between each of the five discourses with their preceding and following narrative sections. Thus, for example, the christology of the infancy narrative in chapters 1-4 is critical for understanding the first discourse, namely, chapters 5-7.[8] Likewise, assuming that Matthew is an expanded rework of Mark, H. B. Green argues, "no enlarged version of Mark could possibly have treated the Passion narrative as an appendix."[9]

H. B. Green puts forth a different proposal: a chiastic structure for Matthew's narrative, thus creating two divisions. In the first half – chapters 1-10 – is the theme of Jesus' proclamation of the kingdom of heaven to the Jews; in the second half – chapters 12-28 – is the theme of Israel's rejection of their Messiah.[10] The centrality of this christological understanding holds true for those who consider chapter 13 and not chapter 11 as the pivotal chapter.[11] In the case of the 'chiastic structure,' there are implications for the grand structure of the Gospel related to portraits of Jesus that find their key in the function and significance of the hinge chapter, whether it is chapter 11 or chapter 13. The theme of Israel's rejection of her messiah remains the same.

Even without the particular literary or topical clues, a number of scholars have attempted to view Matthew's concept of salvation history as critical to the Gospel's composition.[12] Even though it is not strictly a literary analysis, the issue of the salvation history touches upon Matthew's design of the grand

---

[6] M. M. Thompson, "The Structure of Matthew: A Survey of Recent Trends," *Studia Biblica et Theologica* 12 (1982): 200-202; H. B. Green, "The Structure of St. Matthew's Gospel" in *Studia Evangelica* IV, part I (Berlin: Academie-Verlag, 1968), 48; A. Farrer, *St Matthew and St Mark* (London: Dacre, 1954), 177-197, suggests a 'Matthean Hexateuch.'

[7] D. L. Barr, "The Drama of Matthew's Gospel: A Reconsideration of Its Structure and Purpose," *TD* 24 (1976): 35-36; similarly, A. G. van Aarde, "Matthew's Portrayal of the Disciples and the Structure of Matthew 13:53-17:27," *Neot* 16 (1982): 21-74; also, H. J. Bernard Combrink, "The Structure of the Gospel of Matthew as Narrative" *TynBul* 34 (1983): 61-90.

[8] Barr, "The Drama of Matthew's Gospel," 35-36.

[9] Green, "The Structure," 49.

[10] Green, "The Structure," 50-59.

[11] H. J. Bernard Combrink, "The Structure," 71, sees ch. 13 as the hinge point, not ch. 11; also, idem, "The Macrostructure of the Gospel of Matthew," *Neot* 26 (1982): 1-20.

[12] Bauer, *Structure*, 12, lists the names such as W. Trilling, W. G. Thompson, and J. P. Meier.

scheme of his Gospel. For instance, G. Strecker, taking his cue from the work of Hans Conzelmann on Luke, argues that Matthew divided salvation history into three epochs: the age of prophecy, of Jesus, and of the church.[13] Strecker does not, however, pinpoint the exact time of the inauguration of the church in the Gospel, and only vaguely surmises that the church began at the 'end of the life of Jesus.'[14]

Similarly, J. P. Meier argues that for Matthew, the death and resurrection of Jesus marks 'die Wende der Zeit,' that is, the decisive in-breaking of the new aeon.[15] According to Meier, events such as 'the tearing of the temple curtain,' 'the earthquake,' and 'the proleptic resurrection of the dead saints' (27:51b-53) indicate the realization of the apocalyptic terminus in Matt 5:18b-d, which means the cessation of the validity of the Law of Moses and the ushering in of the end time.[16] Besides the unlikely link between the apocalyptic events in 27:51b-53 (cf. Ezek 37:1-15; Zech 14:11-14) and the conditional clauses in 5:18, : ἕως ἂν παρέλθῃ ὁ οὐρανὸς καὶ ἡ γῆ, our study shows that the end time has already dawned at the appearance of Jesus as *Immanuel* (1:23), through whom the eschatological Shepherd begins to seek the lost and heal the sick among his own flock (4:23; 9:35-36). The time of Jesus as the therapeutic Son of David among the lost sheep of the house of Israel is not just 'proleptic' to the new aeon, but it belongs to the eschatological time of restoration.

On the other hand, J. D. Kingsbury proposes that there are only two epochs – those of Israel and those of Jesus, in which the time of the church is subsumed under the time of Jesus, that is, the 'last days' inaugurated by John and Jesus, and which presents a christologically oriented 'three-fold' structure of the Gospel.[17] As to the grand structure of the Gospel, Kingsbury underscores the literary transitional markers in the Gospel at 4:17 and 16:21 (ἀπὸ τότε ἤρξατο ὁ Ἰησοῦς) arguing that the function of this 'fixed formula' is 'to mark the beginning of a new period of time' in the Gospel.[18]

According to his analysis, Matthew arranges his materials to fall into three main sections demarcated by 'superscriptions' (4:17; 16:21; cf. 1:1) in order to inform the reader that the major purpose of his Gospel is to set forth: (a) the genesis and significance of the person of Jesus (1:1-4:16); (b) the nature and effect of his proclamation (4:17-16:20), and (c) the reason and finality of his

---

[13] G. Strecker,"The Concept of History in Matthew," *JAAR* 36 (1967): 219-230; idem, *Der Weg der Gerechtigkeit: Untersuchung zur Theologie des Matthäus* (Göttingen: Vandenhoeck & Ruprecht, 1971), 45-49.

[14] Strecker, *Weg*, 117.

[15] J. P. Meier, *Law and History in Matthew's Gospel: A Redactional Study of Mt. 5:17-48* (Rome: Biblical Institute, 1976), 30-124.

[16] Meier, *Law and History*, 64. Cf. Matt 5:18b-d reads: ἕως ἂν παρέλθῃ ὁ οὐρανὸς καὶ ἡ γῆ, ἰῶτα ἓν ἢ μία κεραία οὐ μὴ παρέλθῃ ἀπὸ τοῦ νόμου, ἕως ἂν πάντα γένηται.

[17] Kingsbury, *Structure*, 31-36.

[18] Ibid., 8.

suffering, death, and resurrection (16:21-28:20).[19] Kingsbury correlates the three-fold structure of the Gospel not only with the Gospel's "chronological scheme that divides history into two epochs [those of Israel and Jesus] after the fashion of prophecy and fulfillment,"[20] but also with his emphatic Son-of-God christology for Matthew's Gospel.[21] Here again, as we have seen in the case of Bacon, we notice the close connection between the grand scheme of the Gospel and its theological/christological implications.

Kingsbury's proposal for the three-fold structure certainly complements Bacon's analysis of the five-fold discourses as he underscores the centrality of the person and mission of Jesus over his teaching material in the Gospel. But, as to the Son of God title, we have already examined that the title hardly explains Matthew's emphatic picture of the therapeutic Son of David. Further, Kingsbury's credence in those literary markers, ἀπὸ τότε ἤρξατο ὁ Ἰησοῦς in 4:17 and 16:21, appears to have received undue emphasis.[22] How can one be sure that those literary markers should bear more weight than those of Bacon's proposal, καὶ ἐγένετο ὅτε ἐτέλεσεν ὁ Ἰησοῦς (7:28; 11:1; 13:53; 19:1; 26:1)?[23] D. R. Bauer finds the more fundamental differences among Bacon's, Kingsbury's, and the others' proposals rather in their methodologies employed respectively:

That is to say, these differences in the understanding of the Gospel's structure are to some degree methodologically determined. Bacon focused upon changes or additions Matthew has made to received traditions (the process of redaction), whereas Jack Dean Kingsbury, who is a chief advocate of the threefold structure of Matthew, stresses the final composition of the work (the product of redaction). The salvation-history proponents generally emphasize specific uniquely Matthean passages, especially 10:5; 15:25; 21:43; 28:16-20.[24]

Once again, Kingsbury's proposal squarely affirms the correlation of the grand structure of the Gospel and its theological implications such as Matthew's concept of salvation-history and the christology of the Gospel.

In the following, we do not intend to suggest another grand scheme for the Gospel. Instead, we hope to summarize how the Davidic Shepherd tradition in

---

[19] Ibid., 36.

[20] Cf. Bauer notes that scholars such as G. Strecker, W. Trilling, R. Walker, W. G. Thompson, and J. P. Meier follow a line of argument for "Matthew's concept of salvation history as the key to the composition of the Gospel."

[21] Kingsbury, *Structure*, 25-37, 40-83.

[22] Luz, *Matthäus*, 1:168, for instance, sees the different function of ἀπὸ τότε in the context of Matt 4:12-17; that is, its function is to link vv. 13-16 (the decisive prerequisite of v. 17) with the 'Hauptaussage' of v. 17, while v.12 prepares the transition.

[23] Davies and Allison, *Matthew*, 1:287, 386-387, noting that ἀπὸ τότε recurs not only in 16:21 but also in 26:16, and ἤρξατο is again used of Jesus in 11:7 and 20 (this last with τότε), denounces this three-fold division analysis (e.g., the thematic unity of Matt 1-2, Matt 5-7); similarly, Gundry, *Matthew*, 10.

[24] Bauer, *The Structure*, 12.

the Gospel can illuminate some aspects of Matthew's narrative structure. For instance, compared to Bacon's theory, our analysis will testify particularly to (i) the significance of the infancy narrative and the passion narrative in view of the whole structure of the Gospel, which Bacon regards merely as Preamble and Epilogue respectively. (ii) The theological implications of the great discourses in the Gospel can be illuminated from the vantage point of the portrait of Jesus as the Davidic Shepherd/Teacher. (iii) Kingsbury/Bauer's view of Matthew's concept of salvation history will be critically assessed in light of the Gospel's interactions with the Davidic Shepherd tradition. With these points in mind, we will now present Matthew's narrative strategy of the Davidic Shepherd tradition.

6.2 The Preamble and Epilogue of the Gospel

The First Gospel opens with the genealogy of Jesus the Anointed, the Son of David (1:1), which implies the Evangelist's intention to invoke all the flavor and connotations attached to the age-old and lingering Davidic expectations at the time of the Evangelist and Jesus. The intensity evoked by this title Son of David, however, is toned down as the Evangelist introduce the history of the sons of Abraham, and passes over the days of the Davidic monarchy and the exile. Jesus the Anointed, introduced as the Son of David in 1:1, arrives in the momentum generated at the end of the exile (1:17). The birth narrative, nevertheless, underscores Jesus, who has been adopted into the Davidic line through Joseph. Still being identified as a son of David, Jesus is now entrusted with two prestigious missions – 'saving his people from their sins' and '*Immanuel* (God with us).'

Jesus' roles and tasks announced in 1:21 and 23 consist of a prelude to the main narrative that elaborates how through Jesus salvation and the divine presence will be experienced by 'his people.' To this prelude, Matthew adds another indication (2:6) when citing from Mic 5:1. This OT text that Matthew claims to have been fulfilled is critical, since it defines clearly – and for the first time in the narrative – in what sense and context Jesus is the Son of David. He is the Son of David as presented in Mic 5:1-4 (cf. Mic 2-5), that is, *as the Davidic Shepherd*, the one who takes upon himself the role of the eschatological Shepherd for Israel.

The Evangelist is aware of and utilizes the full development of the Davidic Shepherd tradition in the OT. He expands and articulates the shepherd images and related motifs drawn from Zechariah 9-14, coalesced in the passion narrative of Matthew 21-27 (21:4-5 [Zech 9:9]; 26:15 [Zech 11:12-13], 26:31 [Zech 13:7]; 27:3-10 [Zech 11:4-6]). Moreover, the shepherd image continues in Matt 26:32 as the smitten shepherd promises, once arisen, to lead the scattered flock . As we have argued, the shepherd image is taken up again in

28:16 and the figure and promise of Jesus in 28:16-20 in fact depicts God establishing his Davidic Shepherd-Appointee/Teacher over one flock of Israel and the nations.

In this respect, both the 'preamble' and 'epilogue' of the Gospel are not to be treated as secondary to the discourses in the main body of the Gospel. Rather, the beginning and closing of the Gospel are endowed with intense images and themes from the post-exilic Davidic Shepherd tradition, resplendent with hope for Israel's restoration, and God's fulfillment of the promise of his Davidic Shepherd. Evidently, taking these Davidic connotations seriously in the beginning and at the end of the Gospel, Bacon's new Moses typology is hard-pressed to find a prominent place in these sections.

6.3 Preaching, Healing, and Teaching

After 2:6, Matthew's quotations from the OT heighten the hope that the exile will end (2:15 [Hos 11:1]; 2:18 [Jer 31:15]), and that ultimately God's presence would return. The results would be the revocation of the curses, the forgiveness of sins, and once again blessings and peace would be bestowed upon their land. Yet how would this happen?

The Evangelist finds the answer in the picture of Jesus as the Davidic Shepherd who accomplishes the mission assigned to the eschatological Shepherd of Ezekiel 34 for the 'harassed and downtrodden flock' (9:36). Jesus' public ministry is summarized in 4:23 and is recapitulated in 9:35 with the identical tripartite mission of teaching, preaching, and healing. In particular, Jesus' healing ministry is tightly bound to the title Son of David – thus the description the therapeutic Son of David (9:27-31; 12:22-24; 15:21-28; 20:29-35; 21:1-16; 22:41-46). Being motivated particularly by compassion for the restoration, Jesus as the eschatological Shepherd seeks the lost, heals the sick, and feed the crowds (9:35-36 [cf. 9:10-13]; 10:1-6; 12:9-12; 14:13-21; 15:24, 29-39; 18:12-14). As a result of Jesus' therapeutic pastoral leadership, the crowds 'follow' him (Matt 4:24; 12:10; 14:14; 19:2; 20:29-34; cf. 9:12-13). Jesus' mission of compassion is that of the eschatological Shepherd – one fundamental reason for God to act and bring about Israel's future restoration compounded with His zeal for His holy name to be recognized by the nations as well as by Israel (cf. Matt 6:9; 28:19).

It is Jesus' authority that amazes the crowds. At the same time, his authority incites hostility from the leaders of the crowds in contexts where the title Son of David is associated with Jesus' healing practices that often take precedent over Pharisaic casuistry (21:23-27; 22:41-45; cf. 8:8; 9:33-34, 12:23-32, 15:23-27). Jesus' authority questioned by the leaders is, in fact, only the flip side of the same coin; the lost and the sick experienced it as God's unfailing compassion.

Jesus' unparalleled compassion and authority likewise indicate his identity as the divinely appointed Davidic Shepherd who takes up the role of the eschatological Shepherd to gather his flock (12:30) and to confront the shepherds of Israel.

In the Gospel, Jesus' authority is testified to not only when he heals the sick but also when he teaches (cf. 7:28-29). How can we explain why Matthew accentuates Jesus' teaching activities, as it is well illustrated by the five (Bacon) or the seven (Green) discourses in the narrative? Bauer makes an intriguing statement on this issue:

> The function of these five discourses within the narrative framework is to point to Jesus' activity of instructing his community, with special reference to the post-Easter existence of the church. Thus, *Matthew incorporates these discourses within his story of Jesus in order to underscore the climax of 28.16-20*, where the exalted Christ is described as being continually present with his community through history, speaking words of instruction and commandment.[25]

The portrait of Jesus as the Teacher with authority explicitly in the time of Jesus and implicitly in the time of the church, makes more sense when we grasp that Jesus is depicted as the Davidic Shepherd-Appointee with the distinctive role of the Teacher/Prince in the midst of his eschatological flock (28:16-20). As Matthew clearly does, teaching is yet another way of expressing the presence of Jesus among his community. Yet, we notice that those discourses (chs. 5-7; 10 [11]; 13; 18; [23] 24-25) in the narrative – each paired with their respective narrative sections (chs. 1-4; 8-9; 12; 15-27; 19-22; 26-28)[26] – likely present the same christological picture of Jesus.

Why is Jesus depicted distinctly as the Teacher in the First Gospel? Does the Evangelist hope to present Jesus as a new Moses in this regard? If it is so, how is this portrait of Jesus as a new Moses supported in narrative sections such as chapters 8-9 and 26-28? The consistency in terms of the christological picture of Jesus both in the discourses and the narrative sections may well be found in the portrait of Jesus as the Davidic Shepherd (e.g., the Christ in Word in chs. 5-7 and the Christ in Deed in chs. 8-9). As the eschatological Shepherd the Son of David distinctly heals and as the Davidic Shepherd-Appointee he is the Teacher *par excellence* for the eschatological flock, i.e., first, the lost house of Israel (10:1-6; 15:24), then the restored Israel, and finally the enjoined nations (28:16-20).

The dual aspects of Jesus' identity as the eschatological Shepherd and as the Davidic Shepherd center around the riddle of the wonder of the crowds contrasted with the distortions of their leaders. The linguistic and thematic comparisons show that the recurring theme of the Shepherd's confrontation with the leaders of the flock is at the bedrock of Matthew's Gospel and Ezekiel's

---

[25] Bauer, *The Structure*, 142 (italics mine).
[26] Barr, "The Drama of Matthew's Gospel," 35-36.

vision. In Ezekiel 34, YHWH's confrontation with the wicked shepherds of Israel is twofold. First, the Shepherd reveres the shepherds' wrongdoings as to the well-being of the flock. Thus, his seeking the lost and healing the sick means the judgement upon their wrongdoings (Ezek 34:1-22). Second, YHWH replaces the wicked shepherds of Israel with the righteous Davidic Shepherd/Teacher who can lead the flock into the way of righteousness (Ezek 34:23-24; Ezek 37:24-25).

Likewise, Matthew's Gospel depicts Jesus' confronting the Pharisees and the leaders of the nation particularly in the context of seeking the sinners and healing the sick. Jesus' healing provokes hostility from the Pharisees. Also, as Matt 9:10-13 demonstrates, Jesus confronts the Pharisees in terms of their teachings and practices (cf. 7:28-29; 23:1-39). These two aspects, healing and teaching, reveals the nature of Jesus' confrontation with the Pharisees, the shepherds of the flock. Eventually this confrontation is resolved with the divine removal of the leadership of the wicked shepherds over Israel (cf. 26:31) and subsequent replacement with the Davidic Shepherd/Teacher (cf. 26:32; 28:16-20), which implies the critical momentum in terms of God's theocratic reign over and in the midst of his own rescued flock.

6.4 The Ezekielian Pattern: Two Shepherds Schema

Regarding the shepherd image, the Son of David, and Jesus' mission of confronting the leaders, seeking the lost, healing the sick, and feeding the hungry (even teaching the gathered, cf. Ezek 37:24b), Matthew's usage of shepherd/sheep image is neither atomistic nor casual. Not only is it evident in linguistic and thematic echoes, but we also find Matthew following Ezekiel's eschatological *pattern* of the theocratic intervention, the coming of the Davidic Appointee, the renewal of the eschatological obedience of God's people, and the return of God's presence in the entire unit of Ezekiel 34-37 (cf. Ezek 40-48).

Jesus' earthly mission in the Gospel is restricted to 'the lost sheep of the house of Israel,' explicit in 10:6, 16; 15:24, while the risen Jesus at the end of the narrative commissions his disciples to go to all the nations (28:18-20). Jesus himself claims that he came to seek the lost sheep (9:10-13; cf. 7:15; 12:11-12; 18:12-14), and he sent his twelve disciples to the lost sheep of the house of Israel, i.e., Israel as the lost sheep that he must bring back (10:1-6). The lost sheep image best represents the utter misery of the community; they are the ones pushed back to the remotest fringe of the community. Ezekiel 34:1-16 presents six different categories related to the sheep – they are in exile; among them, there are the lost and the outcasts with whom YHWH initiates his eschatological visitation. Hence, when Jesus claims to seek the lost sheep (of the house of

Israel), he is taking the required action to proclaim that the eschatological Shepherd's divine intervention has begun.

In turn, this implies the rescue of the flock from the wild beasts – the wicked shepherds and evil spirits (cf. *An. Apoc* 89-90). More significantly, this divine eschatological reversal results in YHWH's replacing Israel's wicked shepherds with his Servant David. Thus the Shepherd and the King (Ezek 34:23-24; 37:24-25) is raised and appointed at the end of the vision (Matt 28:16-20). Ezekiel's pattern clearly divides the role and the mission of the eschatological, compassionate Shepherd for the harassed flock of Israel from the universal vision of the Davidic Shepherd appointed only after YHWH's confrontation with and judgment meted out against the wicked shepherds of Israel is finished (Ezek 34:17-22; cf. Matt 23). The throne of the Davidic Shepherd is established only upon the completion of Israel's rescue from the exile. The return of YHWH the eschatological Shepherd for his downtrodden flock must precede the establishment of the throne of the Davidic Shepherd, 'My Servant David,' 'the Shepherd/King over my flock.'

In this respect, the christology of the Davidic Shepherd and Matthew's theology of mission correspond to each other with critical divergence as well. For Ezekiel, it is YHWH whose mission is to rescue the scattered flock, and it is His David, the Shepherd-Appointee whose mission is to lead the rescued flock on to the path of righteousness by teaching them the laws and decrees of YHWH. For Matthew, however, it is Jesus who assumes both roles of the eschatological compassionate Shepherd of Israel and the Davidic Shepherd/King to be appointed (Matt 28:16-20). The continuity of the pre-Easter mission of the earthly Jesus and that of the post-Easter disciples find common ground in *the christology of Jesus both as the eschatological and Davidic Shepherd* (Two Shepherds schema).

Jesus as the Shepherd-Judge in Matt 25:31-46 lays down the criteria for the final judgment, by which now all the nations will be judged (cf. Ezek 34:17-22). Those criteria reflect the characteristics of the mission of compassion wrought by Jesus as the eschatological Shepherd demonstrated in the midst of the lost sheep of the house of Israel. The implication is that the restoration process must continue now beyond the house of Israel, even among the nations; here, Jesus is envisioned as One Shepherd over all, both Israel and the nations. Yet this picture of the triumphant Shepherd the Judge as the Son of Man is reserved for the future, located at the end of Jesus' final discourse on the Mount of Olives (24:3-25:46).

Seen at the narrative momentum in 25:31-46, Jesus *was* the Shepherd full of compassion sent to İsrael (cf. 9:36; 10:1-6), and he *will be* the Shepherd-Judge who will judge his flock even from all the nations according to the mercy he has testified among the lost sheep of Israel (25:31-46). Yet he is *going to be* the smitten shepherd of YHWH for the completion of the rescue-mission of the lost sheep of the house of Israel in 26:31 (Zech 13:7).

For the passion narrative, the Evangelist seems to instinctively turn to Zechariah's interpretation of the Davidic Shepherd tradition (Zech 9-14). The narrative describes the therapeutic Davidic Shepherd, followed by those who have been healed in 20:29-34, entering into Jerusalem as the humble king in 21:1-5 (Zech 9:9). Then, in 26:31, Jesus is identified as the smitten shepherd of Zech 13:7. Zechariah's interaction with Ezekiel's vision of the Davidic Shepherd underscores how Ezekiel's grand vision should be fulfilled, and his solution is the suffering of the shepherd(s) whose identity is ambiguous followed by the restoration of the blessings beginning with the outburst of the fountain for the house of David (Zech 13:1).

In the First Gospel, the death of Jesus as the smitten shepherd is followed by the eschatological break-through of the risen Shepherd with the risen saints in Matt 27:51b-53. As the graves open, the risen saints enter the holy city as they follow the risen Jesus, echoing the Breaker in Mic 2:12-13 who is none other than the Davidic Shepherd in Mic 5:1-4 (cf. Matt 2:6). The language and the procedure in Matt 27:51b-53 correspond to Ezek 37:12. Even the consequence of these breathtaking events in Matthew 27 and Ezekiel 37 is identical: the nations come to acknowledge God's holy name in Ezek 37:13, and the Roman centurion and those with him confess that Jesus was surely the Son of God in Matt 27:54.

Yet, this is not the end of the vision in Ezekiel 34-37 and Matthew 26-28. Jesus as the risen Shepherd continues to lead his flock after he arose from the dead just as he promises to go ahead of his disciples to Galilee (26:32). Zion, the eschatological locus for God's fulfillment of the promise of Israel's restoration, is split in two in 27:51b, coinciding with the death of Jesus as the smitten shepherd of Israel. The little flock is scattered once again (26:31). The judgment upon the shepherd and thus upon the whole flock is complete. Nevertheless, it is not accurate to say that Zion is 'replaced' by the Mount in Galilee.[27] The risen Jesus, who went ahead of his disciples to Galilee (26:32), is still the Shepherd; he once was the smitten one, but now he is the enthroned Davidic Appointee, the Davidic Shepherd-King over the rescued flock with all authority even over the nations just as it is envisioned in the end of Ezekiel 34 and 37 (cf. Mic 5:3-4).

The expectations for Mount Zion are not bypassed but fulfilled. Yet, features appear that are radically new: now, Jesus himself is the eschatological locus of the divine presence. Jesus is the Prince of the eschatological community in their midst, the Teacher who yet commands them to teach not exactly the laws and decrees of YHWH, but all that he has taught, believed to be the same Law and

---

[27] Donaldson, *Mountain*, 41-48, 171-180, omits the reference of the Mount of Olives split in two (27:51b) as he treats the mountain motif in the Gospel. He treats the mountain motif in terms of the mountain of temptation, teaching, feeding, transfiguration, Olivet discourse, and of commissioning, but excludes 27:51b.

Prophets of old, but fulfilled, taught, and lived through the Shepherd himself (28:18-20). Thus a comparison of Matt 28:18-20 and Ezek 34:23-30/37:24-28 is powerfully striking in that both texts involve all four integral elements of the identical vision: the authority, the teaching, all of the nations, and the divine presence. As foretold in the parable of the sheep and the goats in 25:31-46, Jesus is now proclaimed to be the 'One Shepherd' over all, both the remnant of Israel and the nations.

Above all, the First Gospel closes with Jesus' promise of his perpetual, divine presence with his eschatological community until the age to come truly reaches its consummation. This is exactly how Ezekiel's vision closes in 34 and 37 and likewise in the unit comprised of 40-48. The theme of divine presence or its absence opens and closes both Ezekiel's vision and Matthew's narrative. Ezekiel's solution is YHWH's theocratic pastoral visitation of his own flock, and Matthew saw in Jesus exactly what Ezekiel envisioned; to him, Jesus is *Immanuel*, 'God with us' (1:23).

Probably the implication of the divine presence in this title, *Immanuel*, would not be monolithic. A pattern emerges. For Ezekiel, the motif of divine presence is introduced first by YHWH's visitation to carry out the rescue mission, and next, the Davidic Shepherd's salvific task mostly centered on the sanctification of the people into righteousness. Jesus in Matthew's Gospel is distinctively the compassionate Son of David who seeks and heals, but it is the teaching that is emphasized by the risen Shepherd in 28:16-20 for the rescued community, including those who will be baptized in his holy name.

Our observations refine Kingsbury and Bauer's argument for the two epochs in terms of Matthew's concept of salvation history, i.e., the time of prophecy (the time of the OT) and the time of fulfilment (the eschatological time of Jesus) while the time of the church is subsumed under the time of Jesus.[28] Kingsbury is convinced that "the christology of Matthew, not his ecclesiology [deemed by those who hold to a three epoch salvation history in the Gospel], more than anything else has modeled his concept of the history of salvation." For instance, Kingsbury points out that the inclusio of Jesus' presence between 1:23 and 28:20 informs us that Matthew conceives of the 'time of Jesus' as extending from his birth (1:23) to his parousia (28:20): *"the earthly Jesus and the exalted Jesus are one."*[29] Bauer further elaborates and raises other points in order to support Kingbury's thesis of two epochs by arguing that Jesus as the Son of God in Matt 28:16-20 tightly correlates with the descriptions of Jesus in chapters 1-27, and traces a literary pattern of 'the repetition of comparison' between Jesus and the expectations for the disciples.[30]

---

[28] Kingsbury, *Structure*, 31-39; Bauer, *The Structure*, 45-55, 135-148.
[29] Ibid., 31-33 (italics mine)
[30] Bauer, *The Structure*, 147.

We have already argued that in Matt 28:16-20 the predominant picture is not the Son of God but the Davidic Shepherd-Appointee. Furthermore, while there is continuity in terms of the divine presence both in 1:23 and 28:20, we must not overlook the aspect of progression narrated in the main body of the Gospel. Both at the onset in the infancy narrative and at the conclusion of the Gospel, the Son of God is not the foremost emphasis. What is prominent is the reign of God himself, the Shepherd of Israel mediated through Jesus as the Davidic Shepherd in the earlier chapters and leading up to the passion narrative. In the closing scene of the Gospel, however, we find the predominance of the Davidic-Shepherd Appointee, who is now appointed with authority over the one flock that consists of Israel and the nations.

Matthew's concept of salvation history thus maintains a fundamental continuity in terms of theology, that is, looked at from the view point of God's acts in the history of Israel as well as among the nations. At the same time, the Evangelist reveals that a groundbreaking transition has occurred in the midst of the continuous work of God: God's establishment of his agent, the Davidic Shepherd, followed by his own mission of compassion through Jesus the Son of David. The unequivocal identification of Jesus speaking to his 'church' in 28:16-20 with the Jesus in chapters 1-27, both as the Son of God, certainly misses out the eschatological breakthrough in terms of salvation history as perceived and communicated in Matthew's Gospel.

While the mission of the disciples (thus implicitly the church) is to resemble Jesus in his earthly ministry (cf. Matt 25:31-46), the risen Jesus does not appear to claim that he will take the lead 'ahead of' his community as is implied in 26:32 and 27:51b-53. Rather, he promises to 'be with' his disciples whom he commissioned 'to go' to the nations (28:19-20). The mission of the church will be a mission of compassion, but the 'rescue' mission of the lost house of Israel from the exile is completed. Jesus' leadership as the Shepherd reaches its climax at 27:51b-53. So to speak, after the incident of the risen Jesus' leading the saints into the holy city, he no longer leads the flock in the way he did among the lost flock of Israel; thus his role of YHWH the eschatological Shepherd is completed. The mysterious scene of 27:15b-53 remains a proleptic vision as a profound encouragement for his eschatological one flock of the Jews and Gentiles at Matt 28:18-20.

Yet, as we come to this last scene of the Gospel, the prominent role is laid, rather, upon his disciples. The new eschatological one flock, the church, is placed on the front in the march, ever moving toward all peoples, while the risen Jesus leads them through his words and his presence. The risen One Davidic Shepherd as the Prince/Teacher indwells in the midst of his eschatological community of one flock, Israel and the nations, *as they are ever on the move to the nations*, until the end of the age.

At the end of our research, we observe that the fulfilment of Zechariah's vision of the smitten shepherd in the Gospel (26:31) works in tandem to bring a fuller understanding of Ezekiel's vision of the Davidic Shepherd/Prince (28:16-20) as well as Mic 5:1 near the beginning section of the Gospel in 2:6. Yet, Matthew adopts the distinctively Ezekielian pattern. The compassionate eschatological Shepherd comes first to rescue the flock; afterwards, he appoints his Davidic Shepherd over the one eschatological flock. In the First Gospel, it is 'Jesus, the Anointed, the Son of David' (1:1) who takes upon himself all the distinctive eschatological missions originating in the tradition's various shepherd-figures.

6.5 Conclusion

John P. Heil has argued that the narrative strategy of Matthew's shepherd metaphor is 'guided and unified' by Ezekiel 34.[31] We have seen that certainly not all of the metaphors are unified by Ezekiel 34. Those citations from Mic 5:1 and especially Zechariah 9-14 in the Gospel, as F. Martin observes, function as critical roles in the Evangelist's usage of the shepherd metaphor.[32] The contributions of Micah 2-5 and Zechariah 9-14 to Jesus' and the Evangelist's understanding of salvation history must not be overlooked. We have argued that Matthew's Gospel interacts not only with Ezekiel 34 but also with the entire Davidic Shepherd tradition of the OT, and likewise echoes the tradition in the Second Temple period. This proposal stands on the ground that the Gospel embraces all three major texts from this tradition: Micah 2-5, Ezekiel 34-37, and Zechariah 9-14.

We conclude that Matthew adopts the language and images from these texts, and interprets and modifies 'the pattern' implied in the vision primarily in Ezekiel 34-37. It is not merely Ezekiel 34 but the entire vision of Ezekiel 34-37 that guides and unifies Matthew's shepherd motifs and related themes. This distinctive mission of Jesus to seek the lost is the eschatological signal for the divine reversal of Israel's future restoration and beyond. The therapeutic Son of David need not be seen as Solomon-the-exorcist, for the whole literary and theological framework of Ezekiel 34-37 verifies the puzzling identity of this healing Son of David. He is none other than the Davidic Shepherd who accomplishes the mission of the eschatological Shepherd of Israel. Further, as an Ezekielian pattern (chs. 34-37) is assumed – that the mission of compassion of the eschatological Shepherd precedes his appointment of the Davidic Shepherd/Prince/Teacher (Two Shepherds schema) – the language and themes in Matt 27:51b-53 and 28:16-20 reveal the Evangelist's careful construction of the narrative structure.

---

[31] Heil, "Ezekiel 34," 708.
[32] Martin, "The Image of Shepherd in the Gospel of Saint Matthew," *ScEs* 27 (1975): 261-301.

To sum up, Matthew's emphasis on the title 'Son of David' in the infancy narrative, the Evangelist's distinctive portrait of Jesus as 'the therapeutic Son of David' in the main body of the Gospel, the figure of the smitten shepherd in the passion narrative (26:31-32), and the puzzling background of the portrait of Jesus in the closing scene of 28:16-20 with its correspondence to 1:23 (divine presence), all indicate that the Evangelist shapes his narrative structure as he deeply interacts with the Davidic Shepherd tradition.

Conclusions

1. Aspects of Jesus as the Davidic Shepherd

Matthew's story of Jesus can be read as the story of the Shepherd. Jesus is announced to be the Davidic Shepherd (Matt 2:6). He indeed takes upon himself the role of YHWH as the eschatological Shepherd as he seeks the lost and heals the sick with divine compassion and authority as the main body of the Gospel describes (esp., Matt 8-9).

Further, Jesus is to be the Davidic Shepherd/King/Judge at the end (25:31-46), yet he suffers first as the smitten shepherd (26:31) embodying the whole flock and its wicked shepherds, but only to be raised up with the saints to continue to lead them (27:51b-53). Finally, he is presented as the Davidic Shepherd/Prince/Teacher who will be in the midst of the eschatological flock until the end (28:16-20).

Having traced the Davidic Shepherd tradition from the OT, Judaism and the First Gospel, we have found that Matthew's Gospel presents a significant interpretation of the tradition in light of the person and mission of Jesus among the lost sheep of the house of Israel. Matthew communicates Jesus as the Shepherd, namely, as the eschatological Shepherd (YHWH), the smitten shepherd, the Davidic Shepherd/King/Judge in the future, and currently the Davidic Shepherd Prince/Leader/Teacher for the eschatological one flock comprised of both Israel and the nations. Inevitably, our conclusions touch upon our understanding of various aspects of Matthew's theology. To investigate this area of study and argue for our case while interacting with other views would require writing a book. Thus in the following, we strive to suggest likely implications of our findings concerning the various theological dimensions of the First Gospel.

1.1 The Christology of the Davidic Shepherd

The Ezekiel pattern is critical for solving the riddle of Jesus' identity as the therapeutic Son of David in the Gospel. The association of the Son of David with healing becomes intelligible when we comprehend this figure as the Davidic Shepherd according to the pattern laid out in the Davidic Shepherd tradition. In Matthew's Gospel, the Solomon-exorcist legend falls short; the figure's link with healing and with the messianic connotation of the title Son of David cannot be sustained. In this respect, Jesus is not merely the Son of David reminiscent of the failed Davidic monarchy. The picture of the coming David is profoundly revised, especially through the utilization of the shepherd image.

The absence of any militant connotations in Matthew's use of the title Son of David can thus be explained by its association with shepherd image in the Davidic Shepherd tradition. The Davidic Shepherd image, exhibited in 'the therapeutic Son of David' in the Gospel, has deep roots in Jewish tradition. The particular way the revision of the Davidic Shepherd is formed coincides with Israel's reflections on her failed monarchy, etched in the backdrop of the traumatic exile. Fundamental to this revision is the return of theocracy and the shepherd image for the Davidic Appointee. This means the divine presence is to return and be in the midst of the harassed flock; the Shepherd's eschatological reign is characteristically pastoral and restorative.

In short, the background of Jesus' healing activity as the 'therapeutic Son of David' need neither be drawn from the post-Easter churches nor from the unlikely Jewish legend (Solomon-the exorcist). The association of the Son of David with shepherd imagery provides the vital link between the expected Jewish messiah and the charismatic healing Jesus in the First Gospel. Consequently, by introducing the 'therapeutic Son of David,' Matthew neither "overturns the textuality of Davidic messianism" nor "refashions both the Jewish memory of the past and hope for the future." [1] Rather, the Evangelist faithfully represents Jesus' understanding of himself in terms of the comprehensive shepherd images of the Davidic Shepherd tradition.

Significantly, the association of the Davidic figure with healing activity can be attested in a sampling of texts like Ezek 34; CD-A 13:9; 4Q521 2, 2:12; 4Q504 1-2, 2:14; *Apoc. Ezek.* 5. Various revisions of Davidic expectation that employ shepherd images already occurred in the OT and Judaism and thus predate Jesus and Matthew. To say Matthew subverts Jewish history would therefore be an anachronistic statement.

In past Matthean scholarship, numerous redactional studies anchored their vantage point in the reconstructed 'Matthean community(-ies)' as they attempted to resolve the puzzling association of the [militant] Son of David with the healing Jesus in the Gospel. The result was often a shaky and irreconcilable composite picture of the Isaianic Servant and Solomon-the-exorcist for the therapeutic Son of David. As the proposed model for this link in our thesis, Jesus as the Davidic Shepherd not only solves this puzzling association but also provides a far more coherent explanation of Jesus' mission to the lost sheep of the house of Israel and beyond in the First Gospel.

Despite the richness of the shepherd images for Jesus in the Gospel, the Davidic Shepherd theme may not be 'the overarching' motif Matthew employs in the narrative. Yet it is a prominent one. Through this figure of the Davidic

---

[1] Cf. John M. Jones, "Subverting the Textuality of Davidic Messianism: Matthew's Presentation of the Genealogy and the Davidic Title," *CBQ* 56 (1994): 256-272.

Shepherd – in particular the 'one Davidic Shepherd' over both Israel and the nations – the motif for Jesus as the Son of David, 'My Servant David' (Ezek 34:23; 37:24), surely finds its rightful place at the Gospel's conclusion. Likewise, the attempt to set up a hypothetical antithesis between the Jewish [militant] Son of David and the post-Easter [healing/Isaianic] Christ falters.

The figure of the one Davidic Shepherd over both Israel and the nations in 28:16-20 fills the christological gap that lies between the first coming of the suffering Son of Man (Matt 16:21; 17:22) and the triumphant yet futuristic heavenly Son of Man in his second coming (25:31-46). In this way, Matthew's christology of the Davidic Shepherd places Jesus exactly in between these two advents of the Son of Man.[2] It is also significant that the earthly Son of Man is eventually identified with the smitten shepherd (Matt 26:31; cf. Zech 13:7), while the heavenly Son of Man (Dan 7:12-14) is depicted as the Davidic Shepherd/King/Judge over one flock (Matt 25:31-46; cf. Ezek 34:17-22), but not exactly with the picture of Jesus in Matt 28:16-20. Both figures of the suffering Son of Man and the futuristic Son of Man are closely associated with the shepherd images taken from the OT Davidic Shepherd tradition.

Our proposal also explains the presence of the teaching motif in Matt 28:19-20 and its absence in 10:1-5. In the context of Ezekiel's vision for the restoration and beyond, teaching is a central task reserved for the Davdic Shepherd-Appointee (Ezek 34:23-24; 37:24-25; cf. 20:15). The Davidic Shepherd tradition envisiones God's eschatological people as they are able to follow the law and produce righteousness under the leadership of Jesus as the Davidic Shepherd who will be the Teacher/Prince among them. Jesus, who sent his twelve apostles to the lost sheep of the house of Israel, executes the tasks presumably reserved for YHWH as the eschatological Shepherd who is to rescue his own flock (Matt 10). On the other hand, Jesus in the discourse sections in the Gospel is the Teacher *par excellence* with matchless authority. In fact, the picture of the risen Jesus as the Teacher present in the midst of his disciples is presented at the end of the Gospel (28:19-20). The teaching role of Jesus in the discourses points toward this climactic scene of God's establishing him as the Teacher for his one eschatological flock.

Also, it is noteworthy that the eschatological Teacher, whose main role is to sanctify the eschatological community, is required to be holy, sinless, or righteous. This requirement is widely attested in the Davidic Shepherd tradition (Ezek 36:25; *Ps. Sol.* 17:36; *An. Apoc.* 90:38). Further, the risen Jesus as the Davidic Shepherd is the unique Teacher who commands his under-shepherds (disciples) to teach his Words, not the Law *per se*, but the Law as fulfilled (cf.

---

[2] Cf. U. Luz,"The Son of Man in Matthew: Heavenly Judge or Human Christ?" *JSNT* 48 (1992): 3-21.

5:17-19), filtered (cf. 22:34-40), and lived out (cf. 9:10-13) by the Shepherd himself through his mission of compassion.

After judgment is inflicted upon the smitten shepherd (Matt 26:31), the risen Jesus in 28:16-20 is presented as the justified Shepherd-Teacher in the narrative. His presence in the midst of the eschatological community implies their sanctification through his role as the righteous Prince/Teacher. Both motifs of Jesus saving his people (Matt 1:21) and being 'God-with-us' (1:23) correspond with the picture of Jesus as the eschatological Prince/Teacher in the midst of the ever increasing eschatological community of all nations consisting of both Jews and Gentiles in 28:16-20.

1.2 The Shepherd's Teaching and Mission

Of all the crucial elements of the profile of the eschatological Shepherd, the task of healing the sick as well as seeking the lost characterizes the theological dimension of Jesus' mission as the Davidic Shepherd in the Gospel. The various redemptive and pastoral shepherding tasks belong to YHWH as the eschatological Shepherd, which Jesus in Matthew's Gospel, clearly takes upon himself. The Evangelist communicates Jesus as the promised 'divine' eschatological Shepherd of Israel. The wonder and might wrought by the God of Israel through Jesus the Son of David are witnessed to and confessed by the lost sheep of the house of Israel (Matt 9:34; 12:22-24; 15:31; 22:41-46), and frequently by the Gentile onlookers as well (e.g., Matt 15:21-28; cf. 27:54).

Jesus embodies the compassion of YHWH for his own lost and downtrodden flock. Just as Jesus demonstrated in his mission to the lost sheep of the house of Israel (9:10-13, 36; 10:1-6), the way Jesus summarizes the Law and the Prophets in terms of mercy (Matt 22:37-40) reveals his identity as the compassionate eschatological Shepherd for the harassed sheep. Throughout Matthew's Gospel, there is a consistent emphasis on compassion or mercy. Jesus' identity as the Shepherd sheds light on the coherence between his mission and teaching. Compassion characterizes the mission of Jesus as the eschatological Shepherd (9:9-13).

At the same time, it is the core of the criteria by which the futuristic Son of Man as the Davidic Shepherd/King judges the community of his flock at the end (25:31-46). Likely, it is also the core of the teachings that the one Davidic Shepherd-Appointee commands his disciples/ under-shepherds to teach all nations (28:18-20). In Matthew's Gospel, Jesus' compassion or mercy occupies a central place in both Jesus' teachings and practices as he confronts the Pharisees (9:10-13), in the final judgment to be wrought by the future Davidic Shepherd (25:31-46). Likewise Jesus' compassion or mercy abounds for the eschatological community to be constituted from both Israel and the nations (28:18-20).

1. Aspects of Jesus as the Davidic Shepherd

Jesus healing the sick as the Son of David signals the inauguration of YHWH's eschatological theocratic rule, that is, the kingdom of God, over his own flock, which is also expected to extend to the nations. The 'sinners' (ἁμαρτωλοί) in the Gospel refer to the lost and the outcast within the house of Israel. Jesus confronts the Pharisees, thus threatening them to maintain their popularity over the flock of Israel, the majority of the people. This echoes YHWH's promise to confront the wicked shepherds of Israel (Ezek 34:1-16; cf. Jer 23:1-4). Jesus proclaims the coming of the kingdom (4:23; 9:35), which coincides with the coming of the eschatological Davidic Shepherd(s) for his flock (Matt 2:6; 9:36). The kingdom comes as Jesus seeks the lost sheep of the house of Israel; it implies the arrival of God's eschatological and theocratic rule over his people. For this mission, Jesus fulfills first the role of YHWH the eschatological Shepherd whose main tasks are to seek the lost and heal the sick. The kingdom of God is thus characteristically Davidic in the First Gospel.

Concerning the picture of Jesus as the therapeutic Son of David, we find not so much his humility as his authority with divine compassion. Jesus' compassion in the Gospel is thus bound primarily to his mission to the lost sheep of the house of Israel (cf. 15:21-28). In seeking sinners and healing the sick, Jesus aims primarily to restore justice within the covenantal community, which explains the close connection of Jesus' table fellowship with sinners and his confrontation with the Pharisees as the wicked shepherds of Israel (9:10-13).

Jesus' confrontation with and its consequential reversal of the leadership of Israel, are thus eschatological and final, the inauguration of theocratic rule under the leadership of the Davidic Shepherd. The eventual break between Judaism and the early churches reflected in Matthew's Gospel, therefore, is rooted in Jesus' identity as the eschatological Shepherd and also as the Davidic Shepherd-Appointee. YHWH has finally returned to his own flock to rescue it. Moreover, he will appoint his Servant David as the One Shepherd/Prince/Teacher for his eschatological community to be gathered from all the nations.

This progress in salvation history moving beyond Israel's restoration is propelled by YHWH's own desire for his holy name to be sanctified among the nations. To delve into this aspect requires additional research. Suffice it to say here that the Great Commission in the Gospel (28:18-20) as well as the Lord's prayer (6:9-13) suggest that sanctifying God's holy name is a central concern for Jesus and Matthew, in view of their missionary movement within and beyond Israel.

From the Davidic Shepherd tradition, we attest that these double fundamental motifs, i.e., YHWH's compassion for his people and his zeal for his holy name among the nations, serve as underlying principles for Jesus' mission as the eschatological Shepherd (Matt 10) and the Davidic Shepherd (Matt 28) in the First Gospel.

1.3 The Shepherd's Mission and Salvation-History

The question of Jesus' seemingly contradictory commissions for Israel (10:1-5) as well as for all the nations (28:16-20) in the First Gospel can successfully be explained by the salvation historical scheme of 'Ezekiel's vision' for the eschatological and Davidic Shepherds. As Jesus sends out his twelve disciples to the lost sheep of the house of Israel in Matt 10:1-5, he fulfills the task of YHWH the eschatological Shepherd as envisioned in Ezekiel 34. Compassion – the fundamental divine motivation for the restoration – is underscored in Matt 9:36. Jesus' earthly mission is thus restricted within Israel, among the lost sheep.

On the other hand, the 'all nations' in Matt 28:19-20 refers to the one flock to be led by the Davidic Shepherd. Representative of the remnant of Israel (28:16-17), the eleven are to join the people from all nations. While the compassion motif is particularly associated with YHWH's irresistible, covenantal compassion and faithfulness toward his chosen flock (9:36; 10:1-5), the concern for God's holy name relates itself to God's sovereignty over all nations (28:18-20), and thus is not limited to the house of Israel. These distinctive yet inseparable motivations seem to play respective roles in Jesus' mission – the one for the lost sheep of Israel, and the other for the one flock that consists of both Jews and Gentiles.

Therefore, the fundamental continuity between Jesus' mission to Jews and to Gentiles can be found in the christology of Jesus as both the divine eschatological Shepherd (10:1-6) for Israel and the one Davidic Shepherd for both Israel and the nations (28:16-20). In this way, the christology of Jesus as the Shepherd provides the continuity of the different yet successive stages of Jesus' mission, first to the Jews and then, more inclusively, to all the nations. Not only does the christology of the Shepherd but also the theological pattern or vision for the future of Israel and the nations, as vividly attested in Ezekiel 34 and 37, make the seemingly contradictory attitudes as to the mission in the Gospel fully intelligible.

Nevertheless, an element of discontinuity must be underlined as well. Zechariah's smitten shepherd (13:7; Matt 26:31) comes into the picture of Jesus as the eschatological Shepherd in chapter 10 and the Davidic Shepherd in chapter 28. This motif touches upon one of the most crucial turning-points in view of the narrative structure. This turning-point in terms of Jesus assuming the role of Zechariah's smitten shepherd might underscore Israel's rejection of her messiah as the Son of David.

Further, Jesus views his death, according to Matthew, in Zechariah's scheme of YHWH's judgment upon the shepherd of Israel, which in turn is expected to fulfill Ezekiel's vision of the establishment of Ezekiel's Davidic Shepherd-Appointee. In short, Israel's rejection of her messiah, as narrated in the First Gospel, is subsumed under the grand theological framework of the Davidic

Shepherd tradition. Thus the function of the hinge chapters – whether it be chapter 11 or chapter 13 according to those who suggest a chiastic structure of the Gospel – does not set itself up against the climactic scene of the establishment of the Davidic Shepherd in 28:16-20.

As a consequence of this analysis, the identity of 'the people' in Matt 2:6 may not be monolithic but ambivalent in view of the narrative sequence. While the term primarily refers to the lost sheep of the house of Israel, it also includes all the nations as the story approaches the closing scene. Matthew's narrative suggests that the eschatological 'one flock' under the Davidic Shepherd is to be viewed in light of Israel as the lost sheep. In other words, the eschatological Shepherd's restoration of Israel, in a sense, continues as the one Davidic Shepherd continues to gather his flock through his disciples from all nations (28:16-20) until this Davidic Shepherd-Judge finally judges them according to the criteria (25:31-36) demonstrated in his earthly mission of compassion for the lost sheep of Israel (9:36).

Jesus' twelve disciples sent to Israel in Matt 10:1-5 are now his 'eleven' in 28:16-20 where the risen Shepherd has already passed through the stage of God's judgment upon the shepherds of Israel and thus upon the whole flock. Yet the renewed flock of Israel is open to the nations, since her shepherd is now the one Davidic Shepherd over all. After all, the questions as to the identity, continuity, and radical newness of God's people in the First Gospel are determined and settled by the story of Jesus, who fits the various shepherd images in the Davidic Shepherd tradition.

Matthew's concept of salvation history is threefold: (i) the time of the OT expectation of the coming of the Davidic Shepherd-Appointee (chs. 1-4); (ii) the time of Jesus as the eschatological Shepherd (chs. 5-26:30); and (iii) his Davidic Shepherd-Appointee (26:31-28:20). In Matt 26:31, as Jesus assumes the role of the smitten shepherd of Israel, the flock scatters. From that point forward (26:32), the Evangelist describes the inevitable consequences of the eschatological Shepherd's rescue of his flock and his judgment inflicted on his flock as a whole. The time of the church is glimpsed to begin only at the end of the Gospel (28:18-20).

The time of the church cannot be legitimately identified with that of Jesus in the Gospel unless one carefully considers salvation historical progress according to the Davidic Shepherd tradition, which is characteristically the Ezekiel pattern (Two Shepherds schema). Jesus' earthly mission to Israel is that of YHWH as the eschatological Shepherd while the mission charged upon the disciples at the end of the Gospel is that of the one Davidic Shepherd of all. Certainly, the ecclesiology of Matthew's Gospel is determined by the christology of Jesus as the Shepherd(s) in the context of his mission toward his flock in view of the nations.

2. Concluding Statements

Several topics deserve further research. It would be a worthwhile venture to investigate fully all the texts related to the motif of ἐξουσία in the light of our study of the Shepherd theme. While constraints did not permit us any elaboration, the topic of the role of the Spirit as it is related to the figure of Jesus as the eschatological and Davidic Shepherd(s) is a worthy pursuit. Furthermore, Jesus as the Teacher emerged as a fresh interest as we examined the picture of Jesus as the Davidic Shepherd/Prince especially in the closing section of the Gospel. A fourth remaining task would be a follow-up study of the use of the shepherd images in the rest of the New Testament, which could prove to be complementary, especially the tenth chapter of the Fourth Gospel. Other minor motifs such as the staff or scepter (tribe) and Jesus' casting out demons with 'a word,' in view of the eschatological battle of the flock, may yield interesting results (cf. Rev 2:27-28).

Reflecting on the significance of Ezekiel 34-37 to Matthew's presentation of Jesus, despite the absence of the Evangelist's explicit quotes from Ezekiel, we cannot help but notice the affinity of the spirituality represented both in Ezekiel and Matthew. On the surface, both books appear quite nationalistic. Among the major prophets, Ezekiel can easily be picked up for his strong nationalistic outlook. Similarly, Matthew's Gospel is known as the most Jewish among the Synoptic Gospels.

Nevertheless, nearly every page of the book of Ezekiel resounds with the prophet's heartbeat with the theme, 'for the sake of My holy name (among the nations).' YHWH's zeal for his holy name is a profound motivation that fulfils the restoration of his flock, despite their defilement of the name in the midst of all the nations. YHWH, for his name's sake, restores Israel and will lead the nations to acknowledge his holy name. The vision of the compassionate mission of the eschatological Shepherd for the harassed flock comes splendidly into view, ensconced in the larger vision for YHWH's sovereignty.

Quite similarly, Matthew – as the most Jewish of the Synoptic Gospels – tells the story of the healing 'Son of David' who restricts his mission exclusively to the lost sheep of the house of Israel. Yet, this Shepherd of Israel teaches his disciples to pray, "Our Father, who is in Heaven, hallowed be Thy Name" (6:9-10). After his death and resurrection, Jesus, as the one Davidic Shepherd, commands his restored flock to go out to all the nations, baptizing them in God's name so that they may join the eschatological community under one Shepherd.

In short, Ezekiel and Matthew share a similar spirituality, i.e., one that is a characteristic mixture of nationalism and universalism. The Evangelist of the First Gospel shares the theological framework of the Davidic Shepherd tradition – that resonates with Ezekiel – in his witnessing of Jesus as the 'divine'

2. Concluding Statements

eschatological Shepherd for Israel, and at the same time, the one 'Davidic' Shepherd for both Jews and Gentiles. We trace a definite progress within the narrative, and it is identical with the Ezekiel pattern. The vision of the Davidic Shepherd tradition is an eschatological drama; the Shepherd returns to his own flock; thus, the divine presence is restored (1:23; 28:20), and the Shepherd provides the flock with that which only he can provide – true leadership (2:6; 9:36; 26:31-32; 27:52-53; 28:16-20).

While it is the magnificent story of the eschatological return of the divine presence in the midst of his people, the vision still awaits its full consummation in the future. The eschatological community – the one flock under the Davidic Shepherd, with his presence in their midst – is called to ever move forward the mission of compassion toward all the nations ἕως τῆς συντελείας τοῦ αἰῶνος ("until the consummation of the age," Matt 28:20), that is, until the Shepherd-King comes as the Judge of his flock, perhaps to surprise many (Matt 25:31-46).

Bibliography

Biblical Texts, Primary Sources and Reference Works

Abegg, Martin G., J. E. Bowley and E. M. Cook, eds. *The Dead Sea Scrolls Concordance.* Vol. 1, *The Non-Biblical Texts from Qumran.* Leiden: Brill, 2003.

Aland, K., ed., *Synopsis Quattuor Evangeliorum: Locis parallelis evangeliorum apocryphorum et patrum adhibitis.* Stuttgart: Deutsche Bibelstiftung, 1976.

Aland, K., J. Karavidopoulos, C. M. Martini, and B. M. Metzger, eds. *Novum Testamentum Graece.* 27th ed. Stuttgart: Deutsche Bibelgesellschaft, 1993.

Allegro, John M. *Qumrân Cave 4: I (4Q158-4Q186).* Discoveries in the Judean Desert 5. Oxford: Clarendon, 1968.

Atkinson, K. R. *An Intertextual Study of the Psalms of Solomon.* Studies in the Bible and Early Christianity 49. Lewiston: The Edwin Mellen Press, 2001.

Baumgarten, J. M., and D. R. Schwarz. "Damascus Document." Pages 4-58 in *The Dead Sea Scrolls: Hebrews, Aramaic, and Greek Texts with English Translations.* Edited by J. H. Charlesworth. Vol. 1. Tübingen: Mohr Siebeck, 1994.

Blass, F., and A. Debrunner. *A Greek Grammar of the New Testament and Other Early Christian Literature.* A Translation and Revision of the Ninth-Tenth German Edition Incorporating Supplementary Notes of A. Debrunner by R. W. Funk. Chicago and London: University of Chicago, 1961.

Brook, G. *Qumran Cave 4: XVII.* Discoveries in the Judean Desert 22. Oxford: Clarendon, 1996.

Charlesworth, James H., ed. *The Old Testament Pseudepigrapha.* 2 vols. New York: Doubleday, 1983.

Colson, F. H., G. H. Whitaker, R. Marcus, eds. *Philo.* 10 vols. (with two supplementary volumes). London: Heinemann, 1929-1953.

Corsten, Thomas. *Die Inschriften von Laodikein am Lykos.* Teil I, *Die Inschriften.* IK 49.1. Bonn: Habelt, 1997.

Eckart, K. G. "Das Apokryphon Ezekiel." Pages 45-54 in *Die griechische Baruch-Apocalypse.* Edited by Wolfgang Hage. *Jüdische Schriften aus hellenistisch-römischer Zeit* 5.1. Gütersloher: Gütersloher Verlaghaus Gerd Mohn, 1984.

Elliger, K., and W. Rudolph, eds. *Biblia Hebraica Stuttgartensia.* Editio funditus renovata. Stuttgart: Deutsche Bibelgesellschaft, 1967/1977.

García Martínez, F., and Eibert J. C. Tigchelaar, eds. *The Dead Sea Scrolls Study Edition.* 2 vols. Leiden/Boston: Brill; Grand Rapids: Eerdmans, 2000.

Grenfell, Bernard P., and Arthur S. Hunt, eds. *The Oxyrhychus Papyri.* Part XIII. London, Egypt Exploration Fund: Greco-Roman Branch, 1919.

Holm-Nielsen, Svend. "Die Psalmen Salomos." Pages 49-112 in *Poetische Schrifte.* Edited by W. G. Kümmel et al. *Jüdische Schriften aus hellenistisch-römischer Zeit* 4.2. Gütersloh: Gerd Mohn, 1977.

Levey, S. H. *The Targum of Ezekiel: Translated, with a Critical Introduction, Apparatus, and Notes.* The Aramaic Bible 13. Wilmington, Del.: Michael Glazier, 1987.

Llyewelyn, S. R. *A Review of the Greek Inscriptions and Papyri.* Vol. 9. Grand Rapids: Eerdmans, 2002.

May, H. G., and B. M. Metzger, eds. *The New Oxford Annotated Bible with the Apocrypha. Revised Standard Version Containing the Second Edition of the New Testament and an Expanded Edition of the Apocrypha.* New York: Oxford University Press, 1973.

Metzger, B. M. "The Fourth Book of Ezra." Pages 517-559 in vol. 1 of *The Old Testament Pseudepigrapha.* Edited by J. H. Charlesworth. Garden City, N.Y.: Doubleday, 1985.

Milik, J. T. *The Books of Enoch: Aramaic Fragment of Qumran Cave 4.* Oxford: Clarendon, 1976.

__, and D. Barthélemy, eds. *Qumrân Cave I.* Discoveries in the Judean Desert 1. Oxford: Clarendon Press, 1955.

Olson, Dennis T. "Words of Lights (4Q504-4Q506)." Pages 107-153 in *Pseudepigraphic and Non-Masoretic Psalms and Prayers.* Vol. 4A of *The Dead Sea Scrolls: Hebrews, Aramaic, and Greek Texts with English Translations.* Edited by J. H. Charlesworth. Tübingen: Mohr Siebeck. 1994.

Pritchard, James B., ed. *Ancient Near East Texts Relating to the Old Testament.* Vol. 1, *Anthology of Texts and Pictures.* Princeton: Princeton University Press, 1973.

Puech, Émil. *Qumrân Grotte 4.* Discoveries in the Judean Desert 25. Oxford: Clarendon Press, 1998.

Rahlfs A., ed. *Septuaginta.* Stuttgart: Deutsche Bibelgesellschaft, 1979.

Sanders, J. A. *The Psalms Scroll of Qumran Cave 11 (11QPs^a).* Discoveries in the Judean Desert 4. Oxford: Claredon, 1965.

__. "A Liturgy for Healing the Stricken." Pages 216-233 in *Pesudepigraphic and Non-Masoretic Psalms and Prayers.* Vol. 4A of *The Dead Sea Scrolls: Hebrews, Aramaic, and Greek Texts with English Translations.* Edited by J. H. Charlesworth. Tübingen: Mohr Siebeck. 1994.

Schreiner, J. "Das 4. Buch Esra." Pages 289-412 in *Apokalypsen.* Edited by W. G. Kümmel et al. *Jüdische Schriften aus hellenistisch-römischer Zeit* 5.4. Gütersloher: Gütersloher Verlagshaus Gerd Mohn, 1984.

Shutt, R. J. H. "Letter of Aristeas (Third Century B. C.- First Century A. D.)." Pages 7-34 in *The Old Testament Pseudepigraph.* Edited by J. H. Charlesworth. Vol. 2. N.Y.: Doubleday, 1985.

Sperber, Alexander, ed. *The Bible in Aramaic: The Latter Prophet according to Targum Jonathan.* Vol. 3. Leiden: Brill, 1962.

Thackeray, H. St. J., R. Marcus, A. Wikgren and L. H. Feldman, eds. *Josephus.* 9 vols. Loeb Classical Library. London: Heinemann, 1926-1965.

Uhlig, Sigbert. "Das äthiopische Henochbuch." Pages 461-780 in *Apocalypsen.* Edited by Wolfgang Hage. *Jüdische Schriften aus hellenistisch-römischer Zeit* 5.6. Gütersloher: Gerd Mohn, 1984.

Usher, S., ed. and trans. *Isocrates: Panegyricus and To Nicocles.* Vol. 3 of *Greek Orators.* Warminster, England: Aris & Phillips, 1990.

Wright, R. B. "Psalms of Solomon (First Century B.C.)." Pages 639-670 in *The Old Testament Pseudepigrapha.* Edited by J. H. Charlesworth. Vol. 2. New York: Doubleday, 1985.

Secondary Literature

Abegg, Martin G. "Messianic Hope and 4Q285: A Reassessment." *Journal of Biblical Literature* 113 (1994): 81-91.

__. "Exile and the Dead Sea Scroll." Pages 111-126 in *Exile: Old Testament, Jewish, and Christian Conceptions.* Edited by James M. Scott. Leiden: Brill, 1997.

Abraham, I. *Studies in Pharisaism and the Gospels: First Series.* Cambridge: University Press, 1917.

Ackroyd, Peter R. *Exile and Restoration: A Study of Hebrew Thought of the Sixth Century B.C.* Philadelphia: Westminster, 1968.
Agourides, Savas. "Little Ones' in Matthew." *The Bible Translator* 35 (1984): 329-334.
Albl, Martin C. *"And Scripture Cannot Be Broken": The Form and Function of the Early Christian Testimonia Collections.* Leiden: Brill, 1999.
Allen, L. C. "Structure, Tradition and Redaction in Ezekiel's Death Valley Vision." Pages 127-142 in *Among the Prophets: Language, Image and Structure in the Prophetic Writings.* Edited by Philip R. Davies and David J. A. Clines. Sheffield: JSOT, 1993.
Allison, Dale C. *The End of the Ages Has Come: An Early Interpretation of the Passion and Resurrection of Jesus.* Philadelphia: Fortress, 1985.
___. "The Son of God as Israel: a Note on Matthean Christology." *Irish Biblical Studies* 9 (1987): 74-81.
___. *The New Moses.* Minneapolis: Fortress, 1993.
Anno, Y. "The Mission to Israel in Matthew: The Intention of Matthew 10:5b-6 considered in the Light of the Religio-Political Background." Th.D. diss., Lutheran School of Theology, Chicago, 1984.
Atkinson, K. R. "On the Use of Scripture in the Development of Militant Davidic Messianism at Qumran: New Light from *Psalm of Solomon 17.*" Pages 106-123 in *The Interpretations of Scripture in Early Judaism and Christianity: Studies in Language and Tradition.* Edited by Craig A. Evans. Sheffield: Sheffield, 2000.
Aus, Roger David. *Samuel, Saul and Jesus: Three Early Palestinian Jewish Christian Gospel Haggadoth.* Atlanta: Scholars, 1994.
Bacon, B. W. "The 'Five Books' of Matthew Against the Jews." *The Expositor* 15 (1918): 56-66.
___. "Jesus and the Law: A Study of the First 'Book' of Matthew." *Journal of Biblical Literature* 47 (1928): 203-231.
___. *Studies in Matthew.* New York: Holt, 1930.
Baltzer, D. "Literarkritsche und literarhistorische Anmerkungen zur Heilsprophetie im Ezechiel-Buch." *Bibliotheca ephemeridum theologicarum lovaniensium* 76 (1986): 166-181.
Banks, R. *Jesus and the Law in the Synoptic Tradition.* London: Cambridge University Press, 1975.
Barr, D. L. "The Drama of Matthew's Gospel: A Reconsideration of Its Structure and Purpose." *Theology Digest* 24 (1976): 349-359.
Barth, G. "Matthew's Understanding of the Law." Pages 58-164 in *Tradition and Interpretation in Matthew.* Edited by G. Bornkamm, G. Barth and H. J. Held. Translated by Percy Scott. Philadelphia: Westminster, 1963.
Batto, B. F. "The Covenant of Peace: A Neglected Ancient Near Eastern Motif." *Catholic Biblical Quarterly* 49 (1987): 187-211.
Bauer, David R. *The Structure of Matthew's Gospel: A Study in Literary Design.* Journal for the Study of the New Testament: Supplement Series 30. Sheffield: Almond, 1988.
Baumgarten, A. I. *The Flourishing of Jewish Sects in the Maccabean Era: An Interpretation.* Leiden: Brill, 1997.
Beare, F. W. "The Mission of the Disciples and the Mission Charge: Matthew 10 and Parallels." *Journal of Biblical Literature* 89 (1970): 1-13.
Beaton, Richard. *Isaiah's Christ in Matthew's Gospel.* Cambridge: Cambridge University Press, 2002.
Beker J. Christiaan. "Echoes and Intertextuality: On the Role of Scripture in Paul's Theology." Pages 64-69 in *Paul and the Scriptures of Israel.* Edited by Craig A. Evans and James A. Sanders. Journal for the Study of the New Testament: Supplement Series 83. Sheffield: Sheffield, 1993.

Beutler, Johannes and Robert T. Fortna eds. *The Shepherd Discourse of John 10 and Its Context.* Cambridge: Cambridge University Press, 1991.

Blackburn, Barry. *Theios Aner and the Markan Miracle Traditions.* Wissenschaftliche Untersuchungen zum Alten und Neuen Testament 2/40. Tübingen: Mohr Siebeck, 1991.

Block, D. I. "Bringing Back David: Ezekiel's Messianic Hope." Pages 167-188 in *The Lord's Anointed: Interpretation of Old Testament Messianic Texts.* Edited by P. E. Satterthwaite, Richard E. Hess and Gordon J. Wenham. Carlisle: Paternoster, 1995.

___. *The Book of Ezekiel: Chapters 25-48.* Grand Rapids: Eerdmans,1997.

Boadt, L. "The Function of the Salvation Oracles in Ezekiel 33 to 37." *Hebrew Annual Review* 12 (1990): 1-21.

Böcher, Otto. "Wölfe in Schafspelzen." *Theologische Zeitschrift* 6 (1968): 405-426.

Bock, Darrell L. *Luke.* Vol. 1, Luke 1:1-9:50; vol. 2, Luke 9:51-53. Grand Rapids: Baker, 1996.

Bockmuehl, Markus N. A. *Revelation and Mystery: In Ancient Judaism and Pauline Christianity.* Grand Rapids: Eerdmans, 1997.

Bolyki, János. *Jesu Tischgemeinschaften.* Wissenschaftliche Untersuchungen zum Neuen Testament 2/96. Tübingen: Mohr Siebeck, 1998.

Borg, Marcus J. *Conflict, Holiness and Politics in the Teaching of Jesus.* New York: E. Mellen, 1984.

Bosch, D. *Die Heidenmission in der Zukunftsschau Jesu.* Abhandlungen zur Theologie des Alten und Neuen Testaments 36. Zürich: Zwingli-Verlag, 1959.

Boyarin, Daniel. "Inner Biblical Ambiguity, Intertextuality and the Dialectic of Midrash: The Waters of Marah." *Prooftexts* 10 (1990): 29-48.

Bracewell, R. E. "Shepherd Imagery in the Synoptic Gospels." Ph.D. diss., Southern Baptist Theological Seminary, 1983.

Brandenburger, Egon. *Die Verborgenheit Gottes im Weltgeschehen: Das literarische und theologische Problem des 4 Esrabuches.* Zürich: Theologischer Verlag, 1981.

Brawley, Robert L. "Table Fellowship: Bane and Blessing for the Historical Jesus." *Perspective in Religious Studies* 22 (1995): 13-31.

Breasted, James H. *Egypt.* New York: Russell & Russell, 1962.

Broer, Ingo. "Das Gericht des Menschensohnes über die Völker: Auslegung von Mt 25,31-46." *Bibel und Leben* 11 (1970): 273-295.

Brooke, G. J. "Ezekiel in Some Qumran and New Testament Texts." Pages 317-337 in *The Madrid Qumran Congress.* Edited by J. T. Barrera and L. V. Montaner. Leiden: Brill, 1992.

___. "The Thematic Context of 4Q252." *Jewish Quarterly Review* 85 (1994): 321-343.

___. "4Q252 as Early Jewish Commentary." *Revue de Qumran* 17 (1996): 385-401.

Brooks, Oscar S. "Matthew xxviii 16-20 and the Design of the First Gospel." *Journal for the Study of the New Testament* 10 (1981): 2-18.

Brown, Raymond E. *Birth of the Messiah: A Commentary on the Infancy Narratives in Matthew and Luke.* New York: Doubleday, 1977.

___. "The Gospel of Peter and Canonical Gospel Priority." *New Testament Studies* 33 (1987): 321-343.

___. *The Death of the Messiah: A Commentary on the Passion Narratives in the Four Gospels.* Anchor Bible Reference Library. New York: Doubleday, 1994.

___. "Eschatological Events Accompanying the Death of Jesus: Especially the Raising of the Holy Ones from Their Tombs (Matt 27:51-53)." Pages 43-73 in *Faith and Future: Studies in Christian Eschatology.* Edited by J. P. Galvin. N.Y.: Paulist, 1994.

Brown, Schuyler. "The Two-Fold Representation of the Mission in Matthew's Gospel." *Studia theologica* 31 (1977): 21-32.

___. "The Mission to Israel in Matthew's Central Section." *Zeitschrift für die neutestamentliche Wissenschaft und die Kunde der älteren Kirche* 69 (1978): 215-221.

Bryan, David. *Cosmos, Chaos, and the Kosher Mentality.* Journal for the Study of the Pseudepigrapha: Supplement Series 12. Sheffield: Sheffield, 1995.
Bryan, Steven M. *Jesus and Israel's Traditions of Judgment and Restoration.* Cambridge: Cambridge University Press, 2002.
Bruner, Frederik D. *The Christbook: A Historical/Theological Commentary.* Waco: Word Books, 1987.
___. *Matthew: A Commentary.* Rev. and enl. ed. Vol. 1, *The Christbook: Matthew 1-12*; vol. 2, *The Churchbook: Matthew 13-28.* Grand Rapids: Eerdmans, 2004.
Budd, Philip J. *Numbers.* Word Biblical Commentary 5. Waco: Word, 1984.
Bultmann. Rudolf. *The History of the Synoptic Tradition.* Translated by John Marsh. New York: Harper & Row, 1963.
Burger, Christopher. *Jesus als Davidssohn: Eine traditionsgeschichtliche Untersuchung* Göttingen: Vandenhoeck & Ruprecht, 1970.
Bussby, F. "Did a Shepherd Leave Sheep upon the Mountains or in the Desert?: A Note on Matthew 18:12 and Luke 15:4." *Anglican Theological Review* 45 (1963): 93-94.
Butterworth, M. *Structure and the Book of Zechariah.* Sheffield: JSOT, 1992.
Byrskog, Samuel. *Jesus the Only Teacher: Didactic Authority and Transmission in Ancient Israel, Ancient Judaism and the Matthean Community.* Almqvist & Wilsell International: Stockholm, 1994.
Campbell, Jonathan G. *The Use of Scripture in the Damascus Document 1-8, 19-20.* Berlin: Walter de Gruyter, 1995.
Carson, D. A. "The Jewish Leaders in Matthew's Gospel: A Reappraisal." *Journal of the Evangelical Theological Society* 25 (1982): 161-174.
___. *Matthew.* Expositor's Bible Commentary. Vol. 1, *Matthew 1-12*; vol. 2, *Matthew 13-28.* Grand Rapids: Zondervan, 1995.
Carter, Warren. *Matthew and the Margins.* Journal for the Study of the New Testament: Supplement Series 204. Sheffield: Sheffield University Press, 2000.
___. *Matthew and Empire: Initial Explorations.* Harrisburg, Pa.: Trinity Press International, 2001.
Catchpole, David R. "The Poor on Earth and the Son of Man in Heaven: A Re-Appraisal of Matthew xxv. 31-46." *Bulletin of the John Rylands University Library of Manchester* (1979): 355-397.
Cathacart, K. J., and R. P. Gordon. *The Targum of the Minor Prophet.* Wilmington: Michael Glazier, 1989.
Charette, Blaine. *Restoring Presence: The Spirit in Matthew's Gospel.* Sheffield: Sheffield University Press, 2000.
Charlesworth, J. H. "From Messianology to Christology: Problems and Prospects." Pages 3-35 in *The Messiah: Developments in Earliest Judaism and Christianity.* Edited by J. H. Charlesworth with J. Brownson, M. T. Davis, S. J. Kraftchick and A. F. Segal. Minneapolis: Fortress, 1992.
___. "The Son of David: Solomon and Jesus (Mark 10.47)." Pages 72-87 in *The New Testament and Hellenistic Judaism.* Edited by Peder Borgen and Giversen Søren. Peabody, Mass.: Hendrickson, 1995.
Chaon, Esther G. "4QDIBHAM: Liturgy or Literature?" *Revue de Qumran* 15 (1991): 447-455.
___. "Is *DIVREI HA-ME'OROT* a Sectarian Prayer?" Pages 3-17 in *The Dead Sea Scrolls: Forty Years of Research.* Edited by D. Dimant and U. Rappaport. Leiden: Brill, 1992.
___. "Dibrê Hamme'orot: Prayer for the Sixth Day (4Q504 1-2 v-vi)" Pages 23-27 in *Prayer from Alexander to Constantine: A Critical Anthology.* Edited by Mark Kiley et al. Routledge: London, 1997.
Childs, B. S. *Isaiah.* Louisville: Westminster John Knox, 2001.

Chilton, B. D. "Jesus *ben David*: reflection on the *Davidssohnfrage*" *Journal for the Study of the New Testament* 14 (1982): 88-112.

___. "The Purity of the Kingdom as Conveyed in Jesus' Meals." *Society of Biblical Literature Seminar Papers* 31(1992): 473-488.

Christudhas, M. *The ἐξουσία of Jesus in the Gospel of Matthew*. New Deli: Regency Publications, 2000.

Coffman, Ralph J. "The Historical Jesus the Healer: Cultural Interpretations of the Healing Cults of the Greco-Roman World as the Basis for Jesus Movements." *Society of Biblical Literature Seminar Papers* 32 (1993): 412-441.

Collins, John. J. "Messianism in the Maccabean Period." Pages 97-109 in *Judaisms and Their Messiahs at the Turn of the Christian Era*. Edited by J. Neusner, W. S. Green and E. S. Frerichs. Cambridge: Cambridge University Press, 1987.

___. "A Pre-Christian 'Son of God' Among the Dead Sea Scrolls." *Biblical Research* 9 (1993): 34-38.

___. *The Work of the Messiah*. Discoveries in the Judean Desert 1. Leiden: Brill, 1994.

___. "Teacher and Messiah? The One Who Will Teach Righteousness at the End of Days." Pages 193-210 in *The Community of the Renewed Covenant: The Notre Dame Symposium on the Dead Sea Scrolls*. Edited by E. Ulrich and J. Vanderkam. Notre Dame, Ind.; University of Norte Dame Press, 1994.

___. *The Scepter and the Star: The Messiah of the Dead Sea Scroll*. New York: Doubleday, 1995.

___. "The Expectation of the End in the Dead Sea Scrolls." Pages 74-90 in *Eschatology, Messianism, and the Dead Sea Scrolls*. Edited by C. A. Evans and P. W. Flint. Grand Rapids: Eerdmans, 1997.

___. *The Apocalyptic Imagination*. Grand Rapids: Eerdmans, 1998.

___. "The Nature of Messianism in the Light of the Dead Sea Scrolls." Pages 199-217 in *The Dead Sea Scrolls in Their Historical Context*. Edited by Timothy H. Lim. Edinburgh: T&T Clark, 2000.

Comber, Joseph A. "Critical Notes: The Verb THERAPEUŌ in Matthew's Gospel." *Journal of Biblical Literature* 97 (1978): 431-434.

Combrink, H. J. Bernard. "The Macrostructure of the Gospel of Matthew." *Neotestamentica* 26 (1982): 1-20.

___. "The Structure of the Gospel of Matthew as Narrative." *Tyndale Bulletin* 34 (1983): 61-90.

Cook, S. L. "The Metamorphosis of a Shepherd." *Catholic Biblical Quarterly* 55 (1993): 453-466.

Cooke, G. A. *A Critical and Exegetical Commentary on the Book of Ezekiel*. Edinburgh: T. & T. Clark, 1985.

Cope, O. Larmar. *Matthew: A Scribe for the Kingdom of Heaven*. Catholic Biblical Quarterly Monograph Series 5, Washington, D.C.: The Catholic Biblical Association of America, 1976.

Corley, Kathleen E. "Jesus' Table Practice: Dining with 'Tax Collectors and Sinners,' including Women." *Society of Biblical Literature Seminar Papers* 32 (1993): 444-459.

Cousland, J. R. C. *The Crowds in the Gospel of Matthew*. Leiden: Brill, 2002.

Crossan, J. D. *Four Other Gospels*. Minneapolis: Winston, 1982.

___. *The Historical Jesus: The Life of a Mediterranean Jewish Peasant*. San Francisco: Harper, 1991.

Davenport, G. L. "The 'Anointed of the Lord' in Psalms of Solomon 17." Pages 67-92 in *Ideal Figures in Ancient Judaism*. Edited by J. J. Collins and G. W. Nickelsburg. Society of Biblical Literature Septuagint and Cognate Studies 12. Chico, Calif.: Scholars, 1980.

Davies, Margaret. *Matthew*. Sheffield: JSOT, 1993.

___. "Stereotyping the Other: The 'Pharisees' in the Gospel According to Matthew." Pages 415-532 in *Biblical Studies/Cultural Studies: The Third Sheffield Colloquium*. Edited by J. Cheryl Exum and Stephen D. Moore. Journal for the Study of the Old Testament: Supplement Series 266. Sheffield, 1998.

Davies, Philip R. *The Damascus Covenant*. Journal for the Study of the Old Testament: Supplement Series 25. Sheffield: JSOT, 1982.

___. "Judaism in the Dead Sea Scrolls: The Case of the Messiah." Pages 219-232 in *Dead Sea Scrolls in their Historical Context*. Edited by Timothy H. Lim. Edinburgh: T. & T. Clark, 2000.

___. "The Judaism(s) of the Damascus Document." Pages 27-43 in *Damascus Document: A Centennial of Discovery*. Edited by J. M. Baumgrarten, E. G. Chazon and Avital Pinnack. Leiden: Brill, 2000.

Davies, W. D. and D. C. Allison. *A Critical and Exegetical Commentary on the Gospel According to Saint Matthew*. Vol. 1, *Matthew 1-7*; vol. 2, *Matthew 8-18*; vol. 3, *Matthew 19-28*. International Critical Commentary. Edinburgh: T. and T. Clark, 1988, 1991, 1997.

Davies, W. D., and D. C. Allison. "Matt. 28:16-20: Texts Behind the Text." *Revue d'historie et de philosophie religieuses* 72 (1992): 89-98.

Davies, W. D. *Torah in Messianic Age and/or the Age to Come*. Journal of Biblical Literature Monograph Series 7. Philadelphia: Society of Biblical Literature, 1952.

___. "The Jewish Source of Matthew's Messianism" Pages 494-511 in *The Messiah*. Edited by James H. Charlesworth. Minneapolis: Fortress, 1992.

Davis, P. G. "Divine Agents, Mediators, and New Testament Christology." *Journal of Theological Studies* 45 (1994): 479-503.

Deines, R. *Jüdische Steingefäße und pharisäische Frömmigkeit*. Wissenschaftliche Untersuchungen zum Neuen Testament 2/52. Tübingen: Mohr Siebeck, 1993.

___. *Die Pharisäer: Ihr Verständnis im Spiegel der christlichen und jüdischen Forschung seit Welhausen und Graetz*. Wissenschaftliche Untersuchungen zum Alten und Neuen Testament 101. Tübingen: Mohr Siebeck, 1997.

___. "The Pharisees Between "Judaisms and 'Common Judaism'." Pages 443-504 in *The Complexities of Second Temple Judaism*. Vol. 1 of *Justification and Varigated Nomism*. Edited by D. A. Carson, Peter T. O'Brian, and Mark A. Seifried. Tübingen: Mohr Siebeck, 2001.

De Lacy, D. R. "In Search of a Pharisee." *Tyndale Bulletin* 43 (1992): 353-372.

Dempsey, Carol J. "Micah 2-3: Literary Artistry, Ethical Message, and Some Considerations about the Image of Yahweh and Micah." *Journal for the Study of the Old Testament* 85 (1999): 117-128.

Derby, Josiah. "Prophetic Views of the Davidic Monarcy." *Jewish Biblical Quarterly* 28 (2000): 111-116.

De Roo, J. C. R. "David's Deeds in the Dead Sea Scrolls." *Dead Sea Discoveries* 6 (1999): 44-65.

Derrett, J. D. "Fresh Light on the Lost Sheep and the Lost Coin." *New Testament Studies* 26 (1979): 36-60.

De Jong, M. "Matthew 27:51 in Early Christian Exegesis." *Harvard Theological Review* 79 (1986): 67-79.

Dimant, D. "Qumran Sectarian Literature." Pages 483-550 in *Jewish Writings of the Second Temple Period*. Edited by Michel E. Stone. Compendia rerum iudaicarum ad Novum Testamentum 2:2. Philadelphia: Fortress, 1984.

Dodd, C. H. *According to the Scriptures: The Sub-Structure of the New Testament Theology*. London: SCM, 1952.

Douglas, M. "Deciphering a Meal." *Daedelus* 101 (1972): 61-81.

Donahue, J. R. "Tax Collectors and Sinners: An Attempt at Identification." *Catholic Biblical Quarterly* 33 (1971): 39-61.
Donalson, T. L. *Jesus on the Mountain*. Journal for the Study of the New Testament: Supplement Series 8. Sheffield: JSOT, 1985.
Driver, G. R. *Canaanite Myths and Legends*. Edinburgh: T. & T. Clark, 1956.
Duguid, Iain M. *Ezekiel and the Leaders of Israel*. Leiden: Brill, 1994.
__. "Messianic Themes in Zechariah 9-14." Pages 265-280 in *The Lord's Anointed*. Edited by P. E. Satterthwaite, R. S. Hess and G. J. Wenham. Grand Rapids: Baker, 1995.
Duling, Dennis C. "The Promises to David and Their Entrance into Christianity - Nailing Down a Likely Hypothesis." *New Testament Studies* 20 (1974): 55-77.
__. "Solomon, Exorcism, and the Son of David." *Harvard Theological Review* 68 (1975): 235-252.
__. "The Therapeutic Son of David: An Element in Matthew's Christological Apologetic." *New Testament Studies* 24 (1978): 392-410.
__. "Matthew and the Problem of Authority: Some Preliminary Observations." Pages 59-68 in vol. 3 of *Proceedings*. Eastern Great Lakes Biblical Society, 1983.
__. "Matthew's Plurisignificant 'Son of David' in Social Science Perspective: Kinship, Kingship, Magic and Miracle." *Biblical Theology Bulletin* 22 (1992): 99-116.
Dunn, James D. G. *Christology in the Making: A New Testament Inquiry into the Origins of the Doctrine of the Incarnation*. Philadelphia: Westminster, 1980.
__. "Pharisees, Sinners, and Jesus." Pages 264-289 in *The Social World of Formative Christianity and Judaism*. Edited by Jacob Neusner et al. Philadelphia: Fortress, 1988.
__. "Matthew 12:18/ Luke 11:20-A Word of Jesus?" Pages 29-49 in *Eschatology and the New Testament: Essays in Honor of George Raymond Beasley-Murray*. Edited by W. H. Gloer. Peabody, Mass.: Hendrickson, 1988.
__. *Jesus, Paul and the Law*. Louisville: Westminster, 1990.
__. *Jesus Remembered*. Christianity in the Making. Vol. 1; Grand Rapids: Eerdmans, 2003.
Edwards, James R. "The Authority of Jesus in the Gospel of Mark." *Journal of the Evangelical Theological Society* 37 (1994): 217-233.
Eichrodt, Walter. *Ezekiel: A Commentary*. Philadelphia: Westerminster, 1970.
Eisemann, Moshe. *The Book of Ezekiel: A New Translation with a Commentary Anthologized from Talmudic, Midrash, and Rabbinic Sources*. Brooklyn: Mesorah, 1980.
Eissfeldt, Otto. *The Old Testament: An Introduction*. Translated by Peter R. Ackroyd. New York: Harper & Row, 1972.
Eslin, M. S. "The Five Books of Matthew: Bacon on the Gospel of Matthew." *Harvard Theological Review* 24 (1931): 67-97.
Evans, Craig A. "Listening for Echoes of Interpreted Scripture." Pages 47-51 in *Paul and the Scripture*. Edited by Craig A. Evans and James A. Sanders. Journal for the Study of the New Testament: Supplement Series 83. Sheffield: Sheffield, 1993.
__. "Aspects of Exile and Restoration in the Proclamation of Jesus and the Gospels." Pages 305-315 in *Exile: Old Testament, Jewish, and Christian Conceptions*. Edited by James M. Scott. Leiden: Brill, 1997.
__. "Jesus and the Continuing Exile of Israel." Pages 78-86 in *Jesus and the Restoration of Israel*. Edited by C. C. Newman. Downers Grove: Inter-Varsity, 1999.
Feldman, Louis H. "The Concept of Exile in Josephus." Pages 145-173 in *Exile: Old Testament, Jewish, and Christian Conceptions*. Edited by James M. Scott. Leiden: Brill, 1997.
Findlay, J. A. "The Book of Testimonies and the Structure of the First Gospel." *The Expositor* 20 (1920): 388-400.
Fishbane, Michael. *Biblical Interpretation in Ancient Israel*. Oxford: Clarendon Press, 1985.

Flint, P. W. *The Bible at Qumran: Text, Shape, and Interpretation.* Grand Rapids: Eerdmans, 2001.
France, R. T. *Matthew.* Leicester: IVP, 1985.
Frankemölle, Hubert. *Jahwebund und Kirche Christi: Studien zur Form und Traditionsgeschichte des Evangeliums nach Matthäus.* Neutestamentliche Abhandlungen 10. Münster: Aschendorff, 1974.
Frankfort, Henri. *Cylinder Seals.* London: Gregg Press, 1939.
Freedman, David N. *Micah: A New Translation with Introduction and Commentary.* Anchor Bible. New York: Doubleday, 2000.
Fröhlich, Ida. "Symbolical Language of the Animal Apocalypse (1 Enoch 85-90)." *Revue de Qumran* 14 (1990): 629-636.
Funk, W., and W. R. Hoover, *The Five Gospel.* New York: Polebridge, 1993.
Garland, David. *Reading Matthew.* New York: Crossroad, 1995.
Gerhardsson, Birger. *The Mighty Acts of Jesus According to Matthew.* Lund: C. W. K. Gleerup, 1979.
Gibbs, J. M. "Purpose and Pattern in Matthew's use of the Title 'Son of David'." *New Testament Studies* 10 (1963-1964): 446-464.
___. "Mark 1, 1-15, Matthew 1, 1-4, 16, Luke 1, 1-4, 30, John 1, 1-51: The Gospel Prologues and their Function." Pages 154-188 in vol. 6 of *Studia Evangelica.* Edited by Elizabeth A. Livingstone. Berlin: Akademie Verlag, 1973.
Gibson, J. "HOI TELŌNAI KAI HAI PORNAI." *Journal of Theological Studies* 32 (1981): 429-433.
Gnilka, Joachim. *Das Matthäusevangelium.* 2 vols. Herders theologischer Kommentar zum Neuen Testament. Freiburg/Basel/Wein: Herder,1988-1992.
Goldman, B., ed. *The Discovery of Dura-Europos.* New Haven: Yale University Press, 1979.
Good, Deirdre J. *Jesus the Meek King.* Harrisburg: Trinity Press International, 1999.
Goodman, Martin. "A Note on Josephus, the Pharisees and Ancestral Tradition." *Journal of Jewish Studies* 50 (1999): 17-20.
Gowan, D. E. "The *Exile* in Jewish Apocalyptic." Pages 205-223 in *Scripture in History and Theology: FS J. C. Rylaarsdam.* Edited by A. L. Merrill and T. W. Overholt. Pittsburgh: Pickwick, 1977.
Grassi, J. "Ezekiel xxxvii 1-14 and the New Testament." *New Testament Studies* 11 (1965): 162-164.
Gray, G. *The Least of My Brothers, Matthew 25:31-46: A History of Interpretation.* Society of Biblical Literature Dissertation Series 114. Atlanta: Scholars, 1989.
Green, H. B. "The Structure of St. Matthew's Gospel." Pages 47-59 in vol. 4 of *Studia Evangelica.* Edited by Elizabeth A. Livingstone. Berlin: Academie-Verlag, 1968.
___. "Solomon the Son of David in Matthaean Typology." Pages 227-230 in vol. 7 of *Studia Evangelica.* Edited by Elizabeth A. Livingston. Berlin: Akademie-Verlag, 1982.
Green, William S. "Doing the Text's Work for It: Richard Hays on Paul's Use of Scripture." Pages 58-63 in *Paul and the Scriptures of Israel.* Edited by Craig A. Evans and James A. Sanders. Journal for the Study of the New Testament: Supplement Series 83. Sheffield: Sheffield, 1993.
Greenberg, Moshe. "The Design and Themes of Ezekiel's Program of Restoration." *Interpretation* 38 (1975): 181-216.
___. *Ezekiel 1-20.* Anchor Bible 22. New York: Doubleday, 1983.
Gruen, Erich S. "Hellenistic Kingship: Puzzles, Problems, and Possibilities." Pages 116-125 in *Aspects of Hellenistic Kingship.* Edited by Per Bilde, Troels Engberg-Pedersen, Lise Hannestad and Jan Zahle. Oxford: Aarhus University Press, 1996.
Guelich, Robert A. *Mark 1-8:26.* Word Biblical Commentary 34A. Nashville: Thomas Nelson, 1989.

Gundry, Robert. H. *The Use of the Old Testament in St. Matthew's Gospel*. Leiden: Brill, 1967.
___. *Mark: A Commentary on His Apology for the Cross*. Grand Rapids: Eerdmans, 1993.
___. *Matthew: A Commentary on His Literary and Theological Art*. Grand Rapids: Eerdmans, 1982.
Hagner, Donald A. *Matthew. 1-13*. Word Biblical Commentary 33A. Dallas: Word, 1993.
___. *Matthew 14-28*. Word Biblical Commentary 33B. Dallas: Word, 1995.
Hahn, F. *Mission in the New Testament*. Naperville, Ill.: Alec R. Allenson, 1965.
Hallbäck Geert. "The Fall of Zion and the Revelation of the Law: An Interpretation of 4 Ezra." *Journal for the Study of the Old Testament* 6 (1992): 263-292.
Hals, Ronalds M. *Ezekiel:The Forms of The Old Testament Literature*. Grand Rapids: Eerdmans, 1989.
Hammershaimb, E. *Some Aspects of Old Testament Prophecy from Isaiah to Malachi*. Copenhagen: Rosenkilde og Bagger, 1966.
Hanson, P. D. *The Dawn of Apocalyptic: The Historical and Sociological Roots of Jewish Apocalyptic Eschatology*. Philadelphia: Fortress, 1979.
Harrelson, Walter J. "Messianic Expectation at the Time of Jesus." *St. Luke's Journal of Theology* 32 (1988): 28-42.
Hayes, William C. *The Scepter of Egypt, Part 1*. The Metropolitan Museum of Art, 1953.
Hays, Richard B. *Echoes of Scripture in the Letters of Paul*. New Haven: Yale University Press, 1989.
Head, Peter M. *Christology and the Synoptic Problem*. Cambridge: Cambridge University, 1998.
Heater, Homer. "Matthew 2:6 and Its Old Testament Sources." *Journal of the Evangelical Theological Society* 26 (1983): 395-397.
___. "A Closer Look at Matt 2:6 and Its Old Testament Sources." *Journal of the Evangelical Theological Society* 28 (1985): 47-52.
Heil, John P. "Significant Aspects of the Healing Miracles in Matthew." *Catholic Biblical Quarterly* 41 (1979): 274-287.
___. *The Death and Resurrection of Jesus: A Narrative Critical Reading of Matthew 26-28*. Minneapolis: Fortress, 1991.
___. "Ezekiel 34 and the Narrative Strategy of the Shepherd and Sheep Metaphor in Matthew." *Catholic Biblical Quarterly* 55 (1993): 698-708.
Heintz, J.-G. "Royal Traits and Messianic Figures: A Thematic and Iconographical Approach." Pages 52-66 in *The Messiah: Developments in Earliest Judaism and Christianity*. Edited by J. H. Charlesworth with J. Brownson, M. T. Davis, S. J. Kraftchick and A. F. Segal. Minneapolis: Fortress, 1992.
Held, Heinz J. "Matthew as Interpreter of the Miracle stories." Pages 165-299 in *Tradition and Interpretation in Matthew*. Edited by G. Bornkamm, G. Barth and H. J. Held. Translated by Percy Scott. Philadelphia: Westminster, 1963.
Hempel, C. *The Laws of the Damascus Document: Sources, Tradition and Redaction*. Leiden: Brill, 1998.
Hengel, M., and R. Deines. "E. P. Sanders' 'Common Judaism,' Jesus, and the Pharisees: A Review Article." *Journal of Theological Studies* 46 (1995): 1-70.
Hill, David. *The Gospel of Matthew*. London: Marshall, Morgan and Scott, 1972.
___. "False Prophets and Charismatics: Structure and Interpretation in Matthew 7:15-23." *Biblica* 57 (1976): 327-348.
___. "Matthew 27:51-53 in the Theology of the Evangelist." *Irish Biblical Studies* 7 (1985): 76-87.
Holladay, Carl. *Theios Aner in Hellenistic Judaism: A Critique of the Use of This Category in New Testament Christology*. Missoula: Scholars, 1977.

Hooker, Morna D. "The Beginning of the Gospel." Pages 18-28 in *The Future of Christology*. Edited by Abraham J. Malherbe and Wayne A. Meeks. Minneapolis: Fortress, 1993.

Horsely, R. A. *Jesus and the Spiral of Violence: Popular Jewish Resistance in Roman Palestine*. San Francisco: Harper & Row, 1987.

___. "Social Conflict in the Synoptic Sayings Source Q." Pages 37-52 in *Conflict and Invention*. Edited by J. S. Kloppenborg. Valley Forge: Trinity Press International, 1995.

Hubbard, B. J. *The Matthean Redaction of a Primitive Apostolic Commissioning: An Exegesis of Matthew 28:16-20*. Society of Biblical Literature Dissertation Series 19. Missoula: Scholars, 1974.

Hubbard, Robert L. "יש׳" Pages 556-562 in vol. 2 of *New International Dictionary of Old Testament Theology and Exegesis*. Edited by Willem VanGemeren. Grand Rapids: Zondervan, 1997.

Hultgren, Arland J. *Jesus and His Adversaries: The Form and Function of the Conflict Stories in the Synoptic Tradition*. Minneapolis: Augsburg, 1979.

Hutton, D. "The Resurrection of the Holy Ones (Matt 27:51-53). A Study of the Theology of the Matthean Passion Narrative." Th.D. diss., Harvard Divinity School, 1970.

Jeremias, J. "ποιμήν, ἀρχιποίμην, ποιμαίνω, ποίμνη, ποίμνιον," *Theological Dictionary of the New Testament* 6: 485-502.

___. *Jesus' Promise to the Nations*. 2d ed. London: SCM, 1967.

___. *Jerusalem in the Time of Jesus*. Philadelphia: Fortress, 1969.

___. *New Testament Theology I: The Proclamation of Jesus*. Translated by John Bowden. London: SCM, 1971.

___. *Parables of Jesus*. Rev. ed. London: SCM, 1972.

Jones, D. R. "A Fresh Interpretation of Zechariah IX-XI." *Vetus Testamentum* 12 (1962): 241-259.

Jones, John Mark. "Subverting the Textuality of Davidic Messianism: Matthew's Presentation of Genealogy and the Davidic Title." *Catholic Biblical Quarterly* 56 (1994): 256-272.

Joyce, Paul. *Divine Initiative and Human Response in Ezekiel*. Journal for the Study of the Old Testament: Supplement Series 51. Sheffield: JSOT, 1989.

Kaufmann, Yehezkel. "The Messianic Idea: The Real and the Hidden Son-of-David." *Jewish Biblical Quarterly* 22 (1994): 141-150.

Kautzsch, E. *Gesenius Hebrew Grammar*. Oxford: Clarendon, 1910.

Kearney, Peter J. "He Appeared to 500 Brothers." *Novum Testamentum* 22 (1980): 264-284.

Keener, Craig S. *A Commentary on the Gospel of Matthew*. Grand Rapids: Eerdmans, 1999.

Keesmaat, Sylvia C. "Exodus and the Intertextual Transformation of Tradition in Romans 8:14-30." *Journal for the Study of the New Testament* 54 (1994): 29-56.

Kingsbury, J. D. "The Composition and Christology of Matt 28:16-20." *Journal of Biblical Literature* 93 (1974): 573-584.

___. *Matthew: Structure, Christology, Kingdom*. Minneapolis: Fortress Press, 1975.

___. "The Title, 'Son of David' in Matthew's Gospel." *Journal of Biblical Literature* 95 (1976): 591-602.

___. "Observations on the 'Miracle Chapters' of Matthew 8-9." *Catholic Biblical Quarterly* 40 (1978): 559-573.

Klausner, J. *The Messianic Idea in Israel*. London: Bradford and Dickens, 1956.

Klein, R. W. *Ezekiel: The Prophet and the Message*. Columbia, S.C.: University of South Carolina Press, 1988.

Knibb, M. A. "The Exile in the Literature of the Intertestamental Period." *Heythrop Journal* 17 (1976): 253-279.

___. "The Interpretation of Damascus Document VII, 9b-VIII, 2a and XIX, 5b-14." *Revue de Qumran* 15 (1991): 243-251.

Knowles, M. P. "Moses, the Law, and the Unity of 4 Ezra." *Novum Testamentum* 31 (1989): 257-274.
Köster, H. "σπλαγχνίζομαι" Pages 548-559 in vol. 7 of *Theological Dictionary of the New Testament*. Grand Rapids: Eerdmans, 1971.
Korpel, Marjo Christina Annette. *A Rift in the Clouds: Ugaritic and Hebrew Descriptions of the Divine*. Münster: UGARIT-Verlag, 1990.
Köstenberger, Andreas J. "Jesus the Good Shepherd Who Will Also Bring Other Sheep (John 10:16): The Old Testament Background of the Familiar Metaphor." *Bulletin for Biblical Research* (2002): 67-96.
Kramer, Samuel N. *History Begins at Sumer: Thirty-Nine Firsts in Man's Recorded History*. Philadelphia: University of Pennsylvania Press, 1981.
Kraus, Hans-Joachim. *Psalms 1-59: A Commentary*. Translated by Hilton C. Oswald. Minneapolis: Augsburg, 1988.
Krealing, C. H. *The Synagogue, The Excavations at Dura-Europs: Final Report*, VIII/1. New Haven: Yale University Press, 1956.
Kupp, David D. *Matthew's Emmanuel: Divine Presence and God's People in the First Gospel*. Cambridge: Cambridge University Press, 1996.
Kvalbein, Hans. "Die Wunder der Endzeit: Beobachtungen zu 4Q521 und Matth 11,5p." *New Testament Studies* 43 (1997): 111-125.
__. "Has Matthew Abandoned the Jews?: A Contribution to a Disputed Issue in Recent Scholarship." Pages 45-62 in *Mission of the Early Church to Jews and Gentiles*. Edited by Jostein Ådna and Hans Kvalbein. Tübingen: Mohr Siebeck, 2000.
Laato, Antti. *Josiah and David Redivivus: The Historical Josiah and the Messianic Expectations of Exilic and Postexilic Times*. Stockholm, Sweden: Almqvist & Wiksell, 1992.
Landmesser, C. *Jüngerberufung und Zuwendung zu Gott: Ein exegetischer Beitrag zum Konzept der matthäischen Soteriologie im Anschluß an Mt 9,9-13*. Wissenschaftliche Untersuchungen zum Alten und Neuen Testament 133. Tübingen: Mohr Siebeck, 2001.
Lang, B. "The Roots of the Eucharist in Jesus' Praxis." *Society of Biblical Literature Seminar Papers* 31 (1992): 467-472.
Larkin, Katrina J. A. *The Eschatology of Second Zechariah: A Study of the Formation of Mantological Wisdom Anthology*. Kampen: Kok Pharos, 1994.
Lemke, W. "Life in the Present and Hope for the Future." *Interpretation* 38 (1984): 165-180.
Leske, Adrian M. "The Influence of Isaiah 40-66 on Christology in Matthew and Luke: A Comparison." *Society of Biblical Literature Seminar Papers* 33 (1994): 897-916.
Levenson, Jon D. *Theology of the Program of Restoration of Ezekiel 40-48*. Harvard Semitic Monographs 10. Atlanta: Scholars, 1986.
Levey, S. H. *The Messiah: An Aramaic Interpretation: The Messianic Exegesis of the Targum*. Cincinnati: Hebrew Union College-Jewish Institute of Religion, 1974.
Levine, A. *Numbers 21-36*. Anchor Bible. New York: Doubleday, 2000.
Leyerle, Blake. "Meal Customs in the Greco-Roman World." Pages 29-45 in *Passover and Easter: Origin and History to Modern Times*. Edited by P. F. Bradshaw and L. A. Hoffman. Notre Dame: University of Notre Dame Press, 1999.
Lim, Timothy H. *Pesharim*. London: Sheffield, 2002.
Lindars, B. *New Testament Apologetic: The Doctrinal Significance of the Old Testament Quotations*. Philadelphia: Westminster, 1961.
__. "A Bull, a Lamb and a Word: 1 Enoch XC. 38." *New Testament Studies* 22 (1976): 483-486.
Litwak, Kenneth D. "Echoes of Scripture?: A Critical Survey of Recent Works on Paul's Use of the Old Testament." *Current Research* 6 (1998): 260-288.

Loader, W. R. G. "Son of David, Blindness, Possession, and Duality in Matthew." *Catholic Biblical Quarterly* 44 (1982): 570-585.
Luckenbill, Daniel D. *Ancient Records of Assyria and Babylonia.* Vol. 1, *Historical Records of Assyria from the earliest times to Sargon*; vol. 2, *Historical Records of Assyria from Sargon to the end.* New York: Greenwood Press, 1968.
Lust, J. "Ezekiel Manuscripts in Qumran: A Preliminary Edition of 4QEzek a and b." Pages 90-100 in *Ezekiel and His Book: Textual and Literary Criticism and Their Interrelation.* Edited by J. Lust. Leuven: Leuven University Press, 1986.
__. "Mic 5, 1-3 in Qumran and in The New Testament and Messianism in the Septuagint." Pages 65-88 in *The Scriptures in the Gospels.* Edited by C. M. Tuckett. *Bibliotheca ephemeridum theologicarum lovaniensium* 131. Leuven: Leuven University Press, 1997.
Luz, Ulrich. *Das Evangelum nach Matthäus.* Vol. 1, *Matt. 1-7*; vol. 2, *Matt. 8-20*; vol. 3, *Matt. 21-25*; vol. 4, *Matt. 26-28.* Evangelisch-Katholischer Kommentar zum Neuen Testament. Benziger/Neukirchener: Düsseldorf/Zürich, 1987, 1990, 1997, 2002.
__. *Matthew 1-7: A Commentary.* Translated by W. C. Linss. Continental Commentaries. Minneapolis: Augsburg, 1989.
__. *Matthew 8-20: A Commentary.* Translated by J. E. Crouch. Minneapolis: Fortress, 2001.
__. "Eine thetische Skizze der matthäschen Christologie." Pages 221-235 in *Anfänge der Christologie.* Edited by C. Breytenbach, H. Paulsen and F. Hahn. Göttingen: Vandenhoeck & Ruprecht, 1991.
__. "The Son of Man in Matthew: Heavenly Judge or Human Christ?" *Journal for the Study of the New Testament* 48 (1992): 3-21.
__. *The Theology of the Gospel of Matthew.* Cambridge: Cambridge University Press, 1995.
__. "Has Matthew Abandoned the Jews?: A Response to Hans Kvalbein and Peter Stuhlmacher concerning Matt 28:16-20." Pages 63-68 in *Mission of the Early Church to Jews and Gentiles.* Edited by Jostein Ådna and Hans Kvalbein. Tübingen: Mohr Siebeck, 2000.
Mack, B. L. "Wisdom Makes a Difference: Alternatives to 'Messianic' Configurations." Pages 15-48 in *Judaisms and Their Messiahs at the Turn of the Christian Era.* Edited by Neusner, W. S. Green and E. S. Frerichs. Cambridge: Cambridge University Press, 1987.
Malina, B. J. "The Literary Structure and Form of Mt 28:16-20." *New Testament Studies* 17 (1970): 87-103.
Manus, Chris. "'King-Christology': The Result of a Critical Study of Matt 28:16-20 as an Example of Contextual Exegesis in Africa." *Scriptura* 39 (1991): 25-42.
Mason, R. A. "Some Examples of Inner-biblical Exegesis in Zech IX-XIV." Pages 343-354 in vol. 7 of *Studia Evangelica.* Edited by E. Livingstone. Berlin: Akademie-Verlag, 1982.
Manson, R. *Micah, Nahum, Obadiah.* Sheffield: JSOT, 1991.
Martin, F. "The Image of Shepherd in the Gospel of Saint Matthew." *Science et esprit* 27 (1975): 261-301.
Mays, J. L. *Micah: A Commentary.* Philadelphia: Westminster, 1976.
McNight, Scot. "New Shepherds of Israel: An Historical and Critical Study of Matthew 9:35-11:1." Ph.D. diss., University of Nottingham, 1986.
__. *A New Vision for Israel: The Teachings of Jesus in National Context.* Grand Rapids: Eerdmans, 1999.
Meeks, W. A. *The Prophet-King: Moses Traditions and the Johannine Christology.* Novum Testamentum Supplements 14. Leiden: Brill, 1967.
Meier, J. P. *Law and History in Matthew's Gospel: A Redactional Study of Mt. 5:17-48.* Rome: Biblical Institute Press, 1976.
__. *Matthew.* Wilmington, Del.: Michael Glazier, 1980.
__. *A Marginal Jew: Rethinking the Historical Jesus.* Vol.1, *The Roots of the Problem and the Person*; vol. 2, *Mentor, Message, and Miracles.* New York: Doubleday, 1991, 1994.

___. "The Present State of the 'Third Quest' for the Historical Jesus: Loss and Gain." *Biblica* 80 (1999): 459-487.

___. "The Quest for the Historical Pharisees: A Review of Essay on Roland Deines, Die Pharisaer." *Catholic Biblical Quarterly* 61 (1999): 713-723.

Mercer, A. B. *The Pyramid Texts in Translation and Commentary* IV. New York: Longmans, 1952.

Michel, O. "Der Abschluss des Matthäusevangeliums." *Evangelische Theologie* 10 (1950/1951): 16-26.

Milik, J. T. *Ten Years of Discovery in the Wilderness of Judea*. Naperville, Ill.: Allenson, 1959.

Miller, Jean. *Les Citations d'accomplissement dans l'évangile de Matthieu*. Rome: Editrice Pontificio Istituto Biblico, 1999.

Moessner, David P. *Lord of the Banquet: The Literary and Theological Significance of the Lukan Travel Narrative*. Minneapolis: Fortress, 1989.

Moo, Douglas J. *The Old Testament in the Gospel Passion Narratives*. Sheffield: Almond, 1983.

___. "Jesus and the Authority of the Mosaic Law." *Journal for the Study of the New Testament* 20 (1984): 16-28.

Morgan, T. E. "Is There an Intertext in this Text?: Literary and Interdisciplinary Approaches to Intertextuality." *American Journal of Semiotics* 3, no. 4 (1985): 1-40.

Moxnes, H. "Meals and the New Community in Luke." *Svensk exegetisk årsbok* 51 (1986): 158-167.

Muilenburg, J. "Form Criticism and Beyond." *Journal of Biblical Literature* 88 (1969): 1-13.

Mullins, Terence Y. "Jesus, the 'Son of David.'" *Andrews University Seminary Studies* 29 (1991): 117-126.

Müller, Mogens. "The Theological Interpretation of the Figure of Jesus in the Gospel of Matthew: Some Principal Features in Matthean Christology." *New Testament Studies* 45 (1999): 157-173.

Murphy-O'Connor, J. "The Damascus Document Revisited." *Revue biblique* 92 (1987): 225-245.

Meyers, C. L., and E. M. Myers, *Zechariah 9-14*. Anchor Bible. New York: Doubleday, 1993.

Neale, David A. *None but the Sinners: Religious Categories in the Gospel of Luke*. Journal for the Study of the New Testament: Supplement Series 58. Sheffield: JSOT, 1991.

Neusner, J. *From Politics to Piety: The Emergence of Pharisaic Judaism*. Englewood Cliffs, N.J.: Prentice-Hall, 1973.

___. *Judaic Law from Jesus to the Mishnah: A Systematic Reply to Professor E. P. Sanders*. Atlanta: Scholars, 1993.

___. "Exile and Return as the History of Judaism." Pages 221-238 in *Exile: Old Testament, Jewish, and Christian Conceptions*. Edited by James M. Scott. Leiden: Brill, 1997.

Neyrey, J. H. "The Thematic Use of Isaiah 42, 1-4 in Matthew 12." *Biblica* 63 (1982): 457-473.

___. "Ceremonies in Luke-Acts: The Case of Meals and Table-Fellowship." Pages 361-387 in *The Social World of Luke-Acts: Models for Interpretation*. Edited by J. H. Neyrey. Peabody, Mass.: Hendrickson, 1991.

Nickelsburg, George W. E. *1 Enoch 1: A Commentary on the Book of 1 Enoch, Chapters 1-36; 81-108*. Minneapolis: Fortress, 2001.

Novakovic, Lidija. *Messiah, the Healer of the Sick: A Study of Jesus as the Son of David in the Gospel of Matthew*. Wissenschaftliche Untersuchungen zum Alten und Neuen Testament 2/170. Tübingen: Mohr Siebeck, 2003.

Thompson, J. G. "The Shepherd-Ruler Concept in the OT and Its Application in the NT." *Scottish Journal of Theology* 8 (1955): 406-418.

Oegema, Gerbern S. *The Anointed and his People*. Journal for the Study of the Pseudepigrapha: Supplement Series 27. Sheffield: Sheffield, 1998.
Orlinsky, Harry M. "Nationalism-Universalism in Ancient Israel." Pages 206-236 in *Translating and Understanding the Old Testament*. Edited by Harry T. Frank, W. L. Reed and H. G. May. Nashville: Abingdon, 1970.
Osborne, G. *The Hermeneutical Spiral*. Downers Grove: IVP, 1991.
Paffenroth, Kim. "Jesus as Anointed and Healing Son of David in the Gospel of Matthew." *Biblica* 80 (1999): 547-554.
Patte, Daniel. *The Gospel According to Matthew*. Philadelphia: Fortress, 1987.
Patton, C. "'I Myself Gave Them Laws That Were Not Good': Ezekiel 20 and the Exodus Traditions." *Journal for the Study of the Old Testament* 69 (1996): 73-90.
Perkins, P. "Christology and Mission: Matthew 28:16-20." *Listening* 24 (1989): 302-309.
Perrin, N. *Rediscovering the Teaching of Jesus*. New York: Harper & Row, 1967.
Person, R. F. *Second Zechariah and the Deutoronomic School*. Journal for the Study of the Old Testament: Supplement Series 67. Sheffield: Sheffield, 1993.
Pesch, Rudolf. *Das Markusevangelium*. Freiburg: Herder,1977.
Peters, E. E. *The Harvest of Hellenism*. New York: Simon and Schuster, 1970.
Peterson, W. L. "The Parable of the Lost Sheep in the Gospel of Thomas and the Synoptics." *Novum Testamentum* 23 (1981): 128-147.
Peterson, D. L. *Zechariah 9-14 and Malachi: A Commentary*. Louisville: Westminster John Knox, 1995.
Petrotta, A. J. "A Closer Look at Matt 2:6 and Its Old Testament Sources." *Journal of the Evangelical Theological Society* 28 (1985): 47-52.
Pilch, J. J. "Understanding Biblical Healing: Selecting the Appropriate Model." *Biblical Theology Bulletin* 18 (1988): 60-66.
___. *Healing in the New Testament: Insights from Medical and Mediterranean Anthropology*. Minneapolis: Fortress Press, 2000.
Pomykala, K. E. *The Davidic Dynasty Tradition in Early Judaism: Its History and Significance for Messianism*. SBL Early Judaism and Its Literature 7. Atlanta: Scholars Press, 1995.
Porter, Stanley E. "The Use of the Old Testament in the New Testament: A Brief Comment on Method and Terminology." Pages 79-96 in *Early Christian Interpretation of the Scriptures of Israel*. Edited by C. A. Evans and James A. Sanders. Journal for the Study of the New Testament: Supplement Series 148. Sheffield: Sheffield, 1997.
___. *The Criterion for Authenticity in Historical-Jesus Research: Previous Discussion and New Proposals*. Journal for the Study of the New Testament: Supplement Series 191. Sheffield: Sheffield, 2000.
Puech, Émile. "Une Apocalypse Messianique (4Q521)." *Revue de Qumran* 15 (1991): 475-519.
___. "Messianism, Resurrection, and Eschatology at Qumran and in the New Testament." Pages 235-256 in *The Community of the Renewed Covenant: The Notre Dame Sympium on the Dead Sea Scrolls*. Edited by Eugene Ulrich and James Vanderkam. Notre Dame: University of Notre Dame Press, 1994.
___. *Qumrân Grotte 4*. Discoveries in the Judean Desert 25. Oxford: Clarendon, 1998.
Rabin, Chaim, and Yigael Yadin, eds. *Aspects of the Dead Sea Scrolls*. Scripta Hierosolymitana IV. Jerusalem: Magnes, 1965.
Rajak, Tessea. "Hasmonean Kingship and the Invention of Tradition." Pages 99-115 in *Aspects of Hellenistic Kingship*. Edited by Per Bilde. Oxford: Aarhus University Press, 1996.
Redditt, Paul L. "Israel's Shepherds: Hope and Pessimism in Zechariah 9-14." *Catholic Biblical Quarterly* 51 (1989): 631-642.

Reid, S. B. "The Structure of the Ten Week Apocalypse and the Book of Dream Visions." *Journal for the Study of Judaism in the Persian, Hellenistic, and Roman Periods* 16 (1985): 189-201.
Renz, Thomas. *The Rhetorical Function of the Book of Ezekiel*. Leiden: Brill, 1999.
Repschinski, Boris. *The Controversy Stories in the Gospel of Matthew*. Göttingen: Vandenhoeck & Ruprecht, 2000.
Ribera, Josep. "The Image of Israel According to Targum Ezekiel." Pages 111-121 in *Targumic and Cognate Studies*. Edited by K. J. Cathcart and M. Maher. Journal for the Study of the Old Testament: Supplement Series 230. Sheffield: Sheffield, 1996.
Repschinski, Boris. The *Controversy Stories in the Gospel of Matthew: Their Redaction, Form and Relevance for the Relationship Between the Matthean Community and Formative Judaism.* Göttingen: Vandenhoeck & Ruprecht, 2000.
Riesenfeld, Harald. *The Resurrection in Ezekiel XXXVII and in the Dura-Europos Paintings* Uppsala: A.-B. Lundequistska Bokhandeln, 1948.
Robbins Veron K. *Jesus the Teacher*. Minneapolis: Fortress, 1992.
Rudolf, W. *Haggai, Sacharja 1-8, Sacharja 9-14, Maleachi*. Kommentar zum Alten Testament XIII/4. Gütersloh: Gerd Mohn, 1976.
Russell, E. A. "The Canaanite Woman and the Gospels (Mt 15:2-28; cf. Mk 7:24-30)." *Studia Biblica* (1978): 263-300.
Saldarini, Anthony J. *Pharisees, Scribes, and Sadducees in Palestine Society: A Sociological Approach*. Wilmington: Glazier, 1988.
___. *Matthew's Christian-Jewish Community*. Chicago: University of Chicago Press, 1994.
___. "Understanding Matthew's Vitriol." *Biblical Research* 13 (1997): 32-39, 45.
Sanders, E. P. *Paul and Palestine Judaism: A Comparison of Patterns of Religion*. Philadelphia: Fortress, 1977.
___. "Jesus and the Sinners." *Journal for the Study of the New Testament* 19 (1983): 5-36.
___. *Jesus and Judaism*. Philadelphia: Fortress, 1985.
___. *Jewish Law from Jesus to Mishnah: Five Studies*. Philadelphia: TPI, 1990.
___. *Judaism: Practice and Belief 63 BCE-66CE*. Philadelphia: TPI, 1992.
Sanders, James A. "Paul and the Theology of History." Pages 52-57 in *Paul and the Scriptures of Israel*. Edited by Craig A. Evans and James A. Sanders. Journal for the Study of the New Testament: Supplement Series 83. Sheffield: Sheffield, 1993.
Schaper, Joachim. "The Pharisees." Pages 402-407 in vol. 3 of *The Cambridge History of Judaism: The Early Roman Period*. Edited by W. Horbury, W. D. Davies and J. Sturdy. Cambridge: Cambridge University Press 1999.
Schlatter, A. *Der Evangelist Matthäus*. Stuttgart: Calwer Verlag, 1963.
Schnabel, Eckhard J. *Law and Wisdom from Ben Sira to Paul: A Tradition Historical Enquiry into the Relation of Law, Wisdom and Ethics*. Wissenschaftliche Untersuchungen zum Alten und Neuen Testament 2/16. Tübingen: Mohr Siebeck, 1985.
___. *Urchristliche Mission*. Wuppertal: R. Brockhaus Verlag, 2002.
Schniedewind, W. M. *Society and the Promise to David*. Oxford: Oxford University Press, 1999.
Schroeder, Roy P. "The 'Worthless Shepherd': A Study of Mark 14:27." *Concordia Theological Monthly* 2 (1975): 342-344.
Scott, B. B. *Hear Then the Parable: A Commentary on the Parables of Jesus*. Minneapolis: Fortress Press, 1989.
Senior, Don. *The Passion According to Matthew: A Redactional Study*. Bibliotheca ephemeridum theologicarum lovaniensium 29. Louvain: Louvain Univ. Press, 1975.
___. "The Death of Jesus and the Resurrection of the Holy Ones (MT 27:51-53)." *Catholic Biblical Quarterly* 38 (1976): 312-329.

___. "Matthew's Special Material in the Passion Story: Implications for the Evangelist's Redactional Technique and Theological Perspective." *Ephemerides theologicae lovanienses* 63 (1987): 272-294.

___. "Revisiting Matthew's Special Material in the Passion Narrative." *Ephemerides theologicae lovanienses* 70 (1994): 417-424.

___. *Matthew*. Abingdon New Testament Commentaries. Nashville: Abingdon, 1998.

Shaw, Charles S. *The Speech of Micah: A Rhetorical-Historical Analysis*. Sheffield: JSOT, 1993.

Silva, M. "Ned. B. Stonehouse and Redaction Criticism. Part I: The Witness of the Synoptic Evangelists to Christ." *Westminster Theological Journal* 40 (1977): 77-88.

___. "Part II: The Historicity of the Synoptic Tradition." *Westminster Theological Journal* 41 (1978): 281-303.

Sim, David. C. *Apcoalyptic Eschatology in the Gspel of Matthew*. Society for New Testament Monograph Series 88. Cambridge: Cambridge University Press, 1996.

Skarsaune, O. *The Proof from Prophecy*. Leiden: Brill, 1987.

___. "The Mission to the Jews-a Closed Chapter?: Some Patristic Reflections Concerning 'the Great Commission'." Pages 69-83 in *Mission of the Early Church to Jews and Gentiles*. Edited by Jostein Ådna and Hans Kvalbein. Tübingen: Mohr Siebeck, 2000.

Smith, Dennis E. "The Historical Jesus at Table." *Society of Biblical Literature Seminar Papers* 28 (1989): 466-486.

___. "Table Fellowship as a Literary Motif in the Gospel of Luke." *Journal of Biblical Literature* 106 (1987): 613-638.

___. "Table Fellowship and the Historical Jesus." *Novum Testamentum* (1994): 135-162.

Smith, K. "Matthew 28: Resurrection as Theophany." *Irish Theological Quarterly* 42 (1975): 259-271.

Smith, R. R. *Hellenistic Sculpture*. London: Thames and Hudson, 1991.

Smith, S. H. "The Function of the Son of David Tradition in Mark's Gospel." *New Testament Studies* 42 (1996): 523-539.

Speiser, E. A. *Genesis*. Anchor Bible. Doubleday: New York, 1964.

Stanley, C. D. *Paul and the Language of Scripture: Citation Technique in the Pauline Epistles and Contemporary Literature*. Cambridge: Cambridge University Press, 1991.

Stanton, Graham. "Origin and Purpose of Matthew's Gospel." Pages 1889-1951 in *Aufstieg und Niedergang der römischen Welt* II. 25.3. Edited by W. Hasse. Berlin: De Gruyter, 1985.

___. "Matthew as Creative Interpreter of the Sayings of Jesus." Pages 273-287 in *Das Evangelium und die Evangelien: Vorträge vom Tübinger Symposium, 1982*. Edited by P. Stuhlmacher. Wissenschaftliche Untersuchungen zum Alten und Neuen Testament 28. Tübingen: Mohr Siebeck. 1983.

___. "Matthew's Christology and the Parting of the Ways." Pages 99-116 in *Jews and Christians: The Parting of the Ways A.D. 70 to 135*. Edited by J. D. G. Dunn. Tübingen: Mohr Siebeck, 1992.

___. *A Gospel of a New People: Studies in Matthew*. Louisville: Westminster, 1993.

___. "The Two Parousias of Christ: Justin Martyr and Matthew." Pages 183-195 in *From Jesus to John: Essays on Jesus and New Testament Christology*. Edited by Martinus C. De Boer. Journal for the Study of the New Testament: Supplement Series 84. Sheffield: JSOT, 1993.

Stauffer, Ethelbert. *Jesus and His Story*. Translated by Richard and Clara Winston. New York: Alfred A. Knopf, 1960.

Steck, O. H. *Israel und das gewaltsame Geschick der Propheten: Untersuchungen zur Überlieferung des deuteronomistischen Geschichtsbildes im Alten Testament, Spätjudentum und Urchristentum*. Wissenschaftliche Monographien zum Alten und Neuen Testament 23. Neukirchen-Vluyn: Neukirchner Verlag, 1967.

Stein, R. H. "A Short Note on Mark xiv.28 and xvi.7." *New Testament Studies* 20 (1974): 445-452.
Stendahl, K. *The School of St. Matthew and Its Use of the Old Testament.* Lund: Gleerup, 1968.
Stevenson, R. *The Vision of Transformation: The Territorial Rhetoric of Ezekiel 40-48.* Society of Biblical Literature Dissertation Series 154. Atlanta: Scholars, 1996.
Steudel, A. "אחרית הימים in the Texts from Qumran." *Revue de Qumran* 16 (1993): 225-246.
Stone, M. E. *Fourth Ezra: A Commentary on the Book of Fourth Ezra.* Minneapolis: Augsburg Fortress, 1990.
Strack, H. L., and P. Billerbeck. *Kommentar zum Neuen Testament aus Talmud und Midrasch.* Munich: C. H. Beck, 1926.
Strecker, G. *Der Weg der Gerechitigkeit: Untersuchungen zur Theologie des Matthäus.* Göttingen: Vandenhoeck und Ruprecht, 1966.
___. "The Concept of History in Matthew." *Journal of the American Academy of Religion* 36 (1967): 219-230.
Strong, John. "Ezekiel's Use of Recognition Formula in His Oracles Against the Nations." *Perspectives in Religious Studies* 22 (1995): 115-133.
Stuhlmacher, Peter. "Matt 28:16-20 and the Course of Mission in the Apostolic and Postapostolic Age." Pages 17-43 in *Mission of the Early Church to Jews and Gentiles.* Edited by Jostein Ådna and Hans Kvalbein. Tübingen: Mohr Siebeck, 2000.
Suhl, Alfred. "Der Davidssohn im Matthäus-Evangelium." *Zeitschrift für die neutestamentliche Wissenschaft und die Kunde der älteren Kirche* 59 (1968): 57-81.
Talmon, S. "The Concepts of MĀŠÎAH and Messianism in Early Judaism." Pages 79-115 in *The Messiah: Developments in Earliest Judaism and Christianity.* Edited by J. H. Charlesworth. Minneapolis: Fortress Press, 1992.
Tate, W. Randolph. *Biblical Interpretation: An Integrated Approach.* Rev. and ed. Peabody: Hendrickson, 1997.
Tatum, W. Barnes. "'The Origin of Jesus Messiah' (Matt 1:1, 18a): Matthew's Use of the Infancy Traditions." *Journal of Biblical Literature* 96 (1977): 523-535.
Theissen, Gerd and Annette Merz. *The Historical Jesus: A Comprehensive Guide.* Minneapolis: Fortress, 1996.
Thiselton, Anthony C. *New Horizons in Hermeneutics.* Grand Rapids: Zondervan, 1992.
Thompson, Alden L. *Responsibility for Evil in the Theodicy of IV Ezra: A Study Illustrating the Significance of Form and Structure for the Meaning of the Book.* Society of Biblical Literature Dissertation Series 29. Missoula, Mont.: Scholars, 1977.
Thompson, J. G. "The Shepherd-Ruler Concept in the OT and Its Application in the NT." *Scottish Journal of Theology* 8 (1955): 406-418.
Thompson, M. *Clothed with Christ: The Example and Teaching of Jesus in Romans 12:1-15:13.* Journal for the Study of the New Testament: Supplement Series 59. Sheffield: Sheffield, 1991.
Thompson, Marianne M. "The Structure of Matthew: A Survey of Recent Trends." *Studia Biblica et Theologica* 12 (1982): 195-238.
Thompson, W. G. "Reflections on the Composition of Matt 8,1-9, 34." *Catholic Biblical Quarterly* 33 (1971): 365-388.
Tigchelaar, Eibert J. C. *Prophets of Old and the Day of the End: Zechariah, the Book of Watchers and Apocalyptic.* Leiden: Brill, 1996.
Tiller, Patrick A. *A Commentary of the Animal Apocalypse of 1 Enoch.* SBL Early Judaism and Its Literature 4. Atlanta: Scholars, 1993.
Trilling, W. *Das wahre Israel: Studien zur Theologie des Matthäusevangeliums.* Leipzig: St. Benno-Verlag, 1959.

Tooley, Wilfred. "The Shepherd and Sheep Image in the Teachings of Jesus." *Novum Testamentum* 7 (1964): 15-25.
Trafton, Joseph L. "The Psalms of Solomon in Recent Research." *Journal for the Study of the Pseudepigrapha* 12 (1994): 3-19.
Trautmann, Maria. *Zeichenhafte Handlungen Jesu: Ein Beitrag zur Frage nach dem geschichtlichen Jesus*. Forschung zur Bibel 37. Würzburg: Echter, 1980.
Trotter, Loren L. "'Can This Be the Son of David?'" Pages 82-97 in *Jesus and the Historian*. Edited by F. T. Trotter. Philadelphia: Westminster, 1968.
Troxel, R. L. "Matt 27.51-4 Reconsidered: Its Role in the Passion Narrative, Meaning and Origin." *New Testament Studies* 48 (2002): 30-47.
Uro, R. *Sheep Among the Wolves: A Study of the Mission Instructions of Q*. Helsinki: Suomalainen Tiedeakatemia, 1987.
Van Aarde, A. G. "Matthew's Portrayal of the Disciples and the Structure of Matthew 13:53-17:27." *Neotesamentica* 16 (1982): 21-74.
___. The *Evangelium Infantium*, the Abandonment of Children, and the Infancy Narrative in Matthew 1 and 2 from a Social Scientific Perspective." *Society of Biblical Literature Seminar Papers* 31 (1992): 435-448.
___. *God-With-Us: The Dominant Perspective in Matthew's Story*. Hervormede teologiese studies Supplementum 5. University of Pretoria, Gutenberg. 1994.
Van Buren, E. D. *Symbols of the Gods in Mesopotamian Art*. Rome: Pontificum Institutum Biblicum, 1945.
Vancil, Wayland J. "The Symbolism of the Shepherd in Biblical, Intertestamental, and New Testament Material." Ph. D. diss., Dropsie University, 1975.
Vanderkam, J. "Messianism in the Scrolls." Pages 211-234 in *The Community of the Renewed Covenant*. Edited by E. Ulrich and J. Vanderkam. Notre Dame: University of Notre Dame Press, 1994.
VanGemeren, W. A. *Interpreting the Prophetic Word*. Grand Rapids: Academie, 1990.
Vanhoozer, Kevin J. *Biblical Narrative in the Philosophy of Paul Ricoeur: A Study in Hermeneutics and Theology*. Cambridge: Cambridge University Press, 1990.
___. *Is There a Meaning in This Text?* Grand Rapids: Zondervan, 1998.
Van Groningen, Gerald. *Messianic Revelation in the Old Testament*. Grand Rapids: Baker, 1990.
Vermes, G. *Jesus the Jew*. Philadelphia: Fortress, 1973.
Verseput, Donald J. *The Rejection of the Humble Messianic King: A Study of the Composition of Matthew 11-12*. EHS.T 291. Frankfurt am Main: Peter Lang, 1986.
___. "The Role and Meaning of the 'Son of God' Title in Matthew's Gospel." *New Testament Studies* 33 (1987): 532-556.
___. "Jesus' Pilgrimage to Jerusalem and Encounter in the Temple: A Geographical Motif in Matthew's Gospel." *Novum Testamentum* 36 (1994): 105-121.
___. "The Davidic Messiah and Matthew's Jewish Christianity." *Society of Biblical Literature Seminar Papers* 34 (1995): 102-116.
Via, Dan O. "Structure, Christology, and Ethics in Matthew." Pages 199-215 in *Orientation by Disorientation: Studies in Literary Criticism and Biblical Literary Criticism*. Edited by Richard A. Spencer. Pittsburgh: Pickwick, 1980.
Vledder, Evert-Jan. *Conflict in the Miracle Stories: A Socio-Exegetical Study of Matthew 8 and 9*. Journal for the Study of the New Testament: Supplement Series 152. Sheffield: Sheffield, 1997.
Von Dobbeler, von Alex. "Die Restitution Israels und die Bekehrung der Heiden: Das Verhältnis von Mt 10,5b.6 und Mt 28,18-20 unter dem Aspekt der Komplementarität. Erwägungen zum Standort des Matthäusevangeliums." *Zeitschrift für die neutestamentliche Wissenschaft und die Kunde der älteren Kirche* 91 (2000): 18-44.

Von Rad, G. *Old Testament Theology.* Edinburgh: Oliver and Boyd, 1962.
Von Wahlde, Urban C. "The Relationships between Pharisees and Chief Priests: Some Observations on the Texts in Matthew, John and Josephus." *New Testament Studies* 42 (1996): 506-522.
Wacholder, Ben Zion. "Ezekiel and Ezekielianism as Progenitors of Essenianism." Pages 186-196 in *The Dead Sea Scrolls: Forty Yeas of Research.* Edited by Devorah Dimant and Uriel Rappaport. Brill: Leiden, 1992.
Wagner, J. Ross. *Heralds of the Good News: Isaiah and Paul "In Concert" in the Letter to the Romans.* Leiden: Brill, 2002.
Wainwright, Elaine. "The Matthean Jesus and the Healing of Women." Pages 74-95 in *The Gospel of Matthew in Current Study.* Edited by David E. Aune. Grand Rapids: Eerdmans, 2001.
Walbank, F. W., ed. *The Hellenistic World.* Vol. 7.1 of *The Cambridge Ancient History*, ed. I. E. S. Edward. Cambridge: Cambridge University Press, 1982.
Weaver, Dorothy Jean. *Matthew's Missionary Discourse: A Literary Analysis.* Journal for the Study of the New Testament: Supplement Series 38. Sheffield: Sheffield, 1990.
Weber, K. "The Image of Sheep and Goats in Matthew 25:31-46." *Catholic Biblical Quarterly* 59 (1997): 657-678.
Weinfeld, Moshe. "The Jewish Roots of Matthew's Vitriol." *Biblical Research* 13, no. 3 (1997): 31.
Wenham, D. "The Resurrection Narratives in Matthew's Gospel." *Tyndale Bulletin* 24 (1973): 21-54.
Wenham, Gordon. *Genesis* 16-50. Word Biblical Commentary 2. Nashville: Nelson, 1994.
Wenham, J. W. "When Were the Saints Raised?" *Journal of Theological Studies* 32 (1981): 150-152.
Westerholm, S. "The Law in the Sermon on the Mount: Matt 5:17-48." *Criswell Theological Review* 6 (1992): 43-56.
White, S. A. "A Comparison of the 'A' and 'B' Manuscripts of the Damascus Document." *Revue de Qumran* 48 (1987): 537-553.
Wilcox, Max. "On Investigating the Use of the Old Testament in the New Testament." Pages 231-243 in *Text and Interpretation.* Edited by E. Best and R. Wilson. Cambridge: Cambridge University Press, 1979.
Willett, Tom W. *Eschatology in the Theodicies of 2 Baruch and 4 Ezra.* Journal for the Study of Pseudepigrapha Supplement Series 4. Sheffield: Sheffield Academic Press, 1989.
Wilson, John. *Before Philosophy*, Baltimore: Penguin, 1971.
Winninge, Mikael. *Sinners and the Righteous: A Comparative Study of the Psalms of Solomon and Paul's Letter.* Stockholm: Almqvist & Wiksell International, 1995.
Wise, M. O., and J. D. Tabor, "The Messiah at Qumran." *Biblical Archaeology Review* 18 (1992): 60-61, 65.
Witherington, Ben. *The Christology of Jesus.* Minneapolis: Augsburg Fortress, 1990.
Witherup, R. D. "The Death of Jesus and the Raising of the Saints: Matthew 27:51-54 in Context." Pages 574-585 in *Faith and the Future: Studies in Christian Eschatology.* Edited by J. P. Galvin. New York: Paulist, 1994.
Wolff, Hans W. *Micah: A Commentary.* Translated by Gary Stansell. Minneapolis: Augsburg, 1990.
Wright, N. T. *The New Testament and the People of God.* Minneapolis: Fortress, 1992.
___. *Jesus and the Victory of God.* Minneapolis: Fortress, 1996.
Young, Norman H. "'Jesus and the Sinners: Some Queries." *Journal for the Study of the New Testament* 24 (1985): 73-75.
Zimmerli, Walter. "The Message of the Prophet Ezekiel." *Interpetation* 23 (1969): 131-157.
___. "Knowledge of God According to the Book of Ezekiel." Pages 29-98 in

I Am Yahweh. Edited by Walter Brueggemann. Atlanta: John Knox, 1982.

___. *Ezekiel 2*. Translated by James D. Martin. Philadelphia: Fortress, 1983.

Zimmermann, Frank. "The Language, the Date, and the Portrayal of the Messiah in IV Ezra." *Hebrew Studies* 26 (1985): 203-218.

Zimmermann, J. *Messianische Texte aus Qumran: Königliche, priesterliche und prophetische Messiasvorstellungen in den Schriftfunden von Qumran*. Wissenschaftliche Untersuchungen zum Alten und Neuen Testament 2/104. Tübingen: Mohr Siebeck, 1998.

Reference Index

Contents: 1. Old Testament; 2. New Testament; 3. Apocrypha and Pseudepigrapha; 4. Dead Sea Scrolls; 5. Philo and Josephus; 6. Rabbinic and Other Jewish Texts; 7. Early Christian Writings; 8. Greek Texts.

1. Old Testament

Genesis
12:1-4	342, 357
23:6	156
27:29	78
27:44	71
29:20	71
35:21	175
48:4	78
48:15	25, 26
49:8-12	131
49:10	27, 29, 78, 132, 133, 134, 153, 155, 174, 176, 177, 178, 179, 180, 183, 186
49:24-25	25, 26, *297*
49:26	156

Exodus
1:13-14	63
3:1-10	342
3:10, 11, 17	27, 107, 297
6:6, 13, 26	27, 59, 107, 297
7:5	27, 297
8:8	91
9:16	52
12:4	297
12:33-36	34
12:38	79
12:42	107
13-14	329
13:21	33, 88, 91, 199, 202, 337
14	34, 35
14:19	89
15:1-21	329
17:3	107, 297
19:7	346
19:22, 24	34
19:34	56
21:32	82
23:23	107, 297
23:25	27
28:21	29
29:14	29
32:15-29	59

Leviticus
4:13	46
10:17	74
21:17-18	320
25:43, 46	63
26	50, 52, 130
26:4-13	131

Numbers
4:18	29
7	73
18:12	29
11	59
14:10-19	59
14:29	74
16:31-50	59
16:33	74
20:5	27, 107, 297
24	151
24:17	29, 153, 178, 180, 186
26:55-56	56
27:16-17	26, 33, 199, 297, 337
27:17	25, 173, 180, 205, 209
36:1	73

Deuteronomy
4:38	27, 107, 297
6:23	27, 91, 297
7:6, 14	78
10:8	29
10:14-15	78
14:2	78
14:21	56
17:11-14	165

17:15	183, 184	7	135, 136, 139
20:1	27	7:5-7	297
21:10	27	7:11-16	129, 165, 184
22:7	56	5:2	25, 26, 28, 91, 174, 178
24:17	56	5:20	34
28:25	27	5:24	27
28:57	177	6:8	34
29:7	29	7:5-7	28, 91, 186
32:8	104	7:5-14	135
33:16	156	7:7-8	25, 26, 37, 89
		7:12-16	19, 28, 30, 37, 48, 69, 89, 91, 95, 116, 117, 134, 222, 368, 372
Joshua			
1:1-11	342		
7:9	52	7:14-15	142, 143, 372
22	29	7:14	123, 133
22:14	73	14:17	89
9:15	73	16:23	89
		22:14	35
Judges		23:5	183
2:16	62		
3:10	196	*1 Kings*	
4:4	196	2:17	25
4:14	27	5:1	178
6:15	36	5:9-14	289
10:2-3	196	5:14	78
12:7-14	196	8:27-53	368
15:20	196	8:41	52
16:31	196	11:34	73
26:17-35	232	14:8	68
		14:14	62
1 Samuel		15:3	68
4:18	196	22:17	26, 202, 209
5:2	295	27:27	34
7:6	196		
7:15	196	*2 Kings*	
9:21	36	14:3	68
16:11-13	36	19:9	27
10:20	29	19:35	89
13:14	68		
14:17	89	*1 Chronicles*	
16:1, 2	36	5	29
16:18	36	11:2	26
17:12	36	15:13	34
17:58	36	16:24	78
21:8	25	17:6	26
24:3	177	29:12	178
2 Samuel		*2 Chronicles*	
5:2	176, 177, 179	6:32	52
5:8	320	7:18	178
6-7	177	9:26	178

11:22	156	78:70-72	25, 28
18:16	27, 202, 205, 209	78:71	26
20:6	178	79:9	52
36:23	342, 348, 365	79:13	25
		80	341, 355, 356
Nehemiah		80:1	25, 26
9:6-37	129, 130	80:1-3	28, 346
12:46	36	80:13	34
		87	341, 355
Job		89:3-4	129
13:16	11	89:10	178
31:16-20	226	89:28-37	129
31:31-32	226	89:41	34
		89:72-90:1	104
Psalms		95:7	25, 28
2	28, 89, 134, 135, 137, 362	96	355
		96:2-3	341
2:7-9	35, 108, 116, 118, 222, 348	98:9	119
		100:3	28
2:9	24, 29, 30, 65, 71, 93, 121, 155	106:8	52
		109:1	343
8:5-6	89	109:5-6	341
9:5-6	89	109:21	52
9:8	119	110:1	13, 89
9:17	257	119:41	36, 185
18:14	34	119:176	28
19:13	28	122:4	34
22:12	225	143:11	52
23	25, 28, 52, 290	147:2-3	28
23:1	123		
23:4-6	30, 93, 108, 109, 121	*Proverbs*	
25:11	52	9:17	178
28:8-9	28	23:1	178
28:9	25	30:31	
29:1	300		
31:4	52	*Ecclesiates*	
41:13	299	12:11	31
44:11, 22	28		
45:7	71	*Isaiah*	
46:6	35	2	362
48:8-14	35	2:14	30
59:5	299	3:13	78
67:4	119	6	135
68:35	299	7:20	177
72	76	9:6, 7	30
74:1	25	10:5	108
76:3-6	35	10:21-11:5	137
77:20	25, 26, 28	10:22-27	137
78	29, 31	10:28-32	137
78:52-55	25, 29	11:1-16	30, 136, 137, 139, 152, 208
78:68-72	19, 29, 132		

11:1-5	117, 118, 133, 134, 140, 141, 142, 153, 171	6:13-15	234
		8:8-12	234
11:4	24, 35, 108, 109, 121, 122, 290	10-11	139
		10:21	25
14:1	30	12:10	25
14:5, 29	29	13:17, 20	34
14:9	225	14:7, 21	52
17:12-14	34	22:10-30	65
25:6-10	348	22:22	25, 26
29:5-8	35, 348	23:1-6	19, 24, 25, 29, 40, 58, 60, 79, 91, 113, 133, 134, 136, 152, 271
35	348		
35:4-6	143, 270, 286, 294, 301	23:2	25, 26, 69
40:1-11	30, 32, 34	23:3	32, 69, 297
40:10-11	25, 26, 30, 31, 92, 123, 147, 297	23:4	26, 62, 69
		23:5-6	132, 135
41:3-4	315, 316	23:25	132
42:1-4	286, 287, 301, 313	25:34	29
42:6	79	25:34-36	25
42:18-21	311	29:15	62
43:15	32	30	29, 30, 31, 40, 69, 93, 355
44:28	25		
45:21	36	30:9	129
46	25	30:13-17	297
46:10	36	31:1-37	19, 29, 93, 348
48:9	52	31:6	34
48:20	34	31:7	32, 297
49:1-6	342	31:8-10	26
49:1-12, 22	30, 31	31:10	25, 31, 32, 34, 69, 347
49:8	79	31:12-14	347, 348
49:9-13	25	31:15	186, 280, 378
52:7, 12	32, 34	31:31-34	31, 186
52:11	34	33:12-18	29
52:13-53:12	69	33:15-17	132, 133
53:4	284, 286, 311	33:15	132, 135
53:5-6	316, 347	34:8-11	66
53:7-8	347	49:19	25
55:3	89	50:6	25, 208
56:11	25	50:19	25
58:6-10	226	50:44	25
61	5, 301, 372		
61:1-14	30, 31, 92, 123, 147	*Lamentations*	
61:1-2	143, 270, 294, 295	3:1	108
63:9	36		
63:10-14	30	*Ezekiel*	
63:11	26, 36	1-11	41, 364
66:5	52	12-34	41
		1:30	38
Jeremiah		3:11	41
2:8	25, 29, 69	5:5-6	53
3:15	25, 26	5:13	54
6:3	25		

6	53, 54	22:15	44
7	53, 54, 70	22:16, 22	54
9:4	149, 194	22:25	60
9:10	259	22:27-28	236
10:3-18	39, 369	23:49	54
11	53, 54	24	53, 54
11:12	53	25	53, 54
11:19	55	26	50, 53, 54
11:20	54	29:6	55
12:10	73	29:12	44
12:12	156	29:21	55
12:15	44, 53, 54	30:10	59
12:16-28	43	30:23, 26	44
13:9	54	32:15	54
13:17	41	33	41, 50
13:22-27	234	33:21	38
14:8	54	33:29	54
14:11	54	34-48	40-45
15:7	54	34-39	41, 72
15:45-48	53	34-37	3, 19, 24, 25, 31, 38, 40,
16:1-34	39		41, 43, 46-74, 77, 88,
16:13	62		91, 93, 96, 103, 112,
16:45-48	54		113, 117, 123, 126, 130,
16:62	54		134, 135, 140, 145, 150,
17:1-21	65		151, 170, 172, 198, 214,
17:2-10	60		217, 230, 246, 247, 260,
17:12	73		261, 268, 297, 298, 302,
17:21, 22, 24	50, 54, 55, 118, 348		312, 324, 325, 335, 351,
18:7	226		367, 380, 382, 385, 394
19:1-4	60	34	23, 5,11, 12, 26, 42, 68,
19:10-14	65, 70		69, 79, 80, 104-105,
19:14-15	29		118, 208, 209, 214, 215,
20	50, 52, 53, 55, 130		218, 222, 241, 294, 295,
20:6	66		297, 302, 321, 338, 349,
20:9	55		361, 370, 392
20:12	54	34:1-16	11, 40, 44, 47, 58, 60,
20:14	55		63-65, 74, 80, 87, 110,
20:15	40, 51, 389		117, 130, 133, 147, 151,
20:22	52, 55		210, 215, 216, 217, 222,
20:23, 40	44, 54		230, 238, 239, 244, 259,
20:25	74		263, 264, 272, 273, 297,
20:32-38	59		302, 324, 365, 380
20:34-35	59, 61	34:2-10	25, 32, 47, 57, 91, 276
20:37	70, 107	34:2-3	236, 298
20:39	92	34:2	26, 102, 221
21:10	54	34:4	103, 118, 209, 243, 260,
21:15, 18	70		297, 321, 346
21:25-27	60	34:5	27, 58, 102, 209, 298
21:30-32	65, 73	34:6-7	86, 93, 117, 208, 210,
22:6	60		241, 364
22:7	54	34:7	26, 276

34:8	27, 58, 60, 209, 210, 364	36:10	216
34:10, 11	48, 58, 60, 62, 69, 86, 91, 130, 143, 147, 208, 272, 276, 293, 313	36:16-32	40, 92, 115, 349
		36:20-21	40, 54, 55, 87, 92, 157, 208, 216, 369
34:12, 13	57, 58, 59, 60, 62, 103, 118, 147, 259, 293, 297, 313	36:22-27	269, 312
		36:22	208, 216
		36:23	55, 87, 130, 157, 159
34:11-22	40, 49, 57	36:24-32	49, 74, 75, 158, 338
34:13	34, 59, 62, 200, 240	36:25	120, 142, 389
34:14-15	60, 62, 143, 240, 241, 298	36:26	55, 94
		36:27	46, 52, 94, 110
34:16	58, 118, 209, 211, 243, 259, 260, 272, 297	36:28	54, 70
		36:29	62, 63, 136
34:17-22	60, 74, 80, 221, 222, 223, 225, 226, 227, 229, 230, 231, 232, 380, 381, 389	36:32	216
		36:33-38	75
		36:35	46
		36:37	157
34:18	103, 143, 259	36:36	55
34:20	118	36:37	173, 216
34:21	259	37:1-14	42, 43, 47, 54, 57, 70, 75, 80, 92, 96, 109, 126, 139, 145, 159, 269, 270, 294, 331, 332, 338, 349, 361, 370, 380
34:22	62, 63, 80, 118		
34:23-24	26, 40, 43, 57, 61, 62, 67, 68, 70, 72, 87, 88, 93, 99, 109, 112, 115, 118, 123, 124, 129, 130, 133, 137, 142, 143, 152, 155, 156, 158, 170, 180, 184, 186, 201, 204, 216, 217, 222, 223, 229, 231, 232, 245, 259, 263, 264, 269, 270, 272, 273, 295, 300, 302, 308, 316, 324, 347, 349, 352, 362, 365, 366, 369, 380, 389		
		37:6	55
		37:7	333
		37:11	157, 173
		37:12	61, 200, 332, 333, 336, 340, 382
		37:13-14	62, 332, 333, 335, 336, 340, 382
		37:15-28	42, 43, 45, 47, 70, 75, 79, 80, 82, 93, 132, 139, 155, 160, 191, 196, 215, 224, 297, 349
34:24-32	70, 362	37:15-23	269
34:25-31	118, 131, 349, 350	37:16-17	71, 173
34:26	321	37:18	41
34:27	54, 57, 59, 62, 270	37:19	71
34:30	54, 110, 157, 173, 216, 349, 366, 369	37:23	54, 62, 63, 71
		37:24-25	26, 40, 43, 43, 52, 57, 66-68, 70, 72, 73, 75, 87, 88, 99, 110, 123, 124, 129, 130, 133, 137, 142, 143, 152, 155, 158, 170, 180, 184, 186, 201, 204, 216, 217, 222, 223, 229, 230, 232, 243, 245, 259, 263, 264, 272, 273, 295, 300, 302, 308, 316, 324, 347, 350, 352, 362,
34:31	25, 57, 113, 135, 157, 192, 216, 224, 230, 368		
34:37	57		
34:40	118		
34:41	118		
35	54		
35:15	157, 216		
36	55, 125		
36:1-23	57		
36:5	369		
36:9-11	46		

	366, 369, 380, 389	*Hosea*	
37:26	42, 44, 46, 47, 57, 75,	3:5	40, 62, 69, 129
	78, 94, 118, 130, 367,	6:5	35, 121
	369	6:6	59, 237, 274
37:27, 28	41, 54, 55, 57, 67, 70,	7:4-5	59
	94, 118, 130, 269, 276,	8:3-4	73
	278, 369, 369	11:1	280, 378
38-39	35, 42, 43, 93, 139, 350	13:10	73
38:5	66		
38:23	54	*Joel*	
39:6	54	2:10	331, 333
39:7, 25	40, 54, 92, 208	3:11-12	224
39:21-23	53, 55	4:16	330
40:1	38, 42		
40:28	57	*Amos*	
40:35	351, 369	1:5, 8	29
40-48	3, 38, 41, 43, 70, 72, 73,	3:8	35
	75, 134, 135, 139, 218,	5:26-27	149
	273, 298, 338, 361, 364,	9	135, 152
	380, 383	9:11	36, 135
43:6	42		
43:7-8	40, 54, 60, 92	*Micah*	
44:1-3	263	1:2	78
44:3	73	2-5	3, 19, 31, 37, 88, 113,
44:7, 9	55, 134, 208		135, 184, 246, 325, 334,
44:10	134, 135		335, 377, 385
44:24	73	2:6-11	32
45:7, 8, 9	73, 124, 263, 361	2:12-13	24, 25, 32- 34, 37, 88,
45:16-17	74, 75, 87, 92, 273, 367		91, 108, 184, 185, 199,
46	44, 45		203, 297, 316, 337, 340,
46:8	263		346, 355, 359, 382
46:10	156	3:1-9, 10-12	32
46:18	73	4:1, 2	32
47	42	4:6-5:4	34, 338, 348
47:13-23	56, 124	4:6-8	25, 32, 33, 36, 69, 338
47:22-23	54, 56	4:11-13	32, 34-35, 37, 88, 93,
48:35	41, 45, 94, 135, 269		105,
			108, 109, 115, 135, 142,
Daniel			184, 185, 337, 338, 359,
2:15	132		360
4:14	132	5:1	353
4:22	132	5:1-4	26, 32, 36-37, 79, 90,
4:29	132		93, 110, 113, 115, 118,
5:21	132		125, 139, 142, 145, 152,
5:29	132		153, 174-189, 217, 219,
7:13-14	341, 342, 348, 356		224, 230, 244, 247, 254,
8:1-12	225		270, 297, 313, 348, 360,
8:20-21	225		365, 377
10-11	104	5:2	12, 13, 88, 89, 173, 174,
12	356		176, 188, 244, 263, 282,
11:20	71		311, 347

5:3-4	311, 315, 316, 348, 351, 353, 365, 382	11:10-12	79, 92, 190
		11:11	150, 173
5:5	65, 90, 93, 142, 175, 224	11:12-13	189, 190, 193, 334, 377
5:6	37, 175	11:13	86, 87, 90, 189
5:6-15	217	11:15-17	83, 84, 85, 86, 88, 105, 109, 130, 190, 196, 197
5:15	297		
7:14-15	25, 36, 108	12:1-4	85, 90
7:18	32, 92, 123, 147, 208	12:2-9	35, 78, 83, 84, 88, 90, 113, 115, 191, 198, 202, 203

Nahum

3:18	25	12:8, 9	33, 88, 89, 91, 93, 108, 199, 337

Habakkuk

2:5	78	12:10	82, 86, 87, 88, 92, 123, 126, 147, 324
		13	3, 105, 151

Zephaniah

3:4	234	13:1-2	81, 82, 83, 90, 106, 297, 382
		13:7	12, 13, 26, 83, 84, 85, 87, 149, 150, 173, 189-204, 232, 244, 246, 318, 334, 337, 377, 381, 392

Haggai

2:6	331, 333

Zechariah

1-8	43	13:7-9	88, 191, 200, 337
4:6	81	14:1-3	35
6:9-15	83	14:4-5	331, 333, 334, 336, 337, 340
8:1-6	88, 189		
8:6	189	14:7	113
9-14	3, 5, 19, 31, 76-90, 106, 113, 114, 125, 126, 134, 135, 189, 197, 246, 319, 325, 334, 335, 347, 382, 385	14:1-21	76, 84, 90, 93
		14:21	134

Malachi

9:9-15	76, 81, 84, 87, 90, 93, 113	1:1-4:6	83
		3:4	36
9:9	189, 190, 318, 319, 323, 324, 334, 337, 377, 382		

2. New Testament

9:16	78, 316	*Matthew*	
10:2-3	77, 89, 90, 91, 209, 211, 297	1-10	374
		1-4	374
10:3	25, 190 10:4-17 115, 191	1-2	30, 297, 376
		1:1-4:16	375
10:6	87, 88, 91, 92, 126, 147	1:1	184, 187, 216, 273, 280, 284, 300, 303, 352, 375, 377, 385
11	275		
11:4-7	79, 81, 232	1:1-17	29, 30, 186, 277, 285, 300, 303
11:7	290		
10:10	34	1:5	278
10:30	34	1:6-11	183
11:3-4	190	1:6	289
11:4-16	77, 79, 80, 82, 83, 85, 132, 190, 191, 195, 202, 377	1:17-21	125, 175, 179, 182, 183, 187
		1:17	188, 273, 377
11:5-8	25, 30, 65, 71, 87, 190	1:18-23	216, 279, 300, 363

1:21	188, 210, 260, 273, 277, 363, 364, 371, 377, 390	6:11	262
		6:12	262
1:23	187, 247, 255, 273, 276, 277, 339, 364, 371, 375, 377, 383, 386, 390, 395	6:29	289
		6:32	354
		7:14	244
1:24-25	175	7:15	15, 173, 173, 211, 233-236, 238, 240, 245, 249, 272
2:1-12	182, 354		
2:1-6	283		
2:1-2	175, 186, 187, 211, 358	7:22	278
2:3-4	188, 211, 284	7:23	234
2:5	13	7:25	329
2:6	1, 2, 3, 12, 13, 173, 174-189, 204, 210, 236, 244, 245, 247, 254, 271, 273, 294, 300, 311, 315, 323, 325, 350, 359, 365, 378, 385, 387, 393, 395	7:28-29	273, 376, 379
		7:29	211, 352
		8-9	188, 207, 212, 255, 273, 295, 305, 308, 379, 387
		8:1	211
		8:2	314, 358
		8:6	314
2:8	358	8:11-12	224, 252, 258
2:9-11	175, 186, 358	8:7	290, 296
2:13-15	175, 378	8:8	314, 378
2:15-17	252, 255, 378	8:16	290, 296, 318
2:18	30, 186, 187, 255, 279, 378	8:17	283, 284, 286, 310
		8:25	314
2:19-21	175	8:27	294
3:3	30	8:28-9:26	269, 270, 306-307
3:7	262	9:2	318
3:17	222, 255	9:4-17	269
4:1-11	255, 358	9:6, 8	211, 294, 352
4:12-17	376	9:9-13	15, 19, 147, 208, 212, 217, 248-279, 361, 390, 391
4:15-16	355		
4:15	199, 354		
4:17-16:20	375	9:10	257, 258, 261, 352
4:17	339, 375, 376	9:11	261, 273, 274, 352
4:23-25	1	9:12-13	220, 238, 260, 271, 274, 276, 296, 297, 305, 307, 352, 361
4:23	19, 205, 219, 232, 296, 339, 363, 375, 378, 391		
		9:14-17	255, 306
4:24	296, 297, 318	9:18	358
4:25	211	9:18-26	208, 255, 270, 271, 290
5-7	255, 273, 373, 374, 376, 379	9:23-25	295, 308
		9:27	2, 15, 273, 274, 281, 284, 285, 302, 303, 304-308, 324
5:1	346		
5:3-10	230		
5:7	2, 306		
5:17-48	359, 360	9:27-31	5, 208, 255, 270, 279, 286, 296, 305, 308, 317, 378
5:17-19	390		
5:18	375		
5:23-24	256	9:28	314, 317
5:35	223	9:29	290, 305, 306, 308
5:47	257	9:32-34	208, 212, 255, 271, 283, 284, 294, 299, 305, 306, 308
6:9-10	394		
6:9	19, 278, 358, 378		

9:32	318, 390	11:27	352
9:35-11:1	2	11:28-30	124, 159, 283
9:35-36	147, 204, 209, 214, 255, 311, 323, 375, 378, 392	12-28	374
		12	379
9:35	1, 205, 207, 208, 214, 219, 232, 296, 301, 306, 363, 391	12:1-8	236, 239, 311, 312, 323
		12:2	236
		12:5	236
9:36	1, 2, 3, 12, 15, 19, 173, 188, 193, 205, 205-212, 222, 232, 240, 241, 245, 247, 255, 294, 295, 297, 301, 306, 325, 327, 364, 365, 378, 381, 391, 393, 395	12:7	237, 238, 290, 361
		12:8	236
		12:9-14	1, 173, 245, 311, 312, 378
		12:10	296, 297, 378
		12:11-12	15, 234, 236-239, 240, 249, 272, 319
		12:13	290, 299, 319
9:37-10:1	207, 208	12:14-15	290, 296, 319
9:38	255	12:17-21	283, 286
10	228, 229, 230, 231, 308, 362, 373, 376, 379, 389, 392	12:18-23	287
		12:21	278, 313, 315
10:1-6	187, 204, 214, 218, 219, 227, 232, 233, 241, 247, 273, 311, 313, 314, 356, 362, 363, 379, 381, 390, 392	12:22-30	1, 5, 212, 239, 272, 279, 285, 286, 296, 308, 318, 378, 390
		12:23	15, 281, 284, 285, 302, 303, 308-313, 378
10:1	98, 217, 301, 306, 308, 352	12:24	275, 284, 378
		12:25-32	311, 312, 313
10:2	218	12:27	211
10:3	214	12:30	359
10:5-6	213, 215, 355, 371	12:33	324
10:5	218, 354, 376	12:41-42	224, 289, 290
10:6, 16	1, 12, 15, 19, 98, 173, 193, 205, 208, 212-219, 232, 245, 260, 266, 272, 284, 301, 306, 327	12:48-49	228
		13	373, 374
		13:36-43	224
		13:53	376
10:7	218, 363, 371	14	15, 295
10:8	296, 371	14:13-21	207, 311
10:8-13	226	14:14	1, 2, 206, 296, 297, 298, 306
10:9-10	230		
10:10	226	14:19	298
10:16	218, 232, 233	14:28	314
10:17-18	218, 226, 354	14:30	314
10:25	283	14:33	358
10:34-35	274	15-27	379
10:40-42	226, 230	15	15, 295
11	374, 379	15:1-20	315
11:1	301, 376	15:1-2	265
11:5	304	15:9	212
11:7	376	15:12-28	19, 279, 284
11:12-13	359, 360	15:19	358
11:15	286	15:20	304
11:19	252	15:22	2, 16, 274, 281, 285, 290, 306, 308, 313-316,
11:20	376		

Reference Index 429

	317, 324, 378		297, 316, 317, 318, 319, 321, 334, 378, 382
15:24	12, 15, 173, 187, 204, 205, 213, 214, 232, 233, 245, 247, 260, 266, 272, 284, 311, 315, 327, 378	20:30-31	274, 281, 285, 302, 303, 306, 314, 316-319, 324
		20:32	354
15:21-28	1, 5, 16, 213, 286, 296, 313, 378, 390, 391	20:34	1, 2, 206, 207, 306, 314, 317
15:22	302, 313-319	20:35	354
15:23-27	378	20:46-52	304
15:24	378, 379	21-27	377
15:25	358, 376	21:1-17	321, 378, 382
15:28	290	21:1-12	1, 5, 279, 296, 319, 337
15:29-31	290, 296, 298, 311, 378	21:4-5	189, 222, 323, 337
15:31	285, 299, 390	21:6-9	334
15:32	1, 206, 207, 298, 306	21:9	278, 281, 282, 283, 284, 322, 323
15:32-39	19		
15:34	213	21:10	322
15:39	2	21:11	322
16:1	262	21:12-17	272, 282, 318, 319, 320
16:21-28:20	376	21:14-15	334
16:21	375, 376, 389	21:14	16, 285, 296, 319-323, 324
16:22	314		
17:1	346	21:15	281, 283, 284, 285, 299, 302, 303, 320, 322, 323
17:4	314		
17:5	346	21:23-27	352
17:15	2, 274, 306, 314	21:32-33	257, 258
17:18	290, 296	21:41-46	365
17:22	389	21:43	376
18:1-14	228, 373, 379	21:45	262
18:5	278	22:2	223
18:10	230, 240, 245, 272, 275	22:8-10	354
18:12-14	1, 2, 15, 173, 234, 236 239-243, 245, 249, 272, 275, 294, 311, 378	22:34-40	360, 390
		22:40	238
		22:41-46	5, 279, 286, 296, 299, 300,
18:14	230		
18:17	257		323, 323, 378, 390
18:19-20	277, 279, 368	23	5, 250, 266, 273, 374, 381
18:20	278		
18:21-35	256	23:1-3	274
18:21	314	23:7-10	362
18:23	223	23:13	212
18:24-35	207	23:23	238
18:26	358	23:25	234
18:27	2, 206, 306	23:28	234
18:33	2, 306	23:39	278, 354
19-22	379	24-25	373
19:1	376	24:3	283
19:2	211, 296, 297	24:5	278
19:28	5	24:7	354
20:19	354	24:27	283
20:29-34	1, 2, 5, 16, 211, 286, 296,	24:29	329
		24:37	283

24:39	283		375, 382, 384, 387, 395
25	228, 229, 230, 231	27:54	330, 339, 340, 390
25:31-46	2, 12, 15, 19, 173, 205, 208, 213, 219-232, 240, 245, 284, 316, 327, 354, 361, 370, 372, 381, 383, 389, 390, 393, 395	27:62	262
		27:63-64	283
		28	1, 358
		28:9	358
		28:10	228
25:31-33	223, 224	28:15	354
25:32	1, 2, 220, 221, 359	28:16-20	16, 17, 19, 186, 188, 191, 204, 208, 213, 247, 256, 273, 277, 278, 302, 308, 313, 315, 326, 340-369, 372, 384, 385, 386, 387, 390, 392, 395
25:33	220, 221		
25:34	221		
25:36	220, 228		
25:37-39	226, 228		
25:40	220, 221, 227, 231		
25:42-43	226	28:16-17	345-347, 358
25:43	220	28:18-20	347-369, 381, 383, 391 393
25:44	220, 226		
25:45	221, 227, 229, 231	28:19-20	215-219, 273, 278, 301, 370, 384, 389, 392, 395
25:46	220		
25:51-53	16		
26-28	379	*Mark*	
26:1	376	1:3	30
26:2	284	2:13-17	252, 253
26:14-27:10	203, 204	2:14	252
26:15-37	232	2:15-17	260, 266
26:15	1, 189, 190, 191, 193, 377	2:16	253
		2:23-28	237
26:26-29	190	3:1-6	236
26:31-32	27, 195, 216, 301, 336, 340, 352, 356, 372, 382, 386, 395	3:13-19	213
		3:22-30	308
		6:7-13	213
26:31	1, 2, 12, 13, 173, 189, 191, 189-198, 220, 230, 244, 245, 318, 325, 326, 327, 337, 353, 354, 355, 363, 377, 381, 382, 384, 387, 389, 392, 393, 395	6:34	205
		7:1-5	265
		7:24-30	313
		7:27	314
		7:31-37	290
		8:22-26	304
26:32	198-204, 244, 337, 346, 347, 355, 359, 377, 382, 393	9:27-31	304
		10:46-52	5, 288, 290, 304, 316, 317
26:56, 58	1	10:47-48	288
26:69-75	195	10:52	317
27:3-7	1, 189, 190, 377	11:11	322
27:11	222	12:35	281
27:29	222, 358	13:37	281
27:32-54	339	14:22-25	190
27:37	222, 281, 339	14:26-31	192
27:42	222, 332, 339	14:27	334, 355
27:43	332	14:28	200
27:51-53	16, 17, 19, 191, 200, 204, 295, 326, 327, 328-340, 363, 364, 370, 372,	14:41	257
		15:22	317
		16:7	199

Luke	
3:4	30
5:27-32	253
5:30	252, 253
6:5	236
6:33	257
11:14-15	309
13:10-17	237
13:29	252
15:4-5	235
15:4	234, 242
19:2	260
19:7	260
19:9-10	260
20:41	281
20:44	281

John	
3:1	262
4:26	310
4:29	309
5:1-9	237
7:45-52	262
10	1
10:16	224

Acts	
2:23-24	283
3:13-14	283
20:29	235

Romans	
5:20	51
7:13	51
9:7	13
9:15	206
10:18	13

1 Corinthians	
1:22	186
7:10	14
9:14	14
15:20	330
15:23	283
15:27	13

2 Corinthians	
1:3	206
8:9	283

Galatians	
2:15	257

3:12	13
3:19	51

Philippians	
1:19	11
2:1	206
2:6-11	283
2:9-11	342, 348

1 Thessalonians	
2:19	283
3:13	283
4:15	283
5:23	283

2 Thessalonians	
2:1	283
2:8	283

1 Thimothy	
3:16	348

Hebrews	
1:5-14	348
8	11

James	
5:7-8	283

1 Peter	
3:19	329

2 Peter	
3:4	283

1 John	
2:28	283

Revelation	
2:26	348
2:27-28	71, 394
22:16	186

3. Apocrypha and Pseudepigrapha

Apocryphon of Ezekiel	
5	96, 170, 295, 324, 325, 388

2 Baruch	
40:1	159
72:1-16	159
72:2	159

77:13	206	89:65-66	104
78:7	195	89:72	99
		89:74	100, 103
1 Enoch		89:75	103, 110
1:3-9	331, 333	90:1-28	196, 276
5:4	112	90:3	208, 209
45:3	223	90:6-17	102
51:3	223	90:6-7	268
55:4	223	90:17-18	123, 133, 135, 139, 160, 171, 185, 196, 224, 316, 360, 361, 363, 367, 368
61:8	223		
63:12	112		
79:1-2	112	90:18-19	107, 108, 109, 124, 152, 186
83-90	117		
83:1-2	104	90:19	105, 108
85-90	112	90:20-42	195
85:3	113	90:20-27	100, 103, 104, 107, 217
85:9	113	90:22	104
86:4	97	90:25	106, 109
86:6	98	90:28-29	107
89-90	169, 170, 171, 230, 231, 241, 247, 268, 381	90:33	101
		90:34	109
89:1	113	90:35	100
89:9	101, 113, 236	90:37-39	101, 106-110, 196
89:10-11, 12	98, 101, 113	90:38	102, 110-112, 115, 389
89:12-14	98, 218, 236	99:2	112
89:15-67	105, 105	104:10	112
89:16	98, 101, 107, 217, 236		
89:22	98, 100, 107, 297	*2 Enoch*	
89:25	107, 236	9:1	226
89:26	101, 217	10:5	226
89:28	100, 112		
89:29	101, 217	*4 Ezra*	157-160
89:30	101, 217	1:7-8	115
89:31	98	1:24	160
89:32	100	2:15-32	159
89:36, 38	98	2:21	159
89:39	98	2:34	158, 159, 160, 171, 224, 316
89:40	100		
89:41	100	2:40	160, 170
89:42	98	2:42	160, 170
89:45, 48	98, 99, 100, 111	3:4-27	158
89:46	98, 99, 111, 130, 180, 181, 186	3:20	158
		3:26	158
89:50-58	208	3:28	158
89:51-54	99, 100	3:31	158
89:52	99	4:4	158
89:54-59	104, 196, 209	6:13-16	331, 333
89:55-56	101, 102, 103	6:38-59	158
89:59	103, 104	7:1-16	158
89:59-90:17	100, 105, 107	7:21-24	158
89:61	104	7:26	158
89:64	104	7:28	159

7:29	158	17:34	120, 123, 124
8:35	158	17:34-37	120
12:32	158, 159	17:35	116, 119, 120, 368
12:33-34	159	17:36	119, 120, 122, 142, 171, 363, 389
13:26	159		
13:32	158	17:37	120
13:37	158	17:40-42	118, 224
13:49	159	17:41	119, 123, 124, 360
14:9	158	17:42	119, 120, 123
14:27-36	158	17:43	119, 122
		17:44	116

Jubilees

1:7-18	268	17:45	116, 123
15:32	104	18:1-12	124, 125, 224, 225
23:18-31	195	18:1	116
23:23-24	257	18:2	123
		18:3	116, 123, 124, 224
		18:5	123, 124, 202

Judith

11:19	205, 208	18:6	116
		18:7	116, 123
		18:9	116, 123, 124, 224

Psalms of Solomon

1:1	257	*Sirach*	
2:1-2	257	18:13	97, 360
8:28	195	36:6	195
17:1-18:12	115-125, 133, 134, 139, 140, 152, 160, 165, 169, 171, 182, 185, 247, 266, 273, 276, 316, 361, 367	36:15-16	195, 268
		45:25	97
		47:4-7	165
17	119, 123, 125, 131, 140	*Testament of Benjamin*	
17:3	116	4:4	226
17:4	116, 117, 123		
17:6, 7	117	*Testament of Joseph*	
17:11-14	117, 184	1:4-7	226
17:18-19	117, 118, 184		
17:19-20	117, 184	*Testament of Judah*	
17:19-25	124, 224	24:1	186
17:20	123, 201		
17:20-46	117, 124, 125, 224	*Testament of Levi*	
17:21-25	121, 124	4:1	333
17:21	116-118, 123, 125, 201, 352	18:3	186
17:22-25	119	*Testament of Moses*	
17:23-24	118, 121	4:8-9	195
17:23	120, 121, 122, 123, 126		
17:24	116, 121, 122	*Testament of Solomon*	
17:25	122, 258	1:6-7	289
17:26-46	121, 122, 124, 224	20:1-26:8	289
17:26	118, 313		
17:28	124	*Tobit*	
17:30	124, 225	1:16-17	226
17:31-32	124, 152	4:16	226
17:33	119, 120, 122	13:3	195

4. Dead Sea Scrolls

CD-A

7:18-22	180, 186
11:13-14	237
12:22-13:7	148, 169
13	170, 171, 273
13:7-12	128, 145, 146-149, 293, 308
13:7	147
13:9-10	147, 149, 151, 324, 325
13:9	148, 170, 388
13:11-12	147, 151
13:22	146, 152, 294
14:3-4	57

CD-B

19:7-13	149-151, 171, 172, 180, 194, 196, 244
19:7-9	149, 194, 201, 126, 128, 146
19:10-35	150, 193
19:10-11	149, 150, 194
19:10	147, 153
19:20	150, 194
19:21	150
19:23	150, 194
19:28	150
19:33	194
19:35	150

1QM

7:4-6	320
12:7-9	320

1QS 258

1:22	206
2:1	206
6:12	148
6:20	148
9:11	149, 181

1Q28a (1QSa)

2:8-9	320

1Q28b (1QSb) 127, 141-143, 144, 169, 170, 181, 273, 367

3:1-6	127
4:22-28	127
5	201
5:20-29	140, 144, 145, 307
5:20-23	141, 142
5:21	141, 142, 360
5:22	142, 360
5:24-26	141, 142, 153
5:26-29	142, 153, 208, 360

1Q33 (1QM)

1:3	195
10-11	136
10:3	136
11:7	127, 186

4Q73 126

4Q74 126

4Q161 123, 126, 127, 137-139, 144, 151, 170, 181, 184, 273, 367

2-6, 12	137
2-6, 15	139, 181
2-6, 22	137
5-6, 2-3	137, 138
5-6, 14	138
8-10, 18	137, 153, 201
8-10, 21	138, 181
8-10, 22	139
8-10, 22-23	139

4Q174 123, 126, 127, 134-137, 139, 151, 169, 170, 171, 181, 273, 367

1-2, 1:1-5	134
1-2, 1:3-5	134
1-2, 1:6-9	136
1-2, 1:6	134, 137, 153, 367
1-2, 1:10-11	134
1-2, 1:10-17	136
1-2, 1:11	136
1-2, 1:11-12	135, 363, 367
1-2, 1:13	136, 201
1-2, 1:14	136
1-2, 1:15	136
1-2, 1:18-19	137

4Q175 127

4Q206 (4QEn^e)

2:16	112

13:5 195
13:8 257
14:5-7 268

Reference Index

4Q212		1-2, 4:10	129, 130
4:12-14	268	1-2, 4:14	130, 299
		1-2, 5-6	268
4Q246	127	1-2, 5	130, 295
		1-2, 5:6-14	129
4Q252	123, 126, 127, 131-134,		
	144, 170, 172, 184, 273,	*4Q521*	5, 128, 143-146, 169,
	367		170, 171, 181, 273, 295,
5:1-7	132, 133		367
5:1-3	132	2, 2:1-113	270-271
5:1	155	2, 2:1	143, 144, 145, 294, 307,
5:2	132		352
5:3-4	132, 133, 145, 360	2, 2:2	144, 145, 146
5:5-7	132, 133, 145, 148, 201,	2, 2:3-14	148, 201, 307
	294, 360	2, 2:3-4	146
		2, 2:5-13	144, 145, 146, 153, 325,
4Q285	123, 126, 127, 136, 139-		366
	141, 144, 151, 159, 170,	2, 2:7	143, 294
	181, 184, 273	2, 2:9-11	143, 294, 307
5:2	140	2, 2:10-14	144, 294, 307, 307, 352
5:3-4	140	2, 2:11	143
5:4	139, 140	2, 2:12-14	144, 388
6+4:2	140	2, 2:12	143, 144, 145, 148, 170,
6+4:5	140		324, 388
6+4:6	140	2, 2:13	143, 299
4Q375	127	*4Q541*	127
4Q376	127	*4Q558*	127
4Q377	127	*11Q13*	127
4Q385c	96	*11Q5 (11QPs^a)*	
		28:11	99, 112, 130
4Q386	96	28:11-12a	99, 178, 181, 366
1, 2:1-3	96		

4Q491 127

5. Philo and Josephus

4Q504	128-131, 151, 169, 170,	*Pilo*
	195, 273, 325	*De Agricultura*
1-2, 2:3-14	294	39-66	167
1-2, 2:14	130, 148, 170, 295, 324	39	167, 207
1-2, 3:8	130, 295	41	167, 207
1-2, 3:13	130	46	167
1-2, 4	131, 295	48	167, 207
1-2, 4:2-5:18	129	67	167, 207
1-2, 4:5-14	128-129, 295		
1-2, 4:6	130	*Josephus*	
1-2, 4:7	129, 130		
1-2, 4:6-8	129, 368	- *Antiquities of the Jews*	
1-2, 4:9	130	8:45-49 (8.2.5)	288, 289, 290

8:404	206	34:17	154
9:206	155	34:20	154, 299
12:145	320	34:21	154
13:293-298	261	34:23	155, 156, 157, 299
13.298	262	34:24	154, 157
13:301, 318	165, 166, 207	34:25	156
13:408-410	261	34:30	154, 157
15:417	320	34:31	157
17:41	262	35:15	157
18:2-6	262	36:5	154
20:97-98	338	36:6	154
29:167-172	338	36:9	154, 299
		36:20	154
- Jewish War		36:36	154
1:110-111	261, 262	36:37	154
2:4-5	338	36:21	157
2:128-133	258	36:22	157
2:165	237, 262	36:32	157
2:258-263	338	37:11	157
4:229	329	37:15-23	156
6:288-309	186	37:24	156
		37:25	156
		37:16-19	155, 156
		37:27	154

6. Rabbinic and Other Jewish Texts

Babylonian Talmud

b. Ker. 6b	61, 113	*Tg. Isa.*	
b. Mes. 114b	61, 113	28:1-6	195
b. Pes. 57a	155	53:8	195
b. Sanh. 8a	112		
b. Sotah 22b	267		
b. Yebad m. 61a	61, 113		
b. Yoma 39b	329		

7. Early Christian Writings

Clement Alexandria
Strom 1.9   96

Midrash
Psalm 29:1   308, 324

Didache
16:3   235

Mishnah

m. Bekarot 43b	265
m. Šabb 22.6	237
m. Taharot 7.6	265
m.Yoma 8:6	237

Gospel of Peter
6:21   330

Gospel of Thomas
107   240

Targums

Origen
Homilies on Jeremiah
18:9   96

Tg. Ezek.

34-37	153-157, 169, 170, 171
34:3	154, 299
34:8	154, 299
34:11	299
34:13	154, 299
34:14	154, 299
34:16	154, 299

Manichean Psalmbook
Psalm 239:5-6   96

8: Greek Texts

Aristeas
Ep. Arist.
15   166, 207
187-294   164

Euripides
Orestes 260   316

Homer

- *Iliad*
ii. 78-85   22
ii. 100-108   24
ii. 505-598   24
v.13-14   22
viii.527   316
xi.373   316

- *Odyssey*
xvii.248   316
xxii.35   316

Isocrates
- *Phil.*
107   161 5. 114, 116   163

- *Paneg.*
16   164
22   164
29   163
115   161

- *Ad Nic.* 1-8   163
10-11, 19   163
10-11, 27-29   163
10-11, 32-34   163
10-11, 58   163

Pausanias
Descr. 49:1-2   166, 207

Plato
Statesman
271-272 22

Greek Papiri
P. Oxy 1611. 42-46   168, 170, 172

Author Index

Abegg, M. G., 140, 148
Abraham, I., 259
Ackroyd, P. R., 55, 91, 159
Albl, M. C., 175, 287, 334, 343
Allen, L. C., 46
Allison, D. C., 1, 5, 11, 174-176, 179, 189,
　192, 198-199, 203, 206, 213-215, 220,
　222, 224,226-230, 235-237
Anno, Y., 215
Atkinson, K. R., 116-120, 121, 123, 125,
　127-128, 130, 133, 136-137, 139-141,
　202
Aus, R. D., 331
Bacon, B. W., 373-374, 376-379
Baltzer, D., 50
Banks, R., 359
Barr, D. L., 254, 374, 379
Barth, G., 304, 315, 360
Batto, B. F., 78, 374, 376
Bauer, D. R., 232, 373, 377, 379, 383
Baumgarten, A. I., 249
Baumgarten, J. M., 146, 148-149, 180,
　194, 201, 294, 293
Beare, F. W., 212
Beaton, R., 301, 312
Beker J. C., 12, 17
Beutler, J., 2
Billerbeck, P., 236
Blackburn, B., 291, 292
Blass, F., 303
Block, D. I., 41, 42, 48, 63, 64, 69, 277,
　308
Boadt, L., 40, 42, 43, 47, 50
Böcher, O., 211, 234
Bock, D. L., 252, 309
Bockmuehl, M. A., 113, 140
Bolyki, J., 254
Borg, M. J., 237, 264, 299, 320
Bosch, D., 212
Boyarin, D., 52, 251
Bracewell, R. E., 1, 2
Brandenburger, E., 157, 158
Brawley, R. L., 252
Breasted, J. H., 21

Broer, I., 220
Brooke, G. J., 132, 133, 147
Brooks, O. S., 147, 219
Brown, R. E., 175, 199, 330, 343
Brown, S., 213, 215, 343
Bryan, D., 101, 102, 113, 114, 195, 196
Bryan, S. M., 195, 196, 268
Bruner, F. D., 186
Budd, P. J., 27
Bultmann. R., 252, 253, 291, 341
Burger, C., 4, 281, 285, 291, 292
Bussby, F., 240, 241
Butterworth, M., 84, 85
Byrskog, S., 219, 362, 363
Campbell, J. G., 149
Carson, D. A., 183, 184, 210, 211, 213,
　215, 228, 234, 237, 250, 254, 255,
　267, 280, 305, 314, 322
Carter, W., 182-185, 198, 203, 210, 215,
　224, 235, 240, 299, 306, 309, 310,
　314, 320
Catchpole, D. R., 224, 226
Cathacart, K. J., 333
Charette, B., 280
Charlesworth, J. H., 35, 96, 99, 119, 126,
　129, 141, 142, 146
Charon, E. G., 128, 130, 131
Childs, B. S., 3, 30
Chilton, B. D., 265, 270, 288, 303, 304
Christudhas, M., 352
Coffman, R. J., 29
Collins, J. J., 4, 25, 95-97, 109, 116, 117,
　119, 120, 127, 129, 132, 133, 136,
　138-141, 144, 147, 149, 151, 152,
　159, 180, 277, 286, 294, 308, 318
Comber, J. A., 284, 296, 301
Combrink, H. J. B., 373, 374
Cook, S. L., 190
Cooke, G. A., 38, 42, 44, 46, 51, 59, 60,
　61, 65, 69, 71, 221
Cope, O. L., 287, 312
Corley, K. E., 251
Corsten, T., 166
Cousland, J. R. C., 208, 210-212, 214,

220, 221, 224, 239, 293, 297, 299, 302, 303, 308, 318
Crossan, J. D., 251, 256, 265, 266, 330
Davenport, G. L., 116, 120, 122, 124
Davies, M., 186
Davies, P. R., 181, 194
Davies, W. D., 1, 5, 174-176, 179, 189, 192, 198, 199, 203, 206, 213-215, 220, 222, 226-230, 235-237, 252-255, 260, 262, 263, 280, 288, 304, 309, 311, 314, 317, 319, 322, 355, 358, 359, 371, 373, 376
Debrunner, A., 303
Deines, R., 250, 262, 264, 268, 360
De Lacy, D. R., 250
Dempsey, C. J., 33
Derby, J., 88
De Roo, J. C. R., 131
Derrett, J. D., 240
Dimant, D., 96, 129, 131,
Dodd, C. H., 146, 184, 189, 287, 334
Douglas, M., 189, 265, 306
Donahue, J. R., 259
Donalson, T. L., 299, 343, 346, 346-348, 359, 362, 382
Driver, G. R., 20
Duguid, I. M., 44, 60, 65, 66, 70, 72, 73, 76, 85, 86, 197
Duling, D. C., 4, 5, 28285-287, 289, 290, 296, 299, 300, 306, 309, 313, 317, 319, 323,
Dunn, J. D. G., 254, 257, 264, 265, 268, 280, 281, 309, 363
Eckart, K. G., 96
Eichrodt, W., 39, 42, 56, 59, 210, 221
Eisemann, M., 154
Eissfeldt, O., 159
Eslin, M. S., 373
Evans, C. A., 8, 10, 11, 17, 186, 195, 214, 268, 294, 338
Feldman, L. H., 268
Findlay, J. A., 373
Fishbane, M., 7
Flint, P. W., 126, 147
France, R. T., 198, 206, 208, 210, 214, 215, 223, 241, 306, 310, 314, 322, 364
Frankemölle, H., 177, 212, 315, 364, 365
Frankfort, H., 23
Freedman, D. N., 32, 33
Fröhlich, I., 98, 99, 101

Funk, W., 256
García Martínez, F., 96, 99, 120, 127-129, 131-135, 138-141, 143-150, 175, 178, 180, 190, 293, 294, 308
Garland, D., 197
Gerhardsson, B., 1, 4, 282
Gibbs, J. M., 187, 281, 279, 285
Gibson, J., 257
Gnilka, J., 176, 179, 183, 199, 208, 214, 220-224, 274, 305, 309, 316, 319-321, 330, 332, 342, 346, 355
Goldman, B., 331
Good, D. J., 161, 163
Goodman, M., 261
Gordon, R. P., 177
Gowan, D. E., 268
Grassi, J., 331, 332
Gray, G., 226, 227
Green, H. B., 254, 290, 374, 379
Green, W. S., 121, 129
Greenberg, M., 39, 42-44, 49, 51, 54, 62, 69, 70, 331, 332
Gruen, E. S. 162, 165
Guelich, R. A., 236
Gundry, R. H., 3, 10, 174, 183, 192, 198, 210, 215, 220, 221, 224, 226, 228, 235, 236, 242
Hagner, D. A., 174, 178, 183, 205, 206, 215, 226, 227, 252, 304, 309, 317, 319, 320
Hahn, F., 187, 213, 341, 343
Hallbäck G., 159
Hals, R. M., 39, 41, 42, 58, 69, 70, 71, 73, 263
Hammershaimb, E., 38, 39
Hanson, P. D., 81, 85
Harrelson, W. J., 3, 43
Hayes, W. C., 23
Hays, R. B., 8, 12, 14-17
Head, P. M., 5, 294
Heater, H., 174, 176, 177, 179
Heintz, J.-G., 35
Held, H. J., 304, 315
Hempel, C., 146, 147
Hengel, M., 126, 250
Hill, D., 183, 197, 211, 220, 221, 223, 224, 234, 241, 306, 310, 315, 328, 330
Holladay, C., 292
Hooker, M. D., 187
Horsely, R. A., 256

Hubbard, B. J., 277, 342, 345, 346, 348
Hubbard, R. L., 62
Hultgren, A. J., 253
Hutton, D., 330, 328
Jeremias, J., 20, 26, 69, 192, 206, 212, 220, 240, 241
Jones, D. R., 82
Jones, J. M., 318, 319, 388
Joyce, P., 52, 53
Kaufmann, Y., 38
Kautzsch, E., 48, 49
Kearney, P. J., 330
Keener, C. S., 198, 211, 213, 215, 220-222, 226-229, 234, 237, 240, 252, 258, 260, 280, 304, 305, 309, 314, 316, 317, 320, 321, 329, 342, 351
Keesmaat, S. C., 7, 8, 12, 14, 15
Kingsbury, J. D., 187, 188, 254, 281, 283, 285, 341, 345, 346, 373, 376, 377, 383
Klausner, J., 119
Klein, R. W., 72, 264
Knibb, M. A., 268
Knowles, M. P., 157
Köster, H., 206, 208
Korpel, M. C. A., 20, 23, 27, 30
Köstenberger, A. J., 128, 224
Kramer, S. N., 21
Kraus, H-J., 28
Krealing, C. H., 331
Kupp, D. D., 364, 365, 369
Kvalbein, H., 283, 341, 354-356, 357
Laato, A., 85, 59
Landmesser, C., 221, 272, 273
Lang, B., 275
Larkin, K. J., 78-83, 89, 190
Lemke, W., 63
Leske, A. M., 287, 300
Levenson, J. D., 38, 40, 42, 44, 59, 74
Levey, S. H., 153-157, 216
Levine, A., 27
Leyerle, B., 265
Lim, T. H., 138, 140, 149
Lindars, B., 10, 111-112, 239
Litwak, K. D., 8-10, 12, 17
Loader, W. R. G., 282
Luckenbill, D. D., 21
Lust, J.126, 178, 179, 183
Luz, U., 174-179, 183, 185-187, 192, 193, 197-199, 203, 205, 208, 210, 213-215, 228, 234-236, 238, 240, 242, 284, 285, 290, 303-305, 310, 314, 317, 318, 320, 231, 342-345, 352-357, 362, 364, 367
Mack, B. L., 121
Malina, B. J., 348, 342, 344
Manus, C., 342
Mason, R. A., 79, 85, 197
Manson, R., 32
Martin, F., 1, 3, 6, 385
Mays, J. L., 26, 34, 37, 63, 177
McNight, S., 2, 217, 227, 264, 265, 273
Meeks, W. A., 187, 279, 332
Meier, J. P., 211, 215, 261, 276, 277, 288, 290, 291, 304, 305, 309, 313, 314, 317, 341, 344, 345, 360
Mercer, A. B., 23
Michel, O., 341, 342, 348
Milik, J. T., 97, 131, 141
Miller, J., 177
Moessner, D. P., 266, 275
Moo, D. J., 189, 190, 192, 360
Morgan, T. E., 12, 13
Moxnes, H., 266
Muilenburg, J., 6, 8
Mullins, T. Y., 282
Müller, M., 187, 317
Murphy-O' Connor, J., 149
Meyers, C. L., 77, 78, 79, 82
Myers, E. M., 77, 78, 79, 82
Neale, D. A., 257
Neusner, J., 107, 121, 129, 250, 257, 265, 268, 275
Nickelsburg, G. W. E., 97-116
Novakovic, L., 286, 287, 305, 306, 308, 316, 319,
Oegema, G. S., 116, 117, 125
Orlinsky, H. M., 55
Osborne, G., 6
Paffenroth, K., 290, 311, 323
Patte, D., 198
Patton, C., 51
Perkins, P., 361
Perrin, N., 269, 256
Person, R. F., 83
Pesch, R., 199, 200
Peters, E. E., 160-162
Peterson, W. L., 240
Peterson, D. L., 83, 89
Petrotta, A. J., 174, 184
Pilch, J. J., 272, 306
Pomykala, K. E., 116-126, 127, 129, 132-

Author Index

136, 137, 139-141
Porter, S. E., 8, 12-14, 189, 234
Pritchard, J. B., 20, 24
Puech, É., 138, 143, 271, 294, 307
Rajak, T., 165
Reid, S. B., 104, 105
Renz, T., 39, 41-43, 45-47, 61, 65, 67-71
Repschinski, B., 237, 252, 263
Riesenfeld, H., 331, 333
Robbins V., 7
Rudolph, W., 79
Russell, E. A., 314
Saldarini, A. J., 227, 250, 262, 267, 286, 288, 292
Sanders, E. P., 196, 218, 220, 249, 250, 252, 256, 257, 261, 262, 264, 265, 320
Sanders, J. A., 7, 8, 11, 99, 178, 180
Schaper, J., 263
Schlatter, A., 183, 221, 252, 263
Schnabel, E. J., 112, 119, 166, 208, 218, 341, 354, 355
Schniedewind, W. M., 295
Schroeder, R. P., 191, 193
Scott, B. B., 240, 241
Senior, D., 203, 210, 329, 330, 333
Silva, M., 174
Sim, D. C., 224, 268, 277, 286, 335
Skarsaune, O., 283, 357
Smith, D. E., 251, 257, 265, 266
Smith, R. R., 162
Smith, S. H., 287, 288
Smyth, K., 345
Speiser, E. A., 177
Sperber, A., 154, 156, 216
Stanley, C. D., 8, 13, 14
Stanton, G., 4, 213, 237, 248, 250, 277, 286-281
Stauffer, E., 320
Steck, O. H., 268
Stein, R. H., 203
Stendahl, K., 174, 179
Stevenson, R., 42, 43
Steudel, A., 136, 147
Stone, M. E., 159
Strack, H. L., 236
Strecker, G., 213, 281, 346, 375, 376
Strong, J., 54, 55
Stuhlmacher, P., 119, 213, 341, 354-357
Suhl, A., 302
Talmon, S., 352
Tate, W. R., 7

Tatum, W. B., 280
Thiselton, A. C., 7, 322
Thompson, A. L., 158
Thompson, J. G., 2, 13, 14
Thompson, M. M., 374
Thompson, W. G., 305, 374, 376
Tigchelaar, E. J. C., 84, 89, 96, 99, 129, 134, 135, 143, 146, 147, 150, 178, 180
Tiller, P. A., 98-100, 103, 107, 110-112, 180
Trilling, W., 342
Tooley, W., 2
Trafton, J. L., 116
Trautmann, M., 212
Trotter, L. L., 288
Troxel, R. L., 329, 330, 334, 335, 338, 339
Uro, R., 214, 215, 218
Van Aarde, A. G., 275, 364, 373, 374
Van Buren, E. D., 23
Vancil, W. J., 2, 20, 22-27, 96, 98, 98, 102, 103, 128, 168
Vanderkam, J., 138, 149, 150, 152
VanGemeren, W. A., 41, 62, 277
Vanhoozer, K. J., 7, 39
Van Groningen, G., 39, 40, 62, 68, 69, 89
Vermes, G., 140
Verseput, D. J., 187, 188, 238, 268, 280, 282, 285, 305, 306, 343
Vledder, E.-J., 211, 212, 252
Von Dobbeler, A., 208, 215, 355, 356
Von Rad, G., 68
Von Wahlde, U. C., 262
Wacholder, B. Z., 96, 131, 138
Wagner, J. R., 8, 9, 12, 14
Wainwright, E., 316
Walbank, F. W., 160-163, 166
Weaver, D. J., 211
Weber, K., 221, 225
Weinfeld, M., 267
Wenham, D., 329
Wenham, G., 76, 177
Wenham, J. W., 329
Westerholm, S., 360
White, S. A., 149
Wilcox, M., 10
Willett, T. W., 157, 158
Wilson, J., 23-25
Winninge, M., 257
Wise, M. O., 141, 144
Witherington, B., 211, 237, 304, 309
Witherup, R. D., 338

Wolff, Hans W., 32-34, 36, 177, 185
Wright, N. T., 16, 192, 195, 196, 218, 256, 265, 267, 268, 272, 275, 299
Zimmerli, W., 38, 39, 51, 54-64, 69-71, 74, 113, 147, 242, 263, 293, 331, 332
Zimmermann, F., 127, 128, 131-135, 137, 139, 140, 143, 144, 145, 157
Zimmermann, J., 133, 134, 135, 137, 139, 140, 143, 144, 183, 184, 196, 218, 271, 307, 358

Subject Index

Abraham, 187, 279, 285, 357
Adam, 97, 98, 113, 114, 157, 216
Amazement, 255, 284, 299, 305, 307, 308, 310, 313, 322, 378
Angel, 88-91, 97-99, 103-106, 114-119, 122, 170, 171, 196, 197, 202, 203, 209, 231, 346
Anointed one, 71, 143, 202, 270, 271, 295, 300, 336, 348, 355, 366
Antiochus IV Epiphanes, 161, 162
Atonement, 74, 87, 273, 367
Authority, 20-30, 43-44, 65, 71, 73, 104, 110, 120, 139, 143-148, 151, 153, 169, 171, 178, 181-181, 188, 214-219, 235, 245, 263, 270, 274, 284, 289, 302, 306-308, 310, 312, 319, 321, 326, 342, 433, 348, 350-354, 359, 360-370, 378-379, 382-384
Bar Kokba, 125
Battle [eschatological], 34-37, 59, 83, 88, 93, 105, 108-110, 114, 116, 120, 135-137, 140, 142, 337
Blindness, 282, 295
Branch of David, 30, 60, 108, 123, 127, 132-156, 170, 181, 184, 201, 294, 360, 367
Canaanite woman, 212, 313-315, 358
Chief priests, 188, 191, 211, 262, 300, 320
Child/Children, 187, 275, 280, 283, 290, 302, 303, 319, 322, 325
Christ, 1, 111, 187, 283, 300, 330, 342, 343, 348, 379
Christological titles
- Branch of David, 132
- Christ, 187, 188, 280
- Immanuel, 383
- King, 50, 67, 167, 239
- Lord, 281
- Lord of the sheep, 107
- Messiah, 152, 153
- Prince, 50, 66, 75, 99, 137, 138, 141, 170
- Ruler, 99
- Shepherd, 20, 22, 24, 26, 45, 67, 69, 106, 167
- Servant, 69
- Son of David, 4, 5, 178, 181, 188, 239, 279-282, 284, 285, 287, 296, 300, 305, 306, 308, 323, 377, 378, 388, 389
- Son of God, 283, 376, 339
- Son of Man, 281, 342
Christology, 1, 4, 5, 16-18, 175, 188, 225, 246, 283, 302, 310, 314, 325
Church, 203, 228, 235, 240-241, 248, 251, 267, 283, 287-288, 296, 300, 329, 334, 341, 343, 364, 375, 379, 384, 388, 391, 393
Cleansing of sins, 42, 59, 81, 106
Compassion, 1, 2, 5, 6, 30-32, 60, 65, 69, 75, 77, 87, 88, 91-93, 97, 116, 120, 123-126, 147, 163, 164, 167, 169, 171, 188, 193, 196, 205-214, 222-227, 231-233, 238, 239, 245, 270-274, 284, 285, 293, 294, 298, 299, 306-308, 311, 312, 316-321, 326, 340, 352, 353, 361-367, 378, 389, 381-385, 389-395
Conflict, 87, 92, 95, 125, 137, 182, 190-194, 198, 204, 211, 212, 235, 236, 239, 248-278
Covenant, 28-32, 38-40, 42-92, 99, 123-130, 138, 141-149, 152, 178, 190-199, 204, 229, 249, 256-270, 299, 320, 322, 349, 353, 364
Covenantal nomism, 249
Crowd, 205-214, 220, 233, 239, 252, 255, 260, 270, 281, 284, 287, 292, 296-319, 322, 325, 339, 340, 347, 367, 378, 379
Crucifixion, 284
David
- Davidic dynasty, 30, 45, 89, 95, 125-127, 175, 183, 300
- Davidic expectation, 4, 19, 25, 29-31, 37, 51, 74, 88-89, 93, 95-99, 112, 116, 124, 126-128, 131-137, 145, 146, 149, 151-160, 184, 186, 188, 204, 273, 277, 280, 293, 295, 299, 388
- David's son, 3-6, 16, 67, 179, 183, 184, 187, 188, 201, 222, 239, 248, 254-255, 277-325, 328, 352, 356, 357, 363, 365,

371, 375, 376-394
Death of Jesus, 283, 291, 328, 339, 349, 351, 352, 370, 375, 376, 382
Deuteronomic
- Exodus-Deuteronomic framework, 31, 113, 365
- Deuteronomistic sin-exile-return pattern, 130, 267, 280
Disciples, 191, 194, 198, 199, 208, 211, 213, 226-228, 233, 236, 239, 241, 245, 252, 255, 263, 283, 314, 319, 340-370, 380, 384
- disciples as sheep, 212
- disciples as [under-]shepherds, 196, 218, 231, 273
- discipleship, 198
- Twelve disciples, 195, 196, 212, 215, 217, 218, 301, 340-370
Elijah, 11, 99, 267, 276, 322
Eschatological
- battle / war, 34, 35, 83, 88, 105, 108, 109, 120, 132, 135, 337
- blessings / bliss, 84, 197, 348
- Davidic son, 6, 152, 201, 223, 286
- drama, 106, 108, 133, 188, 360
- gathering, 32, 33, 45, 338
- healer, 149
- judgement / confrontation, 3, 30, 35, 38, 86, 107, 108, 123, 137, 140, 302, 354
- Israel / community, 1, 43, 45, 66, 115, 119, 120, 133, 137, 144, 152, 155, 157, 171, 181, 232-293, 361, 367
- mission, 75, 188, 216, 308, 385
- pattern, 139, 151, 158, 202, 380
- obedience, 169, 380
- Provider, 154, 299
- rest / peace, 160
- restoration, 33, 35, 66, 88, 110, 114, 120, 150, 154, 159, 196, 202, 269, 294, 341, 346, 355, 375
- prophet, 276
- reversal, 75, 91, 381, 385
- Teacher, 124, 133, 136, 171, 360, 363, 367
- Temple / sanctuary, 3, 134, 368
- theocracy, 91, 277, 363, 380
- transformation / sanctification, 101, 135, 137, 142, 196, 217
- white bull, 100, 101, 106, 107, 109, 110, 114, 119, 142, 360
Exile, 19, 26, 30, 34, 39-41, 44, 46, 49, 50, 52, 53, 55, 57, 60, 62, 64, 65, 74, 75, 81, 83, 91-93, 100, 116, 130, 131, 152, 154, 169, 170, 186, 193-198, 208, 209, 223, 255, 267-269, 277, 280, 285, 297, 306, 321, 325, 331, 347, 349, 377, 380, 384
Exorcist, 4, 5, 211, 285-292, 303, 305, 319, 323, 325, 385
Faith, 292, 305, 315, 317
Faithful / unfaithful, 20, 21, 29, 30, 36, 43, 47, 48, 53, 57, 68, 93, 109, 121, 144, 145, 147, 165, 170, 193, 224, 290, 299, 338, 364
Forgiveness, 29, 31, 150, 243, 256, 264, 378
Fulfilment, 57, 187, 210, 361, 383, 384
Formula citation / expression, 8, 12, 13, 174, 189, 365, 375
Genealogy, 187, 300, 377
Gentiles, 79, 101, 124, 142, 165, 185-187, 199, 200, 213, 220, 224, 225, 257, 258, 260, 266, 272, 292, 320
God's zeal for his name, 38, 40, 52, 54, 57, 75, 76, 92, 208, 261, 264-269, 339, 378, 391, 394
Grace, 21, 81, 251, 264
Healing, 1-6, 17, 24, 25, 29, 31, 64, 65, 77, 90-92, 96, 130, 131, 143, 145, 148, 151, 170, 188, 205-222, 236-279, 281-326, 357, 363, 365, 370, 371, 378-380, 388, 391, 394
Jerusalem, 34, 36, 38, 39, 41, 50, 53, 56, 59, 76, 78, 81, 83-89, 95, 99, 102, 106, 116, 117, 124, 125, 131, 157, 182-188, 198, 202, 210, 262, 282, 316-326, 333-338, 344, 353, 365, 382
Jesus
- Jesus' birth, 183, 186, 211, 367, 383
- Jesus' identity, 1, 15, 202, 205, 214, 222-227, 230, 231, 239, 244, 273, 278-280, 293, 299, 301, 305, 307, 309-314, 319, 321-324, 346, 358, 363, 371, 379, 382, 385
- Jesus' mission of compassion, 211, 227, 231, 271, 284, 295, 307, 308, 311, 327, 352, 353, 362, 363, 378, 381, 384, 393, 395
- Jesus' death, 191, 197, 198, 246, 283, 291, 328, 330-339, 349, 351-370, 375, 376, 382
- Jesus' resurrection, 191, 199, 200, 203,

Subject Index

291, 328, 330, 334, 335, 337, 351, 352, 365, 370, 375, 376
Judge
- Shepherd-Judge, 61, 63, 80, 99, 148, 219, 232, 247, 283-285, 298, 310, 354, 359, 361, 379, 371, 381
- Judgement, 28, 30, 59, 65, 82, 90, 91, 104, 109, 114, 115, 118, 139, 147, 150, 154, 196, 259, 272, 324, 349, 352, 353, 360, 380
Hostility, 311, 378, 380
Humility, 284, 312, 365
Infancy narrative, 187, 280, 384, 374, 377, 386
Israel [see *eschatological Israel*]
Kingdom
-of God, 6, 20, 141,143, 212, 246, 270, 278, 298, 342, 359, 364, 374
- of Israel, 20, 46, 68, 70, 80, 99, 116, 121, 165, 302
Kingship, 19, 20-24, 26, 29, 31-36, 42-45, 51, 72, 92, 95, 132, 138, 151, 153, 159-170, 182, 207, 222, 223, 338, 353
Law, 39-58, 71-75, 76, 81, 88, 92, 95, 100, 112, 115-122, 130-137, 142, 152-170, 176, 180, 207, 215, 225, 234-243, 249, 257, 258, 261-265, 273, 290, 300, 312, 320, 338, 348, 350, 356, 359-362, 369, 375, 381, 382
Lead, 26, 35, 51, 59, 62, 77, 93, 174, 200-207, 297, 299, 313, 346, 352, 384, 386
- leader, 27, 29, 32, 33, 37, 40, 45, 50, 60-65, 71-75, 85, 178, 180-184, 210-212, 238, 239, 261, 297, 299, 360, 368, 379
- Pharisee's leadership/Jewish leaders, 261-264, 283, 284, 296, 312
- Shepherd-leadership, 22, 24, 28, 31, 37, 52, 60, 75, 80, 86, 90, 91, 92, 182, 190, 217, 244, 337, 352, 359
Lord
- as a reference to YHWH, 28-35, 42, 42, 48, 50, 53, 54, 56, 61, 69, 71, 76, 91-94, 96, 97, 121, 122, 146, 168-171, 202, 203, 224, 235, 294, 295, 307, 308, 321, 330
- as the Lord of the sheep, 99-106, 196, 197, 209, 217, 231, 297, 360
- as a reference to the Son of David, 313
- as a reference to the Son of Man, 214
- as a reference to Jesus, 282, 285, 300, 302, 306, 314

- as a reference to other gods, 20, 21
Messiah, 4, 5, 223, 224, 277, 279, 281, 282-288, 291-298, 300-311, 315, 317, 318, 322-326, 374
Miracles, 232, 237, 238
Mission, 185, 193, 197, 199, 211-219, 226-233, 243-245, 254, 255-258, 266, 274, 278, 280, 284, 294, 306, 311, 315- 317, 321, 342, 352, 354, 356, 358, 361-371
Monarchy [Davidic], 26, 52, 73-77, 83, 89, 91, 95, 152, 159
- Hellenistic monarchy, 160-167
Moses, 11, 375, 378, 379
Name
- God's zeal for his name, 52-56, 339, 340, 349-371
- divine name, 52, 350, 358
Nathan, 37, 89
Nations, 90-93, 171, 244-246, 326, 379-384, 391-395
Narrative, 3, 5, 9, 16, 18, 77, 182, 184, 186, 191, 204, 210, 248, 252, 310, 372-386
New covenant, 345, 353
Obedience to the law, 94, 356, 360, 362, 370
Passion narrative, 173, 189, 190-193, 204, 246, 252, 334, 335, 337, 347, 352, 356, 374, 377, 382, 386
Pharisees, 211, 212, 217, 234, 236-238, 244, 261, 275, 248-278, 296, 300, 308-313, 319
Peace, 24, 30, 37, 40, 43, 46-48, 57, 76, 79, 86, 90, 94, 114, 129, 142, 159, 163-169
Presence [divine], 2, 6, 32, 34, 41-43, 47, 48, 52, 53, 57, 59, 70, 75-77, 88, 91, 92, 110, 118, 133-135, 137, 152, 153, 172, 230, 236, 280, 297, 302, 315, 321
Preaching, 1, 205, 218, 255, 296, 301, 307, 363, 378, 379
Prince, 30, 43-50, 60, 66, 67, 69, 71-74, 96, 99, 112, 114, 123-156, 169, 170, 176, 178, 180, 181, 183, 186, 197, 201, 216, 217, 245, 263-265, 275, 293-295, 300, 308, 324, 348, 350, 356-367, 369, 370, 379, 382-385
Promise, 6, 10, 15
Provider, 154, 156, 241, 299
Rejection of Jesus, 185, 188, 190
Restoration, 17, 94, 159, 165, 169, 170, 171, 239-241, 243-245, 385
Resurrection of the saints, 269, 331-335,

337, 338, 329, 340, 375
Reversal [Divine], 247-278, 381, 385
Righteousness,380, 381, 383
Rod [the iron], 290
(see also, *scepter* and *staff*)
Sabbath, 236-239, 244, 311, 312
Salvation, 95, 110, 114, 144, 160, 364, 367, 374, 375, 377, 383, 384, 385
Scepter, 23, 25, 27, 28, 35, 71, 72, 108, 109, 116-120, 127, 132, 133, 136, 142, 149, 151, 155, 159, 168, 171, 176, 180, 186
Scheme, 18, 73, 97, 102, 205, 210, 281, 283, 371, 373, 375, 376, 392
Scribes, 175, 188, 211, 262, 263, 320
Servant, 29, 44, 45-48, 64, 66-72, 75, 112, 118, 138, 151, 156, 222, 239, 255, 283, 286, 287, 289, 292, 299-301, 308, 310, 344, 347-353, 358, 364, 381
Sheep
- the six groups of the sheep in exile, 63, 270, 380
- the lost sheep of the house of Israel, 187, 188, 205, 208, 211, 212-219, 227, 229, 230, 232, 233-236, 239, 240, 242, 245, 253, 255, 256, 260, 266, 275, 284, 301, 306, 314-316, 324, 350, 356, 362, 363, 375, 380, 381
Shepherd
- good shepherd, 21, 58, 80, 85, 87, 95, 103, 114, 167, 189, 190, 191, 197, 207, 222
- shepherd's crook (see also *staff*, *scepter* and *rod*), 22-24, 27
- Shepherd-Judge, 219, 222, 227, 230, 231, 247, 361, 381, 393
- smitten shepherd, 191-197, 202
- worthless shepherd, 77, 80, 83, 83, 105, 190, 191, 193
Sinners, 261
Sin(s), 29, 31, 50, 56, 68, 74, 82, 86, 92, 105, 106, 110, 114-117, 130, 134, 150, 151, 158, 194, 197, 258, 273, 274, 363, 377, 378

Sinless Messiah, 120, 363, 364
Solomon, 4, 5, 98, 99, 114-116, 125, 286, 288-292, 303-305, 310, 311, 319, 323, 325, 385, 387, 388
Spirit, 46, 47, 51, 70, 75, 81, 82, 84, 92, 165, 171, 269, 277, 311-313, 338, 363, 394
Staff, 93 (see also *scepter, crook* and *rod*)
Star, 149, 175, 180, 186
Suffering
- of Jesus. 376
- of sheep, 64, 102, 103, 105, 115, 158, 170, 209, 228, 268, 293
- of shepherd, 3, 76, 82, 85, 87, 90, 92, 190, 197, 202, 278, 337, 352, 353, 382
- of servant, 64, 102, 103, 105, 115, 158, 170, 209, 228, 268, 293
- of the Son of God, 338
- of the Son of Man, 284, 289
- of YHWH, 85, 87, 189
Synoptics, 1, 226
Teacher, 119, 120, 123-125, 133, 135-137, 142, 145, 152, 170, 171, 272, 273, 324, 356, 359, 360-367, 377-385, 389-391, 394
Teaching, 2, 273, 301, 350, 378, 390
Temple, 320-322
Theocracy, 74, 75, 91, 169, 188, 273, 277, 302, 324, 388
Titular,
- approach, 141, 153, 302, 325, 343
- use, 26, 31, 69, 90, 141, 295, 303, 321, 322
Truth, 4, 116, 164, 171, 172, 361
Therapeutic Son of David, 5, 15, 16, 319, 322, 325, 326, 375, 376, 378, 382, 385, 386
Unbelief, 284, 296
War (see *battle, eschatological*)
Zion, 28, 32, 34, 35, 45, 50-52, 108, 129, 158-160,318, 388, 346-349, 359, 369, 382

Wissenschaftliche Untersuchungen zum Neuen Testament
Alphabetical Index of the First and Second Series

Ådna, Jostein: Jesu Stellung zum Tempel. 2000. *Volume II/119.*
Ådna, Jostein (Ed.): The Formation of the Early Church. 2005. *Volume 183.*
– and *Kvalbein, Hans* (Ed.): The Mission of the Early Church to Jews and Gentiles. 2000. *Volume 127.*
Alkier, Stefan: Wunder und Wirklichkeit in den Briefen des Apostels Paulus. 2001. *Volume 134.*
Anderson, Paul N.: The Christology of the Fourth Gospel. 1996. *Volume II/78.*
Appold, Mark L.: The Oneness Motif in the Fourth Gospel. 1976. *Volume II/1.*
Arnold, Clinton E.: The Colossian Syncretism. 1995. *Volume II/77.*
Ascough, Richard S.: Paul's Macedonian Associations. 2003. *Volume II/161.*
Asiedu-Peprah, Martin: Johannine Sabbath Conflicts As Juridical Controversy. 2001. *Volume II/132.*
Avemarie, Friedrich: Die Tauferzählungen der Apostelgeschichte. 2002. *Volume 139.*
Avemarie, Friedrich and *Hermann Lichtenberger* (Ed.): Auferstehung – Ressurection. 2001. *Volume 135.*
– Bund und Tora. 1996. *Volume 92.*
Baarlink, Heinrich: Verkündigtes Heil. 2004. *Volume 168.*
Bachmann, Michael: Sünder oder Übertreter. 1992. *Volume 59.*
Bachmann, Michael (Ed.): Lutherische und Neue Paulusperspektive. 2005. *Volume 182.*
Back, Frances: Verwandlung durch Offenbarung bei Paulus. 2002. *Volume II/153.*
Baker, William R.: Personal Speech-Ethics in the Epistle of James. 1995. *Volume II/68.*
Bakke, Odd Magne: 'Concord and Peace'. 2001. *Volume II/143.*
Baldwin, Matthew C.: Whose *Acts of Peter?* 2005. *Volume II/196.*
Balla, Peter: Challenges to New Testament Theology. 1997. *Volume II/95.*
– The Child-Parent Relationship in the New Testament and its Environment. 2003. *Volume 155.*
Bammel, Ernst: Judaica. Volume I 1986. *Volume 37.*
– Volume II 1997. *Volume 91.*
Bash, Anthony: Ambassadors for Christ. 1997. *Volume II/92.*

Bauernfeind, Otto: Kommentar und Studien zur Apostelgeschichte. 1980. *Volume 22.*
Baum, Armin Daniel: Pseudepigraphie und literarische Fälschung im frühen Christentum. 2001. *Volume II/138.*
Bayer, Hans Friedrich: Jesus' Predictions of Vindication and Resurrection. 1986. *Volume II/20.*
Becker, Eve-Marie: Das Markus-Evangelium im Rahmen antiker Historiographie. 2006. *Volume 194.*
Becker, Eve-Marie and *Peter Pilhofer* (Ed.): Biographie und Persönlichkeit des Paulus. 2005. *Volume 187.*
Becker, Michael: Wunder und Wundertäter im früh-rabbinischen Judentum. 2002. *Volume II/144.*
Becker, Michael and *Markus Öhler* (Ed.): Apokalyptik als Herausforderung neutestamentlicher Theologie. 2006. *Volume II/214.*
Bell, Richard H.: The Irrevocable Call of God. 2005. *Volume 184.*
– No One Seeks for God. 1998. *Volume 106.*
– Provoked to Jealousy. 1994. *Volume II/63.*
Bennema, Cornelis: The Power of Saving Wisdom. 2002. *Volume II/148.*
Bergman, Jan: see *Kieffer, René*
Bergmeier, Roland: Das Gesetz im Römerbrief und andere Studien zum Neuen Testament. 2000. *Volume 121.*
Betz, Otto: Jesus, der Messias Israels. 1987. *Volume 42.*
– Jesus, der Herr der Kirche. 1990. *Volume 52.*
Beyschlag, Karlmann: Simon Magus und die christliche Gnosis. 1974. *Volume 16.*
Bittner, Wolfgang J.: Jesu Zeichen im Johannesevangelium. 1987. *Volume II/26.*
Bjerkelund, Carl J.: Tauta Egeneto. 1987. *Volume 40.*
Blackburn, Barry Lee: Theios Ane-r and the Markan Miracle Traditions. 1991. *Volume II/40.*
Bock, Darrell L.: Blasphemy and Exaltation in Judaism and the Final Examination of Jesus. 1998. *Volume II/106.*
Bockmuehl, Markus N.A.: Revelation and Mystery in Ancient Judaism and Pauline Christianity. 1990. *Volume II/36.*
Bøe, Sverre: Gog and Magog. 2001. *Volume II/135.*

Böhlig, Alexander: Gnosis und Synkretismus. Volume 1 1989. *Volume 47* – Volume 2 1989. *Volume 48.*
Böhm, Martina: Samarien und die Samaritai bei Lukas. 1999. *Volume II/111.*
Böttrich, Christfried: Weltweisheit – Menschheitsethik – Urkult. 1992. *Volume II/50.*
Bolyki, János: Jesu Tischgemeinschaften. 1997. *Volume II/96.*
Bosman, Philip: Conscience in Philo and Paul. 2003. *Volume II/166.*
Bovon, François: Studies in Early Christianity. 2003. *Volume 161.*
Brocke, Christoph vom: Thessaloniki – Stadt des Kassander und Gemeinde des Paulus. 2001. *Volume II/125.*
Brunson, Andrew: Psalm 118 in the Gospel of John. 2003. *Volume II/158.*
Büchli, Jörg: Der Poimandres – ein paganisiertes Evangelium. 1987. *Volume II/27.*
Bühner, Jan A.: Der Gesandte und sein Weg im 4. Evangelium. 1977. *Volume II/2.*
Burchard, Christoph: Untersuchungen zu Joseph und Aseneth. 1965. *Volume 8.*
– Studien zur Theologie, Sprache und Umwelt des Neuen Testaments. Ed. by D. Sänger. 1998. *Volume 107.*
Burnett, Richard: Karl Barth's Theological Exegesis. 2001. *Volume II/145.*
Byron, John: Slavery Metaphors in Early Judaism and Pauline Christianity. 2003. *Volume II/162.*
Byrskog, Samuel: Story as History – History as Story. 2000. *Volume 123.*
Cancik, Hubert (Ed.): Markus-Philologie. 1984. *Volume 33.*
Capes, David B.: Old Testament Yaweh Texts in Paul's Christology. 1992. *Volume II/47.*
Caragounis, Chrys C.: The Development of Greek and the New Testament. 2004. *Volume 167.*
– The Son of Man. 1986. *Volume 38.*
– see *Fridrichsen, Anton.*
Carleton Paget, James: The Epistle of Barnabas. 1994. *Volume II/64.*
Carson, D.A., O'Brien, Peter T. and *Mark Seifrid* (Ed.): Justification and Variegated Nomism.
Volume 1: The Complexities of Second Temple Judaism. 2001. *Volume II/140.*
Volume 2: The Paradoxes of Paul. 2004. *Volume II/181.*
Chae, Young Sam: Jesus as the Eschatological Davidic Shepherd. 2006. *Volume II/216.*
Ciampa, Roy E.: The Presence and Function of Scripture in Galatians 1 and 2. 1998. *Volume II/102.*
Classen, Carl Joachim: Rhetorical Criticsm of the New Testament. 2000. *Volume 128.*
Colpe, Carsten: Iranier – Aramäer – Hebräer – Hellenen. 2003. *Volume 154.*

Crump, David: Jesus the Intercessor. 1992. *Volume II/49.*
Dahl, Nils Alstrup: Studies in Ephesians. 2000. *Volume 131.*
Deines, Roland: Die Gerechtigkeit der Tora im Reich des Messias. 2004. *Volume 177.*
– Jüdische Steingefäße und pharisäische Frömmigkeit. 1993. *Volume II/52.*
– Die Pharisäer. 1997. *Volume 101.*
Deines, Roland and *Karl-Wilhelm Niebuhr* (Ed.): Philo und das Neue Testament. 2004. *Volume 172.*
Dettwiler, Andreas and *Jean Zumstein* (Ed.): Kreuzestheologie im Neuen Testament. 2002. *Volume 151.*
Dickson, John P.: Mission-Commitment in Ancient Judaism and in the Pauline Communities. 2003. *Volume II/159.*
Dietzfelbinger, Christian: Der Abschied des Kommenden. 1997. *Volume 95.*
Dimitrov, Ivan Z., James D.G. Dunn, Ulrich Luz and *Karl-Wilhelm Niebuhr* (Ed.): Das Alte Testament als christliche Bibel in orthodoxer und westlicher Sicht. 2004. *Volume 174.*
Dobbeler, Axel von: Glaube als Teilhabe. 1987. *Volume II/22.*
Dryden, J. de Waal: Theology and Ethics in 1 Peter. 2006. *Volume II/209.*
Du Toit, David S.: Theios Anthropos. 1997. *Volume II/91.*
Dübbers, Michael: Christologie und Existenz im Kolosserbrief. 2005. *Volume II/191.*
Dunn, James D.G.: The New Perspective on Paul. 2005. *Volume 185.*
Dunn, James D.G. (Ed.): Jews and Christians. 1992. *Volume 66.*
– Paul and the Mosaic Law. 1996. *Volume 89.*
– see *Dimitrov, Ivan Z.*
–, *Hans Klein, Ulrich Luz* and *Vasile Mihoc* (Ed.): Auslegung der Bibel in orthodoxer und westlicher Perspektive. 2000. *Volume 130.*
Ebel, Eva: Die Attraktivität früher christlicher Gemeinden. 2004. *Volume II/178.*
Ebertz, Michael N.: Das Charisma des Gekreuzigten. 1987. *Volume 45.*
Eckstein, Hans-Joachim: Der Begriff Syneidesis bei Paulus. 1983. *Volume II/10.*
– Verheißung und Gesetz. 1996. *Volume 86.*
Ego, Beate: Im Himmel wie auf Erden. 1989. *Volume II/34.*
Ego, Beate, Armin Lange and *Peter Pilhofer* (Ed.): Gemeinde ohne Tempel – Community without Temple. 1999. *Volume 118.*
– and *Helmut Merkel* (Ed.): Religiöses Lernen in der biblischen, frühjüdischen und früh-christlichen Überlieferung. 2005. *Volume 180.*
Eisen, Ute E.: see *Paulsen, Henning.*
Elledge, C.D.: Life after Death in Early Judaism. 2006. *Volume II/208.*

Ellis, E. Earle: Prophecy and Hermeneutic in Early Christianity. 1978. *Volume 18.*
- The Old Testament in Early Christianity. 1991. *Volume 54.*

Endo, Masanobu: Creation and Christology. 2002. *Volume 149.*

Ennulat, Andreas: Die 'Minor Agreements'. 1994. *Volume II/62.*

Ensor, Peter W.: Jesus and His 'Works'. 1996. *Volume II/85.*

Eskola, Timo: Messiah and the Throne. 2001. *Volume II/142.*
- Theodicy and Predestination in Pauline Soteriology. 1998. *Volume II/100.*

Fatehi, Mehrdad: The Spirit's Relation to the Risen Lord in Paul. 2000. *Volume II/128.*

Feldmeier, Reinhard: Die Krisis des Gottessohnes. 1987. *Volume II/21.*
- Die Christen als Fremde. 1992. *Volume 64.*

Feldmeier, Reinhard and *Ulrich Heckel* (Ed.): Die Heiden. 1994. *Volume 70.*

Fletcher-Louis, Crispin H.T.: Luke-Acts: Angels, Christology and Soteriology. 1997. *Volume II/94.*

Förster, Niclas: Marcus Magus. 1999. *Volume 114.*

Forbes, Christopher Brian: Prophecy and Inspired Speech in Early Christianity and its Hellenistic Environment. 1995. *Volume II/75.*

Fornberg, Tord: see *Fridrichsen, Anton.*

Fossum, Jarl E.: The Name of God and the Angel of the Lord. 1985. *Volume 36.*

Foster, Paul: Community, Law and Mission in Matthew's Gospel. *Volume II/177.*

Fotopoulos, John: Food Offered to Idols in Roman Corinth. 2003. *Volume II/151.*

Frenschkowski, Marco: Offenbarung und Epiphanie. Volume 1 1995. *Volume II/79* – Volume 2 1997. *Volume II/80.*

Frey, Jörg: Eugen Drewermann und die biblische Exegese. 1995. *Volume II/71.*
- Die johanneische Eschatologie. Volume I. 1997. *Volume 96.* – Volume II. 1998. *Volume 110.* – Volume III. 2000. *Volume 117.*

Frey, Jörg and *Udo Schnelle* (Ed.): Kontexte des Johannesevangeliums. 2004. *Volume 175.*
- and *Jens Schröter* (Ed.): Deutungen des Todes Jesu im Neuen Testament. 2005. *Volume 181.*

Freyne, Sean: Galilee and Gospel. 2000. *Volume 125.*

Fridrichsen, Anton: Exegetical Writings. Edited by C.C. Caragounis and T. Fornberg. 1994. *Volume 76.*

Gäbel, Georg: Die Kulttheologie des Hebräerbriefes. 2006. *Volume II/212.*

Gäckle, Volker: Die Starken und die Schwachen in Korinth und in Rom. 2005. *Volume 200.*

Garlington, Don B.: 'The Obedience of Faith'. 1991. *Volume II/38.*
- Faith, Obedience, and Perseverance. 1994. *Volume 79.*

Garnet, Paul: Salvation and Atonement in the Qumran Scrolls. 1977. *Volume II/3.*

Gemünden, Petra von (Ed.): see *Weissenrieder, Annette.*

Gese, Michael: Das Vermächtnis des Apostels. 1997. *Volume II/99.*

Gheorghita, Radu: The Role of the Septuagint in Hebrews. 2003. *Volume II/160.*

Gräbe, Petrus J.: The Power of God in Paul's Letters. 2000. *Volume II/123.*

Gräßer, Erich: Der Alte Bund im Neuen. 1985. *Volume 35.*
- Forschungen zur Apostelgeschichte. 2001. *Volume 137.*

Green, Joel B.: The Death of Jesus. 1988. *Volume II/33.*

Gregg, Brian Han: The Historical Jesus and the Final Judgment Sayings in Q. 2005. *Volume II/207.*

Gregory, Andrew: The Reception of Luke and Acts in the Period before Irenaeus. 2003. *Volume II/169.*

Grindheim, Sigurd: The Crux of Election. 2005. *Volume II/202.*

Gundry, Robert H.: The Old is Better. 2005. *Volume 178.*

Gundry Volf, Judith M.: Paul and Perseverance. 1990. *Volume II/37.*

Häußer, Detlef: Christusbekenntnis und Jesusüberlieferung bei Paulus. 2006. *Volume 210.*

Hafemann, Scott J.: Suffering and the Spirit. 1986. *Volume II/19.*
- Paul, Moses, and the History of Israel. 1995. *Volume 81.*

Hahn, Ferdinand: Studien zum Neuen Testament. Volume I: Grundsatzfragen, Jesusforschung, Evangelien. 2006. *Volume 191.*
- Volume II: Bekenntnisbildung und Theologie in urchristlicher Zeit. 2006. *Volume 192.*

Hahn, Johannes (Ed.): Zerstörungen des Jerusalemer Tempels. 2002. *Volume 147.*

Hamid-Khani, Saeed: Relevation and Concealment of Christ. 2000. *Volume II/120.*

Hannah, Darrel D.: Michael and Christ. 1999. *Volume II/109.*

Harrison; James R.: Paul's Language of Grace in Its Graeco-Roman Context. 2003. *Volume II/172.*

Hartman, Lars: Text-Centered New Testament Studies. Ed. von D. Hellholm. 1997. *Volume 102.*

Hartog, Paul: Polycarp and the New Testament. 2001. *Volume II/134.*

Heckel, Theo K.: Der Innere Mensch. 1993. *Volume II/53.*
- Vom Evangelium des Markus zum viergestaltigen Evangelium. 1999. *Volume 120.*

Heckel, Ulrich: Kraft in Schwachheit. 1993. *Volume II/56.*

- Der Segen im Neuen Testament. 2002. *Volume 150.*
- see *Feldmeier, Reinhard.*
- see *Hengel, Martin.*

Heiligenthal, Roman: Werke als Zeichen. 1983. *Volume II/9.*

Hellholm, D.: see *Hartman, Lars.*

Hemer, Colin J.: The Book of Acts in the Setting of Hellenistic History. 1989. *Volume 49.*

Hengel, Martin: Judentum und Hellenismus. 1969, ³1988. *Volume 10.*
- Die johanneische Frage. 1993. *Volume 67.*
- Judaica et Hellenistica. Kleine Schriften I. 1996. *Volume 90.*
- Judaica, Hellenistica et Christiana. Kleine Schriften II. 1999. *Volume 109.*
- Paulus und Jakobus. Kleine Schriften III. 2002. *Volume 141.*
- and *Anna Maria Schwemer:* Paulus zwischen Damaskus und Antiochien. 1998. *Volume 108.*
- Der messianische Anspruch Jesu und die Anfänge der Christologie. 2001. *Volume 138.*

Hengel, Martin and *Ulrich Heckel* (Ed.): Paulus und das antike Judentum. 1991. *Volume 58.*
- and *Hermut Löhr* (Ed.): Schriftauslegung im antiken Judentum und im Urchristentum. 1994. *Volume 73.*
- and *Anna Maria Schwemer* (Ed.): Königsherrschaft Gottes und himm-lischer Kult. 1991. *Volume 55.*
- Die Septuaginta. 1994. *Volume 72.*
-, *Siegfried Mittmann* and *Anna Maria Schwemer* (Ed.): La Cité de Dieu / Die Stadt Gottes. 2000. *Volume 129.*

Herrenbrück, Fritz: Jesus und die Zöllner. 1990. *Volume II/41.*

Herzer, Jens: Paulus oder Petrus? 1998. *Volume 103.*

Hill, Charles E.: From the Lost Teaching of Polycarp. 2005. *Volume 186.*

Hoegen-Rohls, Christina: Der nachösterliche Johannes. 1996. *Volume II/84.*

Hoffmann, Matthias Reinhard: The Destroyer and the Lamb. 2005. *Volume II/203.*

Hofius, Otfried: Katapausis. 1970. *Volume 11.*
- Der Vorhang vor dem Thron Gottes. 1972. *Volume 14.*
- Der Christushymnus Philipper 2,6-11. 1976, ²1991. *Volume 17.*
- Paulusstudien. 1989, ²1994. *Volume 51.*
- Neutestamentliche Studien. 2000. *Volume 132.*
- Paulusstudien II. 2002. *Volume 143.*
- and *Hans-Christian Kammler:* Johannesstudien. 1996. *Volume 88.*

Holtz, Traugott: Geschichte und Theologie des Urchristentums. 1991. *Volume 57.*

Hommel, Hildebrecht: Sebasmata. Volume 1 1983. Volume 31 – Volume 2 1984. *Volume 32.*

Horbury, William: Herodian Judaism and New Testament Study. 2006. *Volume 193.*

Horst, Pieter W. van der: Jews and Christians in Their Graeco-Roman Context. 2006. *Volume 196.*

Hvalvik, Reidar: The Struggle for Scripture and Covenant. 1996. *Volume II/82.*

Jauhiainen, Marko: The Use of Zechariah in Revelation. 2005. *Volume II/199.*

Jensen, Morten H.: Herod Antipas in Galilee. 2006. *Volume II/215.*

Johns, Loren L.: The Lamb Christology of the Apocalypse of John. 2003. *Volume II/167.*

Joubert, Stephan: Paul as Benefactor. 2000. *Volume II/124.*

Jungbauer, Harry: „Ehre Vater und Mutter". 2002. *Volume II/146.*

Kähler, Christoph: Jesu Gleichnisse als Poesie und Therapie. 1995. *Volume 78.*

Kamlah, Ehrhard: Die Form der katalogischen Paränese im Neuen Testament. 1964. *Volume 7.*

Kammler, Hans-Christian: Christologie und Eschatologie. 2000. *Volume 126.*
- Kreuz und Weisheit. 2003. *Volume 159.*
- see *Hofius, Otfried.*

Kelhoffer, James A.: The Diet of John the Baptist. 2005. *Volume 176.*
- Miracle and Mission. 1999. *Volume II/112.*

Kelley, Nicole: Knowledge and Religious Authority in the Pseudo-Clementines. 2006. *Volume II/213.*

Kieffer, René and *Jan Bergman* (Ed.): La Main de Dieu / Die Hand Gottes. 1997. *Volume 94.*

Kim, Seyoon: The Origin of Paul's Gospel. 1981, ²1984. *Volume II/4.*
- Paul and the New Perspective. 2002. *Volume 140.*
- "The 'Son of Man'" as the Son of God. 1983. *Volume 30.*

Klauck, Hans-Josef: Religion und Gesellschaft im frühen Christentum. 2003. *Volume 152.*

Klein, Hans: see *Dunn, James D.G.*

Kleinknecht, Karl Th.: Der leidende Gerechtfertigte. 1984, ²1988. *Volume II/13.*

Klinghardt, Matthias: Gesetz und Volk Gottes. 1988. *Volume II/32.*

Kloppenborg, John S.: The Tenants in the Vineyard. 2006. *Volume 195.*

Koch, Michael: Drachenkampf und Sonnenfrau. 2004. *Volume II/184.*

Koch, Stefan: Rechtliche Regelung von Konflikten im frühen Christentum. 2004. *Volume II/174.*

Köhler, Wolf-Dietrich: Rezeption des Matthäusevangeliums in der Zeit vor Irenäus. 1987. *Volume II/24.*

Köhn, Andreas: Der Neutestamentler Ernst Lohmeyer. 2004. *Volume II/180.*
Kooten, George H. van: Cosmic Christology in Paul and the Pauline School. 2003. *Volume II/171.*
Korn, Manfred: Die Geschichte Jesu in veränderter Zeit. 1993. *Volume II/51.*
Koskenniemi, Erkki: Apollonios von Tyana in der neutestamentlichen Exegese. 1994. *Volume II/61.*
– The Old Testament Miracle-Workers in Early Judaism. 2005. *Volume II/206.*
Kraus, Thomas J.: Sprache, Stil und historischer Ort des zweiten Petrusbriefes. 2001. *Volume II/136.*
Kraus, Wolfgang: Das Volk Gottes. 1996. *Volume 85.*
Kraus, Wolfgang and *Karl-Wilhelm Niebuhr* (Ed.): Frühjudentum und Neues Testament im Horizont Biblischer Theologie. 2003. *Volume 162.*
– see *Walter, Nikolaus.*
Kreplin, Matthias: Das Selbstverständnis Jesu. 2001. *Volume II/141.*
Kuhn, Karl G.: Achtzehngebet und Vaterunser und der Reim. 1950. *Volume 1.*
Kvalbein, Hans: see *Ådna, Jostein.*
Kwon, Yon-Gyong: Eschatology in Galatians. 2004. *Volume II/183.*
Laansma, Jon: I Will Give You Rest. 1997. *Volume II/98.*
Labahn, Michael: Offenbarung in Zeichen und Wort. 2000. *Volume II/117.*
Lambers-Petry, Doris: see *Tomson, Peter J.*
Lange, Armin: see *Ego, Beate.*
Lampe, Peter: Die stadtrömischen Christen in den ersten beiden Jahrhunderten. 1987, ²1989. *Volume II/18.*
Landmesser, Christof: Wahrheit als Grundbegriff neutestamentlicher Wissenschaft. 1999. *Volume 113.*
– Jüngerberufung und Zuwendung zu Gott. 2000. *Volume 133.*
Lau, Andrew: Manifest in Flesh. 1996. *Volume II/86.*
Lawrence, Louise: An Ethnography of the Gospel of Matthew. 2003. *Volume II/165.*
Lee, Aquila H.I.: From Messiah to Preexistent Son. 2005. *Volume II/192.*
Lee, Pilchan: The New Jerusalem in the Book of Relevation. 2000. *Volume II/129.*
Lichtenberger, Hermann: Das Ich Adams und das Ich der Menschheit. 2004. *Volume 164.*
– see *Avemarie, Friedrich.*
Lierman, John: The New Testament Moses. 2004. *Volume II/173.*
Lieu, Samuel N.C.: Manichaeism in the Later Roman Empire and Medieval China. ²1992. *Volume 63.*

Lindgård, Fredrik: Paul's Line of Thought in 2 Corinthians 4:16-5:10. 2004. *Volume II/189.*
Loader, William R.G.: Jesus' Attitude Towards the Law. 1997. *Volume II/97.*
Löhr, Gebhard: Verherrlichung Gottes durch Philosophie. 1997. *Volume 97.*
Löhr, Hermut: Studien zum frühchristlichen und frühjüdischen Gebet. 2003. *Volume 160.*
– see *Hengel, Martin.*
Löhr, Winrich Alfried: Basilides und seine Schule. 1995. *Volume 83.*
Luomanen, Petri: Entering the Kingdom of Heaven. 1998. *Volume II/101.*
Luz, Ulrich: see *Dunn, James D.G.*
Mackay, Ian D.: John's Raltionship with Mark. 2004. *Volume II/182.*
Maier, Gerhard: Mensch und freier Wille. 1971. *Volume 12.*
– Die Johannesoffenbarung und die Kirche. 1981. *Volume 25.*
Markschies, Christoph: Valentinus Gnosticus? 1992. *Volume 65.*
Marshall, Peter: Enmity in Corinth: Social Conventions in Paul's Relations with the Corinthians. 1987. *Volume II/23.*
Mayer, Annemarie: Sprache der Einheit im Epheserbrief und in der Ökumene. 2002. *Volume II/150.*
Mayordomo, Moisés: Argumentiert Paulus logisch? 2005. *Volume 188.*
McDonough, Sean M.: YHWH at Patmos: Rev. 1:4 in its Hellenistic and Early Jewish Setting. 1999. *Volume II/107.*
McDowell, Markus: Prayers of Jewish Women. 2006. *Volume II/211.*
McGlynn, Moyna: Divine Judgement and Divine Benevolence in the Book of Wisdom. 2001. *Volume II/139.*
Meade, David G.: Pseudonymity and Canon. 1986. *Volume 39.*
Meadors, Edward P.: Jesus the Messianic Herald of Salvation. 1995. *Volume II/72.*
Meißner, Stefan: Die Heimholung des Ketzers. 1996. *Volume II/87.*
Mell, Ulrich: Die „anderen" Winzer. 1994. *Volume 77.*
– see *Sänger, Dieter.*
Mengel, Berthold: Studien zum Philipperbrief. 1982. *Volume II/8.*
Merkel, Helmut: Die Widersprüche zwischen den Evangelien. 1971. *Volume 13.*
– see *Ego, Beate.*
Merklein, Helmut: Studien zu Jesus und Paulus. Volume 1 1987. *Volume 43.* – Volume 2 1998. *Volume 105.*
Metzdorf, Christina: Die Tempelaktion Jesu. 2003. *Volume II/168.*
Metzler, Karin: Der griechische Begriff des Verzeihens. 1991. *Volume II/44.*

Metzner, Rainer: Die Rezeption des Matthäusevangeliums im 1. Petrusbrief. 1995. *Volume II/74.*
– Das Verständnis der Sünde im Johannesevangelium. 2000. *Volume 122.*
Mihoc, Vasile: see *Dunn, James D.G.*
Mineshige, Kiyoshi: Besitzverzicht und Almosen bei Lukas. 2003. *Volume II/163.*
Mittmann, Siegfried: see *Hengel, Martin.*
Mittmann-Richert, Ulrike: Magnifikat und Benediktus. 1996. *Volume II/90.*
Mournet, Terence C.: Oral Tradition and Literary Dependency. 2005. *Volume II/195.*
Mußner, Franz: Jesus von Nazareth im Umfeld Israels und der Urkirche. Ed. von M. Theobald. 1998. *Volume 111.*
Mutschler, Bernhard: Das Corpus Johanneum bei Irenäus von Lyon. 2005. *Volume 189.*
Niebuhr, Karl-Wilhelm: Gesetz und Paränese. 1987. *Volume II/28.*
– Heidenapostel aus Israel. 1992. *Volume 62.*
– see *Deines, Roland*
– see *Dimitrov, Ivan Z.*
– see *Kraus, Wolfgang*
Nielsen, Anders E.: "Until it is Fullfilled". 2000. *Volume II/126.*
Nissen, Andreas: Gott und der Nächste im antiken Judentum. 1974. *Volume 15.*
Noack, Christian: Gottesbewußtsein. 2000. *Volume II/116.*
Noormann, Rolf: Irenäus als Paulusinterpret. 1994. *Volume II/66.*
Novakovic, Lidija: Messiah, the Healer of the Sick. 2003. *Volume II/170.*
Obermann, Andreas: Die christologische Erfüllung der Schrift im Johannesevangelium. 1996. *Volume II/83.*
Öhler, Markus: Barnabas. 2003. *Volume 156.*
– see *Becker, Michael*
Okure, Teresa: The Johannine Approach to Mission. 1988. *Volume II/31.*
Onuki, Takashi: Heil und Erlösung. 2004. *Volume 165.*
Oropeza, B. J.: Paul and Apostasy. 2000. *Volume II/115.*
Ostmeyer, Karl-Heinrich: Kommunikation mit Gott und Christus. 2006. *Volume 197.*
– Taufe und Typos. 2000. *Volume II/118.*
Paulsen, Henning: Studien zur Literatur und Geschichte des frühen Christentums. Ed. von Ute E. Eisen. 1997. *Volume 99.*
Pao, David W.: Acts and the Isaianic New Exodus. 2000. *Volume II/130.*
Park, Eung Chun: The Mission Discourse in Matthew's Interpretation. 1995. *Volume II/81.*
Park, Joseph S.: Conceptions of Afterlife in Jewish Insriptions. 2000. *Volume II/121.*
Pate, C. Marvin: The Reverse of the Curse. 2000. *Volume II/114.*

Peres, Imre: Griechische Grabinschriften und neutestamentliche Eschatologie. 2003. *Volume 157.*
Philip, Finny: The Origins of Pauline Pneumatology. 2005. *Volume II/194.*
Philonenko, Marc (Ed.): Le Trône de Dieu. 1993. *Volume 69.*
Pilhofer, Peter: Presbyteron Kreitton. 1990. *Volume II/39.*
– Philippi. Volume 1 1995. *Volume 87.* – Volume 2 2000. *Volume 119.*
– Die frühen Christen und ihre Welt. 2002. *Volume 145.*
– see *Becker, Eve-Marie.*
– see *Ego, Beate.*
Pitre, Brant: Jesus, the Tribulation, and the End of the Exile. 2005. *Volume II/204.*
Plümacher, Eckhard: Geschichte und Geschichten. 2004. *Volume 170.*
Pöhlmann, Wolfgang: Der Verlorene Sohn und das Haus. 1993. *Volume 68.*
Pokorný, Petr and *Josef B. Souček:* Bibelauslegung als Theologie. 1997. *Volume 100.*
Pokorný, Petr and *Jan Roskovec* (Ed.): Philosophical Hermeneutics and Biblical Exegesis. 2002. *Volume 153.*
Popkes, Enno Edzard: Die Theologie der Liebe Gottes in den johanneischen Schriften. 2005. *Volume II/197.*
Porter, Stanley E.: The Paul of Acts. 1999. *Volume 115.*
Prieur, Alexander: Die Verkündigung der Gottesherrschaft. 1996. *Volume II/89.*
Probst, Hermann: Paulus und der Brief. 1991. *Volume II/45.*
Räisänen, Heikki: Paul and the Law. 1983, ²1987. *Volume 29.*
Rehkopf, Friedrich: Die lukanische Sonderquelle. 1959. *Volume 5.*
Rein, Matthias: Die Heilung des Blindgeborenen (Joh 9). 1995. *Volume II/73.*
Reinmuth, Eckart: Pseudo-Philo und Lukas. 1994. *Volume 74.*
Reiser, Marius: Syntax und Stil des Markusevangeliums. 1984. *Volume II/11.*
Rhodes, James N.: The Epistle of Barnabas and the Deuteronomic Tradition. 2004. *Volume II/188.*
Richards, E. Randolph: The Secretary in the Letters of Paul. 1991. *Volume II/42.*
Riesner, Rainer: Jesus als Lehrer. 1981, ³1988. *Volume II/7.*
– Die Frühzeit des Apostels Paulus. 1994. *Volume 71.*
Rissi, Mathias: Die Theologie des Hebräerbriefs. 1987. *Volume 41.*
Roskovec, Jan: see *Pokorný, Petr.*
Röhser, Günter: Metaphorik und Personifikation der Sünde. 1987. *Volume II/25.*
Rose, Christian: Die Wolke der Zeugen. 1994. *Volume II/60.*

Rothschild, Clare K.: Baptist Traditions and Q. 2005. *Volume 190.*
– Luke Acts and the Rhetoric of History. 2004. *Volume II/175.*
Rüegger, Hans-Ulrich: Verstehen, was Markus erzählt. 2002. *Volume II/155.*
Rüger, Hans Peter: Die Weisheitsschrift aus der Kairoer Geniza. 1991. *Volume 53.*
Sänger, Dieter: Antikes Judentum und die Mysterien. 1980. *Volume II/5.*
– Die Verkündigung des Gekreuzigten und Israel. 1994. *Volume 75.*
– see *Burchard, Christoph*
– and *Ulrich Mell* (Hrsg.): Paulus und Johannes. 2006. *Volume 198.*
Salier, Willis Hedley: The Rhetorical Impact of the Se-meia in the Gospel of John. 2004. *Volume II/186.*
Salzmann, Jorg Christian: Lehren und Ermahnen. 1994. *Volume II/59.*
Sandnes, Karl Olav: Paul – One of the Prophets? 1991. *Volume II/43.*
Sato, Migaku: Q und Prophetie. 1988. *Volume II/29.*
Schäfer, Ruth: Paulus bis zum Apostelkonzil. 2004. *Volume II/179.*
Schaper, Joachim: Eschatology in the Greek Psalter. 1995. *Volume II/76.*
Schimanowski, Gottfried: Die himmlische Liturgie in der Apokalypse des Johannes. 2002. *Volume II/154.*
– Weisheit und Messias. 1985. *Volume II/17.*
Schlichting, Günter: Ein jüdisches Leben Jesu. 1982. *Volume 24.*
Schnabel, Eckhard J.: Law and Wisdom from Ben Sira to Paul. 1985. *Volume II/16.*
Schnelle, Udo: see *Frey, Jörg.*
Schröter, Jens: see *Frey, Jörg.*
Schutter, William L.: Hermeneutic and Composition in I Peter. 1989. *Volume II/30.*
Schwartz, Daniel R.: Studies in the Jewish Background of Christianity. 1992. *Volume 60.*
Schwemer, Anna Maria: see *Hengel, Martin*
Scott, Ian W.: Implicit Epistemology in the Letters of Paul. 2005. *Volume II/205.*
Scott, James M.: Adoption as Sons of God. 1992. *Volume II/48.*
– Paul and the Nations. 1995. *Volume 84.*
Shum, Shiu-Lun: Paul's Use of Isaiah in Romans. 2002. *Volume II/156.*
Siegert, Folker: Drei hellenistisch-jüdische Predigten. Teil I 1980. *Volume 20* – Teil II 1992. *Volume 61.*
– Nag-Hammadi-Register. 1982. *Volume 26.*
– Argumentation bei Paulus. 1985. *Volume 34.*
– Philon von Alexandrien. 1988. *Volume 46.*
Simon, Marcel: Le christianisme antique et son contexte religieux I/II. 1981. *Volume 23.*

Snodgrass, Klyne: The Parable of the Wicked Tenants. 1983. *Volume 27.*
Söding, Thomas: Das Wort vom Kreuz. 1997. *Volume 93.*
– see *Thüsing, Wilhelm.*
Sommer, Urs: Die Passionsgeschichte des Markusevangeliums. 1993. *Volume II/58.*
Souèek, Josef B.: see *Pokorný, Petr.*
Spangenberg, Volker: Herrlichkeit des Neuen Bundes. 1993. *Volume II/55.*
Spanje, T.E. van: Inconsistency in Paul? 1999. *Volume II/110.*
Speyer, Wolfgang: Frühes Christentum im antiken Strahlungsfeld. Volume I: 1989. *Volume 50.*
– Volume II: 1999. *Volume 116.*
Stadelmann, Helge: Ben Sira als Schriftgelehrter. 1980. *Volume II/6.*
Stenschke, Christoph W.: Luke's Portrait of Gentiles Prior to Their Coming to Faith. *Volume II/108.*
Sterck-Degueldre, Jean-Pierre: Eine Frau namens Lydia. 2004. *Volume II/176.*
Stettler, Christian: Der Kolosserhymnus. 2000. *Volume II/131.*
Stettler, Hanna: Die Christologie der Pastoralbriefe. 1998. *Volume II/105.*
Stökl Ben Ezra, Daniel: The Impact of Yom Kippur on Early Christianity. 2003. *Volume 163.*
Strobel, August: Die Stunde der Wahrheit. 1980. *Volume 21.*
Stroumsa, Guy G.: Barbarian Philosophy. 1999. *Volume 112.*
Stuckenbruck, Loren T.: Angel Veneration and Christology. 1995. *Volume II/70.*
Stuhlmacher, Peter (Ed.): Das Evangelium und die Evangelien. 1983. *Volume 28.*
– Biblische Theologie und Evangelium. 2002. *Volume 146.*
Sung, Chong-Hyon: Vergebung der Sünden. 1993. *Volume II/57.*
Tajra, Harry W.: The Trial of St. Paul. 1989. *Volume II/35.*
– The Martyrdom of St.Paul. 1994. *Volume II/67.*
Theißen, Gerd: Studien zur Soziologie des Urchristentums. 1979, ³1989. *Volume 19.*
Theobald, Michael: Studien zum Römerbrief. 2001. *Volume 136.*
Theobald, Michael: see *Mußner, Franz.*
Thornton, Claus-Jürgen: Der Zeuge des Zeugen. 1991. *Volume 56.*
Thüsing, Wilhelm: Studien zur neutestamentlichen Theologie. Ed. von Thomas Söding. 1995. *Volume 82.*
Thurén, Lauri: Derhethorizing Paul. 2000. *Volume 124.*
Tolmie, D. Francois: Persuading the Galatians. 2005. *Volume II/190.*

Wissenschaftliche Untersuchungen zum Neuen Testament

Tomson, Peter J. and *Doris Lambers-Petry* (Ed.): The Image of the Judaeo-Christians in Ancient Jewish and Christian Literature. 2003. *Volume 158.*
Trebilco, Paul: The Early Christians in Ephesus from Paul to Ignatius. 2004. *Volume 166.*
Treloar, Geoffrey R.: Lightfoot the Historian. 1998. *Volume II/103.*
Tsuji, Manabu: Glaube zwischen Vollkommenheit und Verweltlichung. 1997. *Volume II/93.*
Twelftree, Graham H.: Jesus the Exorcist. 1993. *Volume II/54.*
Urban, Christina: Das Menschenbild nach dem Johannesevangelium. 2001. *Volume II/137.*
Visotzky, Burton L.: Fathers of the World. 1995. *Volume 80.*
Vollenweider, Samuel: Horizonte neutestamentlicher Christologie. 2002. *Volume 144.*
Vos, Johan S.: Die Kunst der Argumentation bei Paulus. 2002. *Volume 149.*
Wagener, Ulrike: Die Ordnung des „Hauses Gottes". 1994. *Volume II/65.*
Wahlen, Clinton: Jesus and the Impurity of Spirits in the Synoptic Gospels. 2004. *Volume II/185.*
Walker, Donald D.: Paul's Offer of Leniency (2 Cor 10:1). 2002. *Volume II/152.*
Walter, Nikolaus: Praeparatio Evangelica. Ed. von Wolfgang Kraus und Florian Wilk. 1997. *Volume 98.*
Wander, Bernd: Gottesfürchtige und Sympathisanten. 1998. *Volume 104.*
Watts, Rikki: Isaiah's New Exodus and Mark. 1997. *Volume II/88.*
Wedderburn, A.J.M.: Baptism and Resurrection. 1987. *Volume 44.*
Wegner, Uwe: Der Hauptmann von Kafarnaum. 1985. *Volume II/14.*
Weissenrieder, Annette: Images of Illness in the Gospel of Luke. 2003. Volume II/164.

–, *Friederike Wendt* and *Petra von Gemünden* (Ed.): Picturing the New Testament. 2005. *Volume II/193.*
Welck, Christian: Erzählte ‚Zeichen'. 1994. *Volume II/69.*
Wendt, Friederike (Ed.): see *Weissenrieder, Annette.*
Wiarda, Timothy: Peter in the Gospels. 2000. *Volume II/127.*
Wifstrand, Albert: Epochs and Styles. 2005. *Volume 179.*
Wilk, Florian: see *Walter, Nikolaus.*
Williams, Catrin H.: I am He. 2000. *Volume II/113.*
Wilson, Walter T.: Love without Pretense. 1991. *Volume II/46.*
Wischmeyer, Oda: Von Ben Sira zu Paulus. 2004. *Volume 173.*
Wisdom, Jeffrey: Blessing for the Nations and the Curse of the Law. 2001. *Volume II/133.*
Wold, Benjamin G.: Women, Men, and Angels. 2005. *Volume II/2001.*
Wright, Archie T.: The Origin of Evil Spirits. 2005. *Volume II/198.*
Wucherpfennig, Ansgar: Heracleon Philologus. 2002. *Volume 142.*
Yeung, Maureen: Faith in Jesus and Paul. 2002. *Volume II/147.*
Zimmermann, Alfred E.: Die urchristlichen Lehrer. 1984, ²1988. *Volume II/12.*
Zimmermann, Johannes: Messianische Texte aus Qumran. 1998. *Volume II/104.*
Zimmermann, Ruben: Christologie der Bilder im Johannesevangelium. 2004. *Volume 171.*
– Geschlechtermetaphorik und Gottesverhältnis. 2001. *Volume II/122.*
Zumstein, Jean: see *Dettwiler, Andreas*
Zwiep, Arie W.: Judas and the Choice of Matthias. 2004. *Volume II/187.*

For a complete catalogue please write to the publisher
Mohr Siebeck • P.O. Box 2030 • D–72010 Tübingen/Germany
Up-to-date information on the internet at www.mohr.de